Lecture Notes in Computer S

Commenced Publication in 1973
Founding and Former Series Editors:
Gerhard Goos, Juris Hartmanis, and Jan van Leeuwen

Stavros Konstantinidis (Ed.)

Implementation and Application of Automata

18th International Conference, CIAA 2013
Halifax, NS, Canada, July 16-19, 2013
Proceedings

 Springer

Volume Editor

Stavros Konstantinidis
Saint Mary's University
Department of Mathematics and Computing Science
923 Robie Street
Halifax, NS B3H 3C3, Canada
E-mail: s.konstantinidis@smu.ca

ISSN 0302-9743   e-ISSN 1611-3349
ISBN 978-3-642-39273-3   e-ISBN 978-3-642-39274-0
DOI 10.1007/978-3-642-39274-0
Springer Heidelberg Dordrecht London New York

Library of Congress Control Number: 2013941377

CR Subject Classification (1998): F.1.1-2, F.1, F.2, F.4, E.1, H.3

LNCS Sublibrary: SL 1 – Theoretical Computer Science and General Issues

© Springer-Verlag Berlin Heidelberg 2013

Typesetting: Camera-ready by author, data conversion by Scientific Publishing Services, Chennai, India

Printed on acid-free paper

Springer is part of Springer Science+Business Media (www.springer.com)

Preface

This volume contains the papers presented at the 18th International Conference on Implementation and Application of Automata (CIAA 2013), which was held at Saint Mary's University in Halifax, Nova Scotia, Canada, during July 16–19 2013.

The CIAA conference series is a major international venue for the dissemination of new results in the implementation, application, and theory of automata. The previous 17 conferences were held in the following locations: Porto (2012), Blois (2011), Winnipeg (2010), Sydney (2009), San Francisco (2008), Prague (2007), Taipei (2006), Nice (2005), Kingston (2004), Santa Barbara (2003), Tours (2002), Pretoria (2001), London Ontario (2000), Potsdam (WIA 1999), Rouen (WIA 1998), London Ontario (WIA 1997 and WIA 1996).

The topics of this volume include: complexity of automata, compressed automata, counter automata, dictionary matching, edit distance, homing sequences, implementation, minimization of automata, model checking, parsing of regular expressions, partial word automata, picture languages, pushdown automata, queue automata, reachability analysis for software verification, restarting automata, transducers, tree automata, weighted automata, XML streams.

The submission and refereeing process was supported by the EasyChair conference system. In all, 43 papers were submitted by authors in 22 different countries (from all six habitable continents), including Algeria, Bangladesh, Canada, Czech Republic, Denmark, France, Germany, Italy, Japan, Korea Republic, Poland, Portugal, Russian Federation, Slovakia, and the USA. Each paper was reviewed by at least three Program Committee members. The Program Committee selected 25 regular papers and three short papers for presentation at the conference and publication in this volume. There were three invited talks by Cezar Câmpeanu, Helmut Jürgensen, and Margus Veanes.

We are very thankful to all invited speakers, contributing authors, Program Committee members, and external referees for their valuable contributions toward the realization of CIAA 2013.

We also thank Alfred Hofmann and Anna Kramer of Springer for their guidance during the process of publishing this volume.

We are grateful to (a) the European Association for Theoretical Computer Science (EATCS) for their scientific sponsorship, (b) to Microsoft Research, and the Dean of Science and the Academic Vice President of Saint Mary's University (SMU) for their financial support, and (c) to Destination Halifax for their assistance in arranging the accommodation and excursion venues.

Finally, we are indebted to the Organizing Committee members Casey Meijer and Rose Daurie, as well as to the research students of the department of Mathematics and Computing Science at Saint Mary's University, Halifax.

We are looking forward to the next CIAA in Giessen, Germany.

July 2013   Stavros Konstantinidis

Organization

CIAA 2013 was organized at the Department of Mathematics and Computing Science, Saint Mary's University, Halifax, Canada.

Invited Speakers

Câmpeanu, Cezar	University of PEI, Canada
Jürgensen, Helmut	University of Western Ontario, Canada
Veanes, Margus	Microsoft Research, USA

Program Committee

Brzozowski, Janusz	Waterloo, Canada
Câmpeanu, Cezar	Charlottetown, Canada
Caron, Pascal	Rouen, France
Champarnaud, Jean-Marc	Rouen, France
Domaratzki, Michael	Manitoba, Canada
Drewes, Frank	Umea, Sweden
Han, Yo-Sub	Seoul, Korea
Holub, Jan	Prague, Czech Republic
Holzer, Markus	Giessen, Germany
Ibarra, Oscar	Santa Barbara, California, USA
Ito, Masami	Kyoto, Japan
Jürgensen, Helmut	London, Canada
Kari, Lila	London, Canada
Konstantinidis, Stavros (Chair)	Halifax, Canada
Lombardy, Sylvain	Bordeaux, France
Maletti, Andreas	Stuttgart, Germany
Maneth, Sebastian	New South Wales, Australia
Maurel, Denis	Tours, France
McQuillan, Ian	Saskatoon, Canada
Moreira, Nelma	Porto, Portugal
Okhotin, Alexander	Turku, Finland
Pighizzini, Giovanni	Milan, Italy
Ravikumar, Bala	Rohnert Park CA, USA
Reidenbach, Daniel	Leicestershire, UK
Reis, Rogerio	Porto, Portugal
Salomaa, Kai	Kingston, Canada
Tommasi, Marc	Lille, France

Volkov, Mikhail   Ekaterinburg, Russian Federation
Watson, Bruce   Stellenbosch University, South Africa
Yen, Hsu-Chun   Taipei, Taiwan

External Referees

Amorim, Ivone   Lemay, Aurelien
Baier, Jan   Liu, Jiamou
Barash, Mikhail   Lucanu, Dorel
Becerra Bonache, Leonor   Maia, Eva
Bertoni, Alberto   Malcher, Andreas
Björklund, Henrik   Mignot, Ludovic
Björklund, Johanna   Nagy, Benedek
Boneva, Iovka   Nicart, Florent
Borchert, Bernd   Niehren, Joachim
Broda, Sabine   Nouvel, Damien
Carayol, Arnaud   Oshita, Masaki
Cerno, Peter   Palioudakis, Alexandros
Cleophas, Loek   Polach, Radomir
David, Julien   Quernheim, Daniel
Day, Joel   Rampersad, Narad
Dubernard, Jean-Philippe   Reinhardt, Klaus
Fraser, Gordon   Roche, Abiel
Freydenberger, Dominik D.   Roche-Lima, Abiel
Friburger, Nathalie   Rodaro, Emanuele
Gao, Yuan   Savari, Serap
Groz, Benoît   Schmid, Markus L.
Jeż, Artur   Seki, Shinnosuke
Khoussainov, Bakhadyr   Simjour, Amir
Klimann, Ines   Strauss, Tinus
Kobayashi, Yuji   Tran, Nicholas
Kolpakov, Roman   Wang, Bow-Yaw
Kopecki, Steffen   Young, Joshua
Kourie, Derrick
Kutrib, Martin

Steering Committee

Champarnaud, Jean-Marc   Rouen, France
Holzer, Markus   Giessen, Germany
Ibarra, Oscar   Santa Barbara, California, USA
Maurel, Denis   Tours, France
Salomaa, Kai T. *(Chair)* Kingston, Ontario, Canada
Yen, Hsu-Chun   Taipei, Taiwan

Organizing Committee

Daurie, Rose	Halifax, Canada
Konstantinidis, Stavros	
(Chair)	Halifax, Canada
Meijer, Casey	Halifax, Canada

Sponsors

Academic Vice President, Saint Mary's University
Dean of Science, Saint Mary's University
Destination Halifax
European Association for Theoretical Computer Science
Microsoft Research

Table of Contents

Invited Talks

Regular Papers

Short Papers

Cover Languages and Implementations

Cezar Câmpeanu

Department of Computer Science and Information Technology,
The University of Prince Edward Island, Canada

Abstract. A cover language is a superset of a given language. Deterministic Finite Cover Automata (DFCA) are Deterministic Finite Automata (DFA) accepting finite languages and other words longer than any word in the given language. Some papers from the 60's were constructing DFCAs as a byproduct using ad-hoc procedures, but DFCAs have never been defined until 1998. The notion of Deterministic Finite Cover Automaton, which is based on the concept of similarity relations, was introduced for the very first time at WIA'98, where the authors give the first rigorous formal definition and a clear minimization algorithm.

We will present a survey of the most important results related to cover automata, and will show the importance of the implementation in obtaining, verifying, and solving new results.

A list of open problems related to cover automata, together with some possible approaches will be presented.

Features of the software packages Grail, miniGrail, and Lisa will be exposed. A list of open problems related to challenges and limitations encountered when using software packages implementing automata, languages, and related objects will be shown.

S. Konstantinidis (Ed.): CIAA 2013, LNCS 7982, p. 1, 2013.

Automata for Codes

Helmut Jürgensen

Department of Computer Science
The University of Western Ontario
London, Ontario, Canada, N6A 5B7
helmut@uwo.ca

Abstract. We survey the actual and potential rôles of automata in the modelling of information transmission systems and, in particular, in the encoder, channel and decoder components of such systems. Our focus is on applications of codes in such systems and on the relevance of automaton theoretic methods to these applications. We discuss, for example, the issues of error-detection, fault-tolerance and error-correction for variable-length codes. Beyond reviewing known work in a possibly new setting, we also present some recent results on fault-tolerant decoders for systems in which synchronization errors are likely. We conclude with a kind of research programme, a list of rather general open problems requiring solutions.

1   Information Transmission Systems

In this paper we attempt a survey of the actual and potential rôles of automata in the modelling of information transmission systems and, in particular, in the encoder, channel and decoder components of such systems. We refer the reader to the following sources for general background and details: the books or book chapters [1,3,4,16,45] for variable-length codes; [36,40,42] for block codes; [8,35] for information theory; and [44,46] for automaton theory.

The focus of this paper is on applications of codes in information transmission systems in which automaton theoretic considerations or tools play an essential rôle. For example, we discuss the issues of error-correction, error-detection and fault-tolerance for variable-length codes, issues which are hardly mentioned in most of the literature on these codes, but which are crucial for the application of codes in real-world systems. Beyond reviewing known work in a possibly new setting, we also present some recent results on fault-tolerant decoders for systems in which synchronization errors are likely.

Our survey is subjective and limited. Given the focus of this paper, a large number of otherwise interesting results connecting automata and codes is completely irrelevant; only few results survive the test. We use this opportunity to point out vast areas of research which would deserve attention, provided one accepts that the primary task of codes is to facilitate communication. Occasionally, we may have misjudged the relevance of a result. If so, a better explanation might have helped. Our survey is limited, because we needed to exclude a large

S. Konstantinidis (Ed.): CIAA 2013, LNCS 7982, pp. 2–15, 2013.
© Springer-Verlag Berlin Heidelberg 2013

part of the recent research on codes for modern channels – to explain the physical details of such channels would have taken more space than allowed.

The literature on codes has two nearly disjoint branches originating from the same task: to represent information in a formal or technical setting. Both branches are called *coding theory* by the researchers in the respective field. One of them focuses on codes applied in information transmission systems, while the other one is mainly concerned with the combinatorial and algebraic structure of codes when considered as special types of sets of words.

Consider an *information transmission system* as shown in Fig. 1. A *source* $\mathcal{S}$ sends information to a *receiver* $\mathcal{R}$. The information is transmitted through a *channel* $\mathcal{C}$. During the transmission noise may distort the information. Devices γ and δ with various technical purposes are introduced to make the system work. To keep the presentation simple, we refer to γ as *encoding* and to δ as *decoding,* although other tasks like modulation and demodulation might also be involved. The encoding γ may be *homomorphic* or *sequential.* We assume that the system works in discrete time and with finitely many discrete signals. The output of the source is a potentially bi-infinite sequence of symbols from the *source alphabet* X. The channel transmits symbols from the *channel alphabet* Y, which may be different from the source alphabet. The receiver expects to get the information sent by the source, hence a sequence of symbols[1] from X. We also assume that the system is stationary, that is, that its behaviour is invariant under time shifts.

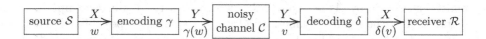

Fig. 1. The standard information transmission model

Before we continue, we introduce some notation. An *alphabet* is a finite non-empty set, the elements of which are called *symbols.* To avoid trivial exceptions, every alphabet considered in this paper is assumed to contain at least two distinct symbols. For a given alphabet X, we consider the set X^+ of finite non-empty words, the set X^ω of right-infinite words[2] and the set X^ζ of bi-infinite words over X. For any alphabet, λ denotes the *empty word* over that alphabet. Let $X^* = X^+ \cup \{\lambda\}$ and $X^\blacklozenge = X^* \cup X^\omega \cup X^\zeta$. For $\eta \in \{*, +, \omega, \zeta, \blacklozenge\}$, an η-word is an element of X^η. By *word* we mean a $*$-word. For a $*$-word w let $\lg(w)$ denote its length. An η-*message* over X is an η-word over X. By *message* we mean an η-message with η chosen as required by the context. The encoding γ transforms

---

[1] It could be sufficient if the receiver gets 'something equivalent' to the message sent. One could also consider additional alphabets for the receiver and the channel output. Nothing essential is gained by such generalizations.

[2] In [16] also left-infinite words are considered. Here they are not needed. There one also distinguishes between word schemata, their instances and words, the latter being equivalences of instances of word schemata. In the present context this distinction is not needed as we consider only stationary systems.

a message w into an *encoded message* $u = \gamma(w)$. As output of the channel one obtains the *transmitted encoded message* v. After decoding one has the *received message* $\delta(v)$.

Requirements 1. *The pair of encoding γ and decoding δ needs to satisfy certain conditions for the system to function, including the following:*

1. *Without noise, there is perfect transmission: $\delta(\gamma(w)) = w$ for all messages w.*
2. *In the presence of noise, transmission is error-free with high probability:*

    $\mathsf{Prob}\big(\delta(v) \neq w \mid w$ *is the source output and v is the channel output*$\big) < \vartheta$

    for some small $\vartheta > 0$.
3. *The channel is used to capacity: Let c be the capacity of the channel C and let h be the average information contents of an output symbol of γ, both measured in bits. Then $c - \varepsilon \leq h < c$ for some small $\varepsilon > 0$.*
4. *If errors cannot be corrected, they should at least be detected.*
5. *Errors should have local effects only: The effect of an error should not propagate through a large part of the message.*
6. *The received message is obtained with little delay: Decoding can start successfully when a bounded part of v is available.*
7. *γ and δ have efficient realizations.*

These requirements form just a small part of a long wish list. They will guide us through the considerations in the present paper.

In what follows, we need some additional notation. We denote finite deterministic acceptors by $A_q = (Q, X, \delta, F)$, where Q is the finite non-empty set of states, X is the input alphabet, $\delta : Q \times X \to Q$ is the – possibly partial – transition function, $F \subseteq Q$ is the set of accepting states and $q \in Q$ is the initial state. The set $L(A_q) = \{w \mid \delta(q, w) \in F\}$ is the language accepted by A. When $q \notin F$, let $L_1(A_q)$ be the set of all those words $w \in L(A_q)$ which have no proper prefix in $L(A_q)$, that is, $L_1(A_q) = L(A_q) \setminus L(A_q)X^+$.

Using the requirements above as a reference, we examine the relevance of automaton theoretic arguments to the usage of codes in information transmission systems from seven points of view: code design according to specifications (Section 2); models for the realization of encodings (Section 3); models for noisy channels (Section 4); error-detection during decoding (Section 5); fault-tolerant decoding (Section 6); error-correcting codes and the realization of error-correction (Section 7); models for the realization of decoding in the absence of noise (Section 8). We conclude with a summary and a wish-list of issues to be addressed in Section 9. Our list of references is extensive, but far from being complete. To keep the survey focused, many relevant details had to be omitted, the choice never being easy.

2   Code Design

Often automata are used to characterize code properties. Turning such characterizations around, one can sometimes use automata to design codes according to a given specification.

For example, to obtain a code with decoding delay 0, a *prefix code*, one can use the following characterization, which, by now, has become folklore.

Proposition 1. *Let* $A_q = (Q, X, \delta, q, F)$ *be a deterministic finite acceptor with* $q \notin F$. *Then* $L_1(A_q)$ *is a prefix code. Conversely, if* $L \subseteq X^+$ *is a non-empty prefix code then there is an deterministic finite acceptor* A_q *such that* $L = L_1(A_q)$.

Similar, but less explicit characterizations exist also for hypercodes [47,48,49] and code classes related to infix or outfix codes [10,12,41], typically in terms of the syntactic monoid of such codes. In addition, many studies concern the automaton theoretic characterization of the set of messages encoded using a given code. To get a fairly comprehensive understanding of the situation we refer to [4,16,45,50]. Further important early studies concerning the connection between automata and codes include [7,30,31,33,38].

Not every characterization of a class of codes by automata lends itself readily to the design and construction of codes according to a given specification. We explain the issue using the example of solid codes.

Recall that a *solid code* over Y is a non-empty set $C \subseteq Y^+$ satisfying the following two conditions (see [16]): (1) if $xuy = v$ with $u, v \in C$ and $x, y \in Y^*$, then $x = y = \lambda$; (2) if $ux, xv \in C$ with $u, v \in X^+$ then $x = \lambda$. The first condition states that C is an *infix code*. The second one states that C is *overlap-free*. Solid codes are revisited in Section 6 below.

Let $A_q = (Q, Y, \delta, F)$ be a deterministic finite acceptor, every state of which is useful and reachable, such that $\delta(f, a)$ is undefined for every $f \in F$ and every $a \in Y$. Consider the *state-pair graph* of A_q defined as follows (see [9]): (1) the set of nodes is $Q \times Q$; (2) there is an edge labelled $a \in Y$ from $(p, r,)$ to (p', r') if and only if $\delta(p, a, p')$ and $\delta(r, a, r')$, where $p, p', r, r' \in Q$.

The following automaton theoretic characterization of the class of solid codes is given in [9].

Proposition 2. *([9], Theorem 9)* $L(A_q)$ *is a solid code, if and only if the state-pair graph of* A_q *has the following two properties:*

1. *For* $p, r \in Q$, *there is no proper path from* (q, p) *to* (f, r) *with* $f \in F$ *unless* $p = q$ *and* $r \in F$.
2. *For* $p, r \in Q$, *there is no proper path from* (q, p) *to* (r, f) *with* $f \in F$ *unless* $p = q$ *and* $r \in F$.

Like many such characterizations, this one of solid codes looks promising until one attempts to apply it to the construction of such codes. Indeed, the automaton theoretic descriptions of classes of codes often only serve as characterizations and are as difficult to use in constructions as the original definitions.

In the case of solid codes, large classes of examples can be constructed by other means than automata [13].

The construction implied by Proposition 1 is a notable exception. As is well-known, it has a host of applications. Below, in Section 6, we present an automaton construction for solid codes, which also meets the expectations expressed above.

3 Encoding

A *homomorphic encoding* of X requires a set[3] $C \subseteq Y^+$ with $|X| = |C|$ and a mapping γ of X onto C, which is then extended to a homomorphism of X^* into Y^*. For unique decodability, the extension must be injective. When there is no noise on the channel this suffices to achieve Requirement 1.1. To satisfy Requirement 1.6, one needs a code with bounded decoding delay, ideally a prefix code. The efficiency of the realization of the encoding depends on the definition of γ. As a theoretical model, a single-state transducer suffices for the encoding. The actual cost of practical realizations is hidden in this model[4].

For *sequential encoding,* on the other hand, the encoder is a *gsm mapping* – or *(sub-)sequential tramsduction* – from X^* to Y^* (see [2, Chapter IV.2] and [4, Chapter 4.3]). Sequential encodings are used in two situations: (1) the size of the input 'alphabet' is unbounded, for example an infinite language[5] over some alphabet Z; (2) the channel has non-zero memory and the noise on the channel depends on the symbols being transmitted. There are many practical situations in which such encodings are used. Here we only mention *convolution codes* for the second kind.

In summary, automata are a useful model for homomorphic encoding only in the form of deterministic transducers. The single-state encoder is of little help. However, as the codes likely to be used are at least prefix codes, an encoder can be derived from the tree representation of the code. On the other hand, for sequential encoding, automata are *the* natural model.

4 Channels

In a very general setting one defines a discrete channel in measure theoretic terms (see [8], for instance). This approach allows one to derive profound results concerning the information transmission properties, capabilities and limitations of channels under extremely weak conditions. For practical applications one will introduce restrictions which match the physical situation at hand and enable technical solutions.

Sometimes it is convenient, to treat channels as stochastic automata [46] or as stochastic transducers [27] with a finite number of states. Such a channel has finite memory and may satisfy additional crucial assumptions, like stationarity or ergodicity, depending on the transition structure of the underlying automaton.

---

[3] As described, this works for finite codes only. For an infinite code C, one would encode the words of an infinite language $L \subseteq X^+$ with $|C| = |L|$ using a mapping γ of L onto C, again extended to a homomorphism. The language L may need to meet certain conditions for such an encoding to be usable.

[4] For block codes, which use words of some fixed length k as input symbols and words of some length ℓ with $\ell > k$ as output words, typical realizations may involve shift registers. This type of implementation relies on the choice of the code as a linear space.

[5] Of course, one could encode Z instead. However, encoding of the words in the language may be more natural and more economical.

In other situations, it is adequate to abstract even further by ignoring probabilities. In this case, the channel could be modelled by a transduction [2] in the
following sense: v is an output of the transduction for input u if and only if the
probability of channel input u resulting in channel output v is significantly greater
than 0. This concept is used frequently in the usual theory of error-detecting
or error-correcting codes. In a more systematic and more rigorous way it is introduced in [20] and [16, pp. 522–525 and 534] and explained in greater detail
in [15,21,22]. To show the connection between automata and channel models, we
sketch the idea: Consider some *basic types of errors,* which might occur during
information transmission. Such types could be, for instance[6], *substitution* σ of a
symbol, *insertion* ι of a symbol, *deletion* δ of a symbol or a combination of these.
Moreover, constraints may be imposed to indicate that a certain error type can
only occur at most m times in L consecutive symbols where $m < L$. Channels
with such specifications are called[7] *SID channels.* How to turn the specification
of an SID channel into a transducer is explained in [24]. In Fig. 2 we show a nondeterministic transducer defined by the binary SID channel $\sigma \odot \delta(1,2)$ in which
at most one substitution or deletion error occurs in every two symbols.

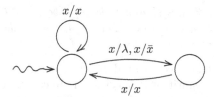

Fig. 2. Transducer for the binary SID channel $\sigma\odot\delta(1,2)$, where $x \in \{0,1\}$ and $\bar{x} = 1-x$

A stochastic transducer modelling the channel with insertions and deletions
of [51] is shown in [27].

5   Error-Detection

An error is detected when the transmitted encoded message v is different from
every possible encoded message. Equivalently, an encoding is *error-detecting,* if
and only if, for every encoded message u, the set of potential channel outputs
for u contains u and no other encoded message [16, Definition 4.1].

Let $W \subseteq X^\eta$ be the set of potential (or highly likely) η-messages, with η as
above. The first of the criteria requires one to decide whether $v \in \gamma(W)$. This
may be very costly or even impossible, even when $W = X^\eta$.

The second criterion is a static condition on the encoding. It states when an
encoding is error-detecting, but does not help with the actual detection of an
error.

---

[6] Which error types are to be considered depends on the physical properties of the
channel.

[7] For brevity, we do not explain the full range of SID channels here. The SID channels
form a small, but important, subset of the set of $\mathcal{P}$-channels [16].

Systematic studies of the error-detection capabilities of languages with respect to SID channels are presented in [25,26,28,29].

6 Fault-Tolerance

Informally, we say that an encoding is *fault-tolerant* when errors have only a bounded effect. For example, consider the following situation: The transmitted encoded message v has a decomposition $v_0 x_1 v_1 x_1 \cdots x_n v_n$ into stretches v_i which contain errors and cannot be decoded and error-free stretches x_i which can be decoded. If, in addition, the decoding of the x_i is not affected by the errors in the other parts, some degree of *fault-tolerance* is achieved. Solid codes can be characterized as the class of codes which afford exactly this kind of fault-tolerance [19]. Fault-tolerance for finite solid codes is realized by deterministic state-invariant transducers without look-ahead. State-invariance means that the transducer's behaviour does not depend on the initial state; lack of look-ahead means that the decoding is output only when a complete code word has been read.

Proposition 3. [16,32,43] *Let X and Y be alphabets and $C \subseteq Y^+$ with $|C| = |X|$. Let γ be a bijection of X onto C. Then C is solid code if and only if there is a state-invariant decoder without look-ahead for γ.*

This characterization of solid codes by decoders has recently been strengthened in two ways, both in the spirit of Proposition 1. Recall that a set $C \subseteq Y^+$ is a *p-infix code* if $Y^* C Y^+ \cap C = \emptyset$ and that the *suffix root* of C is the set $\sqrt[\text{Suff}]{C}$ of words $y \in C$ such that no proper suffix of y is in C.

Proposition 4. [17,18] *Let A_q be a deterministic, possibly infinite, acceptor.*

1. *$L_1(A_q)$ is a prefix code (Proposition 1).*
2. *$\bigcap_{q' \in Q} L_1(A_{q'})$ is an overlap-free p-infix code.*
3. *$\sqrt[\text{Suff}]{\bigcap_{q' \in Q} L_1(A_{q'})}$ is a solid code.*

Thus, as for prefix codes, every deterministic acceptor defines a solid code. Moreover, the construction used by Levenshtein [32] and Romanov [43] in the proof of Proposition 3 can be adapted to infinite, in particular regular, solid codes.

Proposition 5. [17,18] *A non-empty set $C \subseteq Y^+$ is a solid code if and only if there is a deterministic initially connected acceptor A_q such that*

$$C = \sqrt[\text{Suff}]{\bigcap_{q' \in Q} L_1(A_{q'})}.$$

Moreover, if C is a regular set then there is a finite acceptor with this property.

When C is regular, one can transform the acceptor built in the proof of Proposition 5 into a transducer for fault-tolerant deoding [18].

In Propositions 4 and 5, taking the intersection corresponds to state-invariance and taking the suffix root corresponds to lack of look-ahead in Proposition 3.

7   Error-Correction

We now turn to the task of *error-correction*. A message w has been issued, encoded as $u = \gamma(w)$; u has been sent through the channel, and v has been observed as the channel output. Because of noise, v can differ from u. There are three cases:

1. $u = v$: Then v is correct.
2. $u \neq v$, but $v = \gamma(w')$ for some message w': Then v is incorrect; however, the error is not detected.
3. $v \neq \gamma(w')$ for any message w': Then an error is detected.

Usually one assumes that errors are an exception[8], that is,

$$\mathsf{Prob}(\text{output} = u \mid \text{input} = u) > \frac{1}{2}.$$

As a first step, one attempts to recover u from v. In general, this may not be possible with absolute certainty. As a compromise, one attempts to find a word w satisfying the following two conditions:

- $\mathsf{Prob}(w \mid v) \geq \mathsf{Prob}(w' \mid v)$ for all encoded messages w'.
- w is an encoded message.

Here $\mathsf{Prob}(w \mid v)$ and $\mathsf{Prob}(w' \mid v)$ are the posterior probabilities of w and w' having been the input of the channel when v has been observed as its output. Thus, one attempts a *maximum likelihood* correction, leading ultimately to a *maximum likelihood decoding*. In the first case above, this step results in u; in the second case, it results in v; in the third case, the result of this step could be u or another encoded message.

If the channel has been modelled by a stochastic transducer [27], maximum likelihood correction can be achieved by an adaptation of the *Viterbi algorithm* [6,35]. Examples of the application of variants of the Viterbi algorithm to the decoding of variable-length codes or of convolution codes in the presence of synchronization errors include [11,37,39].

In most of the usual theory of error-correcting codes one works with block codes, and error-correction is performed by code words. Moreover, insertions and deletions are not considered. Thus, a channel input and the resulting output have the same lengths. Let $Y = \{0, 1, \dots, q-1\}$ be the channel alphabet, and let $C \subseteq Y^\ell$ be the block code in use with $\ell > 1$ being the code word length. One could build a table of the posterior probabilities $\mathsf{Prob}(u \mid v)$ for $u \in C$ and $v \in Y^\ell$. Typical assumptions about the channel include that it is stationary and memoryless and that $\mathsf{Prob}(x \mid x) = \mathsf{Prob}(y \mid y) > \frac{1}{2}$ for all $x, y \in Y$. An additional typical assumption states that there is a number $p \in (0, 1)$ such that $p = \mathsf{Prob}(x \mid y)$ for all $x, y \in Y$ with $x \neq y$. Clearly, $(q-1)p < \frac{1}{2}$. Thus, if $u \in C$ differs from $v \in Y^\ell$ at r positions, then

$$\mathsf{Prob}(u \mid v) = \big(1 - (q-1)p\big)^{\ell - r} \cdot p^r,$$

---

[8] In addition to this assumption some further, more technical conditions must be met for the present explanation of maximum likelihood correction to be really correct. See [35].

and this probability strictly decreases when r increases. The number r is called the *Hamming distance* d_H between u and v. One can now replace the maximum likelihood correction as follows: Instead of looking for $w \in C$ with $\mathsf{Prob}(w \mid v)$ maximal, one determines $w \in C$ for which $d_H(w, v)$ is minimal.

The set Y^ℓ is a metric space with respect to the Hamming distance. In essence, a part of this space is covered with disjoint circles around the code words, each circle having the same diameter, the diameter chosen maximal. A word inside a circle is assumed to have resulted from the word at the centre. There may be words outside the circles. For these no clear assignment is possible.

Instead of the Hamming distance other distance measures like the *Lee distance* d_L have been used for block codes. Which distance measure is adequate depends on the probabilistic or physical characteristics of the channel. For details of this connection see [16, pp. 595–597] and, in greater detail, [5, pp. 74–77]. In the usual theory of error-correcting codes one then adds further mathematical structure to simplify the calculation of the centres of these circles.

If the channel can delete or insert symbols, one can no longer rely on the input and output having the same lengths. In that situation the distinction between block codes or variable-length codes and even between homomorphic and sequential encodings becomes blurred.

Still, under special assumptions about the physical or stochastic properties of the channel, one can reduce maximum likelihood decoding to minimum distance decoding. This can work, for example, when the *Levenshtein distance* reflects the probabilities in the sense outlined above [34]. In [23], a distance measure D_τ for SID channels is introduced, which specializes to the Hamming distance or various kinds of the Levenshtein distance, depending on the error type τ of the channel. As in the cases of the Hamming distance for block codes, one can relate the error-correction capability of a code C with the minimum distance between distinct code words[9].

Proposition 6. [23,34] *Let C be a finite, non-empty subset of Y^+ and let $\ell = \max\{\lg(x) \mid x \in C\}$.*

1. *If C is a block code and the channel is of type $\tau(m, L)$ with $L \geq \ell$ then C is error-correcting for this channel if and only if $D_\tau(C) > 2m$ [23].*
2. *C is error-correcting for the channel $\iota \odot \delta(m, L)$ with $L \geq \ell$ if and only if $D_{\iota\odot\delta}(C) > 2m$ [34].*
3. *C is error-correcting for the channel $\sigma \odot \iota \odot \delta(m, L)$ with $L \geq \ell$ if and only if $D_{\sigma\odot\iota\odot\delta}(C) > 2m$ [34].*

Greater detail and further references are found in [23]. Proposition 6 expresses the idea, mentioned before, of drawing disjoint circles of maximal diameter around the code words. For error-correction, all words inside such a circle are mapped onto its centre.

In the traditional context of block codes for the substitution channel, minimum-distance decoding is aided by algebraic structure on the codes. For variable-length codes of for channels with insertions and deletions such structure seems

---

[9] Let $C \subseteq Y^+$ and let d be a distance on Y^+. Define the *minimum distance* of C as $d(C) = \min\{d(x,y) \mid x,y \in C, x \neq y\}$. In cases when $d(x,y)$ is undefined, we use the convention that $d(x,y) = \infty$.

not to help. In this situation, the specification of the channel as a transducer could solve the problem. By definition, a code is error-correction for a channel if no two distinct encoded messages can result in the same channel output (see [16]). Thus, ideally, error-correction could be achieved by the inverse of the transduction defining the channel.

Regardless of whether the channel is modelled as a stochastic or a discrete transducer, the delay resulting from the uncertainty about the most likely correction can be a serious problem.

In [14] the error-correction capabilities of solid codes and closely related codes are analysed. The binary solid code $\{0011, 010111\}$ corrects errors in finite messages sent through a channel of type $\delta(1, 6)$, but not in infinite messages. The ternary solid code $\{0022, 02122, 001122\}$ corrects errors in ζ-messages sent through the same channel [15,16].

8 Decoding

In an information transmission system, the decoding δ can be divided into two steps: (1) transformation of the channel output v into an encoded message u'; (2) computation of $\gamma^{-1}(u')$. The former concerns error-handling. The latter is decoding in the absence of noise. For this, one relies primarily on the theory of (sequential) transductions (see [4, Chapter 4]). Both in the noiseless and the noisy case, the decoding (or deciphering) delay is an important issue. Definitions of the *decoding delay* and the *synchronization delay* of codes used with noisy channels can be found in [16, pp. 535–539].

9 Summary

With a focus on the usage of codes in information transmission systems, which questions or results relating automata and codes are useful?

We have examined eight aspects: (1) general requirements; (2) code design; (3) encoding; (4) channels; (5) error-detection; (6) fault-tolerance; (7) error-correction; (8) decoding without noise. The general requirements provide the framework for the investigation. In code design one attempts to construct a code according to a given specification and to prove the correctness of the construction. For encoding a realization is needed, hence also a cost analysis; automata look like the natural candidates to model encoders. For many modern information transmission systems, SID channels seem to model the physical characteristics of the channels involved adequately; in turn, SID channels can be described as non-deterministic or stochastic transducers. Error-detection, fault-tolerance and error-correction capabilities depend on the properties of the channel and of the code; for error-correction one uses either a variant of the Viterbi algorithm or the inverse of the transducer describing the channel; both are based on state-transition models. After error-correction, standard decoding is required which could use the inverse of the transducer implementing the encoding.

There are automata everywhere in an information transmission system!

Research into all aspects of codes for modern noisy channels is still in its early stages. Given the natural presence of automaton models in all parts of an information transmission system, we believe that automaton theoretic methods can be quite useful in addressing not only the theoretical issues concerning such systems, but also the practical implementation problems. With this in mind, we formulate the following wish-list or, rather, research programme:

1. Code properties with respect to errors: Many classes of codes below the prefix codes, even below the bifix codes, have distinct error-handling behaviours. This concerns, for instance, the shuffle hierarchy, solid codes, semaphore codes and types of codes derived from these. Precise descriptions of the error-handling capabilities for these classes of code are needed with respect to SID channels.

2. Code design: 'Useful' characterizations of classes of codes by automata are needed, useful from several points of view: (1) construction of codes according to specification; (2) verification of code properties; (3) translation into efficient implementations.

3. Encoding and decoding: Code-specific or code-class-specific transducer models for encoding and decoding (in the absence of noise) are needed to lead to efficient implementations. Ideally, they would be derived from the code design.

4. Fault-tolerance: Provide a formal, possibly parameterized definition of fault-tolerance. Characterize the codes which achieve fault-tolerance in this general sense. How are these codes related to the solid codes?

5. Error-detection: Under which conditions can error-detection be guaranteed at low cost? Such conditions would concern the SID channel, the message space and the code.

6. Error-correction: Analyse correction strategies. How much of the channel output suffices for the error-correction to succeed with high enough probability? This would determine when the Viterbi algorithm can be reset or when the error-correcting transducer can produce the next output.

7. Correction strategies: Evaluate and compare the performance of error-correction strategies as outlined in Section 7, in particular the variants of the Viterbi algorithm, for various classes of codes low in the code hierarchy and various SID channels.

8. Decoding delay: Find criteria, similar to the ones for the noiseless case, by which the decoding delay of a code with respect to a given SID channel is characterized.

9. Distance measures: Determine methods by which to design codes satisfying a bound on the minimum distance within a given class of codes or by which to compute the minimum distance for a given code.

Partial solutions to some of these problems exist: some of them are mentioned above. With our list we intend to exhibit some of the major problems concerning codes for modern information transmission systems. In the traditional theory of error-correcting codes it took several decades of research from the first formulation of the task to efficient solutions. Regarding variable-length codes and systems with synchronization errors, many similar steps must be made. Analyses of concrete cases and simulations of systems could serve as further guidance.

Given the nature of information transmission systems, automaton models seem to be the natural tool.

Acknowledgments. This research was supported by the Natural Sciences and Engineering Research Council of Canada.

References

1. Béal, M.-P., Berstel, J., Marcus, B.H., Perrin, D., Reutenauer, C.H., Siegel, P.H.: Variable length codes and finite automata. In: Woungang, I., Misra, S., Misra, S.C. (eds.) Selected Topics on Information and Coding Theory. Series on Coding Theory and Cryptology, vol. 7, pp. 505–584. World Scientific, Singapore (2010)
2. Berstel, J.: Transductions and Context-Free Languages. Leitfäden der angewandten Mathematik und Mechanik, LAMM, vol. 38. B. G. Teubner, Stuttgart (1979)
3. Berstel, J., Perrin, D.: Theory of Codes. Academic Press, Orlando (1985)
4. Berstel, J., Perrin, D., Reutenauer, C.H.: Codes and Automata. Encyclopedia of Mathematics and Its Applications, vol. 129. Cambridge University Press, Cambridge (2010)
5. Duske, J., Jürgensen, H.: Codierungstheorie. BI Wissenschaftsverlag, Mannheim (1977)
6. Forney Jr., G.D.: The Viterbi algorithm. Proc. IEEE 61(3), 268–278 (1973)
7. Glebskii, Y.V.: Coding by means of finite automata. Dokl. Akad. Nauk. SSSR 141, 1054–1057 (1961) (in Russian); English translation: Soviet Physics Dokl. 6, 1037–1039 (1962)
8. Guiaşu, S.: Information Theory with Applications. McGraw-Hill, New York (1977)
9. Han, Y.-S., Salomaa, K.: Overlap-free languages and solid codes. Internat. J. Foundations Comput. Sci. 22, 1197–1209 (2011)
10. Ito, M., Jürgensen, H., Shyr, H.J., Thierrin, G.: Outfix and infix codes and related classes of languages. J. Comput. System Sci. 43, 484–508 (1991)
11. Jégou, H., Malinowski, S., Guillemot, C.: Trellis state aggregation for soft decoding of variable length codes. In: Proceedings of the 2005 IEEE Workshop on Signal Processing Systems: Design and Implementation (SIPS 2005), Athens, Greece, November 2-4, pp. 603–608. IEEE, Piscataway (2005)
12. Jürgensen, H.: Syntactic monoids of codes. Acta Cybernet. 14, 117–133 (1999)
13. Jürgensen, H., Katsura, M., Konstantinidis, S.: Maximal solid codes. J. of Automata, Languages and Combinatorics 6, 25–50 (2001)
14. Jürgensen, H., Konstantinidis, S.: Variable-length codes for error correction. In: Fülöp, Z. (ed.) ICALP 1995. LNCS, vol. 944, pp. 581–592. Springer, Heidelberg (1995)
15. Jürgensen, H., Konstantinidis, S.: Error correction for channels with substitutions, insertions, and deletions. In: Chouinard, J.-Y., Fortier, P., Gulliver, T.A. (eds.) Information Theory 1995. LNCS, vol. 1133, pp. 149–163. Springer, Heidelberg (1996)
16. Jürgensen, H., Konstantinidis, S.: Codes. In: Rozenberg, G., Salomaa, A. (eds.) Handbook of Formal Languages, vol. 1, pp. 511–607. Springer, Berlin (1997)
17. Jürgensen, H., Staiger, L.: Fault-tolerant acceptors for solid codes. In: Proceedings of the 14th Journées Montoises of Theoretical Computer Science, Université catholique de Louvain, Belgium, Louvain-la-Neuve, September 11-14, p. 8 (2012); the proceedings were distributed electronically on a memory stick at the conference

18. Jürgensen, H., Staiger, L.: Automata for solid codes. Manuscript in Preparation (2013)

19. Jürgensen, H., Yu, S.S.: Solid codes. J. Inform. Process. Cybernet., EIK 26, 563–574 (1990)

20. Konstantinidis, S.: Error Correction and Decodability. PhD Thesis, The University of Western Ontario, London, Canada (1996)

21. Konstantinidis, S.: Structural analysis of error-correcting codes for discrete channels that involve combinations of three basic error types. IEEE Trans. Inform. Theory 45(1), 60–77 (1999)

22. Konstantinidis, S.: An algebra of discrete channels that involve combinations of three basic error types. Inform. and Comput. 167, 120–131 (2001)

23. Konstantinidis, S.: Relationships between different error-correcting capabilities of a code. IEEE Trans. Inform. Theory 47(5), 2065–2069 (2001)

24. Konstantinidis, S.: Transducers and the properties of error-detection, error-correction, and finite-delay decodability. J. UCS 8(2), 278–291 (2002)

25. Konstantinidis, S., O'Hearn, A.: Error-detecting properties of languages. Theoret. Comput. Sci. 276, 355–375 (2002)

26. Konstantinidis, S., Perron, S., Wilcox-O'Hearn, L.A.: On a simple method for detecting sychronization errors in coded messages. IEEE Trans. Inform. Theory 49(5), 1355–1363 (2003)

27. Konstantinidis, S., Santean, N.: On the definition of stochastic λ-transducers. Internat. J. Comput. Math. 86(8), 1300–1310 (2009)

28. Konstantinidis, S., Silva, P.: Maximal error-detecting capabilities of formal languages. J. of Automata, Languages and Combinatorics 13(1), 55–71 (2008)

29. Konstantinidis, S., Silva, P.V.: Computing maximal error-detecting capabilities and distances of regular languages. Fund. Inform. 101, 257–270 (2010)

30. Levenshtein, V.I.: Самонастраибающиеся автоматы для декодирования сообщений. Dokl. Akad. Nauk. SSSR 141, 1320–1323 (1961); English translation: Self-adaptive automata for decoding messages. Soviet Physics Dokl. 6, 1042–1045 (1962)

31. Levenshtein, V.I.: Об обращении конечных автоматов. Dokl. Akad. Nauk. SSSR 147, 1300–1303 (1962); English translation: The inversion of finite automata. Soviet Physics Dokl. 7, 1081–1084 (1963)

32. Levenshtein, V.I.: Декодирующие абтоматы, инвариантные относительно начального состояния. Problemy Kibernet. 12, 125–136 (1964) (in Russian); Decoding Automata, Invariant with Respect to the Initial State

33. Levenshtein, V.I.[10]: О некоторых сбойствах кодирования и самонастраибающихся абтоматах для декодиробания цообщений. Problemy Kibernet. 11, 63–121 (1964); German translation: Über einige Eigenschaften von Codierungen und von selbstkorrigierenden Automaten zur Decodierung von Nachrichten. Probleme der Kybernetik 7, 96–163 (1966); An English translation is available from the Clearinghouse for Federal Scientific and Technical Information, U. S. Department of Commerce, under the title Some Properties of Coding and Self-Adjusting Automata for Decoding Messages, Problems of Cybernetics, Part II, document AD 667 849; it was prepared as document FTD-MT-24-126-67 by the Foreign Technology Division, U. S. Air Force

---

[10] In the translations the author's name is written as *Löwenstein, Levenstein* and *Levenshteyn*.

34. Levenshtein, V.I.: Двоичные коды ц исправлением выпадений, вставок и замещений символов. Dokl. Akad. Nauk. SSSR 163(4), 845–848 (1965); English translation: Binary codes capable of correcting deletions, insertions, and reversals. Soviet Physics Dokl. 10, 707–710 (1966)

35. MacKay, D.J.C.: Information Theory, Inference, and Learning Algorithms, 6th edn. Cambridge University Press, Cambridge (2007)

36. MacWilliams, F.J., Sloane, N.J.A.: The Theory of Error-Correcting Codes, vol. 2. North-Holland, Amsterdam (1977)

37. Mansour, M.F., Tewfik, A.H.: Convolutional decoding in the presence of synchronization errors. IEEE J. Selected Areas in Communications 28(2), 218–227 (2010)

38. Markov, AI.A.[11]: Non-recurrent coding. Problemy Kibernet. 8, 169–186 (1961) (in Russian); German translation: Nicht rekurrente Codierung. Probleme der Kybernetik 8, 154–175 (1962)

39. Maxted, J.C., Robinson, J.P.: Error recovery for variable length codes. IEEE Trans. Inform. Theory 31(6), 794–801 (1985)

40. Peterson, W.W., Weldon Jr., E.J.: Error-Correcting Codes, 2nd edn. MIT Press, Cambridge (1972)

41. Petrich, M., Thierrin, G.: The syntactic monoid of an infix code. Proc. Amer. Math. Soc. 109, 865–873 (1990)

42. Pless, V.S., Huffman, W.C. (eds.): Handbook of Coding Theory, vol. 2. North-Holland, Elsevier (1998)

43. Romanov, O.T.: Об инвариантных декодирующих автоматах без предвосхищения. Problemy Kibernet. 17, 233–236 (1966) (in Russian); Invariant Decoding Automata without Look-ahead

44. Sakarovitch, J.: Elements of Automata Theory. Cambridge University Press, Cambridge (2009); English translation of Éléments de théorie des automates, Vuibert Informatique, Paris (2003)

45. Shyr, H.J.: Free Monoids and Languages, 3rd edn. Hon Min Book Company, Taichung (2001)

46. Starke, P.H.: Abstrakte Automaten. VEB Deutscher Verlag der Wissenschaften, Berlin (1969); English translation by I. Shepherd: Abstract Automata. North-Holland, Amsterdam (1972)

47. Thierrin, G.: Hypercodes, right convex languages and their syntactic monoids. Proc. Amer. Math. Soc. 83(2), 255–258 (1981)

48. Thierrin, G.: The syntactic monoid of a hypercode. Semigroup Forum 6, 227–231 (1973)

49. Valkema, E.: Syntaktische Monoide und Hypercodes. Semigroup Forum 13(77), 119–126 (1976/1977)

50. Yu, S.-S.: Languages and Codes. Tsang Hai Book Publishing Co., Taichung (2005)

51. Zigangirov, K.Sh: Последовательное декодирование для двоичного канала с выпадениями и вставками. Problemy Peredachi Informatsii 5(2), 23–30 (1969); English translation: Sequential decoding for a binary channel with drop-outs and insertions. Problems Inform. Transmission 5(2), 17–22 (1969)

---

[11] In the German translation the author's name is written as *Markow*.

Applications of Symbolic Finite Automata

Margus Veanes

Microsoft Research
margus@microsoft.com

Abstract. Symbolic automata theory lifts classical automata theory to rich alphabet theories. It does so by replacing an explicit alphabet with an alphabet described implicitly by a Boolean algebra. How does this lifting affect the basic algorithms that lay the foundation for modern automata theory and what is the incentive for doing this? We investigate these questions here. In our approach we use state-of-the-art constraint solving techniques for automata analysis that are both expressive and efficient, even for very large and infinite alphabets. We show how symbolic finite automata enable applications ranging from modern regex analysis to advanced web security analysis, that were out of reach with prior methods.

1 Introduction

Classical automata theory makes two basic assumptions: there is a *finite state space*; and there is a *finite alphabet*. Here we challenge the second assumption by looking at how we can relax it while still maintaining all or most of the benefits of classical automata theory. One of the drawbacks of classical finite state automata is that they do not scale well for large alphabets. Although there are various techniques that address the scalability problem, such as, partial transition functions to avoid irrelevant or unused characters [3,13], integer ranges for succinct representation of contiguous ranges of characters [1], binary decision diagrams for succinct representation of transition functions [7], as well as various extensions with registers such as register automata [10,5] and extended finite automata [12]. Extensions with registers in general lead to infinite state systems or lack of closure properties. There is also research on register automata or automata over data words that focuses on their expressive power and decidability properties [11].

Our interest in this topic originates from the need to support regular expressions in the context of program analysis [17]. Regular expressions or regexes are stated over strings of basic Unicode characters. The runtime representation of characters in modern runtimes like JVM and .NET, as well as in scripting languages like JavaScript, uses the UTF16 encoding. From the point of view of regexes, the alphabet is the set of unsigned integers less than 2^{16} or in other words 16-bit bitvectors. For example the regex character class [\u2639\u263A] matches the symbols ☹ and ☺. Regexes do not directly support symbols in the supplementary Unicode planes (i.e. symbols that are formed from surrogate

S. Konstantinidis (Ed.): CIAA 2013, LNCS 7982, pp. 16–23, 2013.

pairs and whose Unicode code point is $\geq 2^{16}$). For example, the surrogate pair \uD83D\uDE0A that also happens to encode a smiley symbol is treated as two separate characters by a regex, and the regex ^(\uD83D[\uDE00-\uDE4F])*$ matches a string that encodes a sequence of Unicode emoticons [2].[1]

Symbolic Finite Automata or SFAs were introduced, as an extension of classical finite state automata that allows transitions to be labeled with predicates defined in a separate alphabet algebra. The concept of automata with predicates instead of concrete symbols was first mentioned in [19] and was first discussed in [14] in the context of natural language processing. The alphabet theory in SFAs is assumed to be an *effective Boolean algebra*. The main intuition is that an SFA uses an alphabet as a plug-in through an API or interface. The only requirement is that the interface supports operations of a Boolean algebra.

To illustrate the role of the alphabet algebra consider the last regex example above. The predicate 0xDE00 $\leq x \wedge x \leq$ 0xDE4F is an example of such a predicate in a character theory that uses linear arithmetic (modulo-2^{16}, or bitvector arithmetic) and one fixed variable x. We abbreviate it by [\uDE00-\uDE4F] using the standard character class notation of regexes. The following SFA is equivalent to the above regex of emoticons, say $M_{\text{emoticons}}$:

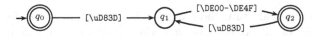

The regex character class [\uDE38-\uDE40] matches the set of low-surrogate halves of a "cat face" emoticon. Suppose we want to construct an SFA that accepts all strings of emoticons that contain no cat face emoticons. One way to do this is to construct the SFA $M_{\text{emoticons}} \times M_{\text{nocats}}$, where M_{nocats} is the SFA:

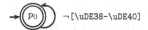

There are many fundamental questions about if and how classical algorithms and techniques can be lifted to SFAs. Some algorithms depend more on the alphabet than others. For example, union of SFAs uses only disjunctions of predicates over characters while intersection uses only conjunctions. Determinization on the other hand needs all Boolean operations. Satisfiability checking of predicates is used to avoid infeasible transitions. Some tradeoffs of the algorithms, when applied to string analysis, are studied in [9]. Minimization of SFAs is studied in [15]. It differs from the classical algorithms [4] with respect to how the alphabet is being used.

---

[1] Emoticons are symbols with code points between 0x1F600 and 0x1F64F. As an example, the surrogate pair \uD83D\uDE0A encodes the Unicode code point 0x1F60A that is the code of a smiley symbol similar to ☺.

Here we discuss basic properties of SFAs, the role of the alphabet, and we describe different applications of SFAs, with a focus on the role of the symbolic alphabet. Two concrete applications are: *regex processing* and *security analysis of string sanitizers*.

2   Effective Boolean Algebras and SFAs

An *effective Boolean algebra* $\mathcal{A}$ has components $(\mathfrak{D}, \Psi, [\![\_]\!], \bot, \top, \vee, \wedge, \neg)$. $\mathfrak{D}$ is an r.e. (recursively enumerable) set of *domain elements*. Ψ is an r.e. set of *predicates* closed under the Boolean connectives and $\bot, \top \in \Psi$. The *denotation function* $[\![\_]\!] : \Psi \rightarrow 2^{\mathfrak{D}}$ is r.e. and is such that, $[\![\bot]\!] = \emptyset$, $[\![\top]\!] = \mathfrak{D}$, for all $\varphi, \psi \in \Psi$, $[\![\varphi \vee \psi]\!] = [\![\varphi]\!] \cup [\![\psi]\!]$, $[\![\varphi \wedge \psi]\!] = [\![\varphi]\!] \cap [\![\psi]\!]$, and $[\![\neg\varphi]\!] = \mathfrak{D} \setminus [\![\varphi]\!]$. For $\varphi \in \Psi$, we write $IsSat(\varphi)$ when $[\![\varphi]\!] \neq \emptyset$ and say that φ is *satisfiable*. $\mathcal{A}$ is *decidable* if $IsSat$ is decidable.

The intuition is that such an algebra is represented programmatically as an API with corresponding methods implementing the Boolean operations and the denotation function. We are primarily going to use two such effective Boolean algebras in the examples, but the techniques in the paper are fully generic.

$\mathbf{2^{bvk}}$ is the powerset algebra whose domain is the finite set BVk, for some $k > 0$, consisting of all nonnegative integers less than 2^k, or equivalently, all k-bit bit-vectors. A predicate is represented by a BDD of depth k.[2] The Boolean operations correspond directly to the BDD operations, $\bot$ is the BDD representing the empty set. The denotation $[\![\beta]\!]$ of a BDD β is the set of all integers n such that a binary representation of n corresponds to a solution of β.

SMT$^{\sigma}$ is the decision procedure for a theory over some sort σ, say integers, such as the theory of integer linear arithmetic. This algebra can be implemented through an interface to an SMT solver. Ψ contains in this case the set of all formulas $\varphi(x)$ in that theory with one fixed free integer variable x. Here $[\![\varphi]\!]$ is the set of all integers n such that $\varphi(n)$ holds. For example, a formula $(x \bmod k) = 0$, say div_k, denotes the set of all numbers divisible by k. Then $div_2 \wedge div_3$ denotes the set of numbers divisible by six.

Extending a given alphabet domain with new characters in the concrete (classical) case is more or less trivial, while in the symbolic case it may not be possible at all or is difficult. We are using the following construct for alphabet extensions.

Definition 1. The *disjoint union* $\mathcal{A}+\mathcal{B}$ of two effective Boolean algebras $\mathcal{A}$ and $\mathcal{B}$, is an effective Boolean algebra where,

---

[2] The variable order of the BDD is the reverse bit order of the binary representation of a number, in particular, the most significant bit has the lowest ordinal.

$$\mathfrak{D}_{A+B} \stackrel{\text{def}}{=} (\mathfrak{D}_A \times \{1\}) \cup (\mathfrak{D}_B \times \{2\});$$

$$\Psi_{A+B} \stackrel{\text{def}}{=} \Psi_A \times \Psi_B;$$

$$[\![\langle \alpha, \beta \rangle]\!]_{A+B} \stackrel{\text{def}}{=} ([\![\alpha]\!]_A \times \{1\}) \cup ([\![\beta]\!]_B \times \{2\})$$

$$\langle \alpha, \beta \rangle \vee_{A+B} \langle \alpha', \beta' \rangle \stackrel{\text{def}}{=} \langle \alpha \vee_A \alpha', \beta \vee_B \beta' \rangle;$$

$$\langle \alpha, \beta \rangle \wedge_{A+B} \langle \alpha', \beta' \rangle \stackrel{\text{def}}{=} \langle \alpha \wedge_A \alpha', \beta \wedge_B \beta' \rangle;$$

$$\neg_{A+B}\langle \alpha, \beta \rangle \stackrel{\text{def}}{=} \langle \neg_A \alpha, \neg_B \beta \rangle;$$

$$\bot_{A+B} \stackrel{\text{def}}{=} \langle \bot_A, \bot_B \rangle;$$

$$\top_{A+B} \stackrel{\text{def}}{=} \langle \top_A, \top_B \rangle.$$

It is straightforward to prove by using distributive laws of intersection and union that the additional conditions of the denotation function hold for the above definition, i.e., that $A+B$ is indeed an effective Boolean algebra. In particular, consider conjunction (we drop the indices of the algebras as they are clear from the context)

$$
\begin{aligned}
[\![\langle \alpha, \beta \rangle \wedge \langle \alpha', \beta' \rangle]\!] &= [\![\langle \alpha \wedge \alpha', \beta \wedge \beta' \rangle]\!] \\
&= [\![\alpha \wedge \alpha']\!] \times \{1\} \cup [\![\beta \wedge \beta']\!] \times \{2\} \\
&= ([\![\alpha]\!] \cap [\![\alpha']\!]) \times \{1\} \cup ([\![\beta]\!] \cap [\![\beta']\!]) \times \{2\} \\
&= ((\underbrace{[\![\alpha]\!] \times \{1\}}_{A}) \cap (\underbrace{[\![\alpha']\!] \times \{1\}}_{A'})) \cup ((\underbrace{[\![\beta]\!] \times \{2\}}_{B}) \cap (\underbrace{[\![\beta']\!] \times \{2\}}_{B'})) \\
&= (A \cap A') \cup (B \cap B') \cup \underbrace{(A \cap B')}_{=\emptyset} \cup \underbrace{(B \cap A')}_{=\emptyset} \\
&= (A \cup B) \cap (A' \cup B') \\
&= [\![\langle \alpha, \beta \rangle]\!] \cap [\![\langle \alpha', \beta' \rangle]\!]
\end{aligned}
$$

Another useful construct when dealing with effective Boolean algebras is *domain restriction*. In SFAs, domain restriction can be used to limit the alphabet to only those characters that matter.

Definition 2. The *domain restriction* of an effective Boolean algebra A with respect to a nonempty r.e. set $V \subseteq \mathfrak{D}_A$, denoted $A{\restriction}V$, is the same effective Boolean algebra as A except that $\mathfrak{D}_{A{\restriction}V} \stackrel{\text{def}}{=} \mathfrak{D}_A \cap V$ and $[\![\psi]\!]_{A{\restriction}V} \stackrel{\text{def}}{=} [\![\psi]\!]_A \cap V$.

It is easy to check that $A{\restriction}V$ is well-defined. In particular, consider disjunction:

$$
\begin{aligned}
[\![\psi \vee \varphi]\!]_{A{\restriction}V} &= [\![\psi \vee \varphi]\!]_A \cap V = ([\![\psi]\!]_A \cup [\![\varphi]\!]_A) \cap V = ([\![\psi]\!]_A \cap V) \cup ([\![\varphi]\!]_A \cap V) \\
&= [\![\psi]\!]_{A{\restriction}V} \cup [\![\varphi]\!]_{A{\restriction}V}
\end{aligned}
$$

and complement:

$$
\begin{aligned}
[\![\neg\psi]\!]_{A{\restriction}V} &= [\![\neg\psi]\!]_A \cap V = (\mathfrak{D}_A \setminus [\![\psi]\!]_A) \cap V = (\mathfrak{D}_A \cap V) \setminus ([\![\psi]\!]_A \cap V) \\
&= \mathfrak{D}_{A{\restriction}V} \setminus [\![\psi]\!]_{A{\restriction}V}
\end{aligned}
$$

Definition 3. A *symbolic finite automaton (SFA)* M is a tuple $(\mathcal{A}, Q, q^0, F, \Delta)$ where $\mathcal{A}$ is an effective Boolean algebra, called the *alphabet*, Q is a finite set of *states*, $q^0 \in Q$ is the *initial state*, $F \subseteq Q$ is the set of *final states*, and $\Delta \subseteq Q \times \Psi_{\mathcal{A}} \times Q$ is a finite set of *moves* or *transitions*.

Elements of $\mathfrak{D}_{\mathcal{A}}$ are called *characters* and finite sequences of characters, elements of $\mathfrak{D}_{\mathcal{A}}^*$, are called *words*; ϵ denotes the empty word. A move $\rho = (p, \varphi, q) \in \Delta$ is also denoted by $p \xrightarrow{\varphi}_M q$ (or $p \xrightarrow{\varphi} q$ when M is clear) where p is the *source* state, denoted $Src(\rho)$, q is the *target* state, denoted $Tgt(\rho)$, and φ is the *guard* or *predicate* of the move, denoted $Grd(\rho)$. A move is *feasible* if its guard is satisfiable. Given a character $a \in \mathfrak{D}_{\mathcal{A}}$, an *a-move* of M is a move $p \xrightarrow{\varphi} q$ such that $a \in [\![\varphi]\!]$, also denoted $p \xrightarrow{a}_M q$ (or $p \xrightarrow{a} q$ when M is clear). In the following let $M = (\mathcal{A}, Q, q^0, F, \Delta)$ be an SFA.

Definition 4. A word $w = a_1 a_2 \cdots a_k \in \mathfrak{D}_{\mathcal{A}}^*$, is *accepted at state p of M*, denoted $w \in \mathscr{L}_p(M)$, if there exist $p_{i-1} \xrightarrow{a_i}_M p_i$ for $1 \leq i \leq k$ where $p_0 = p$ and $p_k \in F$. The *language accepted by M* is $\mathscr{L}(M) \stackrel{\text{def}}{=} \mathscr{L}_{q^0}(M)$.

For $q \in Q$, we use the definitions

$$\overrightarrow{\Delta}(q) \stackrel{\text{def}}{=} \{\rho \in \Delta \mid Src(\rho) = q\}, \quad \overleftarrow{\Delta}(q) \stackrel{\text{def}}{=} \{\rho \in \Delta \mid Tgt(\rho) = q\}.$$

The definitions are lifted to sets in the usual manner. The following terminology is used to characterize various key properties of M. A state p of M is called *partial* if there exists a character a such that there is no a-move from p.

- M is *deterministic*: for all $p \xrightarrow{\varphi} q, p \xrightarrow{\varphi'} q' \in \Delta$, if $IsSat(\varphi \wedge \varphi')$ then $q = q'$.
- M is *complete*: there are no partial states.
- M is *clean*: for all $p \xrightarrow{\varphi} q \in \Delta$, p is reachable from q^0 and $IsSat(\varphi)$,
- M is *normalized*: for all $p, q \in Q$, there is at most one move from p to q.
- M is *minimal*: M is deterministic, complete, clean, normalized, and for all $p, q \in Q$, $p = q$ if and only if $\mathscr{L}_p(M) = \mathscr{L}_q(M)$.[3]

Determinization of SFAs is always possible and is studied in [16]. Completion is straightforward: if M is not complete then add a new state $q_\emptyset$ and the self-loop $q_\emptyset \xrightarrow{\top} q_\emptyset$ and for each partial state q add the move $(q, \bigwedge_{\rho \in \overrightarrow{\Delta}(q)} \neg Grd(\rho), q_\emptyset)$. Observe that completion requires complementation of predicates.

Normalization is obvious: if there exist states p and q and two distinct transitions $p \xrightarrow{\varphi} q$ and $p \xrightarrow{\psi} q$ then replace these transitions with the single transition $p \xrightarrow{\varphi \vee \psi} q$. This does clearly not affect $\mathscr{L}_p(M)$ for any p.

Cleaning amounts to running standard forward reachability that keeps only reachable states, and eliminates infeasible moves. Observe that having infeasible moves $p \xrightarrow{\perp} q$ is semantically useless and may cause unnecessary state space explosion.

---

[3] It is sometimes convenient to define minimality over incomplete SFAs, in which case the *dead-end* state q ($q \neq q^0$ and $\mathscr{L}_q(M) = \emptyset$) is eliminated if it is present.

3 Applications

The development of the theory of symbolic automata has been driven by several concrete practical problems. Here we discuss two such applications. In each case we illustrate what kind of character theory we are working with, and focus on the benefits of the symbolic representation.

3.1 Regex Processing

Practical applications of regular expressions or *regexes* is ubiquitous. What distinguishes practical regexes from schoolbook regular expressions (besides non-regular features that go beyond capabilities of finite state automata representations) are certain constructs that make them appealing (more succinct) than their classical counterparts such as *bounded quantifiers* and *character classes*.

The size of the alphabet is 2^{16} due to the widely adopted UTF16 standard of Unicode characters, e.g., as a somewhat unusual example, the regex `^[\uFF10-\uFF19]$` matches the set of digits in the so-called Wide Latin range of Unicode. We let the alphabet algebra be $\mathbf{2}^{\text{BV}16}$. Let the BDD β_w^7 represent all ASCII word characters (letters, digits, and underscore) as the set of character codes $\{\text{'0'}, \ldots, \text{'9'}, \text{'A'}, \ldots, \text{'Z'}, \text{'\_'}, \text{'a'}, \ldots, \text{'z'}\}$. (We write '0' for the code 48, 'a' for the code 97, etc.) Let also β_d^7 represents the set of all decimal digits $\{\text{'0'}, \ldots, \text{'9'}\}$ and let $\beta_\_$ represent underscore $\{\text{'\_'}\}$. By using the Boolean operations, e.g., $\beta_w^7 \wedge \neg(\beta_d^7 \vee \beta_\_)$ represents the set of all upper- and lower-case ASCII letters. As a regex character class it is expressible as `[\w-[\d_\x7F-\uFFFF]]`.

Regexes are used in many different contexts. A common use of regexes is as a constraint language over strings for *checking* presence or absence of different patterns, e.g., for *security validation* of packet headers in network protocols. Another application, is the use of regexes for *generating* strings that match certain criteria, e.g., for *fuzz testing* applications that use regexes. A further application is *password generation* based on constraints given in form of regexes. Here is a scenario:[4]

1. Length is k and characters are in visible ASCII range: `^[\x21-\x7E]{k}$`
2. There are at least two letters: `[a-zA-Z].*[a-zA-Z]`
3. There is at least one digit: `\d`
4. There is at least one non-word character: `\W`

Consider SFAs for each case and build their product. The product is constructed by using depth-first search. Unsatisfiable predicates are eliminated so that the result is clean. Dead-end states are also eliminated. Random strings accepted by the automaton can be generated uniformly from its minimized or determinized form. Here the canonical structure of BDDs can be exploited to achieve uniformly random selection of characters from predicates.

---

[4] Recall the standard convention: a regex without the start-anchor `^` matches any prefix and a regex without the end-anchor `$` matches any suffix.

3.2   Sanitizer Analysis

Sanitizers are string transformation routines (special purpose encoders) that are extensively used in web applications, in particular as the first line of defense against cross site scripting (XSS) attacks. There are at least three different string sanitizers involved in a single web page (CssEncoder, UrlEncoder, HtmlEncoder) that have very different semantics and sometimes use other basic encoders, e.g., UrlEncoder uses Utf8Encoder as the first step, while the raw input strings are in fact Utf16 encoded during runtime. A large class of sanitizers (including all the ones mentioned above) can be described and analyzed by using *symbolic finite state transducers* (SFTs) [8]. SFAs are used in that context for certain operations over SFTs, for example for checking domain equivalence of SFTs [18].

The character algebra here is modular integer linear arithmetic (or bitvector arithmetic of an SMT solver, the SMT solver used in our implementation is Z3 [6]). The main advantage of this choice is that it makes it possible to seamlessly combine the guards over characters with expressions over *yields* that are the *symbolic outputs* of SFT moves. A concrete example of a yield is the following transformation that takes a character and encodes it as a sequence of other characters:

$$ f: \quad \lambda x.['\&', '\#', (((x \div 10) \bmod 10) + 48), ((x \bmod 10) + 48), ';'] $$

In general, a yield denotes a function from an input character to an output word (of length that is independent of the input character). For example, a yield can be a function $\lambda x.[x, x]$ that duplicates the input character. Thus, an image of an SFTs is not necessarily SFA-recognizable, which is unlike the classical case where the image of a finite state transducer is always regular. In the above example, for example $f('a')$ is the sequence $['\&', '\#', '9', '7', ';']$ (or the string "a"). A typical SFT move ρ looks like:

$$ \rho: \quad q \xrightarrow{(\lambda x.0 < x < 32)/\lambda x.['\&', '\#', (((x \div 10) \bmod 10) + 48), ((x \bmod 10) + 48), ';']} q $$

that is an HtmlEncoder rule for encoding control characters in state q and remaining in that state. For analyzing say idempotence of an encoder with such rules, the encoder is composed with itself. As a result, this leads to more complex guards and outputs of the resulting composed SFT (SFTs are closed under such composition). Imagine for example composing the move ρ with itself, i.e., roughly speaking, feeding the five output characters as its inputs again five times in a row. Then the guard of the composed rule will have subconditions such as $0 < (((x \div 10) \bmod 10) + 48) < 32$ involving potentially nontrivial arithmetic operations. (In this particular case the guard of the composed move will be infeasible.) One task of idempotence checking is *domain equivalence* of SFTs that reduces to *language equivalence* of SFAs whose guards now involve arithmetic operations of the above kind. Domain equivalence of SFTs essentially means that they accept/reject the same input sequences. Note that not all inputs sequences are valid. Perhaps a bit surprising, but even raw input strings may have misplaced characters (e.g. singleton occurrences of surrogates), assuming the standard Utf16 encoding of characters.

References

1. BRICS finite state automata utilities, http://www.brics.dk/automaton/
2. Emoticons, Unicode standard, v. 6.2,
   http://unicode.org/charts/PDF/U1F600.pdf
3. Béal, M.-P., Crochemore, M.: Minimizing incomplete automata. In: 7th International Workshop on Finite-State Methods and Natural Language Processing, pp. 9–16 (2008)
4. Berstel, J., Boasson, L., Carton, O., Fagnot, I.: Minimization of automata. To appear in Handbook of Automata (2011)
5. Bojanczyk, M., Muscholl, A., Schwentick, T., Segoufin, L., David, C.: Two-variable logic on words with data. In: LICS, pp. 7–16 (2006)
6. de Moura, L., Bjørner, N.: Z3: An Efficient SMT Solver. In: Ramakrishnan, C.R., Rehof, J. (eds.) TACAS 2008. LNCS, vol. 4963, pp. 337–340. Springer, Heidelberg (2008)
7. Henriksen, J., Jensen, J., Jørgensen, M., Klarlund, N., Paige, B., Rauhe, T., Sandholm, A.: Mona: Monadic second-order logic in practice. In: Brinksma, E., Steffen, B., Cleaveland, W.R., Larsen, K.G., Margaria, T. (eds.) TACAS 1995. LNCS, vol. 1019, Springer, Heidelberg (1995)
8. Hooimeijer, P., Livshits, B., Molnar, D., Saxena, P., Veanes, M.: Fast and precise sanitizer analysis with Bek. In: USENIX Security (August 2011)
9. Hooimeijer, P., Veanes, M.: An evaluation of automata algorithms for string analysis. In: Jhala, R., Schmidt, D. (eds.) VMCAI 2011. LNCS, vol. 6538, pp. 248–262. Springer, Heidelberg (2011)
10. Kaminski, M., Francez, N.: Finite-memory automata. TCS 134(2), 329–363 (1994)
11. Segoufin, L.: Automata and logics for words and trees over an infinite alphabet. In: Ésik, Z. (ed.) CSL 2006. LNCS, vol. 4207, pp. 41–57. Springer, Heidelberg (2006)
12. Smith, R., Estan, C., Jha, S., Kong, S.: Deflating the big bang: fast and scalable deep packet inspection with extended finite automata. In: SIGCOMM 2008, pp. 207–218. ACM (2008)
13. Valmari, A., Lehtinen, P.: Efficient minimization of DFAs with partial transition functions. In: Albers, S., Weil, P. (eds.) 25th International Symposium on Theoretical Aspects of Computer Science (STACS 2008), Dagstuhl, pp. 645–656 (2008)
14. van Noord, G., Gerdemann, D.: Finite state transducers with predicates and identities. Grammars 4(3), 263–286 (2001)
15. Veanes, M.: Minimization of symbolic automata. Technical Report MSR-TR-2013-48, Microsoft Research (2013)
16. Veanes, M., Bjørner, N., de Moura, L.: Symbolic automata constraint solving. In: Fermüller, C., Voronkov, A. (eds.) LPAR-17. LNCS, vol. 6397, pp. 640–654. Springer, Heidelberg (2010)
17. Veanes, M., de Halleux, P., Tillmann, N.: Rex: Symbolic Regular Expression Explorer. In: ICST 2010, pp. 498–507. IEEE (2010)
18. Veanes, M., Hooimeijer, P., Livshits, B., Molnar, D., Bjørner, N.: Symbolic finite state transducers: Algorithms and applications. In: POPL 2012, pp. 137–150 (2012)
19. Watson, B.W.: Implementing and using finite automata toolkits. In: Extended Finite State Models of Language, pp. 19–36. Cambridge University Press, New York (1999)

Computing Weights

Houda Abbad[1] and Éric Laugerotte[2]

[1] E.E.D.I.S, University of Djillali Liabes, BP 89, 22000 Sidi Bel Abbes, Algeria
houda.abbad@lycos.com
[2] 1 Normandie Univ, France
2 UR, LITIS, F-76000 Rouen, France
eric.laugerotte@univ-rouen.fr

Abstract. This paper introduces an efficient weighted regognition algorithm. It is based on a suitable tree structure called ZPC without building the position automaton. The ZPC-structure results from the compact language and the polynomial structure of weighted expressions. We show that the time complexity of this algorithm is the best one until now.

Keywords: (Partial) Conway semirings, formal power series, rational weighted expressions, weighted recognition.

1 Introduction

Rational weighted expressions are finite representations for a class of formal power series called rational [18,2]. The manipulation of these objects amounts to work with weighted automata [17] by the equivalence between rational series and series that are behaviours of finite state machines.

In this paper, we present a new efficient algorithm for computing weights of words for rational weighted expressions. A classical algorithm consists to build from a rational weighted expression E an equivalent weighted automaton $\mathcal{A} = (\lambda, \mu, \gamma)$ of dimension n whose the behaviour is the expression E, then to evaluate the weight of a word w in $\mathcal{A}$ using the matrix product $\lambda\, \mu(w)\, \gamma$ realized in $O(|w| \times n^2)$. This time complexity is due to the multiplication of a row vector to a matrix for each letter of the word w. Recent research deals with the conversion of rational weighted expressions into weighted automata [16,17]. In [5], a step by step algorithm constructs a position automaton associated to a rational weighted expression in a cubic time w.r.t. the size of this expression. Next, quadratic time algorithms have been proposed using either a generalization of Thompson automata for multiplicities [15], or syntactic tree structures [7]. Thus, the best time to recognize a word w including the construction of the weighted automaton is $O(|w| \times |E|^2)$. However the problem of the weighted recognition is solved in $O(|w| \times |E|)$ by the use of a generalization of Thompson automata without ε-transitions removal [15]. The purpose of this paper is to present a new way to compute the weight of a word w for a rational weighted expression without going through the construction of an equivalent weighted automaton. The

S. Konstantinidis (Ed.): CIAA 2013, LNCS 7982, pp. 24–35, 2013.

best time complexity is preserved with a sharper tree structure [20,7] which is really suited to rational weighted expressions. It generalizes the syntactic tree of classical arithmetical expressions and stands halfway between rational weighted expressions and associated weighted automata.

In the first part, some theoretical notions are reminded. A particular attention is given to the definition of the star operation which must make possible the weighted recognition. Section 3 is dedicated to the study of the compact language which has been characterized for the first time in [7] under a different shape. The polynomial structure of rational weighted expressions is also detailed. In Section 4, a position automaton is constructed from the polynomial structure in a quadratic time w.r.t. the size of the rational weighted expression. Next, the ZPC-structure for multiplicities introduced in [7] is presented. The polynomial structure is also used to build this tree structure in a linear time w.r.t. the size of the rational weighted expression.

Before the conclusion, Section 6 is devoted to a new recognition algorithm in $O(|w| \times |E|)$ of a word w for a rational weighted expression E. It assigns weights to the nodes of the ZPC-structure by means of tree traversals. An overall description of the whole process of computation is described. Proofs of the main results and a detailed example are given in the appendix.

2   Theoretical Background

A semiring $R = (R, +, \cdot, 0, 1)$ is a suitable set of weights for valued graphs and finite state machines such that $(R, +, 0)$ is a commutative monoid and $(R, \cdot, 1)$ is a monoid [11]. Moreover 0 is an absorbing element with respect to multiplication and product distributes over sum. Conway semirings introduced in [8] are equipped with an unary star operation $* : R \to R$ satisfying the sum star identity $(r_1 + r_2)^* = r_1^*(r_2 r_1^*)^*$ and the product star identity $(r_1 r_2)^* = 1 + r_1(r_2 r_1)^* r_2$ for all $r_1, r_2 \in R$. Most important semirings used in computer science and its applications are Conway semirings such that the boolean semiring $\mathbb{B} = (\{0, 1\}, +, \cdot, 0, 1)$ with $0^* = 1^* = 1$, the semiring $\mathbb{N}^\infty = (\mathbb{N} \cup \{\infty\}, +, \cdot, 0, 1)$ with $0^* = 1$ and $a^* = \infty$ for $a \neq 0$, the semiring $\mathbb{R}_+^\infty = (\mathbb{R}_+ \cup \{\infty\}, +, \cdot, 0, 1)$ with $a^* = 1/(1-a)$ for $0 \leq a < 1$ and $a^* = \infty$ for $a \geq 1$, and the tropical semiring $(\mathbb{R}_+ \cup \{\infty\}, \min, +, \infty, 0)$ with $a^* = 0$ for all $a \in \mathbb{R}$. Explicit examples of non Conway semirings using the boolean semiring can be found in [14]. For partial Conway semirings [4], the star operation is only defined on an ideal of R. Furthermore, the sum star and product star identities hold. The semiring $\mathbb{R}_+^p = (\mathbb{R}_+, +, \cdot, 0, 1)$ with $a^* = 1/(1-a)$ for $0 \leq a < 1$ is a partial Conway semiring. It can be noticed that $0^* = 1$ for any (partial) Conway semiring.

Let Σ be an alphabet. The empty word and the length of a word w in the free monoid Σ^* are symbolized respectively by ε and $|w|$. Mappings from Σ^* into a semiring R are called (formal) power series S and they are collected in the set $R\langle\langle \Sigma \rangle\rangle$ [18]. A power series S can be written as $S = \sum_{w \in \Sigma^*} S(w)w$. Polynomials are power series whose the support $\{w \in \Sigma^* : S(w) \neq 0\}$ is finite. The set of polynomials is denoted by $R\langle\Sigma\rangle$. With the addition $(S + T)(w) = S(w) + T(w)$ and the Cauchy product $ST(w) = \sum_{w=uv} S(u)T(v)$, the set $R\langle\langle\Sigma\rangle\rangle$ inherits the structure of a semiring from R. Moreover, if R is a (partial) Conway semiring, then the set $R\langle\langle\Sigma\rangle\rangle$ is again a (partial) Conway semiring [3,4] with the recursive definition of the star operation:

$$S^*(\varepsilon) = S(\varepsilon)^*, \ S^*(w) = S^*(\varepsilon) \sum_{\substack{uv=w \\ u\neq\varepsilon}} S(u)S^*(v) \text{ with } w \neq \varepsilon.$$

Weighted automata are finite-state machines encoded by linear representations [9]. A linear representation $\mathcal{A} = (\lambda, \mu, \gamma)$ with weights in a semiring R is given by an alphabet Σ, a set Q of n states, a mapping μ from Σ into $R^{Q\times Q}$ whose the image of a letter is a transition matrix, the initial vector $\lambda \in R^{1\times Q}$, and the final vector $\gamma \in R^{Q\times 1}$. If we point out the nature of the weights, we just say R-automaton. A state $q \in Q$ is described as an initial (resp. final) state if $(\lambda)_{1q} \neq 0$ (resp. $(\gamma)_{q1} \neq 0$). A transition labelled by the letter $a \in \Sigma$ links the state $p \in Q$ to the state $q \in Q$ if $(\mu(a))_{pq} \neq 0$. The dimension of a weighted automaton $\mathcal{A}$ is its number of states. The mapping μ can be extended as a morphism of monoids from Σ^* to $R^{Q\times Q}$. The behaviour $\|\mathcal{A}\| \in R\langle\langle A\rangle\rangle$ of the weighted automaton $\mathcal{A}$ is the power series defined by $\|\mathcal{A}\|(w) = \lambda\mu(w)\gamma$. The collection of power series that are behaviours of weighted automata is denoted by $\mathrm{REC}(\Sigma, R)$. This set is closed for rational laws: addition, Cauchy product, star, left and right exterior products. Universal constructions are explained in details in [10]. A power series is said rational [2] if it is in the closure of rational laws on letters in Σ and scalars in R. The set of such power series is denoted by $\mathrm{RAT}(\Sigma, R)$. In [3] and more recently in [4] is generalized the Kleene's theorem [13] when R is respectively a Conway semiring or a partial Conway semiring: $\mathrm{RAT}(\Sigma, R) = \mathrm{REC}(\Sigma, R)$. For any semiring R, as the recursive definition of the star operation remains valid in the ideal of proper series S which verify $S(\varepsilon) = 0$, the Schützenberger's theorem holds [19].

The set $\mathrm{RatEx}(\Sigma, R)$ of weighted rational expressions, or again rational R-expressions, over the alphabet Σ with weights in R is the universal free algebra generated by $\Sigma \cup \{0, 1\}$ as constants, the binary operations "+" and ".", the unary operations "*" and $r \in R$ with left and right actions [6,17]. Depending on the properties of the semiring R, the constant term function const of a rational R-expression is inductively computed by:

$$\mathrm{const}(0) = 0 \quad \mathrm{const}(1) = 1 \quad \mathrm{const}(a) = 0$$
$$\mathrm{const}(\mathrm{F} + \mathrm{G}) = \mathrm{const}(\mathrm{F}) + \mathrm{const}(\mathrm{G}) \quad \mathrm{const}(\mathrm{F} \cdot \mathrm{G}) = \mathrm{const}(\mathrm{F})\,\mathrm{const}(\mathrm{G})$$
$$\mathrm{const}(r\,\mathrm{F}) = r\,\mathrm{const}(\mathrm{F}) \quad \mathrm{const}(\mathrm{F}\,r) = \mathrm{const}(\mathrm{F})\,r$$
$$\mathrm{const}(\mathrm{F}^*) = \mathrm{const}(\mathrm{F})^* \text{ if the right-hand side is defined in } R$$

In the domain of the constant term function, a rational power series $\|E\|$ is recursively obtained by:

$$\|0\| = 0 \quad \|1\| = \varepsilon \quad \|a\| = a$$
$$\|\mathrm{F} + \mathrm{G}\| = \|\mathrm{F}\| + \|\mathrm{G}\| \quad \|\mathrm{F} \cdot \mathrm{G}\| = \|\mathrm{F}\|\,\|\mathrm{G}\|$$
$$\|r\,\mathrm{F}\| = r\,\|\mathrm{F}\| \quad \|\mathrm{F}\,r\| = \|\mathrm{F}\|r$$
$$\|\mathrm{F}^*\| = \|\mathrm{F}\|^*$$

A power series S belongs to $\mathrm{RAT}(\Sigma, R)$ if and only if there exits a rational R-expression E in the domain of the constant term function such that $S = \|E\|$ with $S(\varepsilon) = \|E\|(\varepsilon) = \mathrm{const}(E)$. The size and the alphabetical width of a rational R-expression E, denoted by $|E|$ and $|E|_\Sigma$, are the number of nodes in the syntactic tree of E and the number of all occurrences of letters in E. The alphabet $\Sigma_E \subseteq \Sigma$ of E is composed of the letters appearing in E.

In the rest of this paper, from a rational R-expression $E \in \mathrm{RatEx}(\Sigma, R)$, we want to construct a R-automaton whose the behaviour is $\|E\|$. A Kleene theorem must be verified. For this reason, we will consider three types of semirings:

(C)   **The semiring R is a Conway semiring**
The construction will finish for any rational R-expression. The domain of the constant term function is equal to $\mathrm{RatEx}(\Sigma, R)$ because the domain of the star operation for the semiring R is R itself.

(PC)   **The semiring R is a partial Conway semiring**
The construction will finish when rational R-expressions F such that F^* is a subexpression of E are taken in the domain of the constant term function. The scalar $\|F\|(\varepsilon)$ must belong to the domain of the star operation for the semiring R. Otherwise, an impracticable evaluation interrupts the process.

(NPC)   **The semiring R is not a (partial) Conway semiring**
The construction will finish when the constant term of rational R-expressions F such that F^* is a subexpression of E is equal to 0. Otherwise the process is interrupted.

Consequently one of the assumptions (C), (PC) or (NPC) is supposed until the end of this paper. The weight of the word $w \in \Sigma^*$ for the rational R-expression $E \in \mathrm{RatEx}(\Sigma, R)$ is then well-defined as the weight of w for the power series $\|E\|$ which is the behaviour of a R-automaton.

3   Compact Language

In order to construct a R-automaton, the notion of compact language is introduced. This structure is a set of terms retaining almost all the informations encoded by rational R-expressions.

The linearized rational R-expression $\overline{E}^n$ is deduced from the rational R-expression E by ranking every letter occurrence with its position in E starting from the non-negative integer n. In what follows, we denote $\overline{E}^1$ more easily by $\overline{E}$. The letters of $\Sigma_{\overline{E}^n}$ are called positions numbered from n of the rational R-expression E. The letters of $\Sigma_{\overline{E}}$ are merely called positions of the rational R-expression E. Conversely the mapping h from $\Sigma_\infty = \{a_i : a \in \Sigma \text{ and } i \in \mathbb{N}\}$ into Σ associates the letter a to every position a_i. It can be extended as a morphism of algebras from $\mathrm{RatEx}(\Sigma_\infty, R)$ into $\mathrm{RatEx}(\Sigma, R)$ such that $h(\overline{E}^n) = E$. Let A and B be two sets of items $r_1 a_{i_1} r_2 a_{i_2} \cdots a_{i_{m-1}} r_m a_{i_m} r_{m+1}$ where r_j is a scalar in R and a_{i_j} is a position. Products $A \cdot B$, $r A$ and $A r$ are given respectively by the set of items $r_1 a_{i_1} \cdots a_{i_m} t b_{j_1} \cdots b_{j_n} s_{n+1}$ such that t is the evaluation of $r_{m+1} s_1$ for any terms $r_1 a_{i_1} \cdots a_{i_m} r_{m+1} \in A$ and $s_1 b_{j_1} \cdots b_{j_n} s_{n+1} \in B$, the set of items $t\,a_{i_1}\cdots a_{i_m}\,r_{m+1}$ such that t is the evaluation of $r\,r_1$ for any term $r_1\,a_{i_1}\cdots a_{i_m}\,r_{m+1}\in A$, and the set of items $r_1\,a_{i_1}\cdots a_{i_m}\,t$ such that t is the evaluation of $r_{m+1}\,r$ for any term $r_1\,a_{i_1}\cdots a_{i_m}\,r_{m+1}\in A$. The compact language $L(\overline{E}^n)$ over the alphabet $\Sigma_{\overline{E}}$ is inductively defined as follows:

$$L(\overline{0}^n)=\emptyset \qquad L(\overline{1}^n)=\emptyset \qquad L(\overline{a}^n)=\{1\,a_n\,1\}$$
$$L(\overline{F+G}^n)=L(\overline{F}^n)\cup L(\overline{G}^{|F|\Sigma+n})$$
$$L(\overline{F\cdot G}^n)=L(\overline{F}^n)\cdot L(\overline{G}^{|F|\Sigma+n})\cup \mathrm{const}(F)\,L(\overline{G}^{|F|\Sigma+n})\cup L(\overline{F}^n)\,\mathrm{const}(G)$$
$$L(\overline{F^*}^n)=\mathrm{const}(F)^*\left(\bigcup_{i\ge1}\left(L(\overline{F}^n)\,\mathrm{const}(F)^*\right)^i\right)$$
$$L(\overline{r\,F}^n)=r\,L(\overline{F}^n)\qquad L(\overline{F\,r}^n)=L(\overline{F}^n)\,r$$

Terms in $L(\overline{E}^n)$ are in the form $\alpha_1\,a_{i_1}\,\alpha_2\,a_{i_2}\cdots a_{i_{m-1}}\,\alpha_m\,a_{i_m}\,\alpha_{m+1}$ where $a_{i_j}\in\Sigma_{\overline{E}}$ and $i_{m+1}>\cdots>i_2>i_1\ge n$. The compact language $L(\overline{E}^n)$ is equivalent to some polynomials $\mathrm{First}(\overline{E}^n)$, $\mathrm{Last}(\overline{E}^n)$ and $\mathrm{Follow}(\overline{E}^n,a_{i_j})$:

$$\mathrm{First}(\overline{E}^n)=\sum_{\alpha_1 a_{i_1}\in\mathrm{Pref}(L(\overline{E}^n))}\alpha_1\,a_{i_1}$$

$$\mathrm{Last}(\overline{E}^n)=\sum_{a_{i_m}\alpha_{m+1}\in\mathrm{Suff}(L(\overline{E}^n))}\alpha_{m+1}\,a_{i_m}$$

$$\mathrm{Follow}(\overline{E}^n,a_{i_j})=\sum_{\alpha_{j+1}a_{i_{j+1}}\in\mathrm{Fact}(L(\overline{E}^n),a_{i_j})}\alpha_{j+1}\,a_{i_{j+1}}.$$

where $\mathrm{Pref}(L(\overline{E}^n))$, $\mathrm{Suff}(L(\overline{E}^n))$ and $\mathrm{Fact}(L(\overline{E}^n),a_{i_j})$ are respectively the set of prefix of terms in $L(\overline{E}^n)$, the set of suffix, and the set of factors which follow the position a_{i_j}.

Example 1. With the partial Conway semiring $\mathbb{R}_+^P$, let $E=(\frac13 a^*+\frac16 b^*)^*\cdot b$ be a rational $\mathbb{R}_+^P$-expression. The linearized version of E is $\overline{E}=(\frac13 a_1^*+\frac16 b_2^*)^*\cdot b_3$. The weight of the empty word for $\overline{E}$ is $\mathrm{const}(E)=(\frac13 0^*+\frac16 0^*)^*0=2\times0=0$. From the compact language

$$L(\overline{E})=2\bigcup_{i,j>0}\{\tfrac13(1\,a_1\,1)^i\,2,\tfrac16(1\,b_2\,1)^i\,2\}^j\cdot\{1\,b_3\,1\}\cup\{2\,b_3\,1\},$$

we deduce the First, Last and Follow polynomials:

$$\mathrm{First}(\overline{E})=\tfrac23 a_1+\tfrac13 b_2+2\,b_3\qquad \mathrm{Last}(\overline{E})=b_3$$
$$\mathrm{Follow}(\overline{E},a_1)=\tfrac53 a_1+\tfrac13 b_2+2\,b_3\qquad \mathrm{Follow}(\overline{E},b_2)=\tfrac23 a_1+\tfrac43 b_2+2\,b_3$$
$$\mathrm{Follow}(\overline{E},b_3)=0$$

Recursive computations of First, Last and Follow polynomials are suggested:

$$\mathrm{First}(\overline{0}^n)=0\qquad \mathrm{First}(\overline{1}^n)=0\qquad \mathrm{First}(\overline{a}^n)=a_n$$
$$\mathrm{First}(\overline{F+G}^n)=\mathrm{First}(\overline{F}^n)+\mathrm{First}(\overline{G}^{|F|\Sigma+n})$$
$$\mathrm{First}(\overline{F\cdot G}^n)=\mathrm{First}(\overline{F}^n)+\mathrm{const}(F)\,\mathrm{First}(\overline{G}^{|F|\Sigma+n})$$
$$\mathrm{First}(\overline{F^*}^n)=\mathrm{const}(F)^*\,\mathrm{First}(\overline{F}^n)$$
$$\mathrm{First}(\overline{r\,F}^n)=r\,\mathrm{First}(\overline{F}^n)\qquad \mathrm{First}(\overline{F\,r}^n)=\mathrm{First}(\overline{F}^n)$$

$$\mathrm{Last}(\overline{0}^{"}) = 0 \qquad \mathrm{Last}(\overline{1}^{"}) = 0 \qquad \mathrm{Last}(\overline{a}^{"}) = a_n$$
$$\mathrm{Last}(\overline{F+G}^{"}) = \mathrm{Last}(\overline{F}^{"}) + \mathrm{Last}(\overline{G}^{|F|_{\Sigma}+n})$$
$$\mathrm{Last}(\overline{F \cdot G}^{"}) = \mathrm{Last}(\overline{F}^{"}) \, \mathrm{const}(G) + \mathrm{Last}(\overline{G}^{|F|_{\Sigma}+n})$$
$$\mathrm{Last}(\overline{F^*}^{"}) = \mathrm{Last}(\overline{F}^{"}) \, \mathrm{const}(F)^*$$
$$\mathrm{Last}(\overline{r\,F}^{"}) = \mathrm{Last}(\overline{F}^{"}) \qquad \mathrm{Last}(\overline{F\,r}^{"}) = \mathrm{Last}(\overline{F}^{"}) \, r$$

$$\mathrm{Follow}(\overline{0}^{"}, a_j) = 0 \qquad \mathrm{Follow}(\overline{1}^{"}, a_j) = 0 \qquad \mathrm{Follow}(\overline{a}^{"}, a_j) = 0$$
$$\mathrm{Follow}(\overline{F+G}^{"}, a_j) = \mathrm{Follow}(\overline{F}^{"}, a_j) + \mathrm{Follow}(\overline{G}^{|F|_{\Sigma}+n}, a_j)$$
$$\mathrm{Follow}(\overline{F \cdot G}^{"}, a_j) = \mathrm{Follow}(\overline{F}^{"}, a_j) + \mathrm{Last}(\overline{F}^{"})(a_j) \, \mathrm{First}(\overline{G}^{|F|_{\Sigma}+n})$$
$$+ \mathrm{Follow}(\overline{G}^{|F|_{\Sigma}+n}, a_j)$$
$$\mathrm{Follow}(\overline{F^*}^{"}, a_j) = \mathrm{Follow}(\overline{F}^{"}, a_j) + \mathrm{Last}(\overline{F^*}^{"})(a_j) \, \mathrm{First}(\overline{F}^{"})$$
$$\mathrm{Follow}(\overline{r\,F}^{"}, a_j) = \mathrm{Follow}(\overline{F}^{"}, a_j) \qquad \mathrm{Follow}(\overline{F\,r}^{"}, a_j) = \mathrm{Follow}(\overline{F}^{"}, a_j)$$

These recursive computations have been proved for rational weighted expressions in [7] with slightly different definitions.

4   Position Automaton

It will be explained in the following sections how to avoid the conversion of a rational R-expression into a R-automaton in order to speed up the weighted recognition. However a specific construction proving the correctness of our new algorithm is recalled.

Let $E \in \mathrm{RatEx}(\Sigma, R)$ be a rational R-expression. The family composed of polynomials $\mathrm{First}(\overline{E})$, $(\mathrm{Follow}(\overline{E}, a_j))_{a_j \in \Sigma_{\overline{E}}}$ and $\mathrm{Last}(\overline{E})$ is used to construct efficiently a weighted automaton $\mathcal{A}_E = (\lambda_E, \mu_E, \gamma_E)$ in a quadratic time w.r.t. the size of E. It is called the position automaton [5,7] whose behaviour $\|\mathcal{A}_E\|$ is the rational power series $\|E\|$. Let $q_0 \notin \Sigma_{\overline{E}}$ and $Q = \Sigma_{\overline{E}} \cup \{q_0\}$ be the set of states. For each letter $a \in \Sigma$, we associate a transition matrix $\mu_E(a) \in R^{Q \times Q}$ given by:

$$(\mu_E(a))_{pq} = \begin{cases} \mathrm{First}(\overline{E})(q) & \text{if } p = q_0 \text{ and } h(q) = a, \\ \mathrm{Follow}(\overline{E}, p)(q) & \text{if } h(p) \in \Sigma \text{ and } h(q) = a, \\ 0 & \text{otherwise.} \end{cases}$$

In order to complete the definition of $\mathcal{A}_E$, we add:

$$(\lambda_E)_{1q} = \begin{cases} 1 & \text{if } q = q_0, \\ 0 & \text{otherwise,} \end{cases} \qquad (\gamma_E)_{q1} = \begin{cases} \mathrm{const}(E) & \text{if } q = q_0, \\ \mathrm{Last}(\overline{E})(q) & \text{otherwise.} \end{cases}$$

The dimension of the position automaton $\mathcal{A}_E$ is $|E|_{\Sigma} + 1$. The geometry is particular. Only the state q_0 is initial with the weight 1. If $\mathrm{const}(E) \neq 0$, the state q_0 is also final with the weight $\mathrm{const}(E)$. From any state, we can not reach the state q_0. Finally, every incoming transition to a state q is indexed by the letter $h(q)$. The position automaton is then homogeneous. More precisely, the linear representation of the position automaton $\mathcal{A}_E$ is explicited by:

$$\left(\left(1 \; 0_{1 \times \Sigma_{\overline{E}}} \right), \left(\begin{array}{c|c} 0 & v_E(a) \\ \hline 0_{\Sigma_{\overline{E}} \times 1} & M_E(a) \end{array} \right)_{a \in A}, \left(\begin{array}{c} \mathrm{const}(E) \\ c_E \end{array} \right) \right)$$

where $0_{1 \times \Sigma_{\overline{E}}}$ (respectively $0_{\Sigma_{\overline{E}} \times 1}$) is the row (respectively column) vector of size $|E|_\Sigma$ composed only with the weight 0. The row vector $v_E(a)$ indicates either the coefficients of all positions q in $\mathrm{First}(\overline{E})$ such that $h(q) = a$, or 0 otherwise. The column vector c_E encodes the coefficients of positions q in $\mathrm{Last}(\overline{E})$. Whereas, the matrix $M_E(a)$ associates at each pair of positions (p, q) such that $h(q) = a$ the scalar of q in the polynomial $\mathrm{Follow}(\overline{E}, p)$, i.e. $(M_E(h(q)))_{pq} = \mathrm{Follow}(\overline{E}, p)(q)$. We set that $M_E(\varepsilon)$ is equal to the identity matrix in the matrix algebra $R^{\Sigma_{\overline{E}} \times \Sigma_{\overline{E}}}$ and $M_E(w) = M_E(w_1) \cdots M_E(w_n)$ for any word $w = w_1 \cdots w_n$ ($w_i \in \Sigma$). Moreover, we can easily observe that:

$$\mu_E(w) = \left(\begin{array}{c|c} 0 & v_E(w_1) \, M_E(w_2 \cdots w_n) \\ \hline 0_{\Sigma_{\overline{E}} \times 1} & M_E(w) \end{array} \right).$$

Theorem 1 establishes the relation between the position automaton built from a rational R-expression and its associated rational series.

Theorem 1. *Let* $E \in \mathrm{RatEx}(\Sigma, R)$ *be a rational R-expression. The behaviour of the position automaton $\mathcal{A}_E$ is the rational series $\|E\|$.*

Let us mention that a detailed proof of the above theorem is given in [7]. The position automaton is constructed in $O(|E|^2)$ from the rational R-expression E. Consequently, the evaluation of the weight of a word for a rational R-expression amounts to compute it for the corresponding position automaton.

Corollary 1. *The weight of the word $w \in \Sigma_E^*$ for the rational R-expression $E \in \mathrm{RatEx}(\Sigma, R)$ is:* $\|E\|(w) = \lambda_E \, \mu_E(w) \, \gamma_E$.

The proof is straightforward from Theorem 1. The weight is obtained in three steps. First, the rational R-expression is converted into the position automaton. Next a row vector λ is produced by sequential multiplications of the initial vector λ_E by the transition matrices $\mu_E(w_1), \ldots, \mu_E(w_n)$ if $w = w_1 \cdots w_n$ ($w_i \in \Sigma$). At last, the weight results from the product $\lambda \, \gamma_E$. The complexity of this algorithm is $O(|w| \times |E|^2)$.

The following example illustrates the construction of the position automaton and the computation of a weight.

Example 2. We take the partial Conway semiring $\mathbb{R}_+^P$ of Example 1 and the same rational weighted expression $E = \left(\frac{1}{3} a^* + \frac{1}{6} b^* \right)^* \cdot b$. According to the polynomial structure $\mathrm{First}(\overline{E}), (\mathrm{Follow}(\overline{E}, a_j))_{a_j \in \Sigma_{\overline{E}}}$ and $\mathrm{Last}(\overline{E})$, we obtain the following position automaton:

$$\lambda_E = (1 \, 0 \, 0 \, 0) \quad \mu_E(a) = \begin{pmatrix} 0 & 2/3 & 0 & 0 \\ 0 & 5/3 & 0 & 0 \\ 0 & 2/3 & 0 & 0 \\ 0 & 0 & 0 & 0 \end{pmatrix} \quad \mu_E(b) = \begin{pmatrix} 0 & 0 & 1/3 & 2 \\ 0 & 0 & 1/3 & 2 \\ 0 & 0 & 4/3 & 2 \\ 0 & 0 & 0 & 0 \end{pmatrix} \quad \gamma_E = \begin{pmatrix} 0 \\ 0 \\ 0 \\ 1 \end{pmatrix}.$$

where the weight of the word ab for the rational weighted expression E is $\frac{4}{3}$.

Besides the construction of position automata, the polynomial structure consti-
tuted of $\mathrm{First}(\overline{E}),(\mathrm{Follow}(\overline{E},a_j))_{a_j \in \Sigma_{\overline{E}}}$ and $\mathrm{Last}(\overline{E})$ is also employed to build a
tree structure representing the rational R-expression E in a linear time w.r.t. the
size of E.

5   Tree Structure

In [20], the ZPC-structure was proposed to manipulate efficiently boolean rational expressions, then extended to the multiplicity case in [7]. From a rational
R-expression E, this structure is based on two similar trees which are deduced
from the syntactic tree $T(E)$ of E: the First tree $TF(E)$ and the Last tree $TL(E)$.
We present here a slightly different version of the ZPC-structure introduced in
[7]. Indeed, in our definition of the compact language, the scalars are considered
as unary functions rather than constants in order to reach a correct algebraic
formulation. The trees $TF(E)$ and $TL(E)$ encode respectively the First and
the Last polynomials. Some links representing the Follow polynomials connect
$TL(E)$ to $TF(E)$. By handling this structure, it turns out that the weight of a
non-empty word for a rational R-expression E can be computed faster.

A node in a syntactic tree will be written ν and the root will be denoted by
ν_0. The nodes are indexed by either letters or positions, scalars and operators.
In the sequel, the node ν is identified with a position x when x is the label of
ν, knowing that x occurs only one time in the syntactic tree. If the arity of ν
is two, the nodes ν_l and ν_r represent respectively its left and right descendant.
When its arity is one, its unique descendant is ν_s. The relation of descendance
over the nodes of a tree is denoted $\preceq$. It is the transitive closure of the relations
$\nu_l \preceq \nu$, $\nu_r \preceq \nu$ or $\nu_s \preceq \nu$ according to the various cases. When the edges are
labelled by elements of the semiring R, we define the function of cost π. Suppose
that $\nu' \preceq \nu$. The cost $\pi(\nu, \nu')$ is the weight of the path from ν to ν'. Otherwise
$\pi(\nu, \nu') = 0$. We set that $\pi(\nu, \nu) = 1$.

Let two rational R-expressions $F, G \in \mathrm{RatEx}(\Sigma, R)$, a scalar $r \in R$ and a
letter (or a position) $a \in \Sigma$. We detail the inductive construction of the syntactic
tree of a rational R-expression:

$$T(0) = \Omega \qquad T(1) = \Omega \qquad T(a) = \langle a, \Omega \rangle$$
$$T(F+G) = \langle +, T(F), T(G) \rangle \qquad T(F \cdot G) = \langle \cdot, T(F), T(G) \rangle$$
$$T(r\,F) = \langle r, T(F) \rangle \qquad T(F\,r) = \langle \cdot, T(F), T(r\,1) \rangle$$
$$T(F^*) = \langle *, T(F) \rangle$$

The tree Ω is known as the empty tree appearing only when a node does not
have descendants or for the ZPC construction of the weighted expressions 0
and 1. For the construction of the tree $T(F\,r)$, the rational R-expression $F\,r$
is seen as the expression $F \cdot (r\,1)$. Nodes labelled by scalars and stars have
only one subtree while the other operations get two subtrees. For a node ν, the
rational R-expression E_ν denotes the subexpression resulting from ν as root. An
external node indexed by a scalar r represents the rational R-expression $r\,1$. The
coefficient $\mathrm{const}(\nu)$ denotes the constant term $\mathrm{const}(E_\nu)$ of the subexpression E_ν.

The nodes of the First tree TF(E) are symbolized by φ and its edges are directed from the root φ_0 to the external nodes. The tree TF(E) is a copy of the syntactic tree $T(\overline{E})$ with the weight 1 for the edges except some of them. For each non-external node φ labelled by the scalar "r", the edge (φ, φ_s) is endowed with the weight r. When a node φ is labelled by "$\cdot$", we associate the weight $\text{const}(\varphi_l)$ to the edge (φ, φ_r). If a node φ is labelled by "$*$", we assign the weight $\text{const}(\varphi_s)^*$ to the edge (φ, φ_s).

The Last tree TL(E) is also a copy of $T(\overline{E})$ where each node is symbolized with the greek letter ρ. Edges in TL(E) are directed from external nodes to the root ρ_0. Their weights are 1 except some of them. For each node ρ labelled by "$\cdot$", we endow the weight $\text{const}(\rho_r)$ to the edge (ρ_l, ρ). If a node ρ is labelled by "$*$", the weight $\text{const}(\rho_s)^*$ is assigned to the edge (ρ_s, ρ).

The Follow links connect the Last tree TL(E) to the First tree TF(E). Let ρ be a node in TL(E) and φ its corresponding node in TF(E). For each node ρ labelled by "$\cdot$", we set a link from ρ_l to φ_r. For any node ρ labelled by "$*$", we create a link from ρ_s to φ. We denote by Δ the set of Follow links, and by $\Delta_x = \{(\rho, \varphi) \in \Delta | x \preceq \rho\}$ the set of Follow links associated to the position x.

Lemma 1. *Let E be a rational R-expression. At most one Follow link leaves a node ρ of* TL(E).

Example 3. Carrying on with Example 2, the ZPC-structure of the rational $\mathbb{R}^P_+$-expression $E = (\frac{1}{3} a^* + \frac{1}{6} b^*)^* \cdot b$ is given in Fig. 1.

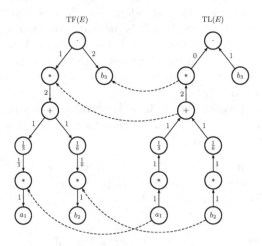

Fig. 1. The ZPC-structure of the rational $\mathbb{R}^P_+$-expression $E = (\frac{1}{3} a^* + \frac{1}{6} b^*)^* \cdot b$.

The construction of the tree TF(E), the tree TL(E) and the Follow links is made in O(E). The number of nodes in TF(E) (resp. TL(E)) is $|E|$. These results appear in [7].

The polynomial structure $\{\text{First}(\overline{E}), (\text{Follow}(\overline{E}, x))_{x \in \Sigma_{\overline{E}}}, \text{Last}(\overline{E})\}$ can be expressed by means of paths in the trees of the ZPC-structure. We establish the relations between the polynomial $\text{First}(\overline{E})$ (resp. $\text{Last}(\overline{E})$, $\text{Follow}(\overline{E}, x)$)) with the tree $\text{TF}(E)$ or the tree $\text{TL}(E)$:

Proposition 1. *Let* E *be a rational R-expression and* x *a position. The following relations are verified:*

$$\text{First}(\overline{E}) = \sum_{x \in \Sigma_{\overline{E}}} \pi(\varphi_0, x)x \qquad \text{Last}(\overline{E}) = \sum_{x \in \Sigma_{\overline{E}}} \pi(\rho_0, x)x$$
$$\text{Follow}(\overline{E}, x) = \sum_{y \in \Sigma_{\overline{E}}} \sum_{(\rho, \varphi) \in \Delta_x} \pi(\rho, x)\, \pi(\varphi, y)y$$

In the following section, a new algorithm is introduced for computing the weight of a word for a rational R-expression with the best time complexity known until now. This time complexity is due to the geometry of the ZPC-structure and its efficient tree traversals.

6   Recognition Algorithm

In order to compute efficiently the weight of a non-empty word $w = w_1 \cdots w_n$ ($w_i \in \Sigma$) for a rational R-expression E, we use here the ZPC-structure of this rational R-expression. The algorithm consists in assigning coefficients $\alpha_\varphi^{(i)}$ (resp. $\beta_\rho^{(i)}$) to each nodes of the tree $\text{TF}(E)$ (resp. $\text{TL}(E)$). In this way, a top-down traversal of $\text{TF}(E)$ and a bottom-up traversal of $\text{TL}(E)$ are performed for each letter w_i ($i = 1, \cdots, n$) of the word w.

Let $w = w_1 \cdots w_n \in \Sigma^* \setminus \{\varepsilon\}$ such that $w_i \in \Sigma$ ($0 < i \leq n$). At the beginning of the recognition process, a particular traversal of $\text{TF}(E)$ is done for the first letter of the word w in order to produce the weights $\alpha_\varphi^{(1)}$:

$$\alpha_\varphi^{(1)} = \begin{cases} 1 & \text{if } \varphi = \varphi_0, \\ 0 & \text{if } \varphi = x \text{ and } h(x) \neq w_1, \\ \alpha_{\varphi'}^{(1)} \pi(\varphi', \varphi) & \text{otherwise.} \end{cases}$$

where φ' is the predecessor of $\varphi \neq \varphi_0$ in the tree $\text{TF}(E)$. Next the computation of the weight $\alpha_\varphi^{(i)}$ is dependent on the weight $\alpha_{\varphi'}^{(i)}$ and the weight $\beta_\rho^{(i-1)}$ if there exists a Follow link $(\rho, \varphi) \in \Delta$:

$$\alpha_{\varphi_0}^{(i)} = \begin{cases} \beta_\rho^{(i-1)} & \text{if } (\rho, \varphi_0) \in \Delta, \\ 0 & \text{otherwise,} \end{cases}$$

$$\alpha_\varphi^{(i)} = \begin{cases} 0 & \text{if } \varphi = x \text{ and } h(x) \neq w_i, \\ \alpha_{\varphi'}^{(i)} \pi(\varphi', \varphi) + \beta_\rho^{(i-1)} & \text{if there exists } (\rho, \varphi) \in \Delta, \\ \alpha_{\varphi'}^{(i)} \pi(\varphi', \varphi) + \beta_{\rho_1}^{(i-1)} + \beta_{\rho_2}^{(i-1)} & \text{if there exists } (\rho_1, \varphi), (\rho_2, \varphi) \in \Delta, \\ \alpha_{\varphi'}^{(i)} \pi(\varphi', \varphi) & \text{otherwise,} \end{cases}$$

where φ' is the predecessor of $\varphi \neq \varphi_0$ in the tree TF(E). The computation of the weight $\beta_\rho^{(i)}$ is dependent on the weights $\beta_{\rho'}^{(i)}$ such that ρ' is a descendant of ρ in the tree TL(E), and the weights $(\alpha_\varphi^{(i)})$ if φ is an external node of TF(E):

$$\beta_x^{(i)} = \alpha_x^{(i)},$$

$$\beta_\rho^{(i)} = 0 \text{ if } \rho \text{ is an external node non labelled by a position,}$$

$$\beta_\rho^{(i)} = \begin{cases} \beta_{\rho_s}^{(i)} \, \pi(\rho_s, \rho) & \text{if } \rho \text{ has one descendant,} \\ \beta_{\rho_l}^{(i)} \, \pi(\rho_l, \rho) + \beta_{\rho_r}^{(i)} & \text{otherwise.} \end{cases}$$

From Proposition 1, the scalar $\alpha_x^{(1)}$ corresponds to the weight of the position x in the polynomial First($\overline{E}$) if $h(x) = w_1$. Moreover we can bring out the coefficients $\alpha_x^{(i)}$ and the coefficients $\beta_x^{(i)}$ $(1 \leq i \leq n)$ where x is a position:

Lemma 2. *The following relations are verified:*

$$\alpha_x^{(1)} = \begin{cases} \text{First}(\overline{E})(x) & \text{if } h(x) = w_1, \\ 0 & \text{otherwise,} \end{cases}$$

$$\alpha_x^{(i)} = \begin{cases} 0 & \text{if } h(x) \neq w_i, \\ \displaystyle\sum_{y \in \Sigma_E} \sum_{\substack{x \preceq \varphi, y \preceq \rho \\ (\rho, \varphi) \in \Delta_y}} \alpha_y^{(i-1)} \, \pi(\rho, y) \, \pi(\varphi, x) & \text{otherwise,} \end{cases}$$

$$\beta_x^{(i)} = \alpha_x^{(i)}.$$

Lemma 2 and Proposition 2 allow to find the weight of a non-empty word for a rational R-expression.

Proposition 2. *The weight of the non-empty word $w = w_1 \cdots w_n$ $(w_i \in \Sigma)$ for the rational R-expression E is:*

$$\|E\|(w) = \beta_{\rho_0}^{(n)}.$$

At this moment, the main result of this work using the ZPC-structure can be expressed:

Theorem 2. *The time complexity of the recognition of a word $w \in \Sigma^*$ for a rational R-expression E is $O(|w| \times |E|)$.*

7   Conclusion

In this paper, we have mentioned the efficiency of the ZPC-structure in order to compute weights of words for rational weighted expressions. Indeed the ZPC-structure can be seen as a convenient generalization of syntactic trees of arithmetical expressions in order to manipulate weighted expressions where the star operator appears. It is the internal structure in our toolbox dedicated for symbolic manipulations of rational weighted expressions and weighted automata [1] which is included in the algebraic combinatorics open-source package `MuPAD-Combinat` [12].

Acknowledgements. We would like to thank Jacques Sakarovitch who put this work in the right way.

References

1. Abbad, H., Laugerotte, É.: Symbolic demonstrations in MuPAD-Combinat. INFOCOMP Journal of Computer Science 7, 21–30 (2008)
2. Berstel, J., Reutenauer, C.: Rational series and their languages. Springer, Berlin (1988)
3. Bloom, S.L., Ésik, Z.: Iteration theories. Springer, Berlin (1993)
4. Bloom, S.L., Ésik, Z., Kuich, W.: Partial Conway and iteration semirings. Fundamenta Informaticae 86, 19–40 (2008)
5. Caron, P., Flouret, M.: Glushkov construction for multiplicities. In: Yu, S., Păun, A. (eds.) CIAA 2000. LNCS, vol. 2088, pp. 67–79. Springer, Heidelberg (2001)
6. Champarnaud, J.M., Duchamp, G.: Derivatives of rational expressions and related theorems. Theoretical Computer Science 313, 31–44 (2004)
7. Champarnaud, J.M., Laugerotte, É., Ouardi, F., Ziadi, D.: From regular weighted expressions to finite automata. International Journal of Foundations of Computer Science 15, 687–699 (2004)
8. Conway, J.C.: Regular algebra and finite machines. Chapman and Hall, London (1971)
9. Droste, M., Kuich, W., Vogler, H.: Handbook of weighted automata. Springer, Berlin (2009)
10. Duchamp, G., Flouret, M., Laugerotte, É., Luque, J.G.: Direct and dual laws for automata with multiplicities. Theoretical Computer Science 267, 105–120 (2001)
11. Golan, J.S.: Semirings and their applications. Kluwer Academic Publishers Dordrecht (1999)
12. Hivert, H., Thiéry, N.M.: MuPAD-Combinat, an open-source package for research in algebraic combinatorics. Séminaire Lotharingien de Combinatoire, 1–70 (2004)
13. Kleene, S.C.: Representation of events in nerve nets and finite automata. Automata Studies, 3–42 (1956)
14. Krob, D.: Models of a K-rational identity system. Journal of Computer and System Sciences 45, 396–434 (1992)
15. Laugerotte, É., Ziadi, D.: Weighted word recognition. Fundamenta Informaticae 83, 277–298 (2008)
16. Lombardy, S., Sakarovitch, J.: Derivatives of rational expressions with multiplicity. Theoretical Computer Science 332, 141–177 (2005)
17. Sakarovitch, J.: Éléments de théorie des automates. Vuibert (2003)
18. Salomaa, A., Soittola, M.: Automata-theoretic aspects of formal power series. Springer, Berlin (1978)
19. Schützenberger, M.P.: On the definition of a family of automata. Information and Control 4, 245–270 (1961)
20. Ziadi, D., Ponty, J.L., Champarnaud, J.M.: Passage d'une expression rationnelle à un automate fini non-déterministe. Bulletin of the Belgian Mathematical Society 4, 177–203 (1997)

Partial Word DFAs*

Eric Balkanski[1], F. Blanchet-Sadri[2], Matthew Kilgore[3], and B.J. Wyatt[2]

[1] Department of Mathematical Sciences, Carnegie Mellon University,
Wean Hall 6113, Pittsburgh, PA 15213, USA
ebalkans@andrew.cmu.edu
[2] Department of Computer Science, University of North Carolina,
P.O. Box 26170, Greensboro, NC 27402–6170, USA
blanchet@uncg.edu, bjwyatt@uncg.edu
[3] Department of Mathematics, Lehigh University, Christmas-Saucon Hall,
14 East Packer Avenue, Bethlehem, PA 18015, USA
m.kilgore0@gmail.com

Abstract. Recently, Dassow et al. connected partial words and regular languages. Partial words are sequences in which some positions may be undefined, represented with a "hole" symbol $\diamond$. If we restrict what the symbol $\diamond$ can represent, we can use partial words to compress the representation of regular languages. Doing so allows the creation of so-called $\diamond$-DFAs which are smaller than the DFAs recognizing the original language L, which recognize the compressed language. However, the $\diamond$-DFAs may be larger than the NFAs recognizing L. In this paper, we investigate a question of Dassow et al. as to how these sizes are related.

1 Introduction

The study of regular languages dates back to McCulloch and Pitts' investigation of neuron nets (1943) and has been extensively developing since (for a survey see, e.g., [7]). Regular languages can be represented by deterministic finite automata, DFAs, by non-deterministic finite automata, NFAs, and by regular expressions. They have found a number of important applications such as compiler design. There are well-known algorithms to convert a given NFA to an equivalent DFA and to minimize a given DFA, i.e., find an equivalent DFA with as few states as possible (see, e.g., [6]). It turns out that there are languages accepted by DFAs that have 2^n states while their equivalent NFAs only have n states.

Recently, Dassow et al. [4] connected regular languages and partial words. Partial words first appeared in 1974 and are also known under the name of strings with don't cares [5]. In 1999, Berstel and Boasson [2] initiated their combinatorics under the name of partial words. Since then, many combinatorial properties and algorithms have been developed (see, e.g., [3]). One of Dassow et al.'s motivations was to compress DFAs into smaller machines, called $\diamond$-DFAs.

---

* This material is based upon work supported by the National Science Foundation under Grant No. DMS–1060775.

S. Konstantinidis (Ed.): CIAA 2013, LNCS 7982, pp. 36–47, 2013.

More precisely, let Σ be a finite alphabet of letters. A *(full) word* over Σ is a sequence of letters from Σ. We denote by Σ^* the set of all words over Σ, the free monoid generated by Σ under the concatenation of words where the empty word ε serves as the identity. A *language L* over Σ is a subset of Σ^*. It is *regular* if it is recognized by a DFA or an NFA. A DFA is a 5-tuple $M = (Q, \Sigma, \delta, q_0, F)$, where Q is a set of states, $\delta : Q \times \Sigma \to Q$ is the transition function, $q_0 \in Q$ is the start state, and $F \subseteq Q$ is the set of final or accepting states. In an NFA, δ maps $Q \times \Sigma$ to 2^Q. We call $|Q|$ the *state complexity* of the automaton. Many languages are classified by this property.

Setting $\Sigma_\diamond = \Sigma \cup \{\diamond\}$, where $\diamond \notin \Sigma$ represents undefined positions or holes, a *partial word* over Σ is a sequence of symbols from $\Sigma_\diamond$. Denoting the set of all partial words over Σ by $\Sigma_\diamond^*$, a *partial language L'* over Σ is a subset of $\Sigma_\diamond^*$. It is regular if it is regular when being considered over $\Sigma_\diamond$. In other words, we define languages of partial words, or partial languages, by treating $\diamond$ as a letter. They can be transformed to languages by using $\diamond$-substitutions over Σ. A $\diamond$-*substitution* $\sigma : \Sigma_\diamond^* \to 2^{\Sigma^*}$ satisfies $\sigma(a) = \{a\}$ for all $a \in \Sigma$, $\sigma(\diamond) \subseteq \Sigma$, and $\sigma(uv) = \sigma(u)\sigma(v)$ for $u, v \in \Sigma_\diamond^*$. As a result, σ is fully defined by $\sigma(\diamond)$, e.g., if $\sigma(\diamond) = \{a, b\}$ and $L' = \{\diamond b, \diamond c\}$ then $\sigma(L') = \{ab, bb, ac, bc\}$. If we consider this process in reverse, we can "compress" languages into partial languages.

We consider the following question from Dassow et al. [4]: Are there regular languages $L \subseteq \Sigma^*, L' \subseteq \Sigma_\diamond^*$ and a $\diamond$-substitution σ with $\sigma(L') = L$ such that the minimal state complexity of a DFA accepting L' or the minimal state complexity of a $\diamond$-DFA accepting L, denoted by $\min_{\diamond\text{-}DFA}(L)$, is (strictly) less than the minimal state complexity of a DFA accepting L, denoted by $\min_{DFA}(L)$? Reference [4, Theorem 4] states that for every regular language L, we have $\min_{DFA}(L) \geq \min_{\diamond\text{-}DFA}(L) \geq \min_{NFA}(L)$, where $\min_{NFA}(L)$ denotes the minimal state complexity of an NFA accepting L, and there exist regular languages L such that $\min_{DFA}(L) > \min_{\diamond\text{-}DFA}(L) > \min_{NFA}(L)$. On the other hand, [4, Theorem 5] states that if $n \geq 3$ is an integer, regular languages L and L' exist such that $\min_{\diamond\text{-}DFA}(L) \leq n + 1$, $\min_{DFA}(L) = 2^n - 2^{n-2}$, $\min_{NFA}(L') \leq 2n + 1$, and $\min_{\diamond\text{-}DFA}(L') \geq 2^n - 2^{n-2}$. This was the first step towards analyzing the sets:

$$D_n = \{m \mid \text{there exists } L \text{ such that } \min_{\diamond\text{-}DFA}(L) = n \text{ and } \min_{DFA}(L) = m\},$$
$$N_n = \{m \mid \text{there exists } L \text{ such that } \min_{\diamond\text{-}DFA}(L) = n \text{ and } \min_{NFA}(L) = m\}.$$

Our paper, whose focus is the analysis of D_n and N_n, is organized as follows. We obtain in Section 2 values belonging to D_n by looking at specific types of regular languages, followed by values belonging to N_n in Section 3. Due to the nature of NFAs, generating a sequence of minimal NFAs from a $\diamond$-DFA is difficult. However, in the case $\min_{DFA}(L) > \min_{\diamond\text{-}DFA}(L) = \min_{NFA}(L)$, we show how to use concatenation of languages to create an L' with systematic differences between $\min_{\diamond\text{-}DFA}(L')$ and $\min_{NFA}(L')$. We also develop a way of applying integer partitions to obtain such values. We conclude with some remarks in Section 4.

2   Constructs for D_n

This section provides some values for D_n by analyzing several classes of regular languages. In the description of the transition function of our DFAs and $\diamond$-DFAs, all the transitions lead to the error state (a sink non-final state) unless otherwise stated. Also, in our figures, the error state and transitions leading to it have been removed for clarity. We will often refer to the following algorithm.

Given a $\diamond$-DFA $M' = (Q', \Sigma_\diamond, \delta', q'_0, F')$ and a $\diamond$-substitution σ, Algorithm 1 gives a minimal DFA that accepts $\sigma(L(M'))$:

- Build an NFA $N = (Q', \Sigma, \delta, q'_0, F')$ that accepts $\sigma(L(M'))$, where $\delta(q, a) = \{\delta'(q, a)\}$ if $a \in \Sigma \setminus \sigma(\diamond)$ and $\delta(q, a) = \{\delta'(q, a), \delta'(q, \diamond)\}$ if $a \in \sigma(\diamond)$.
- Convert N to an equivalent minimal DFA.

First, we look at languages of words of equal length. We give three constructs. The first two both use an alphabet of variable size, while our third one restricts this to a constant k. We prove the second construct which is illustrated in Fig. 1.

Theorem 1. *For $n \geq 1$, $\left\lfloor \frac{n-1}{3} \right\rfloor^2 + \left\lfloor \frac{n-1}{3} \right\rfloor + 2 + (n-1) \bmod 3 \in D_n$.*

Theorem 2. *For $n \geq 1$, if $x = \left\lfloor \frac{\sqrt{1+8(n-1)}-1}{2} \right\rfloor$ then $2^x + n - 1 - \frac{x(x+1)}{2} \in D_n$ for languages of words of equal length.*

Proof. We start by writing n as $n = r + \sum_{i=1}^{x} i$ such that $1 \leq r \leq x+1$ (from the online encyclopedia of integer sequences, x is as stated). Let $M = (Q, \Sigma, \delta, q_0, F)$ be the DFA defined as follows:

- $\{(i,j) \mid 0 \leq i < x,\ 0 \leq j < 2^i,\ (i,j) \neq (x-1,0)\} \cup \{(i,0) \mid x \leq i \leq x+r\} = Q$, $q_0 = (0,0)$, $F = \{(x+r-1, 0)\}$, and $(x+r, 0)$ is the error state;
- $\Sigma = \{a_0, a_1, c\} \cup \{b_i \mid 1 \leq i < x\}$;
- δ is defined as follows:
    - $\delta((i,j), a_k) = (i+1, 2j+k)$ for all $(i,j), (i+1, 2j+k) \in Q$, $a_k \in \Sigma$, $i \neq x-1$, with the exception of $\delta((x-2, 0), a_0) = (x+r, 0)$,
    - $\delta((x-1, i), b_j) = (x, 0)$ for all $(x-1, i) \in Q$, $b_j \in \Sigma$ where the jth digit from the right in the binary representation of i is a 1,
    - $\delta((i, 0), c) = (i+1, 0)$ for $x \leq i < x+r$.

Each word accepted by M can be written in the form $w = ub_ic^{r-1}$, where u is a word of length $x-1$ over $\{a_0, a_1\}$ except for a_0^{x-1}, and b_i belongs to some subset of Σ unique for each u. This implies that M is minimal with $2^x + n - 1 - \frac{x(x+1)}{2}$ states. We can build the minimal equivalent $\diamond$-DFA for $\sigma(\diamond) = \{a_0, a_1\}$, giving $M' = (Q', \Sigma_\diamond, \delta', q'_0, F')$ with n states as follows:

- $\{(i,j) \mid 0 \leq i < x,\ 0 \leq j \leq i,\ (i,j) \neq (x-1,0)\} \cup \{(i,0) \mid x \leq i \leq x+r\} = Q'$, $q'_0 = (0,0)$, $F' = \{(x+r-1, 0)\}$, and $(x+r, 0)$ is the error state;
- δ' is defined as follows:
    - $\delta'((i,0), a_1) = (i+1, i+1)$ for $0 \leq i < x-1$,

- $\delta'((i,j),\diamond)=(i+1,j)$ for all $(i,j)\in Q'\backslash\{(x-2,0)\}$ where $i<x-1$,
- $\delta'((x-1,i),b_{x-i})=(x,0)$ for $1\le i<x$,
- $\delta'((x+i,0),c)=(x+i+1,0)$ for $0\le i<r-1$.

Observe that $L(M')=\{\diamond^{x-i-1}a_1\diamond^{i-1}b_ic^{r-1}\mid 1\le i<x\}$, so $\sigma(L(M'))=L(M)$. Each accepted word consists of a unique prefix of length $x-1$ paired with a unique $b_i\in\Sigma$, and r states are needed for the suffix c^{r-1}, which implies that M' is minimal over all $\diamond$-substitutions. Note that $|Q'|=(\sum_{i=1}^{x}i)+r=n$. □

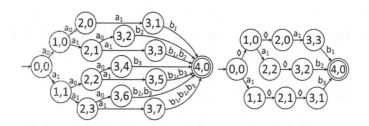

Fig. 1. M (left) and M' (right) from Theorem 2, $n=11, x=4$

Theorem 3. *For $k>1$ and $l,r\ge 0$, let $n=\frac{k(k+2l+3)}{2}+r+2$. Then*

$$2^{k+1}+l(2^k-1)+r\in D_n,$$

for languages of words of equal length.

Next, we look at languages of words of bounded length. The following theorem is illustrated in Fig. 2.

Theorem 4. *For $n\ge 3$, $[n,n+\frac{(n-2)(n-3)}{2}]\subseteq D_n$.*

Proof. Write $m=n+r+\sum_{i=l}^{n-3}i$ for the lowest value of $l\ge 1$ such that $r\ge 0$. Let $M=(Q,\Sigma,\delta,q_0,F)$ be defined as follows:

- $\Sigma=\{a_0,a_r\}\cup\{a_i\mid l\le i\le n-3\}$;
- $Q=\{(i,0)\mid 0\le i<n\}\cup\{(i,j)\mid a_j\in\Sigma$ and $1\le i\le j\}$, $q_0=(0,0)$, $F=\{(n-2,0)\}\cup\{(i,i)\mid i\ne 0,(i,i)\in Q\}$, and $(n-1,0)$ is the error state;
- δ is defined by $\delta((0,0),a_i)=(1,i)$ for all $a_i\in\Sigma$ where $i>0$, $\delta((i,j),a_0)=(i+1,j)$ for all $(i,j)\in Q,i\ne j$, and $\delta((i,i),a_0)=(i+1,0)$ for all $(i,i)\in Q$.

Then $L(M)=\{a_ia_0^{n-3}\mid a_i\in\Sigma\}\cup\{a_ia_0^{i-1}\mid a_i\in\Sigma,i\ne 0\}$. For each $a_i,i\ne 0$, M requires i states. These are added to the error state and $n-1$ states needed for a_0^{n-2}. Thus, M is minimal with m states. Let $M'=(Q',\Sigma_\diamond,\delta',q_0',F')$, where $Q'=\{i\mid 0\le i<n\}$, $q_0'=0$, $F'=\{n-2\}$, and $n-1$ is the error state; δ' is defined by $\delta'(0,\diamond)=1$, $\delta'(0,a_i)=n-1-i$ for all $a_i\in\Sigma,i>0$, and $\delta'(i,a_0)=i+1$ for $1\le i<n-1$. For $\sigma(\diamond)=\Sigma$, we have $\sigma(L(M'))=L(M)$. Furthermore, M' needs $n-1$ states to accept $\diamond a_0^{n-3}\in L(M')$, so M' is minimal with n states. □

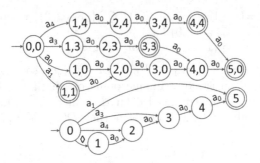

Fig. 2. M (top) and M' (bottom) from Theorem 4, $n = 7$ and $m = 15$ ($l = 3, r = 1$)

Theorem 4 gives elements of D_n close to its lower bound. To find an upper bound, we look at a specific class of machines. Let $n \geq 2$ and let

$$R_n = (\{0, \ldots, n-1\}, \{a_0\} \cup \{(\alpha_i)_j \mid 2 \leq i+2 \leq j \leq n-2\}_\diamond, \delta', 0, \{n-2\}) \quad (1)$$

be the $\diamond$-DFA where $n-1$ is the error state, and δ' is defined by $\delta'(i, \diamond) = i+1$ for $0 \leq i < n-2$ and $\delta'(i, (\alpha_i)_j) = j$ for all $(\alpha_i)_j$. Fig. 3 gives an example when $n = 7$. Set $L_n = \sigma(L(R_n))$, where σ is the $\diamond$-substitution that maps $\diamond$ to the alphabet. Note that R_n is minimal for $L(R_n)$, since we need at least $n-1$ states to accept words of length $n-2$ without accepting longer strings. Furthermore, R_n is minimal for σ, as each letter $(\alpha_i)_j$ encodes a transition between a unique pair of states (i, j). This also implies that R_n is minimal for any $\diamond$-substitution. The next two theorems look at the minimal DFA that accepts L_n. We refer the reader to Fig. 3 to visualize the ideas behind the proofs.

Referring to Fig. 3, in the DFA, each explicitly labelled transition is for the indicated letters. From each state, there is one transition that is not labelled - this represents the transition for each letter not explicitly labelled in a different transition from that state. (For example, from state 0, a_3 transitions to $\{1, 3\}$, a_2 transitions to $\{1, 2\}$, a_4 transitions to $\{1, 4\}$, a_5 transitions to $\{1, 5\}$, and all other letters $a_0, b_3, b_4, b_5, c_4, c_5, d_5$ transition to $\{1\}$). The idea behind the proof of Theorem 6 is that we start with this DFA. We introduce a new letter, "e", into the alphabet and add a new state, $\{2, 3, 4, 5\}$, along with a transition from $\{1, 3\}$ to $\{2, 3, 4, 5\}$ for e. We want to alter the $\diamond$-DFA to accommodate this. So we add a transition for e from 1 to 3 and from 3 to 5 (represented by dashed edges). All other states transition to the error state for e. Now consider the string $a_3 e$. We get four strings that correspond to some partial word that produces $a_3 e$ after substitution: $a_3 e$, $a_3 \diamond$, $\diamond e$, and $\diamond \diamond$. When the $\diamond$-DFA reads the first, it halts in state 5; on the second, it halts in 4; on the third, it halts in 3; and for the fourth, it halts in 2, which matches the added state $\{2, 3, 4, 5\}$. Finally, we need to consider the effect of adding e and the described transitions to the $\diamond$-DFA - does it change the corresponding minimal DFA in other ways? To show that it does not, all transitions with dashed edges in the DFA represent the transitions for e. For example, from state $\{2, 3\}$, an e transitions to $\{3, 4, 5\}$.

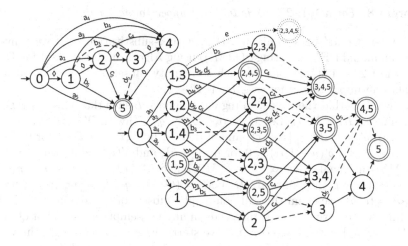

Fig. 3. $\diamond$-DFA R_7 (top if the dashed edges are seen as solid) and minimal DFA for $\sigma(L_7)$ (bottom if the dotted element is ignored and the dashed edges are seen as solid) where $\alpha_0 = a, \alpha_1 = b, \alpha_2 = c, \alpha_3 = d$ and $\sigma(\diamond) = \{a_0, a_2, a_3, a_4, a_5, b_3, b_4, b_5, c_4, c_5, d_5\}$.

Theorem 5. *Let Fib be the Fibonacci sequence defined by $Fib(1) = Fib(2) = 1$ and for $n \geq 2$, $Fib(n+1) = Fib(n) + Fib(n-1)$. Then for $n \geq 1$, $Fib(n+1) \in D_n$.*

Proof. For $n \geq 2$, applying Algorithm 1, convert $M' = R_n$ to a minimal DFA $M = (Q, \Sigma, \delta, q_0, F)$ that accepts L_n, where $Q \subseteq 2^{\{0,\dots,n-1\}}$. For each state $\{i\} \in Q$ for $0 \leq i \leq n - 2$, M requires additional states to represent each possible subset of one or more states of $\{i+1, \dots, n-2\}$ that M' could reach in i transitions. Thus M is minimal with number of states

$$1 + \sum_{i=0}^{n-2} \sum_{j=0}^{\min\{i,n-2-i\}} \binom{n-2-i}{j} = Fib(n+1),$$

where the 1 refers to the error state and where the inside sum refers to the number of states with minimal element i. $\square$

Theorem 6. *For $n \geq 3$, the following is the least upper bound for $m \in D_n$ in the case of languages of words of bounded length:*

$$\sum_{i=0}^{n-1} \binom{n-1-\lceil \log_2 i \rceil}{i}.$$

Our next result restricts the alphabet size to two.

Theorem 7. *For $n \geq 1$, $\dfrac{\lfloor \frac{n}{2} \rfloor (\lfloor \frac{n}{2} \rfloor + 1) + \lfloor \frac{n-1}{2} \rfloor (\lfloor \frac{n-1}{2} \rfloor + 1)}{2} + 1 \in D_n$.*

Finally, we look at languages with some arbitrarily long words.

Theorem 8. *For $n \geq 1$, $2^n - 1$ is the least upper bound for $m \in D_n$.*

Proof. First, let M' be a minimal $\diamond$-DFA with $\diamond$-substitution σ. If we convert this to a minimal DFA accepting $\sigma(L(M'))$ using Algorithm 1, the resulting DFA has at most $2^n - 1$ states, one for each non-empty subset of the set of states in M'. Thus an upper bound for $m \in D_n$ is $2^n - 1$.

Now we show that there exists a regular language L such that $\min_{\diamond\text{-}DFA}(L) = n$ and $\min_{DFA}(L) = 2^n - 1$. Let $M' = (Q', \Sigma_\diamond, \delta', q'_0, F')$ with $Q' = \{0, \ldots, n-1\}$, $\Sigma = \{a, b\}$, $q'_0 = 0$, $F' = \{n-1\}$, and δ' defined by $\delta'(i, \alpha) = i + 1$ for $0 \leq i < n - 1$, $\alpha \in \{\diamond, a\}$; $\delta'(n-1, \alpha) = 0$ for $\alpha \in \{\diamond, a\}$; and $\delta'(i, b) = 0$ for $0 \leq i < n$. Then M' is minimal, since $\diamond^{n-1} \in L(M')$ but $\diamond^i \notin L(M')$ for $0 \leq i < n - 1$. After constructing the minimal DFA $M = (Q, \Sigma, \delta, q_0, F)$ using Algorithm 1 for $\sigma(\diamond) = \{a, b\}$, we claim that all non-empty subsets of Q' are states in Q. To show this, we construct a word that ends in any non-empty subset P of Q'. Let $P = \{p_0, \ldots, p_x\}$ with $p_0 < \cdots < p_x$. We start with a^{p_x}. Then create the word w by replacing the a in each position $p_x - p_i - 1$, $0 \leq i < x$, with b.

We show that w ends in state P by first showing that for each $p_i \in P$, some partial word w' exists such that $w \in \sigma(w')$ and M' halts in p_i when reading w'. First, suppose $p_i = p_x$. Since $|w| = p_x$, let $w' = \diamond^{p_x}$. For w', M' halts in p_x. Now, suppose $p_i \neq p_x$. Let $w' = \diamond^{p_x - p_i - 1} b \diamond^{p_i}$. After reading $\diamond^{p_x - p_i - 1}$, M' is in state $p_x - p_i - 1$, then in state 0 for b, and then in state p_i after reading $\diamond^{p_i}$.

Now suppose a partial word w' exists such that $w \in \sigma(w')$ where M' halts in p for $p \notin P$. Suppose $p > p_x$. Each state $i \in Q'$ is only reachable after i transitions and $|w'| = p_x$, so M' cannot reach p after reading w'. Now suppose $p < p_x$. Then M' needs to be in state 0 after reading $p_x - p$ symbols to end in p, so we must have $w'[p_x - p - 1] = b$. However, $w[p_x - p - 1] = a$, a contradiction.

Furthermore, no states of Q are equivalent, as each word w ends in a unique state of Q. Therefore, M has $2^n - 1$ states, and $2^n - 1 \in D_n$. □

To further study intervals in D_n, we look at the following class of $\diamond$-DFAs. For $n \geq 2$ and $0 \leq r < n$, let

$$R_{n,r}\{s_1, \ldots, s_k\} = (\{0, \ldots, n-1\}, \{a_0, a_1, \ldots, a_k\}_\diamond, \delta', 0, \{n-1\}) \quad (2)$$

be the $\diamond$-DFA where $\{s_1, \ldots, s_k\}$ is a set of tuples whose first member is a letter a_i, distinct from a_0, followed by one or more states in ascending order, and where $\delta'(q, a_i) = 0$ for all (q, a_i) that occur in the same tuple, $\delta'(i, \diamond) = i + 1$ for $0 \leq i \leq n - 2$, $\delta'(n-1, \diamond) = r$, and $\delta'(q, a_i) = \delta'(q, \diamond)$ for all other (q, a_i). Since $R_{n,r}\{\}$ is minimal for any $\diamond$-substitution, and since $\diamond$ and non-$\diamond$ transitions from any state end in the same state, Algorithm 1 converts $R_{n,r}\{\}$ to a minimal DFA with exactly n states. The next result looks at $\diamond$-DFAs of the form $R_{n,r}\{(a_1, 0)\}$.

Theorem 9. *For $n \geq 2$ and $0 \leq i < n$, $n + (n-1)i \in D_n$.*

Proof. Let $a_0 = a$ and $a_1 = b$, let $r = n - i - 1$, let $\sigma(\diamond) = \Sigma = \{a, b\}$, and let $M' = R_{n,r}\{(b, 0)\}$. Using Algorithm 1, let $M = (Q, \Sigma, \delta, \{0\}, F)$ be the minimal DFA accepting $\sigma(L(M'))$. For all words over Σ of length less than n, M must halt in some state $P \in Q$, a subset of consecutive states of $\{0, \ldots, n-1\}$. Moreover, any state $P \in Q$ of consecutive states of $\{0, \ldots, n-1\}$, with minimal element p, is reached by M when reading $b^q a^p$ for some $q \geq 0$. Also, any accepting states in Q that are subsets of $\{0, \ldots, n-1\}$ of size $n-r$ or greater are equivalent, as are any non-accept states that are subsets of size $n-r$ or greater such that the $n-r$ greatest values in each set are identical. This implies that M requires $\sum_{j=n-i}^{n} j$ states for words of length less than n.

For words of length n or greater, M may halt in a state $P \in Q$ that is not a subset of consecutive states of $\{0, \ldots, n-1\}$, as for some $r < p < n-1$, it is possible to have $r, n-1 \in P$ but $p \notin P$. This only occurs when a transition from a state P with $n-1 \in P$ occurs, in which case, M moves to a state P' containing r, corresponding to $\delta'(n-1, \alpha)$ for all $\alpha \in \Sigma_\diamond$. Thus, all states can be considered subsets of consecutive values if we consider r consecutive to $n-1$ or, in other words, if we allow values from $n-1$ to r to "wrap" around to each other. This means that M requires $\sum_{j=1}^{i-1} j$ states for words of length n or greater. Therefore, $\sum_{j=n-i}^{n} j + \sum_{j=1}^{i-1} j = n + (n-1)i \in D_n$.   □

3   Constructs for N_n

Let Σ be an alphabet, and let $\Sigma_i = \{a_i \mid a \in \Sigma\}$ for all integers i, $i > 0$. Let $\sigma_i : \Sigma \to \Sigma_i$ such that $a \mapsto a_i$, and let $\#_j$ be a symbol in no Σ_i, for all i and j. Given a language L over Σ, the *ith product of L* and the *ith #-product of L* are, respectively, the languages

$$\pi_i(L) = \prod_{j=1}^{i} \sigma_j(L), \quad \pi_i'(L) = \sigma_1(L) \prod_{j=2}^{i} \{\#_{j-1}\} \sigma_j(L).$$

In general, we call any construct of this form, languages over different alphabets concatenated with $\#$ symbols, a *#-concatenation*. With these definitions in hand, we obtain our first bound for N_n.

Theorem 10. *For $n > 0$, $\left[n - \left\lfloor \frac{n-1}{3} \right\rfloor, n \right] \subseteq N_n$.*

Proof. Let $L = \{aa, ba, b\}$ be a language over $\Sigma = \{a, b\}$. A minimal NFA recognizing $\pi_i(L)$ is defined as having $2i + 1$ states, $q_0, \ldots, q_{2i}$, with accepting state q_{2i}, starting state q_0, and transition function δ defined by $\delta(q_{2j}, b_{j+1}) = \{q_{2j+1}, q_{2(j+1)}\}$, $\delta(q_{2j}, a_{j+1}) = \{q_{2j+1}\}$, and $\delta(q_{2j+1}, a_{j+1}) = \{q_{2(j+1)}\}$ for $j < i$. It is easy to see this is minimal: the number of states is equal to the maximal length of the words plus one. A minimal $\diamond$-DFA recognizing $\pi_i(L)$ is defined as having $3i + 1$ states, $q_0, \ldots, q_{3i-1}$ and q_{err}, with accepting states q_{3i-1} and q_{3i-2}, starting state q_0, and transition function δ defined as follows:

- $\delta(q_0, b_1) = q_2$, $\delta(q_0, \diamond) = q_1$, and $\delta(q_1, a_1) = q_2$;
- $\delta(q_{3j-1}, a_{j+1}) = q_{3j}$, $\delta(q_{3j-1}, b_{j+1}) = q_{3j+1}$, $\delta(q_{3j}, a_{j+1}) = q_{3(j+1)-1}$, and $\delta(q_{3j+1}, a_{j+1}) = q_{3(j+1)-1}$ for $0 < j < i$;

$$- \delta(q_{3j+1}, a_{j+2}) = \delta(q_{3(j+1)-1}, a_{j+2}) \text{ and } \delta(q_{3j+1}, b_{j+2}) = \delta(q_{3(j+1)-1}, b_{j+2})$$
for $0 < j < i - 1$.

The $\diamond$-substitution corresponds to $\Sigma_1 = \{a_1, b_1\}$ here. This is minimal.

Now, fix n; take any $i \leq \lfloor \frac{n-1}{3} \rfloor$. We can write $n = 3i + r + 1$, for some $r \geq 0$. Let $\{\alpha_j\}_{0 \leq j \leq r}$ be a set of symbols not in the alphabet of $\pi_i(L)$. Minimal NFA and $\diamond$-DFA recognizing $\pi_i(L) \cup \{\alpha_0 \cdots \alpha_r\}$ can clearly be obtained by adding to each a series of states $q'_0 = q_0, q'_1, \ldots, q'_r$, and $q'_{r+1} = q_{2i}$ and $q'_{r+1} = q_{3i-1}$ respectively, with $\delta(q'_j, \alpha_j) = q'_{j+1}$ for $0 \leq j \leq r$. Hence, for $i \leq \lfloor \frac{n-1}{3} \rfloor$, we can produce a $\diamond$-DFA of size $n = 3i + r + 1$ which reduces to an NFA of size $2i + r + 1 = n - i$.   □

Our general interval is based on $\pi'_i(L)$, where no $\diamond$-substitutions exist over multiple Σ_i's. We need the following lemma.

Lemma 1. *Let L, L' be languages recognized by minimal NFAs $N = (Q, \Sigma, \delta, q_0, F)$ and $N' = (Q', \Sigma', \delta', q'_0, F')$, where $\Sigma \cap \Sigma' = \emptyset$. Moreover, let $\# \notin \Sigma, \Sigma'$. Then $L'' = L\{\#\}L'$ is recognized by the minimal NFA $N'' = (Q \cup Q', \Sigma \cup \Sigma', \delta'', q_0, F')$, where $\delta''(q, a) = \delta(q, a)$ if $q \in Q$ and $a \in \Sigma$; $\delta''(q, a) = \delta'(q, a)$ if $q \in Q'$ and $a \in \Sigma'$; $\delta''(q, \#) = \{q'_0\}$ if $q \in F$; and $\delta''(q, a) = \emptyset$ otherwise. Consequently, the following hold:*

1. *For any L, $\min_{NFA}(\pi'_i(L)) = i \min_{NFA}(L)$;*
2. *Let $L_1, \ldots, L_n$ be languages whose minimal DFAs have no error states and whose alphabets are pairwise disjoint, and without loss of generality, let $\min_{DFA}(L_1) - \min_{\diamond\text{-}DFA}(L_1) \geq \cdots \geq \min_{DFA}(L_n) - \min_{\diamond\text{-}DFA}(L_n)$. Then*

$$\min_{\diamond\text{-}DFA}(L_1\{\#_1\}L_2\{\#_2\} \cdots L_n) = 1 + \min_{\diamond\text{-}DFA}(L_1) + \sum_{i=2}^{n} \min_{DFA}(L_i).$$

Theorem 11. *Let L be a language whose minimal DFA has no error state. Moreover, assume $\min_{\diamond\text{-}DFA}(L) = \min_{NFA}(L)$. Fix some n and j, $0 < j \leq \left\lfloor \frac{n - \min_{\diamond\text{-}DFA}(L) - 1}{\min_{DFA}(L)} \right\rfloor$. Then $n - j(\min_{DFA}(L) - \min_{\diamond\text{-}DFA}(L)) - 1 \in N_n$.*

Proof. Since $0 < j \leq \left\lfloor \frac{n - \min_{\diamond\text{-}DFA}(L) - 1}{\min_{DFA}(L)} \right\rfloor$, we can write $n = 1 + \min_{\diamond\text{-}DFA}(L) + j \min_{DFA}(L) + r$ for some r. Then, by Lemma 1(2), this corresponds to $n = \min_{\diamond\text{-}DFA}(\pi'_{j+1}(L) \cup \{w\})$, where w is a word corresponding to an r-length chain of states, as we used in the proof of Theorem 10. We also have $\min_{NFA}(\pi'_{j+1}(L) \cup \{w\}) = (j + 1) \min_{\diamond\text{-}DFA}(L) + r$ using Lemma 1(1) and our assumption that $\min_{\diamond\text{-}DFA}(L) = \min_{NFA}(L)$. Alternatively,

$$\min_{NFA}(\pi'_{j+1}(L) \cup \{w\}) = n - j\left(\min_{DFA}(L) - \min_{\diamond\text{-}DFA}(L)\right) - 1.$$

Our result follows.   □

The above linear bounds can be improved, albeit with a loss of clarity in the overall construct. Consider the interval of values obtained in Theorem 4. Fix an integer x. The minimal integer y such that $x \leq y + \frac{(y-2)(y-3)}{2}$ is clearly $n_x = \left\lceil \frac{3+\sqrt{8x-15}}{2} \right\rceil$, for $x \geq 4$. Associate with x and n_x the corresponding DFAs and $\diamond$-DFAs used in the proof of Theorem 4, i.e., let $L_{n,m}$ be the language in the proof with minimal $\diamond$-DFA size n and minimal DFA size m. If we replace each $\diamond$-transition in the minimal $\diamond$-DFA and remove the error state, we get a minimal NFA of size $n-1$ accepting $L_{n,m}$ (this NFA must be minimal since the maximal length of a word in $L_{n,m}$ is $n-2$). Noting that all deterministic automata in question have error states, we get, using Lemma 1(1), that $\min_{\diamond\text{-}DFA}(\pi'_i(L_{n_x,x})) = n_x + (i-1)(x-1)$ and $\min_{NFA}(\pi'_i(L_{n_x,x})) = i(n_x-1)$. This allows us to obtain the following linear bound.

Theorem 12. *For $n > n_x \geq 4$, $\left[n - (x - n_x) \left\lfloor \frac{n - n_x}{x-1} \right\rfloor - 1, n \right] \subseteq N_n$.*

Proof. For any n and fixed x, write $n = n_x + (i-1)(x-1) + r$, for some $0 \leq r < x - 1$, which is realizable as a minimal $\diamond$-DFA by appending to the minimal $\diamond$-DFA accepting $\pi'_i(L_{n_x,x})$ an arbitrary chain of states of length r, using letters not in the alphabet of $\pi'_i(L_{n_x,x})$, similar to what we did in the proof of Theorem 10. This leads to a minimal NFA of size $i(n_x - 1) + r$, giving the lower bound $n - (x - n_x) \left\lfloor \frac{n - n_x}{x-1} \right\rfloor - 1$ if we solve for i. Anything in the upper bound can be obtained by decreasing i or replacing occurrences of $L_{n_x,x}$ with $L_{n_x,x-j}$ (for some j) and in turn adding additional chains of states of length r, to maintain the size of the $\diamond$-DFA. $\qquad\qquad\qquad\qquad\qquad\qquad\qquad\square$

We can obtain even lower bounds by considering the sequence of DFAs defined in Theorem 8. Recall that for any $n \geq 1$, we have a minimal DFA, which we call M_n, of size $2^n - 1$; the equivalent minimal $\diamond$-DFA, M'_n, has size n. Applying Algorithm 1 to M'_n, the resulting NFA of size n is also minimal. Let $n_0 \geq n_1 \geq \cdots \geq n_k$ be a sequence of integers and consider

$$\min_{\diamond\text{-}DFA} (L(M_{n_0})\{\#_1\}L(M_{n_1})\cdots\{\#_k\}L(M_{n_k})) = 1 + n_0 + \sum_{i=1}^{k}(2^{n_i} - 1), \quad (3)$$

where the equality comes from Lemma 1(2). Iteratively applying Lemma 1 gives

$$\min_{NFA}(L(M_{n_0})\{\#_1\}L(M_{n_1})\cdots\{\#_k\}L(M_{n_k})) = \sum_{i=0}^{k} n_i. \quad (4)$$

To understand the difference between (3) and (4) in greater depth, let us view $(n_1, \ldots, n_k)$ as an integer partition, λ, or as a Young Diagram and assign each cell a value (see, e.g., [1]). In this case, the ith column of λ has each cell valued at 2^{i-1}. Transposing about $y = -x$ gives the diagram corresponding to the transpose of λ, $\lambda^T = (m_1, \ldots, m_{n_1})$, in which the ith row has each cell valued at 2^{i-1}. Note that $m_1 = k$ and there are, for each i, m_i terms of 2^{i-1}. Fig. 4 gives an example of an integer partition and its transpose. Define $\Pi(\lambda^T) = \sum_{i=1}^{n_1} 2^{i-1} m_i = \sum_{i=1}^{k}(2^{n_i} - 1)$ and $\Sigma(\lambda) = \sum_{i=1}^{k} n_i$.

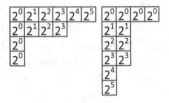

Fig. 4. $\lambda = (6, 4, 1, 1)$ (left) and $\lambda^T = (4, 2, 2, 2, 1, 1)$ (right)

Given this, we can view the language L described in (3) and (4), i.e, $L = L(M_{n_0})\{\#_1\}L(M_{n_1})\cdots\{\#_k\}L(M_{n_k})$, as being defined by the integer n_0 and the partition of integers $\lambda = (n_1, \ldots, n_k)$ with $n_0 \geq n_1$. This gives

$$\min_{\diamond\text{-}DFA}(L) = 1 + n_0 + \Pi(\lambda^T) \quad \text{and} \quad \min_{NFA}(L) = n_0 + \Sigma(\lambda).$$

To further understand this, we must consider the following sub-problem: let $\Pi(\lambda) = n$. What are the possible values of $\Sigma(\lambda)$? To proceed here, we define the sequence p_n recursively as follows: if $n = 2^k - 1$ for some k, $p_n = k$; otherwise, letting $n = m + (2^k - 1)$ for k maximal, $p_n = k + p_m$. This serves as the minimal bound for the possible values of $\Sigma(\lambda)$.

Theorem 13. If $\Pi(\lambda) = n$, then $\Sigma(\lambda) \geq p_n$. Consequently, for all n and $k = \lfloor \log_2(n + 1) \rfloor$, $k + p_n \in N_{1+k+n}$.

Proof. To show that p_n is obtainable, we prove that the following partition, λ_n, satisfies $\Sigma(\lambda) \geq p_n$: if $n = 2^k - 1$ for some k, $\lambda_n = (1^k)$; otherwise, letting $n = m + (2^k - 1)$ for k maximal, $\lambda_n = \lambda_{2^k-1} + \lambda_m$. Here, the sum of two partitions is the partition obtained by adding the summands term by term; (1^k) is the k-tuple of ones. Clearly, for partitions λ and λ', $\Pi(\lambda + \lambda') = \Pi(\lambda) + \Pi(\lambda')$ and $\Sigma(\lambda + \lambda') = \Sigma(\lambda) + \Sigma(\lambda')$. By construction, $\Pi(\lambda_n) = n$ and $\Sigma(\lambda_n) = p_n$. To see this, if $n = 2^k - 1$ for some k, $\Pi(\lambda_n) = \Pi((1^k)) = \Pi((k)^T) = 2^k - 1 = n$ and $\Sigma(\lambda_n) = \Sigma((1^k)) = \sum_{i=1}^{k} 1 = k = p_n$. Otherwise,

$$\Pi(\lambda_n) = \Pi(\lambda_{2^k-1}) + \Pi(\lambda_m) = \Pi((1^k)) + \Pi(\lambda_m) = 2^k - 1 + m = n,$$

$$\Sigma(\lambda_n) = \Sigma(\lambda_{2^k-1}) + \Sigma(\lambda_m) = \Sigma((1^k)) + \Sigma(\lambda_m) = k + p_m = p_n.$$

To show that p_n, or λ_n, is minimal, we can proceed inductively.

From the above, each p_n is obtainable by a partition of size k, where k is the maximal integer with $n \geq 2^k - 1$. Alternatively, $k = \lfloor \log_2(n + 1) \rfloor$. Fixing n, we get $k + p_n \in N_{1+k+n}$. $\square$

4 Conclusion

For languages of words of equal length, Theorem 2 gives the maximum element in D_n found so far and Theorem 3 gives that maximum element when we restrict to a constant alphabet size. For languages with words of bounded length, Theorem 6 gives the least upper bound for elements in D_n based on minimal $\diamond$-DFAs of the form (1) and Theorem 7 gives the maximum element found so far when we restrict to a binary alphabet. For languages with words of arbitrary length, Theorem 8 gives the least upper bound of $2^n - 1$ for elements in D_n, bound that can be achieved over a binary alphabet. We conjecture that for $n \geq 1$, $[n, 2^n - 1] \subseteq D_n$. This conjecture has been verified for all $1 \leq n \leq 7$ based on all our constructs from Section 2.

In Section 3, via products, Theorem 10 gives an interval for N_n. If we replace products with #-concatenations, Theorem 12 increases the interval further. Theorem 13 does not give an interval, but an isolated point not previously achieved. With the exception of this latter result, all of our bounds are linear. Some of our constructs satisfy $\min_{\diamond\text{-}DFA}(L) = \min_{NFA}(L)$, ignoring error states. As noted earlier, this is a requirement for #-concatenations to produce meaningful bounds. Constructs without this restriction are often too large to be useful.

References

1. Andrews, G.E., Eriksson, K.: Integer Partitions. Cambridge University Press (2004)
2. Berstel, J., Boasson, L.: Partial words and a theorem of Fine and Wilf. Theoretical Computer Science 218, 135–141 (1999)
3. Blanchet-Sadri, F.: Algorithmic Combinatorics on Partial Words. Chapman & Hall/CRC Press, Boca Raton (2008)
4. Dassow, J., Manea, F., Mercaş, R.: Connecting partial words and regular languages. In: Cooper, S.B., Dawar, A., Löwe, B. (eds.) CiE 2012. LNCS, vol. 7318, pp. 151–161. Springer, Heidelberg (2012)
5. Fischer, M., Paterson, M.: String matching and other products. In: Karp, R. (ed.) 7th SIAM-AMS Complexity of Computation, pp. 113–125 (1974)
6. Hopcroft, J.E., Motwani, R., Ullman, J.D.: Introduction to Automata Theory, Languages, and Computation, 2nd international edn. Addison-Wesley (2003)
7. Yu, S.: Regular languages. In: Rozenberg, G., Salomaa, A. (eds.) Handbook of Formal Languages, vol. 1, ch. 2, pp. 41–110. Springer, Berlin (1997)

Using Regular Grammars for Event-Based Testing

Fevzi Belli and Mutlu Beyazıt

EIM-E/ADT, University of Paderborn, Paderborn, Germany
{belli,beyazit}@adt.upb.de

Abstract. Model-based testing involves formal models for test generation. This paper suggests regular grammars for event-based modeling. This model, represented in BNF, will then be systematically modified by well-defined mutation operators in order to generate fault models, called *mutants*. Specific algorithms apply to both the model of the system under consideration and the mutants to generate test cases. While existing methods focus on single events the approach introduced in this paper suggests considering event sequences of length $k{\geq}1$, that is, *k-sequences*. The approach also enables to cope with a tough problem encountered in mutation-oriented testing: the elimination of mutants that are equivalent to the original model, and mutants that model the same faults multiple times. These mutants lead to unproductive test suites that cause wasting of resources. The approach proposed devises strategies to exclude the mentioned mutants in that they will not be generated at all.

Keywords: model-based mutation testing, event-based, mutant selection.

1 Introduction and Related Work

Testing is a user-centric technique for quality assurance based on *test cases* that consist of *test inputs* and expected *test outputs*. A *test* invokes the execution, or training of the system under consideration (SUC) using a test case. SUC *succeeds* the test if, upon a test input, the expected test output is produced; otherwise, SUC *fails* the test, which entails the tough *oracle* problem for deriving the expected output. A *set of test cases*, also called *test set/suite*, is generated and executed in the target environment of SUC or an environment which closely resembles the target environment. A *coverage criterion* [16] is used as a stopping condition for testing to provide a measure of test quality of a test suite. This paper prefers the term SUC to "system under test (SUT)" because the approach introduced applies both to a model and an implementation whereas SUT applies to an implementation.

Model-based testing is based on the creation of an abstraction, called *model*, and operating on this model for testing [4]. The use of models has various advantages, such as increased effectiveness and efficiency in terms of fault detection and costs by avoiding the oracle problem if the model is formal in the sense that it can be used to determine the expected test outputs [15].

Model-based mutation testing [9][2][8] includes the additional use of *fault models* in test generation. Fault models are also called *mutants* because they are generated

S. Konstantinidis (Ed.): CIAA 2013, LNCS 7982, pp. 48–59, 2013.
© Springer-Verlag Berlin Heidelberg 2013

from the original model using *mutation operators*. Some mutants can be *equivalent* to the original model. This is a major problem because these mutants do not describe any different behaviors and cause wasting of testing resources [1]; for example, Grün, Schuler and Zeller [10] report that up to 40% of the generated mutants can be equivalent. By using mutants, model-based mutation testing approaches aim to generate test cases which distinguish or discriminate the mutants from the original model, that is, they *kill* the mutants. When such a test case is executed, SUC can be tested as to whether it contains the fault modeled by the mutant or not.

Tests based on the model of SUC form *positive testing* to check whether or not SUC comply with the user expectations, while *negative* testing is based on mutants, or fault models, to check whether or not SUC does not behave as the user does not wish.

Grammar-based testing [14] has already attracted some interest. Mostly, they are used for testing of software termed as *grammarware*, such as compilers, debuggers, code generators, documentation generators [13]. The approaches entail the use of context-free grammars and generation of well-formed inputs to the programs.

In this paper, adopting a model-based mutation testing methodology, a new event-based [7] regular grammar model using event sequences of length $k{\geq}1$ (*k-sequences*) is proposed and related mutation operators to generate fault models are defined. Most importantly, based on certain assumptions about the testing process, mutant selection strategies are devised to exclude mutants that are equivalent to the original model and multiple mutants that model the same faults. The significance of these strategies lies in the fact that we exclude these mutant without even generating them, as opposed to other model-based mutation testing approaches, such as the ones based on [2] or [3], which compare each generated mutant to the original model to determine whether the mutant is equivalent or not. Furthermore, these approaches do not discuss how to avoid the generation of multiple mutants modeling the same faults.

The paper is organized as follows: Section 2 introduces our event-based grammar model and related terminology using examples, Section 3 discusses the mutant selection strategies, Section 4 performs an evaluation, and Section 5 concludes the paper.

2   Formal Grammars for Event-Based Modeling

Basically, an *event* is an externally observable phenomenon, such as an environmental stimulus or a system response, punctuating different stages of the system activity. It is clear that testing activity often disregards the detailed internal behavior of SUC and, hence, can be satisfied by relatively more abstract and simpler representations compared to system development [7]. This is the reason why the approach introduced in this paper chooses formal grammars the elements of which refer to events that are perceivable to the tester to enable him or her to unambiguously decide whether or not SUC passes the test (Oracle problem, see Section 1).

Example 1 (Running Example). Suppose we have 3 events
$$c: \text{copy}, x: \text{cut and } p: \text{paste}.$$
At the beginning, one can perform either c or x. c can be followed by either c, x or p, and x can be followed by either c, x or p. If p is performed after c, it can be followed

by either *c*, *x* or *p*. However, if *p* is performed after *x*, it can only be followed by either *c* or *x*, that is, after cutting and pasting an object, it is not possible to paste it again. One can stop after a *p*.

Fig. 1a represents an event-based directed graph model for Example 1. Such models are popular in testing community [7] and have the same expressiveness as FSA (or FSA with outputs, if output events are also included). Since events are the observable entities in model-based testing and states generally represent the internals of the system, we focus on events and refrain from visualizing states. Therefore, events are placed at the nodes, and the *follows* relation between the events is described using arcs. Pseudo-events *[* and *]* are used to mark start and finish events, respectively. [5]

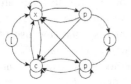   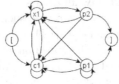

$S \rightarrow c1\ c(c1)\ |\ x1\ c(x1)$
$c(c1) \rightarrow c1\ c(c1)\ |\ x1\ c(x1)\ |\ p1\ c(p1)$
$c(x1) \rightarrow c1\ c(c1)\ |\ x1\ c(x1)\ |\ p2\ c(p2)$
$c(p1) \rightarrow c1\ c(c1)\ |\ x1\ c(x1)\ |\ p1\ c(p1)\ |\ \varepsilon$
$c(p2) \rightarrow c1\ c(c1)\ |\ x1\ c(x1)\ |\ \varepsilon$

(a) Model with ambiguity.   (b) Model with contexted events (No ambiguity).   (c) Grammar which makes use of 1-sequences.

Fig. 1. Event-based models for Example 1

The model in Fig. 1a causes a severe problem: When we say "event *p* in the model," one cannot differentiate which *p* event is referred to. To avoid such ambiguities, we use two types of events. We distinguish events from each other using the *context*s they reside in, and use *contexted events*, such as *{c1, x1, p1, p2}*, as shown in Fig. 1b. We also keep the events as they are visible to the user, that is, *basis events*, such as *{c, x, p}*. Thus, contexted events can be viewed as different instances of basis events.

$S \rightarrow c1\ c1\ c(c1\ c1)\ |\ c1\ x1\ c(c1\ x1)\ |\ c1\ p1\ c(c1\ p1)\ |$
$\quad x1\ c1\ c(x1\ c1)\ |\ x1\ x1\ c(x1\ x1)\ |\ x1\ p2\ c(x1\ p2)$
$c(c1\ c1) \rightarrow c1\ c1\ c(c1\ c1)\ |\ c1\ x1\ c(c1\ x1)\ |\ c1\ p1\ c(c1\ p1)$
$c(c1\ x1) \rightarrow x1\ c1\ c(x1\ c1)\ |\ x1\ x1\ c(x1\ x1)\ |\ x1\ p2\ c(x1\ p2)$
$c(c1\ p1) \rightarrow p1\ c1\ c(p1\ c1)\ |\ p1\ x1\ c(p1\ x1)\ |\ p1\ p1\ c(p1\ p1)\ |\ \varepsilon$
$c(x1\ c1) \rightarrow c1\ c1\ c(c1\ c1)\ |\ c1\ x1\ c(c1\ x1)\ |\ c1\ p1\ c(c1\ p1)$
$c(x1\ x1) \rightarrow x1\ c1\ c(x1\ c1)\ |\ x1\ x1\ c(x1\ x1)\ |\ x1\ p2\ c(x1\ p2)$
$c(x1\ p2) \rightarrow p2\ c1\ c(p2\ c1)\ |\ p2\ x1\ c(p2\ x1)\ |\ \varepsilon$
$c(p1\ c1) \rightarrow c1\ c1\ c(c1\ c1)\ |\ c1\ x1\ c(c1\ x1)\ |\ c1\ p1\ c(c1\ p1)$
$c(p1\ x1) \rightarrow x1\ c1\ c(x1\ c1)\ |\ x1\ x1\ c(x1\ x1)\ |\ x1\ p2\ c(x1\ p2)$
$c(p1\ p1) \rightarrow p1\ c1\ c(p1\ c1)\ |\ p1\ x1\ c(p1\ x1)\ |\ p1\ p1\ c(p1\ p1)\ |\ \varepsilon$
$c(p2\ c1) \rightarrow c1\ c1\ c(c1\ c1)\ |\ c1\ x1\ c(c1\ x1)\ |\ c1\ p1\ c(c1\ p1)$
$c(p2\ x1) \rightarrow x1\ c1\ c(x1\ c1)\ |\ x1\ x1\ c(x1\ x1)\ |\ x1\ p2\ c(x1\ p2)$

(a) Productions.   (b) Directed graph visualization.

Fig. 2. A grammar model for Example 1 which makes use of 2-sequences

Such a model considers the relation between only single events. To define an event-based abstraction which models the occurrences of single events with respect to event sequences of length $k{\geq}1$ (*k-sequences*), we use grammars [12], because a grammar allows use of multiple events in their *productions*.

Fig. 1c shows the productions of a grammar model which describe Example 1 using 1-sequences where $c(r)$ represents the *context of r*. Also, productions can be visualized via directed graphs (See Fig. 1b).

To model Example 1 using 2-sequences, we can make use of the grammar model whose productions are given in Fig. 2. Using the following semantics for productions in Fig. 2, we can say that grammars in Fig. 1c and Fig. 2 model the same system.

• Production $S \rightarrow a\ b\ c(a\ b)$ (or arc $([,\ a\ b))$ means that $a\ b$ is a start 2-sequence.
• Production $c(a\ b) \rightarrow \varepsilon$ (or arc $(a\ b,\ ])$) means that $a\ b$ is a finish 2-sequence.
• Production $c(a\ e) \rightarrow e\ b\ c(e\ b)$ (or arc $(a\ e,\ e\ b)$) means that b follows a.

Having outlined the intuitive background for our event-based grammar model which makes use of k-sequences, now we can give its formal definition.

Definition 1 (k-Sequence Right Regular Grammar (*k-Reg*)). *A k-sequence right RG (k-Reg) (integer $k \geq 1$) is a 6-tuple $G = (E,\ B;\ K;\ C;\ S;\ P)$ where:*

• *E is a finite set of events (or contexted events).*
• *B is a finite set of basis events, which is the set of all visible events under consideration. For $e \in E$, $d(e) \in B$ is the corresponding basis event, and $d(.)$ is the decontexting function.*
• *$K \subseteq E^k$ is a finite set of k-sequences (or terminals). For $r \in K$, $r = r_1...r_k$ and $d(r) = d(r_1)...d(r_k) \in B^k$ is the corresponding basis k-sequence.*
• *C is a finite set of contexts (or nonterminals).*
• *S is the start context (or start symbol).*
• *P is a finite set of productions of the form*

$$Q \rightarrow \varepsilon\ or\ Q \rightarrow r\ c(r)$$

where $Q \in C$ is a context, $r \in K$ is a k-sequence, $c(r) \in C\backslash\{S\}$ is the unique context of r, and ε is the empty string. If $k \geq 2$, for each $c(q) \rightarrow r\ c(r) \in P$

$$q_2...q_k = r_1...r_{k-1}.$$

The semantics of the productions is as follows: For each $c(q) \rightarrow r\ c(r) \in P$, we say that r_k *follows q in the system modeled by grammar G*, that is, $q\ r_k$ is a (k+1)-sequence in the system. Also, for each $S \rightarrow r\ c(r) \in P$, r is a *start k-sequence*, and, for each $c(q) \rightarrow \varepsilon \in P$, q is a *finish k-sequence*.

Productions of a k-Reg are used to derive strings. A *derivation*, denoted by $\Rightarrow_G{}^*$, is a sequence of *derivation steps* each of which is of the form $xQy \Rightarrow_G xRy$ where $x,y \in (C\cup K)^*$ and $Q \rightarrow R \in P$ (We use $\Rightarrow^*$ and $\Rightarrow$ when there is no confusion). Also, the *language* defined by grammar G is the set of strings $L(G) = \{w|\ S \Rightarrow^* w\ (w \in K^*)\}$.

k-Reg Productions can be visualized using directed graphs by labeling the nodes using the k-sequences, and $[$ and $]$. Arcs of the form "$([,\ r)$", "$(r,\ ])$" and "$(q,\ r)$"correspond to the productions of the form "$S \rightarrow r\ c(r)$", "$c(r) \rightarrow \varepsilon$" and "$c(q) \rightarrow r\ c(r)$", respectively. We often use such visualizations to refer to our grammar models.

To associate (contexted) sequences with basis sequences, we extend function $d(.)$ given in Definition 1.

Definition 2 (Decontexting Function). *Given a k-Reg $G = (E,\ B;\ K;\ C;\ S;\ P)$. Let $s = s_1\ s_2\ ... \in E^*$ be an event sequence and X be a set of event sequences. The* *corresponding basis event sequence of s is $d(s) = d(s_1) d(s_2) \ldots \in B^*$ if $s \neq \varepsilon$, and $d(\varepsilon)$ = ε. The corresponding set of basis event sequences of X is $d(X) = \{d(s)| s \in X\}$.*

Example 2 (Decontexted Event Sequences). Consider the 1-Reg in Fig. 1c. for set of event sequences $X = \{c1, c1\ p1, c1\ x1\ p2\}$, $d(X) = \{c, c\ p, c\ x\ p\}$.

For testing, we distinguish between event sequences that can and cannot be obtained using k-Reg productions.

Definition 3 (Event Sequences in a k-Reg). *Given a k-Reg $G = (E, B; K; C; S; P)$. Event sequence s is in grammar G, if there is a derivation of the form $Q \Rightarrow^* xsy$ for some $Q \in C$ and $x,y \in (C \cup E)^*$. A nonempty event sequence s in G is a start [or finish] sequence, if there is a derivation of the form $S \Rightarrow^* s\ Q\ (Q \in C)$ [or $Q \Rightarrow^* s\ (Q \in C)$]. An event sequence which is not in G is also called a faulty event sequence.*

Example 3 (Event Sequences in a 1-Reg). For the 1-Reg in Fig. 1c:
- 2-sequences in $\{c1\ x1, x1\ p2, p1\ p1\}$ are in the grammar, whereas 2-sequences in $\{p2\ p1, p2\ p2\}$ are not.
- $\{c1, x1, c1\ c1, x1\ x1\ p2, c1\ p1\ x1\}$ is a set of start sequences, and $\{p1, p2, p1\ p1, x1\ p2\}$ is a set of finish sequences.

Since we use k-Regs in testing, we often refer to test cases of a k-Reg.

Definition 4 (Test Cases). *Given a k-Reg $G = (E, B; K; C; S; P)$.*
- *An event sequence is a positive test case, if it is a start sequence in G, or it is ε. $T_P(G)$ denotes the set of all positive test cases. A complete event sequence (CES) is a positive test case which is both a start and a finish sequence in G, or it is ε if $\varepsilon \in L(G)$. $T_{CES}(G) = L(G) \subseteq T_P(G)$ denotes the set of all CESs.*
- *An event sequence is a negative test case, if the first event in it is a non-start event or it contains at least one 2-sequence which is not in G. $T_N(G)$ denotes the set of all negative test cases. A faulty complete event sequence (FCES) is a negative test case which either is composed of only a non-start event, or contains only a single 2-sequence which is not in G and it ends with this 2-sequence. $T_{FCES}(G) \subseteq T_N(G)$ denotes the set of all FCESs.*
- *A set of test cases is also called a test set.*

Example 4 (Test Cases of a 1-Reg). For the 1-Reg in Fig. 1c:
- $\{c1, x1\ x1, c1\ p1\ p1\ x1\}$ is a set of positive test cases, and $\{x1\ p2, x1\ x1\ p2, c1\ p1\ p1\ p1\}$ is a set of CESs.
- $\{p1, x1\ p2\ p1\ c1, c1\ x1\ p2\ p2\}$ is a set of negative test cases, and $\{x1\ p2\ p2, c1\ x1\ p2\ p2\}$ is a set of FCESs.

Each event in a given k-Reg is contexted. However, system behaviors are based on basis events, since they correspond to system events as they are visible to the user (Definition 1). Thus, we define the equivalence of two k-Regs as follows.

Definition 5 (Equivalence). *Two k-Regs G and H are equivalent, if $d(T_{CES}(G)) = d(T_{CES}(H))$.*

In practice, it is important that all k-sequences in a k-Reg are utilized.

Definition 6 (Usefulness). *Given a k-Reg G = (E, B; K; C; S; P). A string $z \in (C \cup E)^*$ is useful in grammar G, if $S \Rightarrow^* xzy \Rightarrow^* w$ for some $x,y \in (C \cup E)^*$ and $w \in E^*$. Grammar G is useful, if all k-sequences in K are useful in G.*

Example 5 (A Useful and a Non-useful 1-Reg). k-Regs in Fig. 1c and Fig. 2 are all useful. To obtain a non-useful 1-Reg from Fig. 1c, one can remove $c(p1) \rightarrow \varepsilon$ and $c(p2) \rightarrow \varepsilon$. The resulting grammar does not have any finish events anymore and, therefore, $T_{CES}(G)$ is empty, but it still describes the follows relation correctly.

Deterministic system models help to exclude redundant event sequences.

Definition 7 (Determinism). *A k-Reg G = (E, B; K; C; S; P) is deterministic, if, for each $Q \in C$, there are no two productions $Q \rightarrow q\ c(q) \in P$ and $Q \rightarrow r\ c(r) \in P$ such that $r \neq q$ and $d(r) = d(q)$.*

Example 6 (Test Cases of a Deterministic 1-Reg). 1-Reg obtained from Fig. 1c by including $c(c1) \rightarrow p2\ c(p2)$ is nondeterministic. Positive test cases $s = c1\ c1\ p1$ and $t = c1\ c1\ p2$ are redundant, because $d(s) = d(t) = c\ c\ p$.

Unless noted otherwise, all grammars under our consideration are useful and deterministic k-Regs.

3   Mutant Selection

Event-based fault types can be classified as *missing event* and *extra event*. In *missing event faults*, an event cannot occur after or before a (possibly empty) sequence of events. In *extra event faults*, an event can occur after or before a (possibly empty) sequence of events. To model such faults, *marking* (*mark start*, *mark finish*, *mark non-start* and *mark non-finish*), *insertion* (*insert sequence*, *insert terminal*) and *omission* (*omit sequence* and *omit terminal*) operators can be defined by extending the operators in [8]. These operators also enable to perform small changes or changes local to specific parts of the model so that different mutants do not model many common faults and a modeled fault does not interfere with one another. We use only some of these operators in test generation due to the following assumptions, which are commonly used in event-based testing (See [7] and the references therein).

A1.Events in a test case are executed in the given order; execution of a test case stops when a failure is observed.
A2.A test case can end with any event; it needs not be a finish event.

Thus, for a given k-Reg G, we have the following:

P1. Missing and extra event faults are limited by considering the k-sequences which precede the missing or extra events, ignoring the succeeding k-sequences. Thus, by exercising all (k+1)-sequences in the k-Reg, one can test whether an event is missing after some k-sequence, and, by exercising all relevant faulty k-sequences, one can test whether an event is extra after some k-sequence. (by A1)
P2. Mark non-start, mark non-finish, omit sequence and omit terminal mutants are discarded; they do not contain any (k+1)-sequence that is not in the original model. (due to P1)

P3. Mark finish and mark non-finish mutants do not really correspond to fault models, because every event can be considered as a finish or non-finish event during the testing process. (by A2)

P4. Extra event faults modeled using insert sequence mutants can be modeled using insert terminal mutants.

P5. All negative test cases are FCES. (by A1)

Consequently, for test generation, we do not need to use all types of mutants; we can use the original k-Reg to cover (k+1)-sequences to reveal missing event faults, and mark start and insert terminal mutants to cover faulty (k+1)-sequences to reveal extra event faults. Thus, below, we study only mark start and insert terminal operators to propose related mutant selection strategies.

In the discussion, we consider $G = (E, B; K; C; S; P)$ as the original k-Reg model and G' as a mark start or an insert terminal mutant of G, unless noted otherwise.

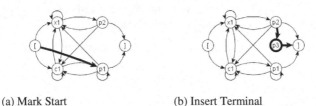

(a) Mark Start   (b) Insert Terminal

Fig. 3. Mutants of the 1-Reg in Fig. 1c (Mutations are drawn in bold)

3.1   Mark Start Mutants

Mark start mutation operators are used to mark k-sequences as start k-sequences. Therefore, mark start mutants are used to model extra start event faults.

Definition 8 (Mark Start). *A mark start (Ms) mutant of G is defined as* $G' = Ms(G, e) = (E, B; K; C; S; P \cup \{S \rightarrow e\ c(e)\})$ *for some* $e \in K$ *such that* $S \rightarrow e\ c(e) \notin P$.

Example 7 (A Mark Start Mutant). Let G be the 1-Reg in Fig. 1c. Fig. 3a shows $Ms(G, p1)$.

The set of all CESs is extended due to the mutation.

Lemma 1 (Set of All CESs of a Mark Start Mutant). *The set of all CESs of G' is given by* $T_{CES}(G') = T_{CES}(G) \cup \{e\ x|\ c(e) \Rightarrow_G^* x\ (x \in E^*)\}$.

Proof: The proof follows from Definition 8.   □

Below, we discuss the equivalence of a mark start mutant to the original k-Reg.

Lemma 2 (Equivalence of a Mark Start Mutant). G' *is not equivalent to G if and only if* $d(X) \setminus d(Y) \neq \emptyset$ *where*

- $X = \{e\ x|\ c(e) \Rightarrow_G^* x\ (x \in E^*)\}$ *and*
- $Y = \{e'\ y|\ S \Rightarrow_G^* e'\ y\ (e' \in K, y \in E^*)$ *where* $e' \neq e$ *and* $d(e') = d(e)\} \subseteq T_{CES}(G)$.

Proof: $T_{CES}(G') = T_{CES}(G) \cup X$ (by Lemma 1). By Definition 5, we have $d(T_{CES}(G')) = d(T_{CES}(G))$ if and only if $d(X) \subseteq d(Y) \subseteq d(T_{CES}(G))$, which completes the proof.   □

Sufficient conditions for usefulness, determinism and nonequivalence of a mark start mutant are outlined in the following.

Theorem 1 (Usefulness of a Mark Start Mutant). *G' is useful, if G is useful.*
Proof: The proof follows from Definition 6 and Definition 8.   □

Theorem 2 (Determinism of a Mark Start Mutant). *G' is deterministic, if G is deterministic and there is no $S \rightarrow e'$ $c(e') \in P$ such that $e' \neq e$ and $d(e') = d(e)$.*

Proof: The proof follows from Definition 7 and Definition 8.   □

Theorem 3 (Nonequivalence of a Mark Start Mutant). *G' is not equivalent to G, if G is useful and there is no $S \rightarrow e'$ $c(e') \in P$ such that $e' \neq e$ and $d(e') = d(e)$.*

Proof: Let X and Y be the sets from Lemma 2. We have:

- $T_{CES}(G) \neq \varnothing$ and $X \neq \varnothing$, since G is useful.
- $Y = \varnothing$, since there is no $S \rightarrow e'$ $c(e') \in P$ such that $e' \neq e$ and $d(e') = d(e)$.

Thus, $d(X) \cap d(Y) = \varnothing$, and G' is not equivalent to G (Lemma 2).   □

Now, we give the mutant selection strategy devised for mark start mutants.

Mark Start Mutant Selection. *For each $G' = Ms(G, e)$, k-sequence e is selected as a mutation parameter if the following hold:*
1. There is no start k-sequence x such that $d(x_1) = d(e_1)$.
1. There is no previously selected parameter y such that $d(y_1) = d(e_1)$.

Let G be a useful and deterministic k-Reg. By Theorem 1, Theorem 2, and Theorem 3, mutants generated from G using the above strategy are useful, deterministic and non-equivalent to G. Also, each of these mutants models a different fault located at the point of mutation: For each $Ms(G, e)$, e_1 (also $d(e_1)$) is an extra start event.

Algorithm 1. Mark Start Mutant Selection

Input: $G = (E, B; K; C; S; P)$ – the k-Reg
Output: M – set of selected mark start mutants
    $M = \varnothing, N = \varnothing$
    for each $b \in B$ **do**
        if there is no $S \rightarrow x$ $c(x) \in P$ such that $d(x_1) = b$ and
        there is no $y \in N$ such that $d(y_1) = b$ **then**
            Select a k-sequence $e \in K$ such that $d(e_1) = b$
            $G' = G, M = M \cup \{Ms(G', e)\}, N = N \cup \{e\}$
        endif
    endfor

The left-out mark start mutants are useful. However, they are either nondeterministic or model previously modeled faults. Some nondeterministic mutants do not model any faults. If they do, these faults are not extra start event faults. Therefore, they can be modeled using insert terminal mutants.

Algorithm 1 generates all mark start mutants using the above strategy. Its runtime complexity is given by $O(|B| |P|)$: (1) The number of mutants generated is bounded by $|B|$, because each mutant represents a different extra start event fault. (2) Each mutant $Ms(G, e)$ can be generated in $O(|P|+|B|) = O(|P|)$ time by checking if there are no start k-sequence x such that $d(x_1) = d(e_1)$ and previously selected mutation parameter y such that $d(y_1) = d(e_1)$, and copying G to modify it.

Example 8 (Mark Start Mutant Selection). Let G be the 1-Reg in Fig. 1c. The only mark start mutant we select is $Ms(G, p1)$. We do not select $Ms(G, c1)$ and $Ms(G, x1)$, because $c1$ and $x1$ are already start events. Furthermore, we do not select $Ms(G, p2)$, because it models the same fault as $Ms(G, p1)$.

3.2   Insert Terminal Mutants

Insert terminal mutation operators are used to add new terminals (k-sequences) by (possibly) connecting them to the existing k-sequences. Therefore, insert terminal mutants are used to model extra event fault where an event follows some k-sequence.

Definition 9 (Insert Terminal). *An insert terminal (It) mutant of G is defined as $G'=$ $It(G, e, U, V) = (E, B; K'; C'; S; P')$ for some $e \notin K$ with $d(e) \in B^k$, $U = \{(a, e)| a \in \{a_1, ..., a_m\} \subseteq K\}$ and $V = \{(e, b)| b \in \{b_1, ..., b_n\} \subseteq K\cup\{e\}\}$, where $K' = K \cup \{e\}$, $C' = C \cup \{c(e)\}$, and $P' = P \cup \{c(e) \to a\ c(a)| (a, e) \in U\} \cup \{c(b) \to e\ c(e)| (e, b) \in V\}$.*

Since we want to generate mutants which contain small number of changes, we limit $|U| = 1$, that is, $U = \{(a, e)\}$. Furthermore, since we use only FCESs in testing, we fix $V=\varnothing$ and insert production $c(e) \to \varepsilon$ for usefulness of k-sequence e.

Example 9 (An Insert Terminal Mutant). Let G be the 1-Reg in Fig. 1c. Fig. 3b shows $It(G, p3, \{(p2, p3)\}, \varnothing)$ where $p3$ is a new contexted p event. Since $V = \varnothing$, $c(p3) \to \varepsilon$ is additionally inserted to preserve usefulness of $p3$.

Lemma 3 (Set of All CESs of an Insert Terminal Mutant). *The set of all CESs of G' is given by $T_{CES}(G') = T_{CES}(G) \cup \{x\ e| S \Rightarrow_G^* x\ c(a)\ (x \in E^*)\}$.*

Proof: The proof follows from Definition 9 considering that $c(e) \to \varepsilon$ is inserted.   □

The following discusses the equivalence of an insert terminal to the original k-Reg.

Lemma 4 (Equivalence of an Insert Terminal Mutant). *G' is not equivalent to G if and only if $d(X) \setminus d(Y) \neq \varnothing$ where*
- *$X = \{x\ e| S \Rightarrow_G^* x\ c(a)\ (x \in E^*)\}$ and*
- *$Y = \{w| w \in T_{CES}(G)$ and w contains e', $e' \in K$, $e' \neq e$ and $d(e') = d(e)\} \subseteq T_{CES}(G)$.*

Proof: $T_{CES}(G') = T_{CES}(G) \cup X$ (by Lemma 3). By Definition 5, we have $d(T_{CES}(G')) = d(T_{CES}(G))$ if and only if $d(X) \subseteq d(Y) \subseteq d(T_{CES}(G))$. This completes the proof.   □

The following give sufficient conditions for usefulness, determinism and nonequivalence of an insert terminal mutant.

Theorem 4 (Usefulness of an Insert Terminal Mutant). *G' is useful, if G is useful.*

Proof: The proof follows from Definition 6, and Definition 9 considering that $c(e) \to \varepsilon$ is inserted.   □

Theorem 5 (Determinism of an Insert Terminal Mutant). *G' is deterministic, if G is deterministic and there is no $c(a) \to e'\ c(e') \in P$ such that $e' \neq e$ and $d(e') = d(e)$.*

Proof: The proof follows from Definition 7 and Definition 9 considering that $c(e) \to \varepsilon$ is also inserted.   □

Theorem 6 (Nonequivalence of an Insert Terminal Mutant). *G' is not equivalent to G, if G is useful and deterministic, and there is no $c(a) \to e'\ c(e') \in P$ such that $e' \neq e$ and $d(e') = d(e)$.*

Proof: Let X and Y be the sets from Lemma 4. We have
- $T_{CES}(G) \neq \emptyset$, since G is useful.
- $X \neq \emptyset$, since G is useful, $U \neq \emptyset$ and $c(e) \rightarrow \varepsilon$ is inserted.
- $d(X) \cap d(Y) = \emptyset$, because each start sequence in deterministic G derived using $m \geq 1$ productions has a unique basis and there is no $c(a) \rightarrow e' c(e') \in P$ such that $e' \neq e$ and $d(e') = d(e)$.

Thus, G' is not equivalent to G (Lemma 4).   □

Now, we give the strategy to select insert terminal mutants.

Insert Terminal Mutant Selection. *For each $G' = It(G, e, \{(a, e)\}, \emptyset)$, 3-tuple $(e, \{(a, e)\}, \emptyset)$ is selected as a mutation parameter if the following hold:*

1. There is no $c(a) \rightarrow x c(x) \in P$ such that $d(x_k) = d(e_k)$.
2. There is no previously selected parameter $(y, \{(a, y)\}, \emptyset)$ such that $d(y_k) = d(e_k)$.

Let G be a useful and deterministic k-Reg. By Theorem 4, Theorem 5 and Theorem 6, mutants generated from G using the above strategy are useful, deterministic and not equivalent to G. Furthermore, each of these mutants models a different fault located at the point of mutation, that is, for each $It(G, e, \{(a, e)\}, \emptyset)$, e_k (also $d(e_k)$) is an extra event that follows k-sequence a.

Algorithm 2. Insert Terminal Mutant Selection

Input: $G = (E, B; K; C; S; P)$ – the k-Reg
Output: M – set of selected insert terminal mutants
$M = \emptyset$
for each $a \in K$ **do**
  $N = \emptyset$
  for each $b \in B$ **do**
    if there is no $c(a) \rightarrow x c(x) \in P$ such that $d(x_k) = b$ and
    there is no $(y, (a, y), \emptyset) \in N$ such that $d(y_k) = b$ **then**
      b' = a new contexted version of b, $e = a_2 \dots a_k b'$
      $G' = G, M = M \cup \{It(G', e, \{(a, e)\}, \emptyset)\}, N = N \cup \{e\}$
    endif
  endfor
endfor

The excluded insert terminal mutants are useful. However, they are either nondeterministic or model previously modeled faults. Some nondeterministic mutants may not model any faults at all. If they do, these faults are not located at the points of mutation. Thus, they are modeled by some other insert terminal mutants that we select.

Algorithm 2 generates all insert terminal mutants using the above strategy. Its runtime complexity is given by $O(|K| |B| |P|)$: (1) The number of mutants generated is bounded by $|K| |B|$, because each mutant represents a different extra event fault. (2) Each mutant $It(G, e, \{(a, e)\}, \emptyset)$ can be generated in $O(|P|+|B|+k) = O(|P|)$ time by checking if there are no $c(a) \rightarrow x c(x) \in P$ such that $d(x_k) = d(e_k)$ and previously selected mutation parameter $(y, \{(a, y)\}, \emptyset)$ such that $d(y_k) = d(e_k)$, preparing e by copying $a_2 \dots a_k$ to append b', and copying G to modify it.

Example 10 (Insert Terminal Mutant Selection). Let G be the 1-Reg in Fig. 1c. We can only use basis terminal p, because c and x can follow all events. The only mutant we select is $It(G, p3, \{(p2, p3\}, \emptyset)$, because only $p2$ is not followed by a p event.

4   Evaluation

To evaluate the proposed mutant selection strategies, we compare the numbers of mutants generated using our approach, grammar-based mutation operators in [14] and graph-based mutation operators in [11] using `Specials` facility of ISELTA as our SUC. ISELTA is a commercial web-portal for marketing touristic services and `Specials` provides addition, editing and deletion of special prices of the marketed services (See [6] for more information on ISELTA and `Specials`). For comparisons, we use k-Reg models of `Specials` for $k=1,2,3,4$.

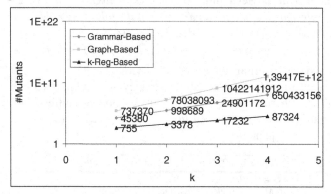

Fig. 4. Total number of mutants generated (Y axis is given in logarithmic scale to increase visibility. Also, the actual mutant numbers are used to label the data points).

Fig. 4 shows that the number of generated mutants can be reduced significantly using the k-Reg-based mutant selection strategies proposed in this paper. Briefly, our results suggest that one can generate ~99% fewer number of mutants.

Threats to Validity. Note that the above evaluation does not consider test generation and execution aspects; therefore, to gain more insight about the effects of mutant selection strategies on the actual performance of the whole testing process, one needs to perform a full-fledged case study including testing simulations. Still, assuming that a similar method is used to generate test cases from each mutant, we can elaborate that the size of the test set generated using our strategies will be significantly smaller.

5   Conclusion

This paper proposes an event-based grammar model to explicitly model the relation between event sequences of length $k\geq1$ and single events, and concentrates on only mutant selection aspects. More precisely, we devise mutant selection strategies to exclude generation of mutant which are equivalent to the original model and multiple mutants which model the same faults based on certain assumptions. Considering that other model-based mutation testing approaches compare each mutant to the original model to determine the equivalence and do not exclude generation of multiple mutants which model the same faults, our results have clear practical benefits, such as increasing the performance by cutting back the number of mutants generated.

For the future work, we plan to consider the way the test execution is performed in a more precise manner or relax the assumptions in Section 3 to propose additional mutant selection strategies and also strategies for other types of mutants.

References

1. Adamopoulos, K., Harman, M., Hierons, R.: How to Overcome the Equivalent Mutant Problem and Achieve Tailored Selective Mutation Using Co-evolution. In: Deb, K., Tari, Z. (eds.) GECCO 2004. LNCS, vol. 3103, pp. 1338–1349. Springer, Heidelberg (2004)
2. Aichernig, B.K.: Mutation Testing in the Refinement Calculus. Formal Aspects of Computing Journal 15(2-3), 280–295 (2003)
3. Ammann, P.E., Black, P.E., Majurski, W.: Using Model Checking to Generate Tests from Specifications. In: Proceedings of the 2nd IEEE International Conference on Formal Engineering Methods (ICFEM 1998), pp. 46–54. IEEE Computer Society, Washington, DC (1998)
4. Beizer, B.: Software Testing Techniques. Van Nostrand Reinhold, New York (1990)
5. Belli, F.: Finite-State Testing and Analysis of Graphical User Interfaces. In: Proceedings of the 12th International Symposium on Software Reliability Engineering (ISSRE 2001), pp. 34–43. IEEE Computer Society, Washington, DC (2001)
6. Belli, F., Beyazıt, M., Güler, N.: Event-Oriented, Model-Based GUI Testing and Reliability Assessment—Approach and Case Study. Advances in Computers 85, 277–326 (2012)
7. Belli, F., Beyazıt, M., Memon, A.: Testing is an Event-Centric Activity. In: Proceedings of the 6th International Conference on Software Security and Reliability (SERE-C 2012), pp. 198–206. IEEE Computer Society, Washington, DC (2012)
8. Belli, F., Budnik, C.J., Wong, W.E.: Basic operations for generating behavioral mutants. In: Proceedings of the 2nd Workshop on Mutation Analysis (MUTATION 2006), pp. 9–18. IEEE Computer Society, Washington, DC (2006)
9. Budd, T.A., Gopal, A.S.: Program Testing by Specification Mutation. Computer Languages 10(1), 63–73 (1985)
10. Grün, B.J.M., Schuler, D., Zeller, A.: The Impact of Equivalent Mutants. In: International Conference on Software Testing, Verification and Validation Workshops (ICSTW 2009), pp. 192–199. IEEE Computer Society, Washington, DC (2009)
11. Hollmann, A.: Model-Based Mutation Testing for Test Generation and Adequacy Analysis. Ph.D. Thesis, University of Paderborn, Paderborn (2011)
12. Hopcroft, J.E., Motwani, R., Ullman, J.D.: Introduction to Automata Theory, Languages and Computation, 3rd edn. Addison-Wesley, Boston (2006)
13. Klint, P., Lämmel, R., Verhoef, C.: Toward an engineering discipline for grammarware. ACM Transactions on Software Engineering and Methodology 14(3), 331–380 (2005)
14. Offutt, A.J., Ammann, P., Liu, L.: Mutation Testing Implements Grammar-Based Testing. In: Proceedings of the 2nd Workshop on Mutation Analysis (MUTATION 2006), pp. 12–21. IEEE Computer Society, Washington, DC (2006)
15. Utting, M., Legeard, B.: Practical Model-Based Testing: A Tools Approach. Morgan Kaufmann Publishers Inc., San Francisco (2006)
16. Zhu, H., Hall, P.A., May, J.H.: Software Unit Test Coverage and Adequacy. ACM Computing Surveys 29(4), 366–427 (1997)

Two-Pass Greedy Regular Expression Parsing⋆

Niels Bjørn Bugge Grathwohl, Fritz Henglein, Lasse Nielsen,
and Ulrik Terp Rasmussen

Department of Computer Science, University of Copenhagen (DIKU)

Abstract. We present new algorithms for producing greedy parses for
regular expressions (REs) in a semi-streaming fashion. Our lean-log algorithm executes in time $O(mn)$ for REs of size m and input strings of
size n and outputs a compact bit-coded parse tree representation. It improves on previous algorithms by: operating in only 2 passes; using only
$O(m)$ words of random-access memory (independent of n); requiring only
kn bits of sequentially written and read log storage, where $k < \frac{1}{3}m$ is
the number of alternatives and Kleene stars in the RE; processing the
input string as a symbol stream and not requiring it to be stored at all.
Previous RE parsing algorithms do not scale linearly with input size, or
require substantially more log storage and employ 3 passes where the first
consists of reversing the input, or do not or are not known to produce
a greedy parse. The performance of our unoptimized C-based prototype
indicates that our lean-log algorithm has also in practice superior performance and is surprisingly competitive with RE tools not performing
full parsing, such as Grep.

1 Introduction

Regular expression (RE) parsing is the problem of producing a parse tree for
an input string under a given RE. In contrast to most regular-expression based
tools for programming such as Grep, RE2 and Perl, RE parsing returns not only
whether the input is accepted, where a substring matching the RE and/or sub-
REs are matched, but a full parse tree. In particular, for Kleene stars it returns
a list of all matches, where each match again can contain such lists depending
on the star depth of the RE.

An RE parser can be built using Perl-style backtracking or general context-
free parsing techniques. What the backtracking parser produces is the *greedy*
parse amongst potentially many parses. General context-free parsing and backtracking parsing are not scalable since they have cubic, respectively exponential
worst-case running times. REs can be and often are grammatically ambiguous
and can require arbitrary much look-ahead, making limited look-ahead context-
free parsing techniques inapplicable. Kearns [1] describes the first linear-time
algorithm for RE parsing. In a streaming context it consists of 3 passes: reverse

⋆ This work has been partially supported by The Danish Council for Independent
Research under Project 11-106278, "Kleene Meets Church: Regular Expressions and
Types". The order of authors is insignificant.

S. Konstantinidis (Ed.): CIAA 2013, LNCS 7982, pp. 60–71, 2013.

the input, perform backward NFA-simulation, and construct parse tree. Frisch and Cardelli [2] formalize greedy parsing and use the same strategy to produce a greedy parse. Dubé and Feeley [3] and Nielsen and Henglein [4] produce parse trees in linear time for fixed RE, the former producing internal data structures and their serialized forms, the latter parse trees in bit-coded form; neither produces a greedy parse.

In this paper we make the following contributions:

1. Specification and construction of symmetric nondeterministic finite automata (NFA) with maximum in- and out-degree 2, whose paths from initial to final state are in one-to-one correspondence with the parse trees of the underlying RE; in particular, the greedy parse for a string corresponds to the lexicographically least path accepting the string.
2. NFA simulation with *ordered state sets*, which gives rise to a 2-pass greedy parse algorithm using $\lceil m \lg m \rceil$ bits per input symbol and the original input string, with m the size of the underlying RE. No input reversal is required.
3. NFA simulation optimized to require only $k \leq \lceil 1/3m \rceil$ bits per input symbol, where the input string need not be stored at all and the 2nd pass is simplified. Remarkably, this *lean-log algorithm* requires fewest log bits, and neither state set nor even the input string need to be stored.
4. An empirical evaluation, which indicates that our prototype implementation of the optimized 2-pass algorithm outperforms also in practice previous RE parsing tools and is sometimes even competitive with RE tools performing limited forms of RE matching.

In the remainder, we introduce REs as types to represent parse trees, define greedy parses and their bit-coding, introduce NFAs with bit-labeled transitions, describe NFA simulation with ordered sets for greedy parsing and finally the optimized algorithm, which only logs join state bits. We conclude with an empirical evaluation of a straightforward prototype to gauge the competitiveness of full greedy parsing with regular-expression based tools yielding less information for Kleene-stars.

2   Symmetric NFA Representation of Parse Trees

REs are finite terms of the form $0, 1, a, E_1 \times E_2, E_1 + E_2$ or E_1^*, where E_1, E_2 are REs.

Proviso. For simplicity and brevity we henceforth assume REs that do not contain sub-REs of the form E^*, where E is nullable (can generate the empty string). All results reported here can be and have been extended to such problematic REs in the style of Frisch and Cardelli [2]. In particular, our implementation BitC handles problematic REs.

REs can be interpreted as types built from singleton, product, sum, and list type constructors [2,5]; see Figure 1. Its structured values $\mathcal{T}[\![E]\!]$ represent the *parse trees* for E such that the regular language $\mathcal{L}[\![E]\!]$ coincides with the strings

$$\begin{aligned}
\mathcal{T}[\![0]\!] &= \emptyset \\
\mathcal{T}[\![1]\!] &= \{()\}, \\
\mathcal{T}[\![a]\!] &= \{a\}, \\
\mathcal{T}[\![E_1 \times E_2]\!] &= \{(V_1, V_2) \mid V_1 \in \mathcal{T}[\![E_1]\!], V_2 \in \mathcal{T}[\![E_2]\!]\}, \\
\mathcal{T}[\![E_1 + E_2]\!] &= \{\mathsf{inl}\ V_1 \mid V_1 \in \mathcal{T}[\![E_1]\!]\} \\
&\quad \cup \{\mathsf{inr}\ V_2 \mid V_2 \in \mathcal{T}[\![E_2]\!]\}, \\
\mathcal{T}[\![E_0^\star]\!] &= \{[V_1, \ldots, V_n] \mid n \geq 0 \wedge \\
&\quad \forall 1 \leq i \leq n.V_i \in \mathcal{T}[\![E_0]\!]\}
\end{aligned}$$

(a) Regular expressions as types.

$$\begin{aligned}
\mathsf{flat}(()) &= \epsilon \\
\mathsf{flat}(a) &= a \\
\mathsf{flat}((V_1, V_2)) &= \mathsf{flat}(V_1)\mathsf{flat}(V_2) \\
\mathsf{flat}(\mathsf{inl}\ V_1) &= \mathsf{flat}(V_1) \\
\mathsf{flat}(\mathsf{inr}\ V_2) &= \mathsf{flat}(V_2) \\
\mathsf{flat}([V_1, \ldots, V_n]) &= \mathsf{flat}(V_1) \ldots \mathsf{flat}(V_n)
\end{aligned}$$

(b) Tree flattening.

$$\begin{aligned}
\mathsf{code}(()) &= \epsilon \\
\mathsf{code}((V_1, V_2)) &= \mathsf{code}(V_1)\,\mathsf{code}(V_2) \\
\mathsf{code}(\mathsf{inl}\ V_1) &= 0\ \mathsf{code}(V_1)
\end{aligned}$$

$$\begin{aligned}
\mathsf{code}(a) &= \epsilon \\
\mathsf{code}([V_1, \ldots, V_n]) &= 0\,\mathsf{code}(V_1) \ldots 0\,\mathsf{code}(V_n)\,1 \\
\mathsf{code}(\mathsf{inr}\ V_2) &= 1\ \mathsf{code}(V_2)
\end{aligned}$$

(c) Bit-coding.

Fig. 1. The type interpretation of regular expressions and bit-coding of parses

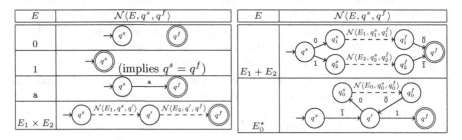

Fig. 2. aNFA construction schema

obtained by flattening the parse trees:

$$\mathcal{L}[\![E]\!] = \{\mathsf{flat}(V) \mid V \in \mathcal{T}[\![E]\!]\}.$$

We recall bit-coding from Nielsen and Henglein [4]. The bit code $\mathsf{code}(V)$ of a parse tree $V \in \mathcal{T}[\![E]\!]$ is a sequence of bits uniquely identifying V within $\mathcal{T}[\![E]\!]$; that is, there exists a function decode_E such that $\mathsf{decode}_E(\mathsf{code}(V)) = V$. See Figure 1 for the definition of code; the definition of decode_E is omitted for brevity, but is straightforward. We write $\mathcal{B}[\![\ldots]\!]$ instead of $\mathcal{T}[\![\ldots]\!]$ whenever we want to refer to the bit codings, rather than the parse trees. We use subscripts to discriminate parses with a specific flattening: $\mathcal{T}_s[\![E]\!] = \{V \in \mathcal{T}[\![E]\!] \mid \mathsf{flat}(V) = s\}$. We extend the notation $\mathcal{B}_s[\![\ldots]\!]$ similarly.

Note that a bit string by itself does not carry enough information to deduce which parse tree it represents. Indeed this is what makes bit strings a compact representation of strings where the underlying RE is statically known.

The set $\mathcal{B}[\![E]\!]$ for an RE E can be compactly represented by *augmented non-deterministic finite automaton (aNFA)*, a variant of enhanced NFAs [4] that has in- and outdegree at most 2 and carries a label on each transition.

Definition 1 (Augmented NFA) An *augmented NFA* (aNFA) is a 5-tuple $M = (Q, \Sigma, \Delta, q^s, q^f)$ where Q is the set of states, Σ is the input alphabet, and q^s, q^f are the starting states. The transition relation $\Delta \subseteq Q \times (\Sigma \cup \{0, 1, \overline{0}, \overline{1}\}) \times Q$ contains directed, labeled transitions: $(q, \gamma, q') \in \Delta$ is a transition from q to q' with label γ, written $q \overset{\gamma}{\longrightarrow} q'$.

We call transition labels in Σ *input labels*; labels in $\{0, 1\}$ *output labels*; and labels in $\{\overline{0}, \overline{1}\}$ *log labels*.

We write $q \overset{p}{\leadsto} q'$ if there is a path labeled p from q to q'. The sequences read(p), write(p) and log(p) are the subsequences of input labels, output labels, and log labels of p, respectively.

We write: J_M for the *join states* $\{q \in Q \mid \exists q_1, q_2. (q_1, \overline{0}, q), (q_2, \overline{1}, q) \in \Delta\}$; S_M for the *symbol sources* $\{q \in Q \mid \exists q' \in Q, a \in \Sigma. (q, a, q')\}$; and C_M for the *choice states* $\{q \in Q \mid \exists q_1, q_2. (q, 0, q_1), (q, 1, q_2) \in \Delta\}$.

If M is an aNFA, then $\overline{M}$ is the aNFA obtained by *flipping* all transitions and exchanging the start and finishing states, that is reverse all transitions and interchange output labels with the corresponding log labels. □

Our algorithm for constructing an aNFA from an RE is a standard Thompson-style NFA generation algorithm modified to accomodate output and log labels:

Definition 2 (aNFA construction) We write $M = \mathcal{N}\langle E, q^s, q^f \rangle$ when M is an aNFA constructed according to the rules in Figure 2.

Augmented NFAs are dual under reversal; that is, flipping produces the augmented NFA for the reverse of the regular language.

Proposition 2.1. *Let $\overline{E}$ be canonically constructed from E to denote the reverse of $\mathcal{L}[\![E]\!]$. Let $M = \mathcal{N}\langle E, q^s, q^f \rangle$. Then $\overline{M} = \mathcal{N}\langle \overline{E}, q^f, q^s \rangle$.*

This is useful since we will be running aNFAs in both forward and backward (reverse) directions.

Well-formed aNFAs—and Thompson-style NFAs in general—are canonical representations of REs in the sense that they not only represent their language interpretation, but their type interpretation:

Theorem 2.1 (Representation). *Given an aNFA $M = \mathcal{N}\langle E, q^s, q^f \rangle$, M outputs the bit-codings of E:*

$$\mathcal{B}_s[\![E]\!] = \{\text{write}(p) \mid q^s \overset{p}{\leadsto} q^f \wedge \text{read}(p) = s\}.$$

3 Greedy Parsing

The *greedy parse* of a string s under an RE E is what a backtracking parser returns that tries the left operand of an alternative first and backtracks to try the right alternative only if the left alternative does not yield a successful parse. The name comes from treating the Kleene star $E^\star$ as $E \times E^\star + 1$, which "greed-ily" matches E against the input as many times as possible. A "lazy" matching interpretation of $E^\star$ corresponds to treating $E^\star$ as $1 + E \times E^\star$. (In practice, multiple Kleene-star operators are allowed to make both interpretations available; e.g. $E\star$ and $E\star\star$ in PCRE.)

Greedy parsing can be formalized by an order $\lessdot$ on parse trees, where $V_1 \lessdot V_2$ means that V_1 is "more greedy" than V_2. The following is adapted from Frisch and Cardelli [2].

Definition 3 (Greedy order) The binary relation $\lessdot$ is defined inductively on the structure of values as follows:

$$
\begin{aligned}
(V_1, V_2) &\lessdot (V_1', V_2') &&\text{if}\quad V_1 \lessdot V_1' \vee (V_1 = V_1' \wedge V_2 \lessdot V_2') \\
\text{inl } V_0 &\lessdot \text{inl } V_0' &&\text{if}\quad V_0 \lessdot V_0' \\
\text{inr } V_0 &\lessdot \text{inr } V_0' &&\text{if}\quad V_0 \lessdot V_0' \\
\text{inl } V_0 &\lessdot \text{inr } V_0' && \\
[] &\lessdot [V_1, \ldots] && \\
[V_1, \ldots] &\lessdot [V_1', \ldots] &&\text{if}\quad V_1 \lessdot V_1' \\
[V_1, V_2, \ldots] &\lessdot [V_1, V_2', \ldots] &&\text{if}\quad [V_2, \ldots] \lessdot [V_2', \ldots]
\end{aligned}
$$

The relation $\lessdot$ is not a total order; consider for example the incomparable elements $(\mathsf{a}, \text{inl } ())$ and $(\mathsf{b}, \text{inr } ())$. The parse trees of any particular RE are totally ordered, however:

Proposition 3.1. *For each E, the order $\lessdot$ is a strict total order on $\mathcal{T}[\![E]\!]$.*

In the following, we will show that there is a correspondence between the structural order on values and the lexicographic order on their bit codings.

Definition 4 For bit sequences $d, d' \in \{0, 1\}^\star$ we write $d \prec d'$ if d is lexicographically strictly less than d'; that is, $\prec$ is the least relation satisfying

1. $\epsilon \prec d$ if $d \neq \epsilon$
2. $b\, d \prec b'\, d'$ if $b < b'$ or $b = b'$ and $d \prec d'$.

Theorem 3.1. *For all REs E and values $V, V' \in \mathcal{T}[\![E]\!]$ we have $V \lessdot V'$ iff $\mathsf{code}(V) \prec \mathsf{code}(V')$.*

Corollary 3.1. *For any RE E with aNFA $M = \mathcal{N}\langle E, q^s, q^f \rangle$, and for any string s, $\min_\lessdot \mathcal{T}_s[\![E]\!]$ exists and*

$$
\min_\lessdot \mathcal{T}_s[\![E]\!] = \mathsf{decode}_E(\min_\prec \{\mathsf{write}(p) \mid q^s \overset{p}{\leadsto} q^f \wedge \mathsf{read}(p) = s\}).
$$

Proof. Follows from Theorems 2.1 and 3.1. □

We can now characterize greedy RE parsing as follows: Given an RE E and string s, find bit sequence b such that there exists a path p from start to finishing state in the aNFA for E such that:

1. $\mathsf{read}(p) = s$,
2. $\mathsf{write}(p) = b$,
3. b is lexicographically least among all paths satisfying 1 and 2.

This is easily done by a backtracking algorithm that tries 0-labeled transitions before 1-labeled ones. It is atrociously slow in the worst case, however: exponential time. How to do it faster?

4 NFA-Simulation with Ordered State Sets

Our first algorithm is basically an NFA-simulation. For reasons of space we only sketch its key idea, which is the basis for the more efficient algorithm in the following section.

A standard NFA-simulation consists of computing $\mathsf{Reach}^*(S, s)$ where

$$\mathsf{Reach}^*(S, \epsilon) = S$$
$$\mathsf{Reach}^*(S, a\, s') = \mathsf{Reach}^*(\mathsf{Reach}(S, a), s')$$
$$\mathsf{Reach}(S, a) = \mathsf{Close}(\mathsf{Step}(S, a))$$
$$\mathsf{Step}(S, a) = \{q' \mid q \in S, q \xrightarrow{a} q'\}$$
$$\mathsf{Close}(S') = \{q'' \mid q' \in S', q' \overset{p}{\rightsquigarrow} q'', \mathsf{write}(p) = \epsilon\}$$

Checking $q^f \in \mathsf{Reach}^*(\{q^s\}, s)$ determines whether s is accepted or not. But how to construct an accepting *path* and in particular the one corresponding to the greedy parse?

We can *log* the set of states reached after each symbol during the NFA-simulation. After forward NFA-simulation, let S_i be the NFA-states reached after processing the first i symbols of input $s = a_1 \ldots a_n$. Given a list of logged state sets, the input string s and the final state q^f, the nondeterministic algorithm Path^* constructs a path from q^s to q^f through the state sets:

$$\mathsf{Path}(S_i, q) = (q', p) \text{ where } q' \in S_i, q' \overset{p}{\rightsquigarrow} q, \mathsf{read}(p) = a_i$$

$$\mathsf{Path}^*(S_0, q) = p' \cdot p \text{ where } (q', p) = \mathsf{Path}(S_0, q), q^s \overset{p'}{\rightsquigarrow} q', \mathsf{read}(p') = \epsilon$$

$$\mathsf{Path}^*(S_i, q) = p' \cdot p \text{ where } (q', p) = \mathsf{Path}(S_i, q), p' = \mathsf{Path}^*(S_{i-1}, q')$$

Calling $\mathsf{write}(\mathsf{Path}^*(S_n, q^f))$ gives a bit-coded parse tree, though not necessarily the lexicographically least.

We can adapt the NFA-simulation by keeping each state set S_i in a particular order: If $\mathsf{Reach}^*(\{q^s\}, a_1 \ldots a_i) = \{q_{i1}, \ldots q_{ij_i}\}$ then order the q_{ij} according to the lexicographic order of the paths reaching them. Intuitively, the highest ranked state in S_i is on the greedy path if the remaining input is accepted from this state; if not, the second-highest ranked is on the greedy path, if the remaining input is accepted; and so on.

The NFA-simulation can be refined to construct properly ordered state sequences instead of sets without asymptotic slow-down. The log, however, is adversely affected by this. We need $\lceil m \lg m \rceil$ bits per input symbol, for a total of $\lceil mn \lg m \rceil$ bits.

The key property for allowing us to list a state at most once in an order state squence is this:

Lemma 4.1. *Let $s, t_1, t_2,$ and t be states in an aNFA M, and let p_1, p_2, q_1, q_2 be paths in M such that $s \overset{p_1}{\rightsquigarrow} t_1, s \overset{p_2}{\rightsquigarrow} t_2,$ and $t_1 \overset{q_1}{\rightsquigarrow} t, t_2 \overset{q_2}{\rightsquigarrow} t.$ If $\mathsf{write}(p_1) \prec \mathsf{write}(p_2)$ then $\mathsf{write}(p_1 q_1) \prec \mathsf{write}(p_2 q_2)$*

Proof. Application of the lexicographical ordering on paths.   □

5   Lean-Log Algorithm

After the ordered forward NFA-simulation with logging, the algorithm Path above can be refined to always yield the greedy parse whend traversing the aNFA in backwards direction. Since the join states J_M of an aNFA M become the choice states $C_{\overline{M}}$ of the reverse aNFA $\overline{M}$ we only need to construct one "direction" bit for each join state at each input string position. It is not necessary to record any states in the log at all, and we do not even have to store the input string. This results in an algorithm that requires only k bits per input symbol for the log, where k is the number of Kleene-stars and alternatives occurring in the RE. It can be shown that $k \leq \frac{1}{3}m$; in practice we can observe $k \ll m$.

Our optimized algorithm is described in Figure 3 below. The forward pass keeps the aNFA and the current character in memory, requiring a $O(m)$ words of random access memory, writing nk bits to the log, and discarding the input string. Finally, the backward pass also requires $O(m)$ words of random access memory and reads from the log in reverse write order. The log is thus a 2-phase stack: In the first pass it is only pushed to, in the second pass popped from.

Both LClose and LStep run in time $O(m)$ per input symbol, hence the forward pass requires time $O(mn)$. Likewise, the backward pass requires time $O(mn)$.

LClose keeps track of visited states and returns the states reached ordered lexicographically according to the paths reaching them. Hence, the following theorem can be proved:

Theorem 5.1. *For any regular expression E and symbol sequence s, if $\mathcal{L}_l =$ LSim(s), and $d =$ LTrace$(\mathcal{L}_l, q^f)$, then* decode$_E(d) = \min_{\lessdot} \mathcal{T}_s[\![E]\!]$.

6   Evaluation

We have implemented the optimized algorithms in C and in Haskell, and we compare the performance of the C implementation with the following existing RE tools:

RE2: Google's RE implementation, available from [6].
Tcl: The scripting language Tcl [7].
Perl: The scripting language Perl [8].
Grep: The UNIX tool grep.
Rcp: The implementation of the algorithm *"DFASIM"* from [4]. It is based on Dubé and Feeley's method [3], but altered to produce a bit-coded parse tree.
FrCa: The implementation of the algorithm*"FrCa"* algorithm used in [4]. It is based on Frisch and Cardelli's method from [2].

In the subsequent plots, our implementation of the lean-log algorithm is referred to as *BitC*.

The tests have been performed on an Intel Xeon 2.5 GHz machine running GNU/Linux 2.6.

$$(Q, L) \oplus (Q', L') = (Q \cdot Q', L \cup L')$$

$$\mathsf{LClose}(q, L) = \begin{cases} ([q], L) & q \xrightarrow{a} q', a \in \Sigma \\ \mathsf{LClose}(q_0, L) \oplus \mathsf{LClose}(q_1, L) & q \xrightarrow{0} q_0, q \xrightarrow{1} q_1 \\ \mathsf{LClose}(q', L \cup \{q' \mapsto t\}) & q \xrightarrow{t} q', t \in \{\overline{0}, \overline{1}\}, q' \notin \mathrm{dom}(L) \\ ([], L) & \text{otherwise} \end{cases}$$

$$\mathsf{LStep}([], a, (Q, L)) = (Q, L)$$

$$\mathsf{LStep}(q \cdot qs, a, (Q, L)) = \begin{cases} \mathsf{LStep}(qs, a, (Q, L) \oplus \mathsf{LClose}(q, L)) & q \xrightarrow{a} q' \\ \mathsf{LStep}(qs, a, (Q, L)) & \text{otherwise} \end{cases}$$

$$\mathsf{LSim}'([], Q, \mathcal{L}) = \begin{cases} \mathcal{L} & \text{if } q^s \in Q \\ \perp & \text{otherwise} \end{cases}$$

$$\mathsf{LSim}'(a \cdot s', Q, \mathcal{L}) = \begin{cases} \mathsf{LSim}(s', Q', L \cdot \mathcal{L}) & \text{if } (Q', L) = \mathsf{LStep}(Q, a, ([], \emptyset)), Q' \neq [] \\ \perp & \text{otherwise} \end{cases}$$

$$\mathsf{LSim}(s) = \mathbf{let}\ (Q_0, L_0) = \mathsf{LClose}(q^s, []) \ \mathbf{in}\ \mathsf{LSim}'(s, Q_0, [L])$$

(a) Forward pass.

$$\mathsf{LTrace}(L \cdot \mathcal{L}, q) = \begin{cases} [] & \text{if } q = q^s \\ \mathsf{LTrace}(L \cdot \mathcal{L}, q') \cdot \gamma & \text{if } q \xrightarrow{\gamma} q', \gamma \in \{0, 1\} \\ \mathsf{LTrace}(\mathcal{L}, q') & \text{if } q \xrightarrow{\gamma} q', \gamma \in \Sigma \\ \mathsf{LTrace}(L \cdot \mathcal{L}, q') & \text{if } q \xrightarrow{L[q]} q', L[q] \in \{\overline{0}, \overline{1}\} \end{cases}$$

(b) Backward pass.

Fig. 3. Forward and backward pass algorithm

6.1   Pathological Expressions

To get an indication of the "raw" throughput for each tool, $a^\star$ was run on sequences of as (Figure 4a). (Note that the plots use log scales on both axes, so as to accommodate the dramatically varying running times.) Perl outperforms the rest, likely due to a strategy where it falls back on a simple scan of the input instead. FrCa stores each position in the input string from which a match can be made, which in this case is every position. As a result, FrCa uses significantly more memory than the rest, causing a dramatic slowdown.

The expression $(a|b)^\star a(a|b)^n$ with the input $(ab)^{n/2}$ is a worst-case for DFA-based methods, as it results in a number of states exponential in n. Perl has been omitted from the plots, as it was prohibitively slow. Tcl, Rcp, and Grep all perform orders of magnitude slower than FrCa, RE2, and BitC (Figure 4b), indicating that Tcl and Grep also use a DFA for this expression. If we fix n to 25, it becomes clear that FrCa is slower than the rest, likely due to high memory consumption as a result of its storing all positions in the input string (Figure 4c). The asymptotic running times of the others appear to be similar to each other, but with greatly varying constants.

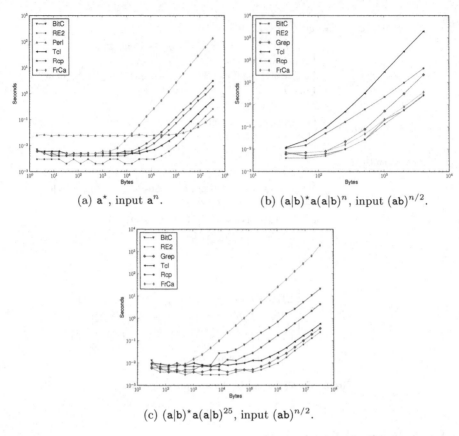

(a) $\mathtt{a}^\star$, input $\mathtt{a}^n$.

(b) $(\mathtt{a}|\mathtt{b})^\star\mathtt{a}(\mathtt{a}|\mathtt{b})^n$, input $(\mathtt{ab})^{n/2}$.

(c) $(\mathtt{a}|\mathtt{b})^\star\mathtt{a}(\mathtt{a}|\mathtt{b})^{25}$, input $(\mathtt{ab})^{n/2}$.

Fig. 4

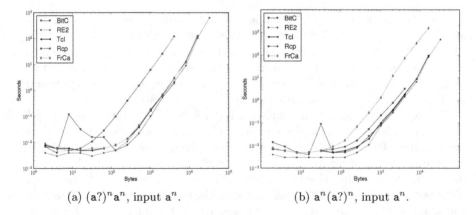

(a) $(\mathtt{a}?)^n\mathtt{a}^n$, input $\mathtt{a}^n$.

(b) $\mathtt{a}^n(\mathtt{a}?)^n$, input $\mathtt{a}^n$.

Fig. 5

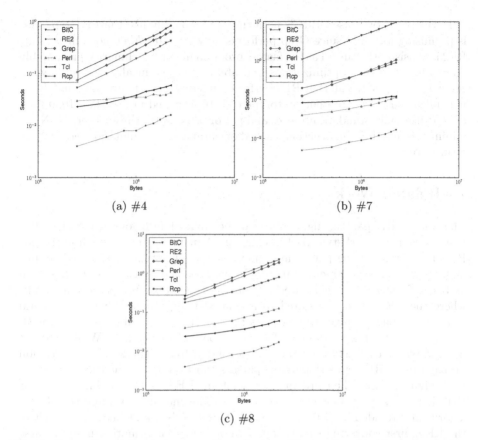

(a) #4   (b) #7

(c) #8

Fig. 6. Comparison using various e-mail expressions. The numbers in the captions refer to the regexes in [9, Table 1].

For the backtracking worst-case expression $(a?)^n a^n$ in Figure 5a, BitC performs roughly like RE2.[1] In contrast to Rcp and FrCa, which are both highly sensitive to the *direction* of non-determinism, BitC has the same performance for both $(a?)^n a^n$ and $a^n (a?)^n$ (Figure 5b).

6.2   Practical Examples

We have run the comparisons with various "real-life" examples of REs taken from [9], all of which deal with expressions matching e-mail addresses. In Fig. 6b, BitC is significantly slower than in the other examples. This can likely be ascribed to heavy use of bounded repetitions in this expression, as they are currently just rewritten into concatenations and Kleene stars in our implementation.

---

[1] The expression parser in BitC failed for the largest expressions, which is why they are not on the plot.

In the other two cases, BitC's performance is roughly like that of Grep. This is promising for BitC since Grep performs only RE *matching*, not full *parsing*. RE2 is consistently ranked as the fastest program in our benchmarks, presumably due to its aggressive optimizations and ability to dynamically choose between several strategies. Recall that RE2 performs greedy leftmost subgroup matching, not full parsing. Our present prototype of BitC is coded in less than 1000 lines of C. It uses only standard libraries and performs no optimizations such as NFA-minimization, DFA-construction, cached or parallel NFA-simulation, etc. This is future work.

7   Related Work

The known RE parsing algorithms can be divided into four categories. The first category is Perl-style backtracking used in many tools and libraries for RE subgroup matching [10]; it has an exponential worst case running time, but always produces the greedy parse and enables some extensions to REs such as backreferences. Another category consists of context-free parsing methods, where the RE is first translated to a context-free grammar, before a general context-free parsing algorithm such as Earleys [11] using cubic time is applied. An interesting CFG method is derivatives-based parsing [12]. While efficient parsers exist for subsets of unambiguous context-free languages, this restriction propagates to REs, and thus these parsers can only be applied for subsets of unambiguous REs. The third category contains RE scalable parsing algorithms that do not always produce the greedy parse. This includes NFA and DFA based algorithms provided by Dubé and Feeley [3] and Nielsen and Henglein [4], where the RE is first converted to an NFA with additional information used to parse strings or to create a DFA preserving the additional information for parsing. This category also includes the algorithm by Fischer, Huch and Wilke [13]; it is left out of our tests since its Haskell-based implementation often was not competitive with the other tools. The last category consists of the algorithms that scale well and always produce greedy parse trees. Kearns [1] and Frisch and Cardelli [2] reverse the input; perform backwards NFA-simulation, building a log of NFA-states reached at each input position; and construct the greedy parse tree in a final forward pass over the input. They require storing the input symbol plus m bits per input symbol for the log. This can be optimized to storing bits proportional to the number of NFA-states reached at a given input position [4], although the worst case remains the same. Our lean log algorithm uses only 2 passes, does not require storing the input symbols and stores only $k < \frac{1}{3}m$ bits per input symbol in the string.

References

1. Kearns, S.M.: Extending Regular Expressions. PhD thesis, Columbia University (1990)
2. Frisch, A., Cardelli, L.: Greedy Regular Expression Matching. In: Díaz, J., Karhumäki, J., Lepistö, A., Sannella, D. (eds.) ICALP 2004. LNCS, vol. 3142, pp. 618–629. Springer, Heidelberg (2004)
3. Dubé, D., Feeley, M.: Efficiently Building a Parse Tree From a Regular Expression. Acta Informatica 37(2), 121–144 (2000)
4. Nielsen, L., Henglein, F.: Bit-coded Regular Expression Parsing. In: Dediu, A.-H., Inenaga, S., Martín-Vide, C. (eds.) LATA 2011. LNCS, vol. 6638, pp. 402–413. Springer, Heidelberg (2011)
5. Henglein, F., Nielsen, L.: Regular expression containment: Coinductive axiomatization and computational interpretation. In: Proc. 38th ACM SIGACT-SIGPLAN Symposium on Principles of Programming Languages (POPL). SIGPLAN Notices, vol. 46, pp. 385–398. ACM Press (January 2011)
6. Cox, R.: RE2, https://code.google.com/p/re2/
7. Ousterhout, J.: Tcl: An Embeddable Command Language. In: Proc. USENIX Winter Conference, pp. 133–146 (January 1990)
8. Wall, L., Christiansen, T., Orwant, J.: Programming Perl. O'Reilly Media, Incorporated (2000)
9. Veanes, M.V.M., de Halleux, P., Tillmann, N.: Rex: Symbolic Regular Expression Explorer. In: Proc. 3d Int'l Conf. on Software Testing, Verification and Validation, Paris, France. IEEE Computer Society Press (April 6-10 2010)
10. Cox, R.: Regular Expression Matching can be Simple and Fast
11. Earley, J.: An Efficient Context-Free Parsing Algorithm. Communications of the ACM 13(2), 94–102 (1970)
12. Might, M., Darais, D., Spiewak, D.: Parsing with derivatives: a functional pearl. In: ACM SIGPLAN Notices, vol. 46, pp. 189–195. ACM (2011)
13. Fischer, S., Huch, F., Wilke, T.: A Play on Regular Expressions: Functional Pearl. In: Proc. of the 15th ACM SIGPLAN International Conference on Functional Programming, ICFP 2010, pp. 357–368. ACM, New York (2010)

Universal Witnesses for State Complexity of Basic Operations Combined with Reversal*

Janusz Brzozowski[1] and David Liu[2]

[1] David R. Cheriton School of Computer Science, University of Waterloo,
Waterloo, ON, Canada N2L 3G1
brzozo@uwaterloo.ca

[2] Department of Computer Science, University of Toronto
Toronto, ON, Canada M5S 3G4
liudavid@cs.toronto.edu

Abstract. We study the state complexity of boolean operations, concatenation, and star, with one or two of the argument languages reversed. We derive tight upper bounds for the symmetric differences and differences of such languages. We prove that the previously discovered bounds for union, intersection, concatenation and star of such languages can all be met by the recently introduced universal witness and its variants.

Keywords: basic operation, boolean operation, regular language, reversal, state complexity, universal witness.

1 Introduction

For background on state complexity see [2,3,11]. The *state complexity of a regular language* is the number of states in the minimal deterministic finite automaton (DFA) recognizing the language. The *state complexity of an operation* on regular languages is the maximal state complexity of the result of the operation as a function of the state complexities of the arguments. We refer to state complexity simply as *complexity*.

The state complexity of basic operations combined with reversal was studied in 2008 by Liu, Martin-Vide, A. Salomaa, and Yu [9]. For regular languages $K, L \subseteq \Sigma^*$ with state complexities m and n, the basic operations considered in [9] were union $(K \cup L)$, intersection $(K \cap L)$, product (catenation or concatenation) (KL), and star (L^*), combined with reversal (L^R). It was shown that $(2^m-1)(2^n-1)+1$ is a tight upper bound for $(K \cup L)^R = K^R \cup L^R$ and $(K \cap L)^R = K^R \cap L^R$. Gao and Yu [8] found the tight upper bound $m2^n - (m-1)$ for $K \cup L^R$ and $K \cap L^R$. It was also proved in [9] that $3 \cdot 2^{m+n-2} - (2^n-1)$ is an upper bound for $(KL)^R = L^R K^R$, but the question of tightness was left open. Cui, Gao, Kari and Yu [5] answered this question positively, and also showed that $3 \cdot 2^{m+n-2}$ is an upper bound for

---

* This work was supported by the Natural Sciences and Engineering Research Council of Canada under grant No. OGP0000871, and was done while the second author was at the University of Waterloo.

S. Konstantinidis (Ed.): CIAA 2013, LNCS 7982, pp. 72–83, 2013.

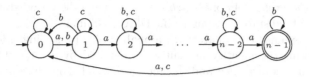

Fig. 1. DFA $\mathcal{U}_n(a, b, c)$ of a complex language $U_n(a, b, c)$

$K^R L$. In another paper [6], they proved that $(m - 1)2^n + 2^{n-1} - (m - 1)$ is a tight upper bound for KL^R. Gao, K. Salomaa, and Yu [7] demonstrated that 2^n is a tight upper bound for $(L^*)^R = (L^R)^*$. Thus eight basic operations with reversal added have been considered so far.

There are two usual steps in finding the state complexity of an operation. First, establish an upper bound; then, find a *stream* $(L_n \mid n \geqslant k)$ of languages, where k is some small positive integer, to act as *witnesses* to show that the bound is tight. In the literature witnesses for binary operations have usually been two distinct streams.

The family of DFAs $\mathcal{U}_n(a, b, c) = (Q, \Sigma, \delta, q_0, F)$ for $n \geqslant 3$ illustrated in Fig. 1 and the corresponding language stream $(U_n(a, b, c) \mid n \geqslant 3)$ were proposed in [3] as the "universal witness."

Throughout this paper we use the notation $w :_{\mathcal{D}} t$ to indicate that the map $\delta(\cdot, w)$ corresponding to word $w \in \Sigma^*$ causes the transformation t on the set of states of the DFA $\mathcal{D}$, omitting the $\mathcal{D}$ if it is clear from the context. The inputs to $\mathcal{U}_n(a, b, c)$ perform the following transformations on $Q = \{0, \ldots, n - 1\}$: the *cycle* of all n states, $a : (0, \ldots, n - 1)$; the *transposition* of 0 and 1 (leaving other states unchanged), $b : (0, 1)$; and the *singular* transformation sending state $n - 1$ to state 0 (and not affecting any other states), $c : (n - 1 \to 0)$. It is well known that the inputs of $\mathcal{U}_n(a, b, c)$ perform all n^n possible transformations of states, and also that the state complexity of the reverse of $U_n(a, b, c)$ is 2^n; the latter result follows by a theorem from [10].

In [3] Brzozowski defined two languages K and L over Σ to be *permutationally equivalent* if one can be obtained from the other by permuting the letters of the alphabet. The permutationally equivalent language of $U_n(a, b, c)$ obtained by interchanging a and b is denoted by $U_n(b, a, c)$. The restriction of the language to alphabet $\{a, b\}$ is denoted by $U_n(a, b)$. These definitions and notation are extended to DFAs in the natural way.

It was proved in [3] that the bound mn for union, intersection, difference $(K \setminus L)$ and symmetric difference $(K \oplus L)$ is met by $(U_m(a, b, c) \mid m \geqslant 3)$ and $(U_n(b, a, c) \mid n \geqslant 3)$. The bound $2^{n-1} + 2^{n-2}$ for star is met by $U_n(a, b)$, and the bound $(m - 1)2^n + 2^{n-1}$ for product, by $U_m(a, b, c)U_n(a, b, c)$. This justifies the use of $U_n(a, b, c)$ as a "universal witness" for the basic operations.

For some operations we require extensions of the universal witness stream. A *dialect* of $U_n(a, b, c)$ is the language of any DFA with three inputs a, b, and c, where a is a cycle as above, b is the transposition of *any* two states (p, q), and c is a singular transformation $(r \to s)$ sending *any* state r to *any* state $s \neq r$. The corresponding *dialect DFAs* of $\mathcal{U}_n(a,b,c)$ always have the initial state 0, but an arbitrary set of final states, as long as the DFA is minimal. The universal witness has also been extended to quaternary alphabets [3], by adding a fourth input d which performs the identity permutation, denoted by $d : 1_Q$. The concepts of permutational equivalence and dialect are extended in the obvious way to quaternary languages and DFAs.

In this paper, we extend the notion of basic operations from [9] by including difference and symmetric difference. Altogether, we study the following 13 operations with these basic operations and reversal:

$$K \cup L^R, \quad K \cap L^R, \quad K \setminus L^R, \quad L^R \setminus K, \quad K \oplus L^R,$$
$$K^R \cup L^R, \quad K^R \cap L^R, \quad K^R \setminus L^R, \quad K^R \oplus L^R,$$
$$KL^R, \quad K^R L, \quad K^R L^R \text{ and } (K^R)^*.$$

Our contributions are as follows:

1. We derive the bounds $m2^n - (m-1)$ for $K_m \setminus L_n^R$ and $L_n^R \setminus K_m$ and the bound $m2^n$ for $K_m \oplus L_n^R$, and show that these and the known bounds for $K_m \cup L_n^R$ and $K_m \cap L_n^R$ are met by the two identical streams $U_m(a,b,c)$ and $U_n(a,b,c)$. This reduces the size of the alphabet for union and intersection from four in [8] to three.

2. We derive the bounds $(2^m - 1)(2^n - 1) + 1$ for $K_m^R \setminus L_n^R$ and 2^{m+n-1} for $K_m^R \oplus L_n^R$, and show that these and the known bounds for $K_m^R \cup L_n^R$ and $K_m^R \cap L_n^R$ are met by two streams, $U_{\{0,2\},m}(a,b,c)$ and $U_{\{1,3\},n}(b,a,c)$, where the set of final states in $\mathcal{U}_{\{0,2\},m}(a,b,c)$ (respectively $\mathcal{U}_{\{1,3\},n}(b,a,c)$) is $\{0,2\}$ (respectively $\{1,3\}$).

3. We prove that the known bound for $K_m L_n^R$ is met by two identical streams of languages $U_m(a,b,c)$ and $U_n(a,b,c)$. As a reviewer pointed out, this result shows that $U_m(a,b,c)$ can be used as a witness for the language LL^R, which to our knowledge has not been explicitly studied before.

4. We show that the known bound for $K_m^R L_n$ is met by two permutationally equivalent dialects of $U_n(a,b,c,d)$.

5. We prove that the known bound for $(K_m L_n)^R = L_n^R K_m^R$ is met by two permutationally equivalent streams $U_m(a,b,c,d)$ and $U_n(d,c,b,a)$. Our proof is considerably simpler than the one in [5].

6. We note that the original proof in [7] for the bound on $(K^R)^*$ uses a dialect of $U_n(a,b,c)$, and point out that the known bound is met by $U_n(a,b,c)$ with final state 0.

In obtaining the results above, we prove Conjectures 1–4, 8, 11, and 14 of [3].

Boolean operations with one and two reversed arguments are considered in Sections 2 and 3. Product and star are examined in Section 4, and Section 5 concludes the paper. Omitted proofs can be found in [4].

2   Boolean Operations with One Reversed Argument

Gao and Yu [8] showed using quaternary witnesses that the complexities of $K_m \cup L_n^R$ and $K_m \cap L_n^R$ are both $m2^n - (m-1)$. These results are improved and extended here as follows: (1) *ternary* alphabets suffice, (2) the *same* language stream can be used for K_m and L_n for both union and intersection, (3) the same language stream is also a witness for two difference operations and symmetric difference, and (4) the bound for symmetric difference is $m2^n$.

First we derive upper bounds for two differences and for symmetric difference.

Proposition 1. *Let K_m and L_n be two regular languages with complexities m and n. Then the complexities of $K_m \setminus L_n^R$ and $L_n^R \setminus K_m$ are at most $m2^n - (m-1)$, and that of $K_m \oplus L_n^R$ is at most $m2^n$.*

Proof. Let $\mathcal{D}_1 = (Q_1, \Sigma, \delta_1, q_1, F_1)$ and $\mathcal{D}_2 = (Q_2, \Sigma, \delta_2, q_2, F_2)$ be the minimal DFAs of K_m and L_n. Let $\mathcal{N}_2 = \mathcal{D}_2^R$ be the NFA obtained by interchanging the sets of initial and final states and reversing all transitions in $\mathcal{D}_2$. Then $\mathcal{N}_2$ accepts the reversed language L_n^R. Let $\mathcal{R}_2$ be the DFA obtained from $\mathcal{N}_2$ by the subset construction; since the reverse of $\mathcal{N}_2$ is deterministic, $\mathcal{R}_2$ is minimal, by a theorem from [1]. Finally, let $\mathcal{P}$ be the direct product DFA of $\mathcal{D}_1$ and $\mathcal{R}_2$. The states of $\mathcal{P}$ are of the form (i, S) where $i \in Q_1$ and $S \subseteq Q_2$, and hence there are at most $m2^n$ reachable states. With appropriate assignments of final states, $\mathcal{P}$ can accept the languages $K_m \setminus L_n^R$, $L_n^R \setminus K_m$, and $K_m \oplus L_n^R$. This proves the bound for $K_m \oplus L_n^R$.

For any input $x \in \Sigma$, any state $(i, \emptyset)$ is mapped to $(j, \emptyset)$ for some j; similarly (i, Q_2) is mapped to (j, Q_2) for some j, because $\delta_2(q, x)$ is defined for all $q \in Q_2$ and $x \in \Sigma$. So for $K_m \setminus L_n^R$, all m states of the form (i, Q_2) are non-final and indistinguishable, and the same is true for $L_n^R \setminus K_m$ with the states $(i, \emptyset)$. So for $K_m \setminus L_n^R$ and $L_n^R \setminus K_m$ there are at most $m2^n - (m-1)$ distinguishable states. □

We now prove that these bounds and the corresponding ones for union and intersection are met by the universal witness streams. For states i and j of a DFA and $w \in \Sigma^*$, we use the notation $i \xrightarrow{w} j$ to denote that state j is reached from state i by word w. Let $K \circ L$ denote any one of the boolean operations $K \cup L$, $K \cap L$, $K \setminus L$ and $K \oplus L$.

Theorem 1 ($K \circ L^R$). *For $m, n \geqslant 3$, the complexities of the four languages $U_m(a,b,c) \cup (U_n(a,b,c))^R$, $U_m(a,b,c) \cap (U_n(a,b,c))^R$, $U_m(a,b,c) \setminus (U_n(a,b,c))^R$, and $(U_n(a,b,c))^R \setminus U_m(a,b,c)$ are all $m2^n - (m-1)$, and that of $U_m(a,b,c) \oplus (U_n(a,b,c))^R$ is $m2^n$.*

Proof. Let $\mathcal{D}_1 = \mathcal{U}_m(a,b,c)$ and $\mathcal{D}_2 = \mathcal{U}_n(a,b,c)$; the various related automata are defined as in the proof of Proposition 1. The problem is illustrated in Fig. 2

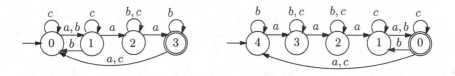

Fig. 2. DFA $\mathcal{D}_1 = \mathcal{U}_4(a,b,c)$ and NFA $\mathcal{N}_2 = \mathcal{D}_2^R = (\mathcal{U}_5(a,b,c))^R$

for $m = 4$ and $n = 5$. We show that all $m2^n$ states of $\mathcal{P}$ of the form (i, S) with $i \in Q_1$ and $S \subseteq Q_2$ are reachable.

The initial state is $(0, \{n-1\})$, and $(0, \{n-1\}) \xrightarrow{c} (0, \emptyset) \xrightarrow{a^i} (i, \emptyset)$ for all $0 \leqslant i \leqslant m-1$. The word ab sends 0 to 0 in $\mathcal{D}_1$ and acts as $ab :_{\mathcal{N}_2} (n-1, n-2, \ldots, 2, 0)$. Therefore all states $(0, \{j\})$, $j \neq 1$ are reachable from $(0, \{n-1\})$ by repeated applications of ab. If $n \nmid m$, then $(0, \{1 + m \pmod{n})\}) \xrightarrow{a^m} (0, \{1\})$. If $n \mid m$, then $(0, \{0\}) \xrightarrow{a^{m-1}} (m-1, \{1\}) \xrightarrow{c} (0, \{1\})$. For $i > 0$, $(0, \{i + j \pmod{n})\}) \xrightarrow{a^i} (i, \{j\})$. Hence all states (i, S), $|S| \leqslant 1$, are reachable.

Consider the reachability of a state (i, S) with $|S| = k$ for some $k \geqslant 2$. The words in $\{a, b\}^*$ can perform all permutations of states in $\mathcal{D}_1$ and $\mathcal{N}_2$. Therefore it suffices to show the reachability of all states (i, S), where $|S| = k$ and $0, n-1 \in S$. Fix such a state (i, S). If $i < m-1$, then $(i, S \setminus \{n-1\}) \xrightarrow{c} (i, S)$. For $(m-1, S)$ there are three cases:

1. $m \nmid n$. $((m-1-n \pmod{m}), S) \xrightarrow{a^n} (m-1, S)$.
2. $m = n = 3$. Since $a^2ba :_{\mathcal{D}_1} (1, 2)$ and $a^2ba :_{\mathcal{N}_2} (0, 2)$ and $0, 2 \in S$, we have $(1, S) \xrightarrow{a^2ba} (2, S)$.
3. $m \mid n$, $n \geqslant 4$. Consider the transposition $a^2ba^{n-2} :_{\mathcal{N}_2} (2, 3)$, and define $S' \subseteq Q_2$ such that $S \setminus \{n-1\} \xrightarrow{ca^2ba^{n-1}} S'$. Since $0 \in S$, S' contains 0; hence applying c adds $n-1$. Moreover, since $m \mid n$, this word also acts as $a^2ba^{n-2} :_{\mathcal{D}_1} (m-2, m-1)$. It follows that $(m-2, S') \xrightarrow{ca^2ba^{n-2}} (m-1, S)$.

Therefore by induction on $|S|$, all $m2^n$ states are reachable, and it remains to find the number of pairwise indistinguishable states for each operation.

We claim that if $S, T \subseteq Q_2$ are distinct states of $\mathcal{R}_2$, then there is an input which takes this pair of states to Q_2 and $\emptyset$. Without loss of generality, let $k \in S \setminus T$; by applying a cyclic shift a^k we may assume that $0 \in S \setminus T$. Then applying ca^{n-1} results in two states S_1 and T_1 with $0, 1 \in S_1 \setminus T_1$. Repeating this transformation $n-1$ times maps S to Q_2 and T to $\emptyset$.

Sets Q_2 and $\emptyset$ are mapped to themselves under all inputs. Also, Q_2 is final and $\emptyset$ non-final in $\mathcal{R}_2$. Therefore the states (i, Q_2) and $(j, \emptyset)$ are distinguishable for the boolean operations as follows:

1. $K_m \cup L_n^R$, $L_n^R \setminus K_m$, and $K_m \oplus L_n^R$: apply a^k, $k \notin \{m-1-i, m-1-j\}$, to send i and j to non-final states.
2. $K_m \cap L_n^R$: apply a^{m-1-i} so that i gets mapped to a final state.
3. $K_m \setminus L_n^R$: apply a^{m-1-j} so that j gets mapped to a final state.

Thus any two states (i, S) and (j, T) with $S \neq T$ are distinguishable for all five boolean operations. Now consider states of the form (i, S) and (j, S), $i < j$.

Case 1. $S = \emptyset$. All states $(i, \emptyset)$ are non-final and indistinguishable for $K_m \cap L_n^R$ and $L_n^R \setminus K_m$. For the other three boolean operations, apply a^{m-1-j} to get the distinguishable states $(k, \emptyset)$, $(m-1, \emptyset)$, $k \neq m-1$.

Case 2. $S \neq \emptyset$, $0 \notin S$. Note that $ba :_{\mathcal{D}_1} (0, 2, 3, \ldots, m-1)$ and $ba :_{\mathcal{N}_2} (n-1, n-2, \ldots, 1)$. Since $i \neq j$, at least one of i and j is not equal to 1. Therefore we can apply $(ba)^d$ for some d so that the states become $(m-1, S')$, (k, S')

where S' is non-final, and $k \neq m-1$. This distinguishes the states for $K_m \cup L_n^R$, $K_m \oplus L_n^R$, and $K_m \setminus L_n^R$. For the other two operations, apply a cyclic shift a^r so that S is mapped to some S'' and $0 \in S''$, and the pair of states is now in Case 3.

Case 3. $S \neq Q_2, 0 \in S$. Again, apply $(ba)^p$ for some p so that the states become $(m-1, S')$, (k, S'), S' is final, and $k \neq m-1$. This distinguishes the states for $K_m \cap L_n^R$ and $L_n^R \setminus K_m$. For the other three operations, apply a cyclic shift a^r so that S is mapped to S'', and $0 \notin S''$, so that Case 2 now applies.

Case 4. $S = Q_2$. All states (i, Q_2) are final for $K_m \cup L_n^R$ and non-final for $K_m \setminus L_n^R$, and indistinguishable in either case. For the other three boolean operations, apply a^{m-1-j} to get the states (k, Q_2), $(m-1, Q_2)$, $k \neq m-1$. This distinguishes the states.

For symmetric difference, all $m2^n$ states are distinguishable. For the other operations, exactly m states are indistinguishable. □

3 Boolean Operations with Two Reversed Arguments

Note that $(K \circ L)^R = K^R \circ L^R$ for all four boolean operations. Liu, Martin-Vide, A. Salomaa, and Yu [9] showed that $(2^m - 1)(2^n - 1) + 1$ is a tight upper bound for $K^R \cup L^R$ and $K^R \cap L^R$, and that the bound is met by ternary witnesses. We first derive upper bounds for difference and symmetric difference.

Proposition 2. *Let K_m and L_n be two regular languages with complexities m and n. Then the complexity of $K_m^R \setminus L_n^R$ is at most $(2^m - 1)(2^n - 1) + 1$, and the complexity of $K_m^R \oplus L_n^R$ is at most 2^{m+n-1}.*

Proof. Let $\mathcal{D}_1 = (Q_1, \Sigma, \delta_1, q_1, F_1)$ and $\mathcal{D}_2 = (Q_2, \Sigma, \delta_2, q_2, F_2)$ be the minimal DFAs of K_m and L_n. As in Proposition 1, we apply the standard subset construction to the NFAs $\mathcal{N}_1$ and $\mathcal{N}_2$ obtained by reversing $\mathcal{D}_1$ and $\mathcal{D}_2$, and then construct their direct product DFA $\mathcal{P}$. The states of $\mathcal{P}$ are of the form (S, T) where $S \subseteq Q_1$ and $T \subseteq Q_2$; hence $\mathcal{P}$ has at most 2^{m+n} states.

For $K_m^R \setminus L_n^R$, all states of the form $(\emptyset, T)$ and (S, Q_2) are non-final. Moreover, because $\mathcal{D}_2$ is complete, every input leads to a state of the same form. Therefore these states are indistinguishable. As there are $(2^m - 1)(2^n - 1)$ states *not* of this form, $\mathcal{P}$ has at most $(2^m - 1)(2^n - 1) + 1$ distinguishable states.

For $K_m^R \oplus L_n^R$, we note that (S, T) is final if and only if $(\overline{S}, \overline{T})$ is final, where $\overline{S} = Q_1 \setminus S$ and $\overline{T} = Q_2 \setminus T$. Let $S \subseteq Q_1$ be a subset of states of $\mathcal{N}_1$; apply $x \in \Sigma$ to get a state S'. Then $i \in S'$ if and only if $\delta_1(i, x) \in S$. It follows that S and $\overline{S}$ are mapped to a pair $S', \overline{S}'$, i.e., complementary states are mapped to complementary states in $\mathcal{N}_1$ and $\mathcal{N}_2$. Therefore complementary states are indistinguishable, so $\mathcal{P}$ has at most 2^{m+n-1} distinguishable states. □

Next, we require a result concerning $\mathcal{U}_m(a, b, c)$ and $\mathcal{U}_n(b, a, c)$. The NFAs $\mathcal{N}_1 = (\mathcal{U}_4(a, b, c))^R$ and $\mathcal{N}_2 = (\mathcal{U}_5(b, a, c))^R$ are shown in Fig. 3.

Theorem 2. *For $m, n \geq 3$, the complexities of $(U_m(a, b, c))^R \cup (U_n(b, a, c))^R$, $(U_m(a, b, c))^R \cap (U_n(b, a, c))^R$ and $(U_m(a, b, c))^R \setminus (U_n(b, a, c))^R$ are $(2^m - 1) \cdot$*

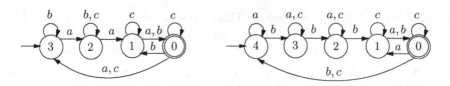

Fig. 3. NFAs $\mathcal{N}_1 = (\mathcal{U}_4(a,b,c))^R$ and $\mathcal{N}_2 = (\mathcal{U}_5(b,a,c))^R$

$(2^n - 1) + 1$, *whereas that of* $(U_m(a,b,c))^R \oplus (U_n(b,a,c))^R$ *is* 2^{m+n-1}, *except when* $m = n = 4$; *then the first three complexities are 202 and the fourth is 116.*

Proof. Let $\mathcal{D}_1 = (Q_1, \Sigma, \delta_1, 0, \{m-1\}) = \mathcal{U}_m(a,b,c)$ and $\mathcal{D}_2 = (Q_2, \Sigma, \delta_2, 0, \{n-1\}) = \mathcal{U}_n(b,a,c)$, and define the related automata as in Proposition 2.

We show that all 2^{m+n} states of $\mathcal{P}$ are reachable unless $m = n = 4$. The initial state is $(\{m - 1\}, \{n - 1\})$. From this state, $(\emptyset, \emptyset)$ is reached by c. Also, $(\{m - 1\}, \{n - 1\}) \xrightarrow{bc} (\emptyset, \{n - 2\}) \xrightarrow{b^{n-2-j} \pmod n} (\emptyset, \{j\})$. Similarly, $(\{m - 1\}, \{n-1\}) \xrightarrow{aca^{m-2-i} \pmod m} (\{i\}, \emptyset)$. For $i, j \geq 2$, $(\{m - 1\}, \{n - 1\}) \xrightarrow{a^{m-1-i}b^{n-1-j}} (\{i\}, \{j\})$. For the other four states, we have the transformations $(\{2\}, \{2\}) \xrightarrow{a^2 b} (\{1\}, \{1\}) \xrightarrow{a} (\{0\}, \{0\})$ and $(\{2\}, \{2\}) \xrightarrow{ab^2} (\{1\}, \{0\}) \xrightarrow{a} (\{0\}, \{1\})$. So all states (S, T) with $|S|, |T| \leq 1$ are reachable.

Consider state $(\{i\}, T)$ with $|T| = k$, $k \geq 2$. As in the proof of Theorem 1, to prove reachability of all such states we need only consider the case $0, n - 1 \in T$. For $1 \leq i \leq m - 1$, $(\{i\}, T \setminus \{n-1\}) \xrightarrow{c} (\{i\}, T)$, while $(\{1\}, T) \xrightarrow{a^2} (\{m - 1\}, T)$ and $(\{2\}, T) \xrightarrow{a^2} (\{0\}, T)$. By induction on k, all states $(\{i\}, T)$ are reachable.

Now suppose all states of the form (S, T) are reachable for $|S| = k$, for some $k \geq 1$. We want to show how to reach all states (S, T) where $|S| = k+1$. Again, it suffices to consider only the case where $0, m - 1 \in S$. As before, $S \setminus \{m-1\} \xrightarrow{c} S$. If 0 and $n - 1$ are both in T or both not in T, then $(S', T) \xrightarrow{c} (S, T)$. Moreover, using arguments symmetric to the ones above, (S, T) is reachable for all $|T| \leq 1$. For the other T, we divide the problem into two cases.

Case 1. m is odd. Let $w \in \{a, b\}^*$ be a permutation of states on $\mathcal{N}_1$ and $\mathcal{N}_2$. We show how to construct another word which performs the same transformation as w on $\mathcal{N}_2$, but maps S to itself in $\mathcal{N}_1$. To do this, make three changes to w:

   (i) Add a^{m-1} to the beginning of w.
   (ii) Replace all instances of a in w by a^m.
   (iii) Add a^{m+1} to the end of w.

Call the resulting word w'. Because m is odd and $a^2 :_{\mathcal{N}_2} \mathbf{1}_{Q_2}$, w' is the same transformation as w on $\mathcal{N}_2$. Consider applying w' to S. Change (i) maps S to some S' with $0, 1 \in S'$ (as we assume $0, m - 1 \in S$). Both a^m and b map S' to itself; so the transformation caused by change (ii) maps S' to itself. Finally, change (iii) is the inverse of (i), mapping S' back to S.

For any state $T \subseteq Q_2$, with $|T| \geqslant 2$, there is a permutation word $w \in \{a,b\}^*$ which maps some T' to T, where $0, n-1 \in T'$. Using the above construction, $(S', T') \xrightarrow{w'} (S, T)$. Therefore all 2^{m+n} states are reachable for odd m. Since the two NFAs are symmetric, if n is odd we can repeat the argument above, interchanging the roles of $\mathcal{N}_1$ and $\mathcal{N}_2$.

Case 2. m and n are both even. Let $T = \{0, t_1, \ldots, t_l\}$, $0 < t_1 < \cdots < t_l < n-1$. Let $j = n - 1 - t_l$, $T' = \{0, t_1 + j, \ldots, t_{l-1} + j, n-1\}$, and $w = (ab)^j$. Since $ab :_{\mathcal{N}_2} (n-1, n-2, \ldots, 1)$, we have $T' \xrightarrow{w} T$. We may assume that j is even, as if j is odd, the same transformation can be caused by $w = (ab)^{j+n-1}$ (as ab is a cycle of length $n-1$). Define the words $tr_i = a^i ba^{m-i}$ as the transpositions $tr_i :_{\mathcal{N}_1} (i, i+1)$. In $\mathcal{N}_2$, they are equivalent to b if i is even, and aba if i is odd.

First suppose $k \geqslant 3$, so that $S \setminus \{0, m-1\} \neq \emptyset$, and let i be the minimal element of this set. If $i = 1$, then define $w' = (tr_{m-1} tr_0)^{j/2}$; this word is equivalent to w in $\mathcal{N}_2$, and maps S to itself (since $0, 1, m-1 \in S$). Therefore $(S, T') \xrightarrow{w'} (S, T)$, and since $0, n-1 \in T'$, (S, T') and (S, T) are reachable. From states of this form where $0 \in T$, any T can be reached by applying cyclic shifts b^j, which map S to itself, as $0, 1 \in S$. Now suppose $i > 1$. If i is even, let $w = tr_{i-1}(tr_{m-1})^{n-1}$. Then there exists S' with $0, i-1, m-1 \in S'$ and $|S'| = |S|$ such that $S' \xrightarrow{w} S$. Moreover, w acts as $(aba)^n : \mathbf{1}_{Q_2}$ on $\mathcal{N}_2$, so $(S', T) \xrightarrow{w} (S, T)$ for all $T \subseteq Q_2$. If i is odd, let $w = tr_{i-1} tr_{i-2} tr_{i-1} (tr_{m-1})^{n-1}$, which acts as the transformation $(i-2, i)(0, m-1)$ on $\mathcal{N}_1$ and $(ab)^2$ in $\mathcal{N}_2$. Since $n-1$ is odd, applying w^{n-1} is equivalent to w on $\mathcal{N}_1$, while $w^{n-1} :_{\mathcal{N}_2} \mathbf{1}_{Q_2}$ (as ab causes a cycle of length $n-1$). Then $(S', T) \xrightarrow{w^{n-1}} (S, T)$ for some S' containing $0, i-2$, and $m-1$. We apply an inductive argument on i to show the reachability of states (S, T) where $|S| \geqslant 3$.

To complete the induction on $k = |S|$, we need to now handle the case $k = 2$. Suppose $S = \{0, m-1\}$. If $m \geqslant 6$, applying a^2 does not change T, but maps $S \xrightarrow{a^2} S' = \{m-2, m-3\}$; thus $0, 1, m-1 \notin S'$. Reachability for all states of the form $S' \cup T$ uses the same argument as the case $0, 1, m-1 \in S$. Since n is even, $(S', T) \xrightarrow{a^{m-2}} (S, T)$, and all of these states are reachable as well. By symmetry, all of the above arguments apply when $n \geqslant 6$.

This completes the reachability proof of all cases except $m = n = 4$. Computation shows that only 232 of the possible 256 states are reachable.

Next we examine the distinguishability of the reachable states. Let (S_1, T_1) and (S_2, T_2) be two distinct states of $\mathcal{P}$, with $S_1 \neq S_2$. We may apply a cyclic shift b^k if necessary so that for each $i = 1, 2$, either (1) $T_i \in \{\emptyset, Q_2\}$, or (2) $T_i \neq \{n-1\}$ and $T_i \neq Q_2 \setminus \{n-1\}$. Applying a cyclic shift a^l if necessary, we may assume that $0 \in S_1 \setminus S_2$. As in Theorem 1, we map S_1 to Q_1 and S_2 to $\emptyset$ by applying $(ca^{m-1})^{m-2}$.

If the T_i are $\emptyset$ or Q_2, this transformation leaves them unchanged. Otherwise, by the above condition, they are mapped neither to $\emptyset$ nor to Q_2. Therefore we can map any pair of states (S_1, T_1) and (S_2, T_2), $S_1 \neq S_2$, to (Q_1, T_1'), $(\emptyset, T_2')$

with $T_i' \in \{\emptyset, Q_2\} \iff T_i \in \{\emptyset, Q_2\}$ for $i = 1, 2$. A similar claim holds for the case $T_1 \neq T_2$ by switching the a's and b's.

We now consider each of the boolean operations separately.

Union. States (Q_1, T) and (S, Q_2) are final and indistinguishable for all possible S and T. Now consider the $(2^m - 1)(2^n - 1)$ states not containing Q_1 or Q_2. By the claim above, and since the two DFAs are symmetric, we can reduce all pairs to the form (Q_1, T_1), $(\emptyset, T_2)$, where $T_1, T_2 \neq Q_2$. These states are distinguishable by applying a cyclic shift b^k mapping T_2 to a non-final state.

Intersection. The states $(\emptyset, T)$ and $(S, \emptyset)$ are non-final and indistinguishable for all possible S and T. By the above claim again, all other states (not containing a $\emptyset$) can be reduced to the case (Q_1, T_1), $(\emptyset, T_2)$, $T_1, T_2 \neq \emptyset$. Mapping T_1 to a final state using a cyclic shift will distinguish the states.

Difference. We consider the operation $U_m(a, b, c)^R \setminus U_n(b, a, c)^R$. The indistinguishable states are those of the form $(\emptyset, T)$ and (S, Q_2), which are all non-final. States (Q_1, T_1) and $(\emptyset, T_2)$ $(T_1, T_2 \neq Q_2)$ are distinguished by shifting T_1 to a non-final state, and (S_1, Q_2), $(S_2, \emptyset)$ $(S_1, S_2 \neq \emptyset)$ are distinguished by shifting S_2 to a final state.

Symmetric Difference. We note that (S, T) is final if and only if $(\overline{S}, \overline{T})$ is final, where the bar denotes complementation. Moreover, if two states are complementary, then they are mapped to complementary states under any input. Therefore (S, T) and $(\overline{S}, \overline{T})$ are indistinguishable. This leads to a maximum of 2^{n+m-1} distinguishable states.

For any state (S, T), either S or $\overline{S}$ contains 0. To complete the proof, we only need to show that all states of the form (S, T) with $0 \in S$ are distinguishable. Let (S_1, T_1) and (S_2, T_2) be two such states. If $T_1 = T_2$, then $S_1 \neq S_2$, there exists $k \in S_1 \oplus S_2$, and hence a^k distinguishes the states. If $T_1 \neq T_2$, by applying b^2 if necessary, we may assume that there exists $k \in \{0, \ldots, n - 2\}$ such that $k \in T_1 \oplus T_2$. By applying ca^{m-1}, we may assume that $0, 1 \in S_1 \cap S_2$. This does not change the fact that T_1 and T_2 are distinct, by the above assumption on k. Applying b^k for $k \in T_1 \oplus T_2$ distinguishes the two states.   □

Theorem 2 shows that $\mathcal{U}_m(a, b, c)$ and $\mathcal{U}_n(b, a, c)$ are witnesses for every case except $m = n = 4$. We show next that this result can be improved if the initial states of the witnesses are modified. For $m \geqslant 3$, let $\mathcal{U}_{\{0,2\}, m}(a, b, c)$ be the DFA obtained from $\mathcal{U}_m(a, b, c)$ by changing the set of final states to $\{0, 2\}$. For $n \geqslant 4$, let $\mathcal{U}_{\{1,3\}, n}(b, a, c)$ be the DFA obtained from $\mathcal{U}_n(b, a, c)$ by changing the set of final states to $\{1, 3\}$, and for $n = 3$, use $\mathcal{U}_{\{1\}, n}(b, a, c)$ with final state 1.

Theorem 3 ($K^R \circ L^R$). *Let* $K_m = U_{\{0,2\}, m}(a, b, c)$ *and* $L_n = U_{\{1,3\}, n}(b, a, c)$ *for* $n \geqslant 4$ *and let* $L_3 = U_{\{1\}, 3}(b, a, c)$. *For* $m, n \geqslant 3$, *the complexities of* $K_m^R \cup L_n^R$, $K_m^R \cap L_n^R$, *and* $K_m^R \setminus L_n^R$ *are* $(2^m - 1)(2^n - 1) + 1$, *but that of* $K_m^R \oplus L_n^R$ *is* 2^{m+n-1}.

Proof. If it is not the case $m = n = 4$, then by Theorem 2, it suffices to show that state $(\{m - 1\}, \{n - 1\})$ is reachable from the initial state of the NFA. If $n = 3$, the initial state is $(\{0, 2\}, \{1\})$, and $(\{0, 2\}, \{1\}) \xrightarrow{ab^2c} (\{1\}, \{1\}) \xrightarrow{a^2b^2}$

$(\{m-1\}, \{n-1\})$. Suppose $n \geqslant 4$. The initial state is $(\{0,2\}, \{1,3\})$. Apply the following: $(\{0,2\}, \{1,3\}) \xrightarrow{ac} (\{1\}, \{0,3,n-1\}) \xrightarrow{a^3} (\{m-2\}, \{1,3,n-1\})$. If $n = 4$, then $n-1 = 3$, and we can apply $(\{m-2\}, \{1,3\}) \xrightarrow{c} (\{m-2\}, \{1\}) \xrightarrow{b^2 a^{m-1}} (\{m-1\}, \{n-1\})$. If $n > 4$, then apply $(\{m-2\}, \{1,3,n-1\}) \xrightarrow{cb^2c} (\{m-2\}, \{1\}) \xrightarrow{b^2 a^{m-1}} (\{m-1\}, \{n-1\})$.

For every case except $m = n = 4$, this shows that all states are reachable. When $m = n = 4$, one can verify through explicit enumeration that the states unreachable from $(\{3\}, \{3\})$ are exactly the states reached from $(\{0,2\}, \{1,3\})$ by words in $\{a,b\}^*$. Therefore in this case all states are reachable as well. Finally, distinguishability follows from the proof of Theorem 2. □

4 Product and Star

4.1 The Language KL^R

Cui, Gao, Kari and Yu showed in [6] that the complexity of KL^R is $(m-1)2^n + 2^{n-1} - (m-1)$, with ternary witnesses. We now prove that the bound can also be met by one stream. As a corollary to this theorem, we get that the universal witness $U_n(a,b,c)$ acts as a witness for the operation LL^R.

Theorem 4. *For $m, n \geqslant 3$, the complexity of the product $U_m(a,b,c)(U_n(a,b,c))^R$ is $(m-1)2^n + 2^{n-1} - (m-1)$.*

Proof. Let $\mathcal{D}_1 = (Q_1, \Sigma, \delta_1, q_0, \{q_{m-1}\})$ and $\mathcal{D}_2 = (Q_2, \Sigma, \delta_2, 0, \{n-1\})$ be the minimal DFAs of $U_m(a,b,c)$ and $U_n(a,b,c)$, where $Q_1 = \{q_0, \dots, q_{m-1}\}$ and $Q_2 = \{0, \dots, n-1\}$. Let $\mathcal{N}_2$ be $\mathcal{D}_2^R$, and let $\mathcal{N}$ be the NFA for the product of $\mathcal{D}_1$ and $\mathcal{N}_2$, obtained by adding an ε-transition from q_{m-1} to $n-1$, where ε is the empty word, and making q_{m-1} non-final. This is illustrated in Fig. 4.

Use the subset construction on $\mathcal{N}$ to get a DFA $\mathcal{P}$ with states of the form $\{q_i\} \cup S$, where $q_i \in Q_1$ and $S \subseteq Q_2$. Any state must either not contain q_{m-1}, or contain both q_{m-1} and $n-1$. There are $(m-1)2^n$ states of the former type, and 2^{n-1} states of the latter. We will show that all of these states are reachable.

Set $\{q_0\}$ is initial, $\{q_0\} \xrightarrow{a^i} \{q_i\}$ for $i \leq m-2$, and $\{q_{m-2}\} \xrightarrow{a} \{q_{m-1}\} \cup \{n-1\}$. Also, $\{q_{m-1}\} \cup \{n-1\} \xrightarrow{a} \{q_0\} \cup \{n-2\}$, and from there all other states $\{q_0\} \cup \{j\}$ are reached by repeated applications of ab, except $j = 1$. If $n \nmid m$, then $\{q_0\} \cup \{1+m \pmod{n}\} \xrightarrow{a^m} \{q_0\} \cup \{1\}$. If $n \mid m$, then $\{q_0\} \cup \{0\} \xrightarrow{a^{m-1}c} \{q_0\} \cup \{1\}$.

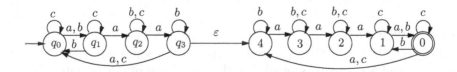

Fig. 4. NFA $\mathcal{N}$ for $U_4(a,b,c)(U_5(a,b,c))^R$

Finally, $\{q_i, j\}$ is reached from $\{q_0, i+j \pmod{n}\}$ by a^i. So all states $\{q_i\} \cup S$, where $i < m-1$ and $|S| \leqslant 1$ are reachable.

States $\{q_{m-1}\} \cup S$ with $|S| = k \geqslant 2$ and $n-1 \in S$ can be reached by a from some state $\{q_{m-1}\} \cup S'$, with $|S'| = k-1$. From these states, all states $\{q_0\} \cup S$ with $|S| = k$, $n-2 \in S$ are reached by a. Since $k \geqslant 2$, repeated applications of the cycle $ab :_{\mathcal{N}_2} (n-1, n-2, \ldots, 2, 0)$ reaches all states $\{q_0\} \cup S$, $|S| = k$. From these states, repeated applications of a reach all states $\{q_i\} \cup S$, $|S| = k$ and $i < m-1$. By induction, all of the required states are reachable.

For distinguishability, note that all m states of the form $\{q_i\} \cup Q_2$ are final and indistinguishable. Consider $\{q_i\} \cup S$ and $\{q_j\} \cup T$ with $S \neq T$. If $k \in S \oplus T$, then a^k distinguishes the two states. Now consider the pair $\{q_i\} \cup S$, $\{q_j\} \cup S$, $S \neq Q_2$. Let $k \notin S$, and apply a^k to get $\{q_{i'}\} \cup S'$, $\{q_{j'}\} \cup T'$. If $S' \neq T'$, then by the previous argument the states are distinguishable. Otherwise, $S' = T'$ and $0 \notin S'$. Note that $ba :_{\mathcal{D}_1} (q_0, q_2, q_3, \ldots, q_{m-1})$, and maps only 0 to 0 in $\mathcal{N}_2$. Since $i \neq j$, at least one of i, j is not equal to 1. Then by applying some $(ba)^d$ if necessary, we may assume that $i < m-2$, $j = m-2$. Apply a to get states $\{q_{i+1}\} \cup T$, $\{q_{m-1}\} \cup T \cup \{n-1\}$, where $n-1 \notin T$. Since these states contain different subsets of Q_2, they are distinguishable by the previous argument. □

4.2 The Language $K^R L$

Let $\mathcal{V}_n(a, b, c, d) = (Q_{\mathcal{V}}, \Sigma, \delta_{\mathcal{V}}, 0, \{n-1\})$, where $Q = \{0, \ldots, n-1\}$, $a : (0, \ldots, n-1)$, $b : (n-2, n-1)$, $c : (n-1 \to n-2)$, and $d : \mathbf{1}_{Q_n}$. Let $V_n(a, b, c, d)$ be the language of $\mathcal{V}_n(a, b, c, d)$.

It was shown in [5] by Cui, Gao, Kari and Yu that $3 \cdot 2^{m+n-2}$ is a tight bound for $K_m^R L_n$. The permutationally equivalent dialects $V_m(a, b, c, d)$ and $V_n(d, c, b, a)$ work here:

Theorem 5 ($K_m^R L_n$). *The complexity of the product* $(V_m(a, b, c, d))^R V_n(d, c, b, a)$ *is* $3 \cdot 2^{m+n-2}$ *for* $m, n \geqslant 3$.

4.3 The Language $(KL)^R = L^R K^R$

Let $\mathcal{U}_n(a, b, c, d) = (Q, \Sigma, \delta_{\mathcal{U}}, 0, \{n-1\})$, where $a : (0, \ldots, n-1)$, $b : (0, 1)$, $c : (n-1 \to 0)$, and $d : \mathbf{1}_Q$. Let $U_n(a, b, c, d)$ be the language of $\mathcal{U}_n(a, b, c, d)$.

It was shown by Cui, Gao, Kari, and Yu [5] that quaternary witnesses meet the bound $3 \cdot 2^{m+n-2} - 2^n + 1$ for $(K_m L_n)^R$. The languages $U_m(a, b, c, d)$ and $U_n(d, c, b, a)$ work here:

Theorem 6 ($L_n^R K_m^R$). *The complexity of* $(U_n(d, c, b, a))^R (U_m(a, b, c, d))^R$ *is* $3 \cdot 2^{m+n-2} - (2^n - 1)$, *for* $m, n \geqslant 3$.

4.4 Reverse of Star

Note that $(L^*)^R = (L^R)^*$. The star of the reverse was studied by Gao, K. Salomaa, and Yu [7], who showed that the complexity of this operation is 2^n. The witness $\mathcal{U}_{\{0\}, n}(a, b, c)$, which is $\mathcal{U}_n(a, b, c)$ with final state set changed to $\{0\}$ works here. The proof is the same as that in [7].

Theorem 7 $((L^*)^R)$. *For $n \geqslant 3$, the complexity of $((U_{\{0\},n}(a, b, c))^*)^R$ is 2^n.*

5   Conclusions

We have proved that the universal witnesses $U_n(a, b, c)$ and $U_n(a, b, c, d)$, along with their permutational equivalents $U_n(b, a, c)$ and $U_n(d, c, b, a)$, and dialects $U_{\{0,2\},m}(a, b, c)$, $U_{\{1,3\},n}(a, b, c)$, $U_{\{0\},n}(a, b, c)$, $V_m(a, b, c, d)$ and $V_n(d, c, b, a)$ suffice to act as witnesses for all the state complexity bounds involving binary boolean operations, product, star and reversal. We have shown that it is efficient to consider all four boolean operations together. Lastly, the use of universal witnesses and their dialects simplified many proofs, and allowed us to utilize the similarities in the witnesses.

Acknowledgment. We thank Baiyu Li for careful proofreading. We thank our reviewers for their helpful feedback and corrections, and pointing out the extension of the universal witness to the operation LL^R.

References

1. Brzozowski, J.: Canonical regular expressions and minimal state graphs for definite events. In: Proc. Symposium on Mathematical Theory of Automata. MRI Symposia Series, vol. 12, pp. 529–561. Polytechnic Institute of Brooklyn, N.Y. (1963)
2. Brzozowski, J.: Quotient complexity of regular languages. J. Autom. Lang. Comb. 15(1/2), 71–89 (2010)
3. Brzozowski, J.: In search of most complex regular languages. In: Moreira, N., Reis, R. (eds.) CIAA 2012. LNCS, vol. 7381, pp. 5–24. Springer, Heidelberg (2012)
4. Brzozowski, J., Liu, D.: Universal witnesses for state complexity of basic operations combined with reversal (July 2012), http://arxiv.org/abs/1207.0535
5. Cui, B., Gao, Y., Kari, L., Yu, S.: State complexity of combined operations with two basic operations. Theoret. Comput. Sci. 437, 82–102 (2012)
6. Cui, B., Gao, Y., Kari, L., Yu, S.: State complexity of two combined operations: catenation-star and catenation-reversal. Int. J. Found. Comput. Sc. 23(1), 51–66 (2012)
7. Gao, Y., Salomaa, K., Yu, S.: The state complexity of two combined operations: star of catenation and star of reversal. Fund. Inform. 83(1-2), 75–89 (2008)
8. Gao, Y., Yu, S.: State complexity of combined operations with union, intersection, star, and reversal. Fund. Inform. 116, 1–14 (2012)
9. Liu, G., Martin-Vide, C., Salomaa, A., Yu, S.: State complexity of basic language operations combined with reversal. Inform. and Comput. 206, 1178–1186 (2008)
10. Salomaa, A., Wood, D., Yu, S.: On the state complexity of reversals of regular languages. Theoret. Comput. Sci. 320, 315–329 (2004)
11. Yu, S.: State complexity of regular languages. J. Autom. Lang. Comb. 6, 221–234 (2001)

Trimming Visibly Pushdown Automata

Mathieu Caralp, Pierre-Alain Reynier, and Jean-Marc Talbot

Laboratoire d'Informatique Fondamentale de Marseille
UMR 7279, Aix-Marseille Université & CNRS, France

Abstract. We study the problem of trimming visibly pushdown automata (VPA).
We first describe a polynomial time procedure which, given a visibly pushdown
automaton that accepts only well-nested words, returns an equivalent visibly
pushdown automaton that is trimmed. We then show how this procedure can be
lifted to the setting of arbitrary VPA. Furthermore, we present a way of building,
given a VPA, an equivalent VPA which is both deterministic and trimmed.

1 Introduction

Visibly pushdown automata (VPA) are a particular class of pushdown automata defined
over an alphabet split into call, internal and return symbols [2,3][1]. In VPA, the stack behavior is driven by the input word: when reading a call symbol, a symbol is pushed onto
the stack, for a return symbol, the top symbol of the stack is popped, and for an internal
symbol, the stack remains unchanged. VPA have been applied in research areas such as
software verification (VPA allow one to model function calls and returns, thus avoiding
the study of data flows along invalid paths) and XML documents processing (VPA can
be used to model properties over words satisfying a matching property between opening
and closing tags).

Languages defined by visibly pushdown automata enjoy many properties of regular
languages such as (effective) closure by Boolean operations and these languages can always be defined by a deterministic visibly pushdown automaton. However, VPA do not
have a unique minimal form [1]. Instead of minimization, one may consider trimming as
a way to deal with smaller automata. Trimming a finite state automaton amounts to removing useless states, *i.e.* states that do not occur in some accepting computation of the
automaton: every state of the automaton should be both reachable from an initial state,
and co-reachable from a final state. This property is important from both a practical and
a theoretical point of view. Indeed, most of the algorithmic operations performed on
an automaton will only be relevant on the trimmed part of that automaton. Removing
useless states may thus avoid the study of irrelevant paths in the automaton, and speed
up the analysis. From a theoretical aspect, there are several results holding for automata
provided they are trimmed. For instance, the boundedness of finite-state automata with
multiplicities can be characterized by means of simple patterns for trimmed automata
(see [13,9]). Similarly, Choffrut introduced in [8] the twinning property to characterize sequentiality of (trimmed) finite-state transducers. This result was later extended to
weighted finite-state automata in [5]. Both of these results have been extended to visibly

---

[1] These automata were first introduced in [4] as "input-driven automata".

S. Konstantinidis (Ed.): CIAA 2013, LNCS 7982, pp. 84–96, 2013.

pushdown automata and transducers in [6] and [11] respectively, requiring these objects to be trimmed.

While trimming finite state automata can be done easily in linear time by solving two reachability problems in the graph representing the automaton, the problem is much more involved for VPA (and for pushdown automata in general). Indeed, in this setting, the current state of a computation (called a configuration) is given by both a "control" state and a stack content. A procedure has been presented in [12] for pushdown automata. It consists in computing, for each state, the regular language of stack contents that are both reachable and co-reachable, and use this information to constrain the behaviors of the pushdown automaton in order to trim it. This approach has however an exponential time complexity.

Contributions. In this work, we present a procedure for trimming visibly pushdown automata. The running time of this procedure is bounded by a polynomial in the size of the input VPA. We first tackle the case of VPA that do only recognize so-called *well-nested* words, *i.e.* words which have no unmatched call or return symbols. This class of VPA is called well-nested VPA, and denoted by wnVPA. We actually present a construction for reducing wnVPA, *i.e* ensuring that every run starting from an initial configuration can be completed into an accepting run. As we consider well-nested VPA, one can consider the "dual" of the automaton (reads the word from right to left), and apply the reduction procedure on it, yielding a trimming procedure. In a second step, we address the general case. To do so, we present a construction which modifies a VPA in order to obtain a wnVPA. This construction has to be reversible, in order to recover the original language, and to be compatible with the trimming procedure. In addition, we also design this construction in such a way that it allows to prove the following result: given a VPA, we can effectively build an equivalent VPA which is both deterministic and trimmed.

Organization of the paper. In Section 2 we introduce useful definitions. We address the case of well-nested VPA in Section 3 and the general case in Section 4. We consider the issue of determinization in Section 5. Due to lack of space, some proofs are omitted but can be found in [7].

Related models. VPA are tightly connected to several models:

Context-free grammars: it is well-known that pushdown automata are equivalent to context-free grammars. This observation yields the following procedure for trimming pushdown automata [2]. One can first translate the automaton into an equivalent context-free grammar, then eliminate from this grammar variables generating the empty language or not reachable from the start symbol, and third convert the resulting grammar into the pushdown automaton performing its top-down analysis. This construction has a polynomial time complexity but, in this form, it does not apply to VPA. Indeed, the resulting pushdown automaton may not satisfy the condition of visibility as the third step may not always produce rules respecting the constraints on push and pop operations associated with call and return symbols.

---

[2] We thank Géraud Sénizergues for pointing us this construction.

Tree automata: by the standard interpretation of XML documents as unranked trees, VPA can be understood as acceptors of unranked tree languages. It is shown in [2] that they actually do recognize precisely the set of regular (ranked) tree languages, using the encoding of so-called *stack-trees*, which is similar to the first-child next-sibling encoding (fcns for short). Trimming ranked tree automata is standard (and can be performed in linear time), and one can wonder whether this approach could yield a polynomial time trimming procedure for VPA. Actually, going through tree automata would not ease the construction of a trimmed VPA. Indeed, trimming the fcns encoding of a wnVPA, and then translating back the result into a wnVPA yields an automaton which is reduced but not trimmed (this is intuitively due to the fact that the fcns encoding realizes a left-to-right traversal of the tree). Moreover, this construction does not ensure a bijection between accepting runs, a property that is useful when moving to weighted VPA.

Nested word automata [3]: this model is equivalent to that of VPA. One could thus rephrase our constructions in this context, and obtain the same results.

2   Definitions

Words and well-nested words. A *structured alphabet* Σ is a finite set partitioned into three disjoint sets Σ_c, Σ_r and Σ_ι, denoting respectively the *call*, *return* and *internal* alphabets. We denote by Σ^* the set of words over Σ and by ϵ the empty word.

The set of *well-nested* words Σ_{wn}^* is the smallest subset of Σ^* such that $\varepsilon \in \Sigma_{wn}^*$, $\Sigma_\iota \subseteq \Sigma_{wn}^*$ and for all $c \in \Sigma_c$, all $r \in \Sigma_r$, all $u, v \in \Sigma_{wn}^*$, $cur \in \Sigma_{wn}^*$ and $uv \in \Sigma_{wn}^*$.

Given a family of elements $e_1, e_2, \ldots, e_n$, we denote by $\Pi_{i=1}^n e_i$ the concatenation $e_1 e_2 \ldots e_n$. The length of a word u is denoted by $|u|$.

Visibly pushdown automata (VPA). Visibly pushdown automata are a restriction of pushdown automata in which the stack behavior is imposed by the input word. On a call symbol, the VPA pushes a symbol onto the stack, on a return symbol, it must pop the top symbol of the stack, and on an internal symbol, the stack remains unchanged. The only exception is that some return symbols may operate on the empty stack

Definition 1 (Visibly pushdown automata). A *visibly pushdown automaton* (VPA) on finite words over Σ is a tuple $A = (Q, I, F, \Gamma, \delta)$ where Q is a finite set of states, $I \subseteq Q$ is the set of initial states, $F \subseteq Q$ the set of final states, Γ is a finite stack alphabet, $\delta = \delta_c \uplus \delta_r \uplus \delta_r^\perp \uplus \delta_\iota$ the (finite) transition relation, with $\delta_c \subseteq Q \times \Sigma_c \times \Gamma \times Q$, $\delta_r \subseteq Q \times \Sigma_r \times \Gamma \times Q$, $\delta_r^\perp \subseteq Q \times \Sigma_r \times \{\perp\} \times Q$, and $\delta_\iota \subseteq Q \times \Sigma_\iota \times Q$.

For a transition $t = (q, a, x, q')$ from δ_c, δ_r or $\delta_r^\perp$ or $t = (q, a, q')$ in δ_ι, we denote by source(t) and target(t) the states q and q' respectively, and by letter(t) the symbol a.

A *stack* is a word from Γ^* and we denote by $\perp$ the empty word on Γ. A *configuration* of a VPA is a pair $(q, \sigma) \in Q \times \Gamma^*$.

Definition 2. A *run of A on a word $w = a_1 \ldots a_l \in \Sigma^*$ over a sequence of transitions* $(t_k)_{1 \leq k \leq l}$ *from a configuration (q, σ) to a configuration (q', σ') is a finite sequence of symbols and configurations $\rho = (q_0, \sigma_0) \Pi_{k=1}^l (a_k (q_k, \sigma_k))$ such that $(q, \sigma) = (q_0, \sigma_0)$, $(q', \sigma') = (q_l, \sigma_l)$, and, for each $1 \leq k \leq l$, there exists $\gamma_k \in \Gamma$ such that either:*

- $t_k = (q_{k-1}, a_k, \gamma_k, q_k) \in \delta_c$ and $\sigma_k = \sigma_{k-1}\gamma_k$, or
- $t_k = (q_{k-1}, a_k, \gamma_k, q_k) \in \delta_r$ and $\sigma_{k-1} = \sigma_k\gamma_k$, or
- $t_k = (q_{k-1}, a_k, \bot, q_k) \in \delta_r^\bot$, $\sigma_{k-1} = \sigma_k = \bot$, or
- $t_k = (q_{k-1}, a_k, q_k) \in \delta_\iota$ and $\sigma_k = \sigma_{k-1}$.

We denote by $\mathsf{Run}_w(A)$ the set of runs of A over the word w. Note that a run for the empty word is simply any configuration. We write $(q, \sigma) \xrightarrow{w} (q', \sigma')$ when there exists a run over w from (q, σ) to (q', σ'). We may omit the superscript w when irrelevant.

Given two runs $\rho_i = (q_0^i, \sigma_0^i)\Pi_{k=1}^{\ell_i}(a_k^i(q_k^i, \sigma_k^i))$ for $i \in \{1, 2\}$, we can consider the concatenation $\rho_1\rho_2$ of these runs, provided that $(q_{\ell_1}^1, \sigma_{\ell_1}^1) = (q_0^2, \sigma_0^2)$, defined as
$\rho_1\rho_2 = (q_0^1, \sigma_0^1)\Pi_{k=1}^{\ell_1}(a_k^1(q_k^1, \sigma_k^1))\Pi_{k=1}^{\ell_2}(a_k^2(q_k^2, \sigma_k^2))$,

Initial (resp. final) configurations are configurations of the form $(q, \bot)$, with $q \in I$ (resp. (q, σ) with $q \in F$). A run is *initialized* if it starts in an initial configuration and it is *accepting* if it is initialized and ends in a final configuration. We denote by $\mathsf{ARun}_w(A)$ the set of accepting runs of A over the word w. The set of all accepting runs of A is denoted $\mathsf{ARun}(A)$. A word is accepted by A iff there exists an accepting run of A on it. The language of A, denoted $L(A)$, is the set of words accepted by A.

Definition 3. *A VPA* $A = (Q, I, F, \Gamma, \delta)$ *is*

- deterministic *if I is a singleton and for all q in Q, for all c in Σ_c, for all i in Σ_ι, for all r in Σ_r,*
  - *there exists at most one rule of the form (q, c, γ, q') in δ_c, of the form (q, i, q') in δ_ι and of the form $(q, r, \bot, q')$ in $\delta_r^\bot$,*
  - *for all γ in Γ, there exists at most one rule of the form (q, r, γ, q') in δ_r*
- co-deterministic *if F is a singleton and for all q' in Q, for all c in Σ_c, for all i in Σ_ι, for all r in Σ_r,*
  - *for all γ in Γ, there exists at most one rule of the form (q, c, γ, q') in δ_c.*
  - *there exists at most one rule of the form (q, r, γ, q') in δ_r, of the form (q, i, q') in δ_ι and of the form $(q, c, \bot, q')$ in $\delta_\iota^\bot$.*

(Co)-reduced and trimmed VPA. A configuration (q, σ) is *reachable from a configuration* (q', σ') if there exists a word u in Σ^* such that $(q', \sigma') \xrightarrow{u} (q, \sigma)$.

We say that a configuration (q, σ) is *reachable* (resp. *co-reachable*) if there exists an initial (resp. a final) configuration κ such that (q, σ) is reachable from κ (resp. κ is reachable from (q, σ)).

Definition 4. *Let A be a VPA. Let us consider the three following conditions :*

 (i) every reachable configuration is co-reachable.
 (ii) for every configuration (p, σ) and every reachable and final configuration κ' of A such that κ' is reachable from (p, σ), then (p, σ) is reachable.
 (iii) for every state q, there exists an accepting run going through a configuration (q, σ).

We say that the automaton A is reduced *(resp.* co-reduced*) if it fulfills conditions (i) and (iii) (resp. (ii) and (iii)),* trimmed *if it is both reduced and co-reduced, and* weakly reduced *if it fulfills condition (i).*

Observe that condition (ii) looks more complicated than the property stating that every co-reachable configuration is reachable. Indeed, unlike in finite state-automata, the presence of a stack requires one to focus on reachable final configurations, and not to consider arbitrary final configuration. However, we will see that for VPA accepting only well-nested words, this condition is equivalent to the simpler one.

Observe that condition (iii) simply corresponds to the removal of states that are useless, which can easily be done in polynomial time.[3]

2.1 Well-Nested VPA (wnVPA)

A VPA A is said to be *well-nested* if $L(A) \subseteq \Sigma_{\mathsf{wn}}^*$. The class of well-nested VPA is denoted by wnVPA. Well-nested VPA enjoy some good properties; we describe some of them here.

Remark 1. If A is well-nested then its final configurations (f, σ) (with f a final state) are reachable only if $\sigma = \bot$.

The property of being co-reduced can be rephrased when considering wnVPA

Proposition 1. *Let A be a* wnVPA *satisfying condition (iii), then A is* co-reduced *iff every configuration that can reach some configuration $(f, \bot)$ with $f \in F$ is reachable.*

Proof. Let $f \in F$. By condition (iii), there exists $\sigma \in \Gamma^*$ such that configuration (f, σ) appears on some accepting run ρ, and by Remark 1, $\sigma = \bot$. Thus the set of reachable final configurations of A is equal to $\{(f, \bot) \mid f \in F\}$.   □

Let $\Sigma = (\Sigma_c, \Sigma_r, \Sigma_\iota)$. The dual alphabet of Σ (denoted dual(Σ)) is the alphabet $(\Sigma_r, \Sigma_c, \Sigma_\iota)$. Roughly speaking call and return symbols are switched.

Let w be a word in Σ_{wn}^*. We define dual(w) as the mirror image of w. We naturally extend this notion to languages L such that $L \subseteq \Sigma_{\mathsf{wn}}^*$.

Definition 5. *Let $A = (Q, I, F, \Gamma, \delta)$ be a* wnVPA *over some alphabet Σ. Its dual VPA denoted* dual$(A) = (Q', I', F', \Gamma', \delta')$ *over the alphabet* dual(Σ) *is given by $Q' = Q$, $I' = F$, $F' = I$, $\Gamma' = \Gamma$, and $\delta'_c = \{(q, r, \gamma, q') \mid (q', r, \gamma, q) \in \delta_r\}$, $\delta'_r = \{(q, c, \gamma, q') \mid (q', c, \gamma, q) \in \delta_c\}$, $\delta'_\iota = \{(q, i, q') \mid (q', i, q) \in \delta_\iota\}$, and $\delta'^\bot_r = \emptyset$.*

It is easy to prove that :

Proposition 2. *Let A be a* wnVPA*. Then*

- *For all $w \in \Sigma_{\mathsf{wn}}^*$, there exists a bijection between* $\mathsf{Run}_w(A)$ *and* $\mathsf{Run}_{\mathsf{dual}(w)}$ *(dual(A)) which induces a bijection between* $\mathsf{ARun}_w(A)$ *and* $\mathsf{ARun}_{\mathsf{dual}(w)}$(dual$(A)$).
- *A is reduced iff* dual(A) *is co-reduced.*
- *A is deterministic iff* dual(A) *is co-deterministic.*

---

[3] For any state q, one can build in polynomial time from A a word automaton over the alphabet Γ whose language is empty iff no configuration of the form (q, σ) is reachable.

3 Trimming Well-Nested VPA

Let A be a wnVPA on a structured alphabet Σ. We present the construction of the VPA $\text{trim}_{\text{wn}}(A)$, which recognizes the same language, and in addition is trimmed. First we define the reduced VPA $\text{reduce}(A)$ which is equivalent to A.

3.1 Construction of wreduce(A) and reduce(A)

Consider a wnVPA $A = (Q, I, F, \Gamma, \delta)$. We describe the construction of the VPA $\text{wreduce}(A)$, which is weakly reduced. The VPA $\text{reduce}(A)$ is then obtained by removing useless states of $\text{wreduce}(A)$, as explained in the previous section.

We first consider the set $\text{WN} = \{(p,q) \in Q \times Q \mid \exists (p, \perp) \to (q, \perp) \in \text{Run}(A)\}$. This set can be computed in quadratic time as the least one satisfying

- $\{(p,p) \mid p \in Q\} \subseteq \text{WN}$,
- if $(p, p') \in \text{WN}$ and $(p', p'') \in \text{WN}$, then $(p, p'') \in \text{WN}$
- if $(p, q) \in \text{WN}$, and $\exists (q, i, q') \in \delta_\iota$, then $(p, q') \in \text{WN}$
- if $(p, q) \in \text{WN}$ and $\exists (p', c, \gamma, p) \in \delta_c, (q, r, \gamma, q') \in \delta_r$, then $(p', q') \in \text{WN}$

Definition 6. *For any* wnVPA $A = (Q, I, F, \Gamma, \delta)$, *we define the* wnVPA $\text{wreduce}(A)$ *as* $(Q', I', F', \Gamma', \delta')$ *where* $Q' = \text{WN}$, $I' = \text{WN} \cap (I \times F)$, $F' = \{(f, f) \mid f \in F\}$, $\Gamma' = \Gamma \times Q$, *and* δ' *is defined by its restrictions on call, return* [4] *and internal symbols respectively (namely* δ'_c, δ'_r *and* δ'_ι):

- $\delta'_c = \{((p,q), c, (\gamma, q), (p', q')) \mid (p, q), (p', q') \in Q', (p, c, \gamma, p') \in \delta_c,$
  $\qquad\qquad \exists r \in \Sigma_r, \exists s \in Q, (q', r, \gamma, s) \in \delta_r \text{ and } (s, q) \in Q'\}$
- $\delta'_r = \{((q', q'), r, (\gamma, q), (p, q)) \mid (q', q'), (p, q) \in Q', (q', r, \gamma, p) \in \delta_r\}$
- $\delta'_\iota = \{((p, q), a, (p', q)) \mid (p, q), (p', q) \in Q', (p, a, p') \in \delta_\iota\}$

Intuitively, the states (and the stack) of $\text{wreduce}(A)$ extend those of A with an additional state of A. This extra component is used by $\text{wreduce}(A)$, when simulating a run of the VPA A, to store the state that the run should reach to

$$(p', \sigma.\gamma) \xrightarrow{w_1} (q', \sigma.\gamma)$$
$$c \nearrow \qquad\qquad\qquad \searrow r$$
$$(p, \sigma) \qquad\qquad\qquad (s, \sigma) \xrightarrow{w_2} (q, \sigma)$$

Fig. 1. Construction of call transitions

pop the symbol on top of the stack. To obtain a weakly reduced VPA, we require for the call transitions the existence of a matching return transition that allows one to reach the target state q. This condition is depicted on Figure 1, and we give an example of the construction in Figure 2.

3.2 Properties of wreduce(A) and reduce(A)

We consider the projection π from configurations of $A_{\text{wred}} = \text{wreduce}(A)$ to configurations of A obtained by considering the first component of states, as well as the first component of stack symbols. By definition of A_{wred}, each transition of A_{wred} is associated with a unique transition of A, we also denote by π this mapping. One can easily prove (see [7]), that π maps runs of A_{wred} onto runs of A.

The constructions wreduce and reduce have the following properties:

---

[4] As the language is well-nested, we do not consider return transitions on the empty stack.

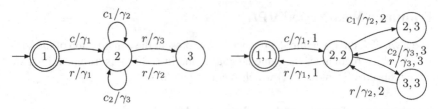

Fig. 2. On the left a VPA A, on the right reduce(A). There exists an initialized run of A over cc_1c_1 which cannot be completed into an accepting run. This run is no longer present in reduce(A).

Theorem 1. *Let A be a wnVPA, and let $A_{\mathsf{wred}} = \mathsf{wreduce}(A)$ and $A_{\mathsf{red}} = \mathsf{reduce}(A)$. A_{wred} and A_{red} can be built in polynomial time, and satisfy:*

(1) *there exist bijections between $\mathsf{ARun}(A_{\mathsf{wred}})$ and $\mathsf{ARun}(A)$, and $\mathsf{ARun}(A_{\mathsf{red}})$ and $\mathsf{ARun}(A)$ and thus in particular $L(A) = L(A_{\mathsf{wred}}) = L(A_{\mathsf{red}})$,*
(2) *A_{wred} is weakly reduced, and A_{red} is reduced,*
(3) *if A is co-reduced, then A_{red} is co-reduced,*
(4) *if A is co-deterministic, then A_{wred} and A_{red} are co-deterministic.*

Proof (Sketch). As explained above, the mapping π induces a mapping from runs of A_{wred} to runs of A. It is easy to verify that this mapping preserves the property of being accepting. In addition, one can prove by induction on the structure of the underlying word that it is both injective and surjective when restricted to the set $\mathsf{ARun}(A_{\mathsf{wred}})$. This yields a bijection between $\mathsf{ARun}(A_{\mathsf{wred}})$ and $\mathsf{ARun}(A)$. The bijection between $\mathsf{ARun}(A_{\mathsf{wred}})$ and $\mathsf{ARun}(A_{\mathsf{red}})$ is trivial as by definition A_{red} only differs from A_{wred} by states that do not appear in accepting runs.

The proof that A_{wred} is weakly reduced proceeds by induction on the size of the stack of the reachable configuration (p, σ) under consideration. It is then an immediate consequence that A_{red} is reduced.

By Proposition 1, to prove Property (3), we only have to prove that every configuration of A_{wred} co-reachable from a final configuration with an empty stack is reachable. This is done by induction on the size of the stack of this configuration. Last, the proof of Property (4) is done by an inspection of the transitions of A_{wred}. □

3.3 From Reduced to Trimmed

A construction for the co-reduction. Given a wnVPA, we can perform the following composition of constructions: coreduce = dual ∘ reduce ∘ dual. As a consequence of Proposition 2 and Theorem 1, the construction coreduce yields an equivalent wnVPA which is co-reduced.

Trimming. We define the construction $\mathsf{trim}_{\mathsf{wn}}$ as $\mathsf{trim}_{\mathsf{wn}} = \mathsf{coreduce} \circ \mathsf{reduce}$. Property (3) of Theorem 1 entails that the construction coreduce preserves the reduction, and we thus obtain the following result:

Theorem 2. *Let A be a wnVPA, and $A_{\mathsf{trim}} = \mathsf{trim}_{\mathsf{wn}}(A)$. A_{trim} is trimmed and can be built in polynomial time. Furthermore there exists a bijection between $\mathsf{ARun}(A)$ and $\mathsf{ARun}(A_{\mathsf{trim}})$, and thus in particular $L(A) = L(A_{\mathsf{trim}})$.*

Remark 2. The construction of co-reduction could also be presented explicitly. It would consist in adding an extra component into states and stack symbols, as done for the construction reduce, but representing the state reached when the top symbol of the stack was pushed. The same approach allows to present explicitly the construction trim$_{wn}$.

4   General Case

In order to trim a VPA A over some alphabet Σ, we will first build a wnVPA extend(A) over a new alphabet. In a second step, we trim the VPA extend(A) using the procedure described in the previous section for wnVPA. Last, we construct from the resulting wnVPA a VPA, which recognizes the language $L(A)$, and which is still trimmed. It is far from being trivial to propose procedures for transforming a VPA into wnVPA, and back, which are compatible with the notion of being trimmed. In addition, we will address the property of determinism in the next section, and our constructions should also be compatible with that issue.

4.1   Constructing Well-Nested Words from Arbitrary Words

Let $\Sigma = \Sigma_c \uplus \Sigma_r \uplus \Sigma_\iota$ be a structured alphabet. We introduce the structured alphabet $\Sigma^{\text{ext}} = \Sigma_c^{\text{ext}} \uplus \Sigma_r^{\text{ext}} \uplus \Sigma_\iota^{\text{ext}}$ defined by $\Sigma_c^{\text{ext}} = \Sigma_c$, $\Sigma_r^{\text{ext}} = \Sigma_r \uplus \{\bar{r}\}$, and $\Sigma_\iota^{\text{ext}} = \Sigma_\iota \uplus \{i_r \mid r \in \Sigma_r\}$, where $\bar{r}$ and $\{i_r \mid r \in \Sigma_r\}$ are fresh symbols.

We define inductively the mapping ext which transforms a word over Σ into a well-nested word over Σ^{ext} as follows, given $a \in \Sigma_\iota$, $r \in \Sigma_r$ and $c \in \Sigma_c$:

$$\text{ext}(aw) = a \cdot \text{ext}(w), \qquad \text{ext}(rw) = i_r \cdot \text{ext}(w),$$

$$\text{ext}(cw) = \begin{cases} cw_1 r \cdot \text{ext}(w_2) & \text{if } \exists w_1 \in \Sigma_{\text{wn}}^* \text{ such that } w = cw_1 rw_2, \\ c \cdot \text{ext}(w) & \text{otherwise.} \end{cases}$$

For example, ext($rccar$) = $i_r ccar\bar{r}$ with $c \in \Sigma_c$, $r \in \Sigma_r$ and $a \in \Sigma_\iota$. The mapping ext replaces every return r on empty stack by the internal symbol i_r, and adds a suffix of the form $\bar{r}^*$ in order to match every unmatched call. As a consequence, ext(w) is a well-nested word over the alphabet Σ^{ext}. We extend the function ext to languages in the obvious way.

4.2   Reduction to wnVPA

From VPA to wnVPA .... We present the construction extend which turns a VPA over Σ into a wnVPA over Σ^{ext}. Intuitively, when firing a call transition, the automaton non-deterministically guesses whether this call will be matched or not. Then, if a call is considered as not matching, the automaton completes it by using a transition over $\bar{r}$ at the end of the run. This is done by adding a suffix to the VPA which reads words from $\bar{r}^+$. Moreover the construction replaces the returns on empty stack by internals, this is done by memorizing in the state the fact that the current stack is empty or not.

We let T denote the set of symbols $\{\bot, \top, \circ\}$. Intuitively, $\bot$ means that the stack is empty, $\circ$ means that the stack contains only useless symbols (*i.e.* symbols associated with calls that will be unmatched), and $\top$ means that the stack contains some useless symbols (possibly none) plus symbols which will be popped. Furthermore we consider a fresh final state denoted by $\bar{f}$. This state is used to pop all the useless symbols.

Definition 7. *Let $A = (Q, I, F, \Gamma, \delta)$ be a VPA over an alphabet Σ, we define the VPA* $\text{extend}(A) = (Q', I', F', \Gamma', \delta')$ *over the alphabet Σ^{ext}, where $Q' = (Q \times T) \cup (\{\bar{f}\} \times \{\bot, \circ\})$, $I' = I \times \{\bot\}$, $F' = (F \cup \{\bar{f}\}) \times \{\bot\}$, $\Gamma' = \Gamma \times T$, and δ' is given by:*

$\delta'_c = \{((p,t), c, (\gamma, t), (q, x)) \mid (p, c, \gamma, q) \in \delta_c \text{ and either } x = \top \text{ or } (x = \circ \wedge t \neq \top)\}$

$\delta'_r = \{((p, \top), r, (\gamma, t), (q, t)) \mid (p, r, \gamma, q) \in \delta_r\} \cup$
$\qquad \{((p, \circ), \bar{r}, (\gamma, t), (\bar{f}, t)) \mid p \in F \cup \{\bar{f}\}, \gamma \in \Gamma, t \neq \top\}$

$\delta'_\iota = \{((p, t), a, (q, t)) \mid (p, a, q) \in \delta_\iota\} \cup \{((p, \bot), i_r, (q, \bot)) \mid (p, r, \bot, q) \in \delta_r\}$

Theorem 3. *Let A be a VPA. Then for all words $w \in \Sigma^*$, there exists a bijection between $\text{ARun}_w(A)$ and $\text{ARun}_{\text{ext}(w)}(\text{extend}(A))$.*

... and back. We present now the construction allowing to go from a wnVPA on Σ^{ext} to the "original" VPA on Σ. It is not always possible to find such a VPA, we thus introduce the property of being retractable.

Definition 8. *Let Σ be an alphabet and $A = (Q, I, F, \Gamma, \delta)$ be a VPA over the alphabet Σ^{ext}. We define two subsets of Q as follows:*
$\text{trap}(A) = \{q \in Q \mid \exists p, (p, \bar{r}, \gamma, q) \in \delta_r\}$
$\text{border}(A) = \{p \notin \text{trap}(A) \mid \exists t \in \delta \text{ such that } \text{source}(t) = p \text{ and } \text{target}(t) \in \text{trap}(A)\}$
Elements of $\text{border}(A)$ are called border states *of A.*

Definition 9. *Let Σ be an alphabet and $A = (Q, I, F, \Gamma, \delta)$ be a VPA over the alphabet Σ^{ext}. Then A is said to be* retractable *if:*

(i) *There exists a VPA B over Σ such that $L(A) = \text{ext}(L(B))$,*
(ii) *We have $\text{trap}(A) \cap I = \emptyset$,*
(iii) *For all transitions t in δ such that $\text{source}(t) \notin \text{trap}(A)$, if $\text{letter}(t) = \bar{r}$, then $\text{target}(t) \in \text{trap}(A)$, otherwise $\text{target}(t) \notin \text{trap}(A)$.*
(iv) *For all transitions t in δ such that $\text{source}(t) \in \text{trap}(A)$, then $\text{letter}(t) = \bar{r}$, and $\text{target}(t) \in \text{trap}(A)$.*
(v) *For each initialized run of A which ends in a border state there exists a unique run ρ' over $\bar{r}^+$ such that $\rho\rho'$ is an accepting run.*

Intuitively, a VPA which is retractable has two components: the first (before entering $\text{trap}(A)$) can read words over $(\Sigma^{\text{ext}} \setminus \{\bar{r}\})^*$ whereas the second reads words of the form $\bar{r}^+$. Note that the only way to go from a state not in $\text{trap}(A)$ to a state in $\text{trap}(A)$ is to use a transition which leaves a border state by $\bar{r}$. We give an example of these properties in Figure 3. We naturally have:

Lemma 1. *Let A be a VPA, then $\text{extend}(A)$ is retractable.*

We now define the converse of the function extend, named retract:

Definition 10. *Let $A = (Q, I, F, \Gamma, \delta)$ be a retractable VPA over the alphabet Σ^{ext}, we define the VPA $\text{retract}(A) = (Q', I', F', \Gamma, \delta')$ over the alphabet Σ by $Q' = Q \setminus \text{trap}(A)$, $I' = I$, $F' = (F \setminus \text{trap}(A)) \cup \text{border}(A)$, and the set of transition rules $\delta' = \delta'_c \uplus \delta'_r \uplus \delta'^{\bot}_r \uplus \delta'_\iota$ is defined by: $\delta'_c = \delta_c$, $\delta'_r = \{t \in \delta_r \mid \text{letter}(t) \neq \bar{r}\}$, $\delta'^{\bot}_r = \{(p, r, \bot, q) \mid (p, i_r, q) \in \delta_\iota\}$, and $\delta'_\iota = \{t \in \delta_\iota \mid \text{letter}(t) \in \Sigma_\iota\}$.*

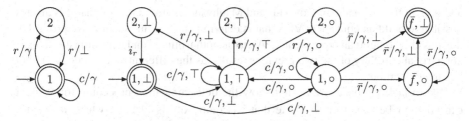

Fig. 3. At left, the VPA A with $L(A) = (crr)^*c^*$, at right the VPA $A_{ext} = \text{extend}(A)$ with $L(A_{ext}) = \{(cri_r)^* c^k \bar{r}^k \mid k \in \mathbb{N}\}$. $\text{border}(A_{ext}) = \{(1, \circ)\}$, $\text{trap}(A_{ext}) = \{(\bar{f}, \circ), (\bar{f}, \bot)\}$.

This construction is very simple, it replaces all the internal transitions over i_r by return transitions on empty stack over symbol r, and removes the return transitions over $\bar{r}$. Note that the final states of $\text{retract}(A)$ are the final states of A which are not in $\text{trap}(A)$ and the border states of A. We list below important properties of retract:

Theorem 4. *Let A be a retractable VPA on Σ^{ext}, we have:*

(i) *for any word $w \in \Sigma^*$, there exists a bijection between $\text{ARun}_w(\text{retract}(A))$ and $\text{ARun}_{ext(w)}(A)$, and thus in particular $L(A) = \text{ext}(L(\text{retract}(A)))$,*

(ii) *if A is trimmed, then so is $\text{retract}(A)$.*

In addition, the retractability is preserved by the trimming procedure described in the previous section for well-nested VPA:

Theorem 5. *Let A be a VPA, then the VPA $\text{trim}_{wn}(\text{extend}(A))$ is retractable.*

4.3 Trimming VPA

We consider the construction trim defined by $\text{trim}(A) = \text{retract} \circ \text{trim}_{wn} \circ \text{extend}(A)$, and state its main properties:

Theorem 6. *Let A be a VPA on the alphabet Σ, and let $A_{trim} = \text{trim}(A)$. The VPA A_{trim} can be built in polynomial time, and satisfies:*

(i) *there is a bijection between $\text{ARun}(A)$ and $\text{ARun}(A_{trim})$, and so $L(A) = L(A_{trim})$,*

(ii) *A_{trim} is trimmed.*

Proof. First, by Theorem 5, the VPA $\text{trim}_{wn} \circ \text{extend}(A)$ is retractable, and thus $\text{trim}(A)$ is well-defined. Then, the first property follows from the fact that such bijections exist for the constructions extend, trim_{wn} and retract. The second property is a consequence of Theorems 2 and 4.(ii). □

5 Deterministic Trimmed VPA

We have proven in the previous section that it is always possible, given a VPA, to build an equivalent VPA (*i.e.* that recognizes the same language) which is trimmed. In addition, in the original paper of Alur and Madhusudan, it was proven that it is always possible to build an equivalent VPA that is deterministic. In this section, we prove that it is possible to build an equivalent VPA that is both trimmed and deterministic. This is not a trivial corollary of the two previous results, as the different constructions can not be directly combined.

Due to lack of space, we do not show the determinization procedure for VPA, it can however be found in [2]. Note that its complexity is $O(2^{n^2})$, where n denotes the number of states of the input VPA. We denote by determinize the procedure obtained by applying the construction of [2], followed by the removal of useless states (according to property (iii) of Definition 4), which can be performed in polynomial time.

5.1   Determinization Preserves Reduction and Retractability

We start by proving that the construction determinize preserves the properties of being weakly reduced and of being retractable. In the sequel, we let A be a VPA and we let $A_{det} = $ determinize(A). The following results are proved in [2]:

Theorem 7. A_{det} *is a deterministic* VPA, *and* $L(A) = L(A_{det})$.

Lemma 2. *Let* $w \in \Sigma^*$. *If there exists an initialized run* ρ' *of* A_{det} *on* w *(not necessarily accepting), then there exists an initialized run* ρ *of* A *on* w.

As a corollary we prove in [7]:

Proposition 3. *If* A *is weakly reduced (resp. retractable), then* A_{det} *is weakly reduced (resp. retractable).*

5.2   Construction of a Deterministic Trimmed VPA

We consider the following composition of the different constructions presented before:

$$\text{det-trim} = \text{retract} \circ \text{coreduce} \circ \text{determinize} \circ \text{reduce} \circ \text{extend}$$

We claim that this composition allows one to build an equivalent VPA that is both deterministic and trimmed, as stated in the following theorem:

Theorem 8. *Let* A *be a* VPA. *The* VPA det-trim(A) *is deterministic, trimmed, and satisfies* $L(A) = L(\text{det-trim}(A))$.

Proof. We first let $A_1 = $ reduce $\circ$ extend(A). By Theorems 1 and 3, A_1 is weakly reduced, and recognizes the language ext$(L(A))$. In addition, we prove in [7] that A_1 is retractable. Consider now $A_2 = $ determinize(A_1). We have $L(A_2) = L(A_1)$ and, by Proposition 3 and as the construction determinize includes the removal of useless states, the VPA A_2 is deterministic, reduced and retractable. Consider now $A_3 = $ coreduce(A_2). By Theorem 1, properties (3) and (4) and by Proposition 2, the construction coreduce preserves the properties of being reduced, and of being deterministic. In addition, we prove in [7] that this construction also preserves the property of being retractable. We thus conclude that A_3 is retractable, trimmed, deterministic, and satisfies $L(A_3) = L(A_1) = \text{ext}(L(A))$. To conclude, it remains to observe that the construction retract preserves the determinism, and to use Theorem 4.   □

6 Conclusion

We introduced a series of constructions to trim a VPA. For each of these constructions, there exist projections of transitions of the obtained VPA onto those of the original one, which yield bijections between the accepting runs of the two VPA's. As a corollary, our constructions can be lifted to weighted VPA (VPA equipped with a labeling function of transitions, such as visibly pushdown transducers).

Our trimming procedure doesn't preserve the deterministic nature of the input VPA. We have however presented an alternative method to simultaneously trim and determinize a VPA, the complexity of this method being exponential. One can wonder whether a deterministic VPA can be trimmed with a polynomial time complexity, preserving its deterministic nature. The answer to this question is negative, and can be obtained using the family of languages $L_N = \{(c_1 + c_2)^k c_1 (c_1 + c_2)^N r^{N+k+1} \mid k \in \mathbb{N}\}$, with $N \in \mathbb{N}$, as a counter-example.

As future work, we plan to study the complexity of determining whether a VPA is trimmed. Another perspective is to implement our constructions in libraries for nested words such as [10].

Acknowledgments. This work has been supported by the project ECSPER (JC09_472677), funded by the ANR, and by the PEPS project SOSP, funded by the CNRS.

References

1. Alur, R., Kumar, V., Madhusudan, P., Viswanathan, M.: Congruences for Visibly Pushdown Languages. In: Caires, L., Italiano, G.F., Monteiro, L., Palamidessi, C., Yung, M. (eds.) ICALP 2005. LNCS, vol. 3580, pp. 1102–1114. Springer, Heidelberg (2005)
2. Alur, R., Madhusudan, P.: Visibly Pushdown Languages. In: STOC, pp. 202–211 (2004)
3. Alur, R., Madhusudan, P.: Adding Nesting Structure to Words. JACM 56(3), 1–43 (2009)
4. von Braunmühl, B., Verbeek, R.: Input-driven Languages are Recognized in log n Space. In: Karpinski, M. (ed.) FCT 1983. LNCS, vol. 158, pp. 40–51. Springer, Heidelberg (1983)
5. Buchsbaum, A.L., Giancarlo, R., Westbrook, J.: On the Determinization of Weighted Finite Automata. SIAM J. Comput. 30(5), 1502–1531 (2000)
6. Caralp, M., Reynier, P.-A., Talbot, J.-M.: Visibly Pushdown Automata with Multiplicities: Finiteness and K-Boundedness. In: Yen, H.-C., Ibarra, O.H. (eds.) DLT 2012. LNCS, vol. 7410, pp. 226–238. Springer, Heidelberg (2012)
7. Caralp, M., Reynier, P.-A., Talbot, J.-M.: A Polynomial Procedure for Trimming Visibly Pushdown Automata. Technical Report hal-00606778, HAL, CNRS, France (2013)
8. Choffrut, C.: Une Caractérisation des Fonctions Séquentielles et des Fonctions Sous-Séquentielles en tant que Relations Rationnelles. Theor. Comput. Sci. 5(3), 325–337 (1977)
9. De Souza, R.: Étude Structurelle des Transducteurs de Norme Bornée. PhD thesis, ENST, France (2008)

10. Driscoll, E., Thakur, A., Reps, T.: OpenNWA: A Nested-Word Automaton Library. In: Madhusudan, P., Seshia, S.A. (eds.) CAV 2012. LNCS, vol. 7358, pp. 665–671. Springer, Heidelberg (2012)
11. Filiot, E., Gauwin, O., Reynier, P.-A., Servais, F.: Streamability of Nested Word Transductions. In: FSTTCS. LIPIcs, vol. 13, pp. 312–324. Schloss Dagstuhl - Leibniz-Zentrum fuer Informatik (2011)
12. Girault-Beauquier, D.: Some Results About Finite and Infinite Behaviours of a Pushdown Automaton. In: Paredaens, J. (ed.) ICALP 1984. LNCS, vol. 172, pp. 187–195. Springer, Heidelberg (1984)
13. Mandel, A., Simon, I.: On Finite Semigroups of Matrices. Theor. Comput. Sci. 5(2), 101–111 (1977)

A Uniformization Theorem
for Nested Word to Word Transductions

Dmitry Chistikov and Rupak Majumdar

Max Planck Institute for Software Systems (MPI-SWS)
Kaiserslautern and Saarbrücken, Germany
{dch,rupak}@mpi-sws.org

Abstract. We study the class of relations implemented by nested word to word transducers (also known as visibly pushdown transducers). We show that any such relation can be uniformized by a functional relation from the same class, implemented by an unambiguous transducer. We give an exponential upper bound on the state complexity of the uniformization, improving a previous doubly exponential upper bound. Our construction generalizes a classical construction by Schützenberger for the disambiguation of nondeterministic finite-state automata, using determinization and summarization constructions on nested word automata. Besides theoretical interest, our procedure can be the basis for synthesis procedures for nested word to word transductions.

Keywords: uniformization, transduction, nested word, visibly pushdown language.

1 Introduction

A central result in the theory of rational languages is the *unambiguity theorem* [1,2], which states that every rational function can be implemented by an unambiguous transducer, that is, a transducer that has at most one successful run on any input. Schützenberger [3] gave an elegant and direct proof of the unambiguity theorem, by showing, for any rational function, a matrix representation which can be made unambiguous. Sakarovitch [2] subsequently showed that the construction of Schützenberger can be used as the foundation for several results in the theory of rational functions, most notably the rational uniformization theorem and the rational cross-section theorem (see [4,5,2]).

In more detail, the construction of Schützenberger starts with a (possibly nondeterministic) finite-state transducer $\mathcal{T}$, and constructs a new transducer with the property that any input string has at most one successful run. The construction performs a cross product of the subset automaton with the original automaton, and shows that certain edges in the cross product can be removed to obtain unambiguity, while preserving the language. As a simple consequence, the input–output relation that relates two words (u, w) if the transducer $\mathcal{T}$ can output w on input u, can be *uniformized*: for every u in the domain of $\mathcal{T}$, we can pick a unique w that is related to it (the rational *uniformization theorem*).

S. Konstantinidis (Ed.): CIAA 2013, LNCS 7982, pp. 97–108, 2013.

In this paper, we present an extension of the construction by Schützenberger to transducers from *nested words* to words. A nested word consists of a word over an alphabet, together with a *matching relation* that relates "call" positions in the word with corresponding "return" positions. Nested word automata [6] were introduced as a subclass of pushdown automata which are expressive enough for a large number of application domains but nevertheless retain many closure and decidability properties of regular languages. Nested word automata distinguish between "call," "return," and "internal" positions of the word, and, informally, push a symbol on the runtime stack on a call letter, pop a symbol on a return letter, and do not touch the stack on an internal letter.

These automata were extended to nested word *transducers* by Raskin and Servais [7] (under the name "visibly pushdown transducers"). The definition of nested word transducers in [7] allows transitions with ε-marking, but ensures that no transition can read and write symbols of different types (call, return, and internal symbols, in terms of visibly pushdown automata). We study a model of nested-word to word transducers considered in [8,9], in which the output word carries no structural information (i.e., call, return, or internal). That is, our transducers transform nested words into "normal" (linear) words.

Our main construction generalizes Schützenberger's construction to give an unambiguous automaton that is at most a single exponential larger than the input automaton. Our construction relies on the standard determinization construction for nested word automata from [6], as well as on a *summarization* of a nested word automaton, which captures the available properly-nested computation paths between two states. We show how to prune the product of these automata, in analogy with [3,2], to get an unambiguous automaton.

As a consequence of our construction, we obtain a uniformization theorem for nested-word to word transducers: any relation defined by such a transducer can be uniformized by a functional (i.e., single-valued) relation implemented by an unambiguous transducer. For functional nested-word to word transductions this yields an unambiguity theorem: every single-valued transduction can be implemented by a unambiguous transducer. The increase in the number of states is at most exponential, which improves a doubly exponential construction of [10] based on a notion of look-ahead. Note that for several other classes of algebraic (i.e., pushdown) transductions, uniformization theorems were obtained in [11].

In this extended abstract we focus on the case of nested words without unmatched calls, which we call *closed* words. Our construction for this special case captures main technical ideas and can be extended with an auxiliary transformation to fit the general case. Another variant of our construction, also not discussed here, applies to so-called weakly hierarchical nested word automata from [6,12], which are restricted to record only the current state on the stack.

Besides theoretical interest, our results provide an easily-implementable disambiguation construction. Since uniformization results are at the core of reactive synthesis techniques [13], our construction can form the basis of implementations of synthesis procedures for nested-word to word transductions.

The structure of the current paper is as follows. After introducing necessary definitions, we describe three basic constructions on nested word automata in Section 3. We show how to combine them to obtain a specific disambiguation of an arbitrary automaton (the *Schützenberger construction*) in Section 4. The proofs are delayed until Section 6, while in Section 5 we discuss how to use the construction to obtain uniformization and unambiguity theorems. Short notes sketching the extension to the general (non-closed-word) case can be found in relevant parts of Sections 4 and 5.

2 Nested Words and Transducers

A *nested word* of length k over an alphabet Σ is a pair $u = (x, \nu)$, where $x \in \Sigma^k$ and ν is a *matching relation* of length k, that is, a subset $\nu \subseteq \{-\infty, 1, \ldots, k\} \times \{1, \ldots, k, +\infty\}$ such that, first, if $\nu(i, j)$ holds, then $i < j$; second, for $1 \leq i \leq k$ each of the sets $\{j \mid \nu(i, j)\}$ and $\{j \mid \nu(j, i)\}$ contains at most one element; third, whenever $\nu(i, j)$ and $\nu(i', j')$, it cannot be the case that $i < i' \leq j < j'$. We assume that $\nu(-\infty, +\infty)$ never holds.

If $\nu(i, j)$, then the position i in the word u is said to be a *call*, and the position j a *return*. All positions from $\{1, \ldots, k\}$ that are neither calls nor returns are *internal*. A call (a return) is *matched* if ν matches it to an element of $\{1, \ldots, k\}$ and *unmatched* otherwise.

We shall call a nested word *closed* if it has no unmatched calls, and *well-matched* if it has no unmatched calls and no unmatched returns. We denote the set of all nested words over Σ by Σ^{*n}, the set of all closed nested words by Σ^{*c}, and the set of all well-matched words by Σ^{*w}. Observe that $\Sigma^{*w} \subsetneq \Sigma^{*c} \subsetneq \Sigma^{*n}$.

The family of all non-empty (word) languages over the alphabet Δ is denoted by $\mathcal{L}(\Delta^*)$.

Define a *(nested-word to word) transducer* over the input alphabet Σ and output alphabet Δ as a structure $\mathcal{T} = (Q, P, \delta, Q^i, Q^f, P^i)$, where:

- Q is a finite non-empty set of (linear) states,
- P is a finite set of hierarchical states,
- $\delta = (\delta^{\mathsf{call}}, \delta^{\mathsf{int}}, \delta^{\mathsf{ret}})$, where
  - $\delta^{\mathsf{int}} \subseteq Q \times \Sigma \times Q \times \mathcal{L}(\Delta^*)$ is a set of internal transitions,
  - $\delta^{\mathsf{call}} \subseteq Q \times \Sigma \times Q \times P \times \mathcal{L}(\Delta^*)$ is a set of call transitions,
  - $\delta^{\mathsf{ret}} \subseteq P \times Q \times \Sigma \times Q \times \mathcal{L}(\Delta^*)$ is a set of return transitions,
- $Q^i \subseteq Q$ and $Q^f \subseteq Q$ are sets of initial and final linear states, and
- $P^i \subseteq P$ is a set of initial hierarchical states.

A *path* through a transducer $\mathcal{T}$ driven by an input word $u = (a_1 \ldots a_k, \nu) \in \Sigma^{*n}$ is a sequence of alternating linear states and transitions of $\mathcal{T}$, where ith transition leads from the current linear state to the next one, carries the letter $a_i \in \Sigma$, and has type chosen according to the matching relation ν; furthermore, for every pair of matched call and return, hierarchical states encountered in the corresponding transitions are required to be the same (we say that they are *sent* and *received* along the hierarchical edges).

The path is *successful* if it starts in a state from Q^i, ends in a state from Q^f, and all states received along the hierarchical edges in unmatched returns belong to P^i. Note that, following [6], we impose no requirement on unmatched calls (our automata are called *linearly accepting*). A (*successful*) *computation* consists of a (successful) path and a sequence of words w_i taken from languages $\ell_i \in \mathcal{L}(\Delta^*)$ in the transitions of the path in the same order. The concatenation of these words gives a word $w \in \Delta^*$, which is said to be *output* by the transducer.

We say that the transducer $\mathcal{T}$ *implements* the transduction $T \subseteq \Sigma^{*n} \times \Delta^*$ that contains each pair (u, w) if and only if there exists a successful computation of $\mathcal{T}$ driven by the input u and having the output w. When we are only interested in the behaviour of $\mathcal{T}$ on closed words, we say that $\mathcal{T}$ *weakly implements* $T^c = T \cap (\Sigma^{*c} \times \Delta^*)$.

3 Automata and Auxiliary Constructions

A *nested word automaton* (NWA, or simply an *automaton*) $\mathcal{A}$ is defined similarly to a transducer, with the only difference that it has no output, that is, all $\mathcal{L}(\Delta^*)$ factors are dropped. Words $u \in \Sigma^{*n}$ carried by successful paths are said to be *accepted* by $\mathcal{A}$, and the automaton itself is then said to *recognize* the language $L \subseteq \Sigma^{*n}$ of all such words. Two automata are *equivalent* if they recognize the same language.

We call $\mathcal{A}$ (*weakly*) *unambiguous* if it has at most one successful path for each (closed) word $u \in \Sigma^{*n}$ ($u \in \Sigma^{*c}$). As usual, $\mathcal{A}$ is *deterministic* if, first, each of the sets Q^i and P^i contains at most one element and, second, for every $q \in Q$, $a \in \Sigma$, and $p \in P$, each of the three sets δ^{int}, δ^{call}, and δ^{ret} contains at most one tuple of the form (q, a, q', p'), (q, a, q', p), and (p, q, a, q'), respectively. Every deterministic automaton is unambiguous, but not vice versa. Also recall that a linear state q of an automaton $\mathcal{A}$ is called *accessible* if there exists a path in $\mathcal{A}$ starting in some linear state $q_0 \in Q^i$ and ending in q, in which all states received along the hierarchical edges in unmatched returns belong to P^i.

We now define three auxiliary constructions for nested word automata. Their combination, as shown in Sections 4 and 5, can be used to obtain uniformization and unambiguity theorems for nested-word to word transductions.

Determinization. Every nested word automaton is known to be equivalent to a deterministic one, by a variant of the well-known subset construction [6],[1] which can be traced to a 1985 paper [14]. Given an automaton $\mathcal{A} = (Q, P, \delta, Q^i, Q^f, P^i)$, construct the automaton with the following components:

- the set of linear states is $2^{Q \times Q}$ (in this context pairs $(q, q') \in Q \times Q$ are called *summaries* and understood as available properly-nested path fragments between pairs of states);
- the set of hierarchical states is $\{p'_0\} \cup (2^{Q \times Q} \times \Sigma)$, where the state p'_0 is new;

---

[1] Note that the construction described in the print version of [6] is flawed, so we refer the reader to the electronic document available on the Web.

- for every input letter a and every state $S \subseteq Q \times Q$, transitions lead from S to states S' defined as follows:
  - for an internal transition, S' contains all summaries (q, q'') such that there exists a summary $(q, q') \in S$ and an internal transition $(q', a, q'') \in \delta^{\text{int}}$;
  - for a call transition, S' contains all summaries (q'', q''), for which there exists a summary $(q, q') \in S$ and a call transition $(q', a, q'', p) \in \delta^{\text{call}}$; along the hierarchical edge the pair (S, a) is sent;
  - for a return transition upon the receipt of a hierarchical state $H = (S^0, b)$, where $S^0 \subseteq Q \times Q$, the state S' contains all summaries (q_0, q'') such that there exist summaries $(q_0, q_1) \in S^0$ and $(q, q') \in S$, a call transition $(q_1, b, q, p) \in \delta^{\text{call}}$ and a return transition $(p, q', a, q'') \in \delta^{\text{ret}}$ with a matching $p \in P$;
  - for a return transition upon the receipt of a hierarchical state p_0', the state S' contains all summaries (q, q'') such that there exists a summary $(q, q') \in S$ and a return transition $(p_0, q', a, q'') \in \delta^{\text{ret}}$ with some $p_0 \in P^{\text{i}}$;
- the only initial linear state is $\{(q_0, q_0) \mid q_0 \in Q^{\text{i}}\}$ (here we deviate from [6], where the set $Q^{\text{i}} \times Q^{\text{i}}$ is used), and an arbitrary linear state S is final whenever it contains some pair (q, q') with $q' \in Q^{\text{f}}$;
- the only initial hierarchical state is p_0'.

The accessible part of this automaton is called the *determinization* of $\mathcal{A}$ and denoted $\mathcal{A}_{\text{det}}$. It is deterministic and can be proved equivalent to $\mathcal{A}$.

Summarization. We also introduce another auxiliary automaton, closely related to that of $\mathcal{A}_{\text{det}}$. This automaton will keep track of summaries instead of single states, similarly to $\mathcal{A}_{\text{det}}$, but still rely on nondeterminism rather than subset construction to mimic the behaviour of $\mathcal{A}$. In short, for any transition in $\mathcal{A}_{\text{det}}$, if it leads from a state containing a summary (q, q') to some state S', this automaton will have transitions from (q, q') to all summaries that can use (q, q') as a witness for their inclusion in S'. More formally, this summary-tracking automaton has the following components:

- the set of linear states is $Q \times Q$;
- the set of hierarchical states is $(\{p_0'\} \times \delta^{\text{ret}}) \cup ((Q \times Q) \times \delta^{\text{call}})$, where the state p_0' is new;
- for every input letter a and every state $(q, q') \in Q \times Q$, transitions from (q, q') are defined as follows:
  - every internal transition $(q', a, q'') \in \delta^{\text{int}}$ of $\mathcal{A}$ is translated here into an internal transition leading to (q, q'');
  - every call transition $t = (q', a, q'', p) \in \delta^{\text{call}}$ is translated into a call transition leading to (q'', q''), with $((q, q'), t)$ sent along the hierarchical edge;
  - every pair of call and return transitions $t = (q_1, b, q, p) \in \delta^{\text{call}}$ and $(p, q', a, q'') \in \delta^{\text{ret}}$ with matching $p \in P$ is translated, for all $q_0 \in Q$, into return transitions to (q_0, q''), depending on the state $((q_0, q_1), t)$ received along the hierarchical edge;

- every return transition $t = (p_0, q', a, q'') \in \delta^{\text{ret}}$ with $p_0 \in P^i$ is also translated into a return transition to (q, q'') with (p'_0, t) received along the hierarchical edge;
- initial linear states are (q_0, q_0), for all $q_0 \in Q^i$, and the set of final linear states is $Q \times Q^f$;
- the set of initial hierarchical states is $\{p'_0\} \times \delta^{\text{ret}}$.

The accessible part of this automaton is called the *summarization* of $\mathcal{A}$ and denoted $\mathcal{A}_{\text{sum}}$. This automaton is also easily shown to be equivalent to $\mathcal{A}$.

Note that we could also define $\mathcal{A}_{\text{sum}}$ in a more natural way, with the set of hierarchical states $\{p'_0\} \cup ((Q \times Q) \times \Sigma)$, similarly to $\mathcal{A}_{\text{det}}$. However, we need to tie transitions of $\mathcal{A}_{\text{sum}}$ to individual transitions of $\mathcal{A}$, as seen in the proof of Lemma 7 in Section 6 below.

Product. Let us define the product of two nested word automata $\mathcal{A}_1 = (Q_1, P_1, \delta_1, Q_1^i, Q_1^f, P_1^i)$ and $\mathcal{A}_2 = (Q_2, P_2, \delta_2, Q_2^i, Q_2^f, P_2^i)$ in the standard way:

- sets of linear and hierarchical states are $Q_1 \times Q_2$ and $P_1 \times P_2$, respectively;
- whenever δ_1 and δ_2 contain transitions from q_1 to q'_1 and from q_2 to q'_2, respectively, having the same type and carrying the same letter $a \in \Sigma$, this results in a transition from (q_1, q_2) to (q'_1, q'_2) of the same type carrying $a \in \Sigma$; the state sent or received along the hierarchical edge (if any) is the ordered pair of the hierarchical states carried by these transitions;
- sets of initial and final linear states are $Q_1^i \times Q_2^i$ and $Q_1^f \times Q_2^f$;
- the set of initial hierarchical states is $P_1^i \times P_2^i$.

The accessible part of this automaton is called the *product* of $\mathcal{A}_1$ and $\mathcal{A}_2$ and denoted $\mathcal{A}_1 \times \mathcal{A}_2$.

4   The Schützenberger Construction for NWA

We now describe a special disambiguation construction for nested word automata, generalizing a construction due to Schützenberger [3,2].

First suppose that $\mathcal{B}$ and $\mathcal{A}$ are nested word automata, and there exists a mapping μ that takes states and transitions of $\mathcal{B}$ to states and transitions of $\mathcal{A}$, preserving letters and transition types, in such a way that the image of a successful path is always a successful path. Then we say that the automaton $\mathcal{B}$ is *morphed* into $\mathcal{A}$ by μ (this word usage roughly corresponds to that in [2,5]). We shall sometimes decorate symbols denoting attributes of $\mathcal{B}$ with the prime ', to distinguish these attributes from those of $\mathcal{A}$.

We now define a special operation eliminating some transitions of an automaton. Suppose that $\mathcal{B}$ is morphed into $\mathcal{A}$ by μ, and δ' is some subset of $\mathcal{B}$'s transitions. For every state s' of $\mathcal{B}$, consider its image $\mu(s')$ in $\mathcal{A}$ and take an arbitrary transition t arriving at $\mu(s')$. Denote the set of all transitions arriving at s' by $\text{IN}(s')$. If the set $\mu^{-1}(t) \cap \text{IN}(s') \cap \delta'$ has cardinality 2 or greater, i.e., if the inverse image $\mu^{-1}(t)$ contains more than one transition from δ' arriving at s', then transform the automaton $\mathcal{B}$ by eliminating all of these transitions but one, arbitrarily chosen. Repeat this procedure for all states s' of $\mathcal{B}$ and all transitions t arriving at their images $\mu(s')$. We shall say that the resulting automaton is obtained from $\mathcal{B}$ by *weeding* transitions δ' with respect to the mapping μ.

Now everything is ready for our generalization of the Schützenberger construction to nested word automata. Suppose we are given an automaton $\mathcal{A}$. First construct the automaton $\mathcal{A}_{\mathsf{sum}} \times \mathcal{A}_{\mathsf{det}}$, where $\mathcal{A}_{\mathsf{sum}}$ and $\mathcal{A}_{\mathsf{det}}$ are the summarization and determinization of $\mathcal{A}$. Let the mapping π take linear states and transitions of $\mathcal{A}_{\mathsf{sum}} \times \mathcal{A}_{\mathsf{det}}$ to their projections in $\mathcal{A}_{\mathsf{det}}$. Weed the sets of internal and return transitions of $\mathcal{A}_{\mathsf{sum}} \times \mathcal{A}_{\mathsf{det}}$ with respect to π (note that all call transitions are left intact), and then for every final state S of $\mathcal{A}_{\mathsf{det}}$, make all but one state in $\pi^{-1}(S)$ non-final (this remaining final state should have the form $((q, q'), S)$, where (q, q') is final in $\mathcal{A}_{\mathsf{sum}}$). Denote the obtained automaton by $\mathcal{S}$.

Theorem 1. *The automaton $\mathcal{S}$ is weakly unambiguous, equivalent to and morphed into $\mathcal{A}$.*

The proof of this theorem is given in Section 6. Before that, in Section 5, we show that $\mathcal{S}$ can be used to obtain a uniformization of any relation weakly implemented by a transducer whose underlying automaton is $\mathcal{A}$. (Recall that this relation is defined as a subset of $\Sigma^{*c} \times \Delta^*$.)

Note that if we cannot disregard non-closed words, then an auxiliary step is needed. Roughly speaking, if in $\mathcal{S}$ we additionally separate matched calls from unmatched calls and weed the latter in the same fashion as earlier, then the obtained automaton will be unambiguous, equivalent to and morphed into $\mathcal{A}$. We leave a detailed discussion of this construction until the full version of the paper, and only note that this transformation involves at most a constant-factor increase in the number of states and transitions.

Proof Idea. The "morphism" part is easy, and the main challenge is the weak unambiguity and equivalence to $\mathcal{A}$. We borrow the overall strategy from [2,5], but prefer not to hide the needed properties behind the framework of covering of automata, and instead make the main steps explicit to achieve more clarity.

Consider the mapping π defined above. For every final state S of $\mathcal{A}_{\mathsf{det}}$, exactly one of the states in its inverse image $\pi^{-1}(S)$ is final. Now for every successful path in $\mathcal{A}_{\mathsf{det}}$ terminating in S we construct its inverse image in $\mathcal{S}$. If we show that at all states every arriving transition in $\mathcal{A}_{\mathsf{det}}$ has a non-empty inverse image in $\mathcal{A}_{\mathsf{sum}} \times \mathcal{A}_{\mathsf{det}}$ (this is almost *in-surjectivity* in terms of [2,5]), then our definition of weeding will ensure that at each state exactly one option remains in $\mathcal{S}$ (*in-bijectivity*). As a result, every successful path in $\mathcal{A}_{\mathsf{det}}$ has exactly one counterpart in $\mathcal{S}$. Since the automaton $\mathcal{A}_{\mathsf{sum}} \times \mathcal{A}_{\mathsf{det}}$ only accepts words accepted by $\mathcal{A}$, the equivalence follows; and the unambiguity then follows from that of $\mathcal{A}_{\mathsf{det}}$.

There are, however, two issues with this strategy. Note that the concatenation of correct path segments is not necessarily a correct path segment, unlike in the standard finite-state machines. To this end, we choose not to weed call transitions, in order to ensure that the path can be correctly prolonged on each step, with matching hierarchical states sent and received along hierarchical edges. This explains why our construction of $\mathcal{S}$ guarantees only weak unambiguity.

At this point we also brush upon the more subtle second issue, which is the reason for our having defined and taken $\mathcal{A}_{\mathsf{sum}}$ in the product, as opposed to just using $\mathcal{A}$ as in the original construction. Imagine we did otherwise, and consider some return transition from (q', S') to (q'', S'') in $\mathcal{S}$ encountered while constructing the inverse image of some path in $\mathcal{A}_{\mathsf{det}}$. Note that the specific transition from q' to q'' is in fact chosen at the weeding stage and fixes some specific state $p \in P$ received along the hierarchical edge.

Now if S' contains two summaries (q_1, q') and (q_2, q'), we cannot know which of q_1 and q_2 we will hit after extending the path backwards until the matching call. But then it may well be the case that none of the call transitions arriving in this state q_i under the appropriate input letter carries the hierarchical state p. This means that the weeding of return transitions would change the recognized language, which is undesirable. Introducing the automaton $\mathcal{A}_{\mathsf{sum}}$, which keeps track of summaries, resolves this issue, as seen from Lemmas 5 and 6 in Section 6.

5   Uniformization of Transductions

Return to the problem of uniformizing an arbitrary nested-word to word transduction T. Recall that a relation $U \subseteq A \times B$ is a *uniformization* of a relation $T \subseteq A \times B$ if U is a subset of T, single-valued as a transduction, and has the same domain, that is, if $U \subseteq T$ and for every $u \in A$, the existence of a $w \in B$ such that $(u, w) \in T$ implies that there is exactly one $w' \in B$ such that $(u, w') \in U$.

In this section, we show how to use the Schützenberger construction to obtain a uniformization of an arbitrary nested-word to word transduction T. Take any transducer $\mathcal{T}$ implementing T, and denote by $\mathcal{A}$ its underlying automaton, that is, one obtained by removing the output labels from transitions. Transform $\mathcal{A}$ into $\mathcal{S}$ as described in Section 4. By the "morphism" part of Theorem 1, $\mathcal{S}$ is morphed into $\mathcal{A}$ by some mapping ρ, so that each transition t' in $\mathcal{S}$ is projected onto a single transition $\rho(t')$ in $\mathcal{A}$. Take the output label $\ell \subseteq \Delta^*$ of $\rho(t')$ in $\mathcal{T}$ and specify any single word $w \in \ell$ as the output of t'. The automaton $\mathcal{S}$ is thus transformed into a transducer $\mathcal{U}$, which is claimed to satisfy our needs.

Theorem 2. *Let the transducer $\mathcal{T}$ weakly implement a transduction T. Then the transducer $\mathcal{U}$ weakly implements a transduction U, which is a uniformization of T.*

Proof. We use the "equivalence" and "weak unambiguity" parts of Theorem 1. First observe that any nested word u belongs to the domain of T (or, respectively, U) if and only if it is is accepted by $\mathcal{A}$ (or, respectively, $\mathcal{S}$). It then follows from the "equivalence" part that T and U have the same domain within Σ^{*n}. Second, if (u, w') and (u, w'') with $w' \neq w''$ are in U, then $\mathcal{U}$ has at least two successful paths driven by u. This is impossible for any closed word u by the "weak unambiguity" part of Theorem 1.   □

We say that a relation $U \subseteq \Sigma^{*c} \times \Delta^*$ is a *weak uniformization* of a transduction $T \subseteq \Sigma^{*n} \times \Delta^*$ if U is a uniformization of the transduction $T^c = T \cap (\Sigma^{*c} \times \Delta^*)$.

Corollary. *Any nested-word to word transduction T implemented by a transducer with n linear states has a weak uniformization implemented by a transducer with at most $n^2 2^{n^2-1}$ linear and $n\, 2^{n^2-1}\, |\delta^{\mathsf{call}}| + |\delta^{\mathsf{ret}}|$ hierarchical states.*

We conclude this section with a series of remarks. First, the statement of Theorem 1 can actually be augmented with the following observation. The transduction U implemented by $\mathcal{U}$ not only has the property that U^c is a uniformization of T^c, but also satisfies $U \subseteq T$ and has the same domain as T. In other words, for all non-closed words u, the existence of a $w \in \Delta^*$ such that $(u, w) \in T$ implies that there is at least one $w' \in \Delta^*$ such that $(u, w') \in U$, and all such pairs (u, w') from U also belong to T.

Second, an additional refinement of the construction of $\mathcal{S}$, as sketched in Section 4, can be used to obtain a stronger version of Theorem 2, giving a uniformization instead of a weak uniformization. The proof of the stronger version repeats the proof above almost literally.

Finally, note that for a single-valued transduction T, the statement above gives an unambiguity theorem: the construction defines a transducer implementing T, whose underlying automaton is (weakly) unambiguous.

6   Proof of Theorem 1

In this section we demonstrate that the automaton $\mathcal{S}$ constructed in Section 4 has the desired properties: weak unambiguity, equivalence to the original automaton $\mathcal{A}$, and the property of being morphed into $\mathcal{A}$. We first prove an important property of accessible states in the automaton $\mathcal{A}_{\mathsf{sum}} \times \mathcal{A}_{\mathsf{det}}$.

Lemma 1. *If a state $((q, q'), S)$ is accessible in $\mathcal{A}_{\mathsf{sum}} \times \mathcal{A}_{\mathsf{det}}$, then (q, q') belongs to S and S is accessible in $\mathcal{A}_{\mathsf{det}}$.*

Proof. It is clear that no state of the form $((q, q'), S)$ can be accessible in $\mathcal{A}_{\mathsf{sum}} \times \mathcal{A}_{\mathsf{det}}$ unless the state S is accessible in $\mathcal{A}_{\mathsf{det}}$. The fact that $(q, q') \in S$ is then proved by induction on the length of the path leading from the initial state of $\mathcal{A}_{\mathsf{det}}$ to S. Indeed, $\{(q_0, q_0) \mid q_0 \in Q^i\}$ is the only initial state of $\mathcal{A}_{\mathsf{det}}$ and the states (q_0, q_0), $q_0 \in Q^i$, are initial in $\mathcal{A}_{\mathsf{sum}}$. This forms the induction base. Further, suppose that $((q_1, q_2), S)$ is accessible and a transition leads from this state to some state $((q, q'), S')$. Then $(q_1, q_2) \in S$ and, by the definition of $\mathcal{A}_{\mathsf{sum}} \times \mathcal{A}_{\mathsf{det}}$, there are transitions from S to S' and from (q_1, q_2) to (q, q'). It remains to use the fact that whenever a transition leads in $\mathcal{A}_{\mathsf{sum}}$ from (q_1, q_2) to (q, q'), any corresponding transition in $\mathcal{A}_{\mathsf{det}}$ from a state containing (q_1, q_2) can only lead to a state containing (q, q').   □

Lemma 1 shows that every accessible state of $\mathcal{A}_{\mathsf{sum}} \times \mathcal{A}_{\mathsf{det}}$ can be regarded as a set of summaries S with a distinguished element $(q, q') \in S$. From now on we shall only use the states of $\mathcal{A}_{\mathsf{sum}} \times \mathcal{A}_{\mathsf{det}}$ satisfying the conclusion of this lemma. For our second lemma, recall that by π we denote the mapping taking every state of $\mathcal{A}_{\mathsf{sum}} \times \mathcal{A}_{\mathsf{det}}$ to its projection in $\mathcal{A}_{\mathsf{det}}$.

Lemma 2 (in-surjectivity). *For every accessible state $((q, q'), S)$ of $\mathcal{A}_{\mathsf{sum}} \times \mathcal{A}_{\mathsf{det}}$ and every transition t arriving at its projection S in $\mathcal{A}_{\mathsf{det}}$, there exists at least one transition in $\mathcal{A}_{\mathsf{sum}} \times \mathcal{A}_{\mathsf{det}}$ whose projection is t.*

Proof. First apply Lemma 1 to note that $(q, q') \in S$. Let the transition t arriving at S depart from a state S^0. We claim that there exists a summary $(q_1, q_2) \in S^0$ and a transition from (q_1, q_2) to (q, q') in $\mathcal{A}_{\mathsf{sum}}$ such that the inverse image $\pi^{-1}(t)$ in $\mathcal{A}_{\mathsf{sum}} \times \mathcal{A}_{\mathsf{det}}$ contains a transition from $((q_1, q_2), S^0)$ to $((q, q'), S)$.

Indeed, since $(q, q') \in S$, it follows that there exists a summary $(q_1, q_2) \in S^0$ that guaranteed the inclusion of (q, q') into S by the definition of $\mathcal{A}_{\mathsf{det}}$. The definition of $\mathcal{A}_{\mathsf{sum}}$ then ensures the existence of a transition of the same type from this (q_1, q_2) to (q, q'). The rest follows by the definition of $\mathcal{A}_{\mathsf{sum}} \times \mathcal{A}_{\mathsf{det}}$. □

Lemma 3 (partial in-bijectivity). *For every state $((q, q'), S)$ of $\mathcal{S}$ and every internal or return transition t arriving at its projection S in $\mathcal{A}_{\mathsf{det}}$, there exists a unique transition in $\mathcal{S}$ whose projection is t.*

Proof. Follows from Lemma 2 and the definitions of $\mathcal{S}$ and weeding. □

Lemma 4. *The automaton $\mathcal{S}$ is weakly unambiguous.*

Proof. Consider any closed word $u \in \Sigma^{*c}$ accepted by $\mathcal{S}$ and assume for the sake of contradiction that there are two different successful paths driven by u. Observe that the projection by means of π of a successful path in $\mathcal{S}$ is a successful path in $\mathcal{A}_{\mathsf{det}}$. Since the automaton $\mathcal{A}_{\mathsf{det}}$ is deterministic and, therefore, unambiguous, it follows that both paths in $\mathcal{S}$ are projected onto a single path in $\mathcal{A}_{\mathsf{det}}$.

Consider the last position in this pair of paths, on which these paths disagree, that is, $((q_1, q_1'), S) \neq ((q_2, q_2'), S)$ (note that the projection onto $\mathcal{A}_{\mathsf{det}}$ should be the same, hence the common component S). The successors of these two states are the same in both paths, and the following segments up to the end of the paths also coincide (both paths end in the same final state of $\mathcal{S}$, since for each S at most one state from $\pi^{-1}(S)$ is final). Suppose that our chosen pair of paths leads from this pair of states to some state $((q, q'), S')$.

Now consider the transition t leading from S to S' in the (single) induced path in $\mathcal{A}_{\mathsf{det}}$. This transition has at least two elements in its inverse image $\pi^{-1}(t)$, one of them departing from $((q_1, q_1'), S)$ and another from $((q_2, q_2'), S)$. By the partial in-bijectivity given by Lemma 3, this transition t can only be a (matched) call transition. We shall show, however, that this conclusion leads to a contradiction.

Indeed, consider the tails of the paths, which coincide by our initial assumptions. It follows that these tails take the same transition at the return position matching the call in question (recall that our word is closed, so all calls are matched). This transition receives some state $((q_0, q_0'), t_0, S^0, a)$ along the hierarchical edge, and so the call in question can only originate at the state $((q_0, q_0'), S^0)$. This contradicts the availability of the choice between $((q_1, q_1'), S)$ and $((q_2, q_2'), S)$ and establishes that one of the paths must be invalid. □

Our next goal is to show that every successful path in $\mathcal{A}_{\mathsf{det}}$ has an inverse image in $\mathcal{S}$ that is also a successful path. To prove this fact, we need an auxiliary lemma that describes the behaviour of $\mathcal{A}_{\mathsf{sum}}$ on well-matched words.

Lemma 5. *Suppose that the automaton $\mathcal{A}_{\mathsf{sum}}$ can be driven by some well-matched word $u \in \Sigma^{*w}$ from a state (q_0, q_1) to a state (q_2, q_3). Then $q_0 = q_2$.*

Proof. Any well-matched word is a concatenation of internal transitions and well-matched fragments enclosed in matched pairs of symbols. By the definition of $\mathcal{A}_{\mathsf{sum}}$, internal transitions do not change the first component of a summary, and every matched return resets it to the state just before the matching call. □

Lemma 6. *The automaton $\mathcal{S}$ is equivalent to $\mathcal{A}$.*

Proof. Since the projection of a successful path through $\mathcal{S}$ is a successful path through $\mathcal{A}_{\mathsf{det}}$, it is sufficient to demonstrate that $\mathcal{S}$ always accepts whenever $\mathcal{A}_{\mathsf{det}}$ accepts. Consider a successful path through $\mathcal{A}_{\mathsf{det}}$ and construct a path through $\mathcal{S}$ as follows. First consider the destination of the path in $\mathcal{A}_{\mathsf{det}}$ and take the (unique) final state in its inverse image. Second, reconstruct a path through $\mathcal{S}$ by the following procedure. On each step, given a state of $\mathcal{S}$ and a transition t arriving at its projection in $\mathcal{A}_{\mathsf{det}}$, choose a transition from the inverse image $\pi^{-1}(t)$ to obtain the previous state in the path through $\mathcal{S}$. Lemmas 2 and 3 reveal that this procedure will indeed yield a path-like sequence in $\mathcal{S}$. Since all summaries from the initial state of $\mathcal{A}_{\mathsf{det}}$ are initial states of $\mathcal{A}_{\mathsf{sum}}$, this sequence will begin in an initial state of $\mathcal{S}$. However, we also need to show that this sequence will indeed be a correct path through $\mathcal{S}$, that is, states sent and received along the hierarchical edges will match.

Consider a fragment of the input word of the form $\langle bua \rangle$, where $u \in \Sigma^{*w}$ (here and below angle brackets decorate call and return positions). Suppose that the automaton $\mathcal{A}_{\mathsf{det}}$ is driven by $a\rangle$ from a state S to a state S', and the reconstructed path in $\mathcal{S}$ distinguishes summaries $(q, q') \in S$ and $(q_0, q'') \in S'$, with $\mathcal{A}_{\mathsf{sum}}$ driven by $a\rangle$ from the former to the latter. Also suppose that at this point a transition receiving a state $((q_0, q_1), t_0, S^0, b)$ along the hierarchical edge is chosen in $\mathcal{S}$. Now assume that the matching call drives $\mathcal{A}_{\mathsf{det}}$ from S^0 to some state S''. We need to show that when the reconstruction of the path in $\mathcal{S}$ reaches this call, the state S'' will have been mapped to the state $((q, q), S'')$ and a transition to this state from $((q_0, q_1), S^0)$ will be available, with $((q_0, q_1), t_0, S^0, b)$ sent along the hierarchical edge.

The second of this claims is relatively straightforward. Indeed, by the definitions of $\mathcal{A}_{\mathsf{det}}$ and $\mathcal{A}_{\mathsf{sum}}$, a transition from $((q, q'), S)$ to $((q_0, q''), S')$ driven by $a\rangle$ upon the receipt of $((q_0, q_1), t_0, S^0, b)$ along the hierarchical edge is witnessed by a pair of call and return transitions of the form $t_0 = (q_1, b, q, p) \in \delta^{\mathsf{call}}$, $(p, q', a, q'') \in \delta^{\mathsf{ret}}$ with matching $p \in P$. Since a call transition from S^0 to S'' is available in $\mathcal{A}_{\mathsf{det}}$ with an input letter b, it follows that a call transition with the same input letter leads from $((q_0, q_1), S^0)$ to $((q, q), S'')$ in $\mathcal{S}$.

Now turn to the first of the claims. Recall that we are now dealing with a fragment of the input word of the form $\langle bua \rangle$ with $u \in \Sigma^{*w}$. Observe that our entire argument can be interpreted as a proof that the reconstructed sequence segment induces a correct path segment in $\mathcal{A}_{\mathsf{sum}}$. We now use induction to prove this fact. The base case corresponds to words containing internal symbols only, and does not require any analysis. The inductive step corresponds to words of the form $\langle bua \rangle$ with $u \in \Sigma^{*w}$, as specified earlier. Here, since just before the input symbol $a\rangle$ the automaton $\mathcal{A}_{\mathsf{sum}}$ is set to the state (q, q') and the word u is well-matched, one concludes with the help of the inductive hypothesis and Lemma 5 that after reading the symbol $\langle b$ the automaton $\mathcal{A}_{\mathsf{sum}}$ must have been in some state of the form $(q, \bar{q})$, where $\bar{q} \in Q$. Since this state $(q, \bar{q})$ is taken from S'', and the state S'' is the destination of some call transition in $\mathcal{A}_{\mathsf{det}}$, it follows from the definition of $\mathcal{A}_{\mathsf{det}}$ that $\bar{q} = q$. This concludes the proof of Lemma 6.   □

Lemma 7. *The automaton $\mathcal{S}$ is morphed into $\mathcal{A}$.*

Proof. Removing $\mathcal{A}_{\mathsf{det}}$-components maps the states of $\mathcal{S}$ into the set of states of $\mathcal{A}_{\mathsf{sum}}$. Transitions are mapped to transitions of $\mathcal{A}_{\mathsf{sum}}$ accordingly. It remains to observe that $\mathcal{A}_{\mathsf{sum}}$ is itself morphed into $\mathcal{A}$, for every transition of $\mathcal{A}_{\mathsf{sum}}$ can be mapped into a specific transition of $\mathcal{A}$.   □

Theorem 1 follows from Lemmas 4, 6 and 7.

References

1. Kobayashi, K.: Classification of formal languages by functional binary transducers. Information and Control 15, 95–109 (1969)
2. Sakarovitch, J.: A construction on finite automata that has remained hidden. Theoretical Computer Science 204, 205–231 (1998)
3. Schützenberger, M.P.: Sur les relations rationnelles entre monoïdes libres. Theoretical Computer Science 3, 243–259 (1976)
4. Eilenberg, S.: Automata, Languages, and Machines, vol. A. Academic Press (1974)
5. Sakarovitch, J.: Elements of Automata Theory. Cambridge University Press (2009)
6. Alur, R., Madhusudan, P.: Adding nesting structure to words. Journal of the ACM 56(3), 16 (2009), revised version available online at
   http://robotics.upenn.edu/~alur/Jacm09.pdf
7. Raskin, J.-F., Servais, F.: Visibly pushdown transducers. In: Aceto, L., Damgård, I., Goldberg, L.A., Halldórsson, M.M., Ingólfsdóttir, A., Walukiewicz, I. (eds.) ICALP 2008, Part II. LNCS, vol. 5126, pp. 386–397. Springer, Heidelberg (2008)
8. Staworko, S., Laurence, G., Lemay, A., Niehren, J.: Equivalence of deterministic nested word to word transducers. In: Kutyłowski, M., Charatonik, W., Gębala, M. (eds.) FCT 2009. LNCS, vol. 5699, pp. 310–322. Springer, Heidelberg (2009)
9. Filiot, E., Raskin, J.-F., Reynier, P.-A., Servais, F., Talbot, J.-M.: Properties of visibly pushdown transducers. In: Hliněný, P., Kučera, A. (eds.) MFCS 2010. LNCS, vol. 6281, pp. 355–367. Springer, Heidelberg (2010)
10. Filiot, E., Servais, F.: Visibly pushdown transducers with look-ahead. In: Bieliková, M., Friedrich, G., Gottlob, G., Katzenbeisser, S., Turán, G. (eds.) SOFSEM 2012. LNCS, vol. 7147, pp. 251–263. Springer, Heidelberg (2012)
11. Konstantinidis, S., Santean, N., Yu, S.: Representation and uniformization of algebraic transductions. Acta Informatica 43, 395–417 (2007)
12. Alur, R., Madhusudan, P.: Adding nesting structure to words. In: Ibarra, O.H., Dang, Z. (eds.) DLT 2006. LNCS, vol. 4036, pp. 1–13. Springer, Heidelberg (2006)
13. Thomas, W.: Facets of synthesis: revisiting Church's problem. In: de Alfaro, L. (ed.) FOSSACS 2009. LNCS, vol. 5504, pp. 1–14. Springer, Heidelberg (2009)
14. Von Braunmühl, B., Verbeek, R.: Input driven languages are recognized in $\log n$ space. Annals of Discrete Mathematics 24, 1–20 (1985)

Towards Nominal Context-Free
Model-Checking*

Pierpaolo Degano, Gian-Luigi Ferrari, and Gianluca Mezzetti

Dipartimento di Informatica — Universitá di Pisa
{degano,giangi,mezzetti}@di.unipi.it

Abstract. Two kinds of automata are introduced, for recognising regular and context-free nominal languages. We compare their expressive power with that of analogous proposals in the literature. Some properties of our languages are proved, in particular that emptiness of a context-free nominal language L is decidable, and that the intersection of L with a regular nominal language is still context-free. This paves the way for model-checking systems against access control properties in the nominal case, which is our main objective.

A Philippe Darondeau, amico schivo e carissimo, P.

Introduction

Languages over an infinite alphabet are receiving growing interest, see e.g. [3,21]. They are used to formalise different aspects of many real computational systems, in particular of those concerning the usage of potentially unboundedly many resources that are dynamically created, accessed, and disposed. Examples can be found in XML schemas, in the Web (URLs), in security protocols (e.g. nonces and time-stamps), in virtualised resources of Cloud systems, in mobile and ubiquitous scenarios, etc [23,8,4]. Also foundational calculi for concurrent and distributed systems faced the similar problem of handling unboundedly many fresh (or restricted) names, under the term *nominal* languages [22,5,13,17,18].

The literature mainly reports on various kinds of regular languages over infinite alphabets and on their recognizers [16,2,7]. Also context-free languages over infinite alphabets have been investigated [9,5,19,20]. Indeed, even the following simple example, as well as the recursive patterns of PCRE [15], shows that the patterns of dynamical resource usage have an intrinsic context-free nature.

```
let rec exec() =
    if(...)
        let socket = newsocketfromenv();
        send(socket);
        exec();
        release(socket);
    else ...
```

---

* This work has been partially supported by the MIUR project *Security Horizons*, and by IST-FP7-FET open-IP project ASCENS.

S. Konstantinidis (Ed.): CIAA 2013, LNCS 7982, pp. 109–121, 2013.

The ML-like script above is an abstract dispatcher of tasks on sockets. The execution environment yields a new socket that is *fresh*, so to guarantee exclusive access. Then a sequence of actions exec occurs (we omit them below), and eventually the socket is released. An example of a trace generated during a run is new(s1)new(s2)new(s3)...release(s3)release(s2)release(s1). Now, forgetting the actions new, release and only keeping the names of the sockets (taken from an infinite alphabet), we get a word ww^R, where the symbols in w are all different. For the sake of simplicity, we omit below actions and we only consider resources; actions, hence data words are dealt with in Examples 2 and 7 showing that only simple extensions are required in the general case.

In the line of [10], we pursue here our investigation on developing a foundational model for nominal languages. Our main interest is in statically verifying nominal regular properties of systems, in particular safety properties. Systems are modelled as nominal context-free languages, and verification is done via model-checking. In this paper, we propose an effective model that characterises a novel class of nominal languages, including the one of [9] and expressing, e.g. the traces ww^R above — because of this ability in dealing with "balanced" parenthesis, we shall call these nominal languages *context-free*, and *regular* those that cannot. Additionally, our model can express traces ww^R where the restriction that the resources of w are different is relaxed at wish, yet keeping freshness. This makes the binding name-resource more flexible, in the spirit of dynamic allocation and garbage collection of variable-location typical of programming languages.

To make our model-checking procedure effective, we need to establish the conditions under which the emptiness of the intersection of a nominal context-free language with a nominal regular one is decidable. Preliminary to that is defining a group of nominal automata with increasing expressive power. Besides regular vs. context free, we also consider a disposal mechanism that supports "garbage collection" of symbols thus allowing to re-use disposed symbols, so having, e.g., the above nominal languages of words ww^R with/without replication of symbols.

We proved that regular nominal languages are not closed under complement and (full) concatenation, but they are under union and also under intersection, provided that symbol re-use is forbidden. In that case the language of *all* the strings over an infinite alphabet is however *not* regular, while it is in the general case. We also establish relations with proposals in the literature: without disposal our class of regular languages includes that of Usage Automata (UA) [2] and is incomparable with Variable Automata (VFA) [14] and Finite-memory Automata (FMA) [16]. With the disposal mechanism instead, VFA languages are included in our class, and we conjecture that the same holds for FMA.

Our class of context-free nominal languages is only closed under union and a restricted form of concatenation. Without disposal, ours are equivalent to Usages [2] and become more powerful than these and than *quasi context-free languages* (QCFL) [9] with the possibility of re-using disposed symbols.

As said above, our main goal is proving nominal properties of nominal models, in particular regular safety properties expressing secure access to resources [10].

Table 1. Notation

Math

$i,j \in \underline{r} = \{i \mid 1 \leq i \leq r\}$	set of indices in $\mathbb{N}$, the natural numbers
$L \nleqq L'$	incomparable sets: $L \nsubseteq L'$ and $L \nsupseteq L'$

Words

$\Sigma = \Sigma_s \cup \Sigma_d,$	alphabet with Σ_s *finite* set of *static* symbols and Σ_d disjoint
$(\{?,\top\} \cup \mathbb{N}) \cap \Sigma = \emptyset$	*countably infinite* set of *dynamic* symbols.
$a,b \in \Sigma;\quad w \in \Sigma^*$	symbols in Σ and words, where ε is the empty string
$w[i], \lvert w \rvert, w^R, \lVert w \rVert$	i-th symbol, length, reverse, and set of symbols of w

Automata

$q \in Q$	state of an automaton
$\sigma \in \Sigma_s \cup \underline{r} \cup \{\varepsilon, \top\}$	input symbol in a transition label
$Z \in \Sigma_s \cup \underline{r} \cup \{\varepsilon, ?\}$	stack read symbol in a transition label
$\zeta \in (\Sigma_s \cup \underline{r})^*; z \in \Sigma_s \cup \underline{r}$	stack write symbols in a transition label
$\Delta \in \{i+, i-\} \cup \{\varepsilon\}$	m-register update in a transition label
S	a stack, $\_$ is the empty one
N, M	m-registers, $\_$ is the empty one
$\lVert N \rVert$	set of (dynamic) symbols in N
C, ρ	a configuration, and a run $C_1 \to \cdots \to C_k$
$R, A; L(R), L(A)$	a FSNA,PSNA automaton and their language
$\mathcal{L}(\text{FSNA}), \mathcal{L}(\text{VFA}), \ldots$	set of languages accepted by FSNA, VFA, $\ldots$ automata

Standard and efficient (automata-based) model-checking techniques require that the emptiness of context-free nominal languages is decidable, and that the intersection of a property and a model is still context-free. We here consider the nominal languages accepted by automata without disposal, and prove that both requirements above hold. We also conjecture that intersecting a regular and a context-free language results in a context-free one, provided that at most one of their recognizers uses a disposal mechanism.

The paper is organised as follows. Notation and abbreviations are summarised in Tab. 1. For lack of space, we occasionally only give an intuition of why our results hold; all the proofs are in www.di.unipi.it/~mezzetti. Sect. 1 introduces Finite State Nominal Automata that express nominal properties richer than those considered in [10]; also, we compare these regular nominal languages (with and without disposal) with other proposals in the literature. Then Sect. 2 defines Pushdown Nominal Automata, studies some language theoretic properties and compares them with existing formalisms. Sect. 3 contains our main achievements: decidability of the emptiness of a context-free nominal language (without disposal) and feasibility of model-checking them against regular nominal properties.

1 Finite State Nominal Automata

Our automata accept languages over infinite alphabets, partitioned in a finite set of *static* symbols and an infinite set of *dynamic* symbols, representing resp. the resources known before program execution and those created at run-time.

Table 2. Operations s-push, s-pop, s-top on m-registers

An *m-register* is a pair $N = \langle x \in \{0,1\}, S \rangle$, where x is the *activation state* of the stack S

s-push$(a, \langle x, S \rangle) = \langle 1, push(a, S) \rangle$ s-top$(\langle 1, S \rangle) = top(S)$	s-pop$(\langle x, S \rangle) = \begin{cases} \langle 0, pop(S) \rangle & \text{if } S \neq \_ \\ \langle 0, S \rangle & \text{if } S = \_ \end{cases}$

A finite nominal automaton is a *non-deterministic finite state automaton* (FSA) enriched with a fixed number of *mindful registers*. These registers are components of the configurations that store dynamic symbols, and are accessed and manipulated while recognizing a string. (See Tab. 1 for notation used below.)

Definition 1 (Finite State Nominal Automata)
A finite state nominal automaton (FSNA) is $R = \langle Q, q_0, \Sigma, \delta, r, F \rangle$ where:

- *Q is a finite set of states, $q_0 \in Q$ $-F \subseteq Q$ is the set of final states*
- *$\Sigma = \Sigma_s \cup \Sigma_d$ is a partitioned alphabet (Σ_s is finite and Σ_d denumerable)*
- *r is the number of mindful registers (m-registers for short)*
- *δ is a relation between pairs (q, σ) and (q', Δ), with $\sigma \neq \top$. For brevity, we write $q \xrightarrow[\Delta]{\sigma} q' \in \delta$ whenever $(q, \sigma, q', \Delta) \in \delta$*

A configuration is a triple $C = \langle q, w, [N_1, \ldots, N_r] \rangle$ where q is the current state, w is the word to be read and $[N_1, \ldots, N_r]$ is an array of r m-registers with symbols in Σ. The configuration is final if $\langle q_f \in F, \varepsilon, [N_1, \ldots, N_r] \rangle$.

An m-register N is actually a stack plus a tag, recording if it is either active or not. The operations on N are almost standard (see Tab. 2), but additionally N becomes inactive after a s-pop, while a s-push makes it active (note that s-popping an empty m-register results in a no-operation). The operation s-top yields a value only if the m-register is active, otherwise it is undefined, as well as when the m-register is empty.

If $\sigma \in \Sigma_s$, just as in FSA, a transition $q \xrightarrow[\Delta]{\sigma} q'$ checks whether the current symbol in the input string is σ. Instead, if $\sigma \in \mathbf{r}$, the symbol to be read is s-top(N_σ), i.e. the top of the σth m-registers (analogously to [16]).

An m-register is at the same time updated according to Δ. There are three cases: $\Delta = \varepsilon$ then nothing has to be done; $\Delta = i+$ then a *fresh* dynamic symbol is s-pushed on the i-th m-register; $\Delta = i-$ then the i-th m-register is s-popped. A symbol is *fresh* if it does not appear in any of the r m-registers.

The application of a transition is detailed in the following definition:

Definition 2 (Recognizing Step)
Given an FSNA R, a step $\langle q, w, [N_1, \ldots, N_r] \rangle \rightarrow \langle q', w', [N'_1, \ldots, N'_r] \rangle$ occurs iff there exists a transition $q \xrightarrow[\Delta]{\sigma} q' \in \delta$ s.t. both conditions hold:

1. $\begin{cases} \sigma = \varepsilon \Rightarrow w = w' \text{ and} \\ \sigma = i \Rightarrow w = s\text{-}top(N_i)w' \text{ and} \\ \sigma \in \Sigma_s \Rightarrow w = \sigma w' \end{cases}$

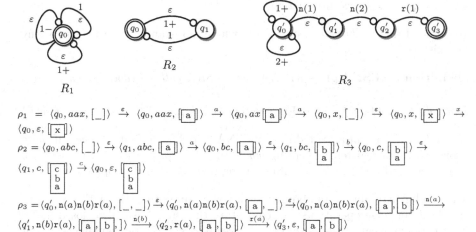

$\rho_1 = \langle q_0, aax, [\_] \rangle \xrightarrow{\varepsilon} \langle q_0, aax, [\boxed{\text{a}}] \rangle \xrightarrow{a} \langle q_0, ax\,[\boxed{\text{a}}] \rangle \xrightarrow{a} \langle q_0, x, [\_] \rangle \xrightarrow{\varepsilon} \langle q_0, x, [\boxed{\text{x}}] \rangle \xrightarrow{x}$
$\langle q_0, \varepsilon, [\boxed{\text{x}}] \rangle$

$\rho_2 = \langle q_0, abc, [\_] \rangle \xrightarrow{\varepsilon} \langle q_1, abc, [\boxed{\text{a}}] \rangle \xrightarrow{a} \langle q_0, bc, [\boxed{\text{a}}] \rangle \xrightarrow{\varepsilon} \langle q_1, bc, [\boxed{\substack{\text{b}\\\text{a}}}] \rangle \xrightarrow{b} \langle q_0, c, [\boxed{\substack{\text{b}\\\text{a}}}] \rangle \xrightarrow{\varepsilon}$
$\langle q_1, c, [\boxed{\substack{\text{c}\\\text{b}\\\text{a}}}] \rangle \xrightarrow{c} \langle q_0, \varepsilon, [\boxed{\substack{\text{c}\\\text{b}\\\text{a}}}] \rangle$

$\rho_3 = \langle q'_0, \text{n}(a)\text{n}(b)\text{r}(a), [\_, \_] \rangle \xrightarrow{\varepsilon} \langle q'_0, \text{n}(a)\text{n}(b)\text{r}(a), [\boxed{\text{a}}, \_] \rangle \xrightarrow{\varepsilon} \langle q'_0, \text{n}(a)\text{n}(b)\text{r}(a), [\boxed{\text{a}}, \boxed{\text{b}}] \rangle \xrightarrow{\text{n}(a)}$
$\langle q'_1, \text{n}(b)\text{r}(a), [\boxed{\text{a}}, \boxed{\text{b}},] \rangle \xrightarrow{\text{n}(b)} \langle q'_2, \text{r}(a), [\boxed{\text{a}}, \boxed{\text{b}}] \rangle \xrightarrow{\text{r}(a)} \langle q'_3, \varepsilon, [\boxed{\text{a}}, \boxed{\text{b}}] \rangle$

Fig. 1. Three examples of FSNA R_i and of their runs ρ_i. The automaton R_1 accepts Σ^*; note that the dynamic symbol x can be any symbol in Σ_d, even a, because the m-register is empty when x is s-pushed and there is no restriction on its freshness. The automaton R_2 accepts L_0 in Ex. 4; and R_3 accepts strings aba ($a \neq b$).

$$2. \begin{cases} \Delta = i+ \Rightarrow N'_i = \text{s-push}(b, N_i) \wedge \forall j. b \notin \|N_j\| \wedge \forall j\ (j \neq i).N_j = N'_j\ and \\ \Delta = i- \Rightarrow N'_i = \text{s-pop}(N_i) \wedge \forall j\ (j \neq i).N_j = N'_j\ and \\ \Delta = \varepsilon \Rightarrow \forall j.N_j = N'_j \end{cases}$$

Finally, the language accepted by R is
$$L(R) = \{w \in \Sigma^* \mid \langle q_0, w, [\_, \dots, \_] \rangle \to^* C_k, \text{with } C_k \text{ final}\}$$

Some examples follow.

Example 1. The FSNA R_1 in Fig. 1 non-deterministically accepts $\Sigma^* \in \mathcal{L}(\text{FSNA})$.

Example 2. The FSNA R_3 in Fig. 1 on data words represents a property for the **new-release** traces of the introduction, in the default-accept paradigm [1]. (The symbols **new**(a), **release**(a) are abbreviated by $\text{n}(a), \text{r}(a)$.) The automaton accepts the unwanted traces where a second socket is created ($\text{n}(2)$) before having released the last one created ($\text{r}(1)$). Dealing with data words, σ assumes the form $\text{n}(u), \text{r}(u), u \in \underline{\text{r}} \cup \Sigma_s$. For readability, we only mention the sockets in the m-registers, omitting the actions n, r.

Example 3. Let $L_r = \{ww^R \in \Sigma^*_d \mid |w| = r \text{ and } \forall i, j. w[i] \neq w[j]\}$ then no FSNA R with less that r states and r m-registers accepts L_r. Indeed, a standard argument on FSA proves that r states are required. Assume now that R has less than r registers and accepts ww^R. By the pigeonhole principle, there is at least one symbol of w, say a, s.t. $\forall i.\ a \neq \text{s-top}(N_i)$ when w has been read. Since $a \in \|w\|$, a needs to be s-pushed while traversing w^R, but it is fresh so it can be replaced by any other (fresh) different symbol, which makes R to accept also ww', where $w' \neq w^R$: contradiction.

We now define a sub-class of FSNA where no transitions can s-pop any m-registers.

Definition 3 (FSNA$_+$). *An FSNA$_+$ is a FSNA with no edge $q \xrightarrow[i-]{\sigma} q'$.*

Example 4. The FSNA$_+$ R_2 in Fig. 1 accepts $L_0 = \{w \in \Sigma_d^* \mid \forall i \neq j. w[i] \neq w[j]\}$.

Example 5. Σ^* is not accepted by any FSNA$_+$, as Σ_d is infinite. Indeed, if there is one with r m-registers, accepting ww with $|w| = \|\|w\|\| = r+1$, the word ww is not accepted if $a \in \|w\|$ and, after having read w, $\forall i. a \neq$ s-top(N_i).

The two kinds of automata above enjoy a few closure properties with respect to standard language operations. We conjecture that FSNA are also closed under intersection; concatenation also preserves regularity, if the two languages do not share any dynamic symbol.

Theorem 1 (Closures)

	∪	∩	‾	•	*
FSNA	✓	?	✗	✓	✓
FSNA$_+$	✓	✓	✗	✗	✗

We now compare the expressive power of FSNA and of FSNA$_+$ with that of other models for regular nominal languages, namely VFA [14], FMA [16] and UA [2].

Theorem 2 (Comparison)

- $\mathcal{L}(FSNA) \supset \mathcal{L}(VFA) \supset \mathcal{L}(UA)$ – $\mathcal{L}(FSNA) \supset \mathcal{L}(FMA)$
- $\mathcal{L}(FSNA_+) \supset \mathcal{L}(UA)$ – $\mathcal{L}(FSNA_+) \not\subseteq \mathcal{L}(VFA)$ – $\mathcal{L}(FSNA_+) \not\subseteq \mathcal{L}(FMA)$

2   Pushdown Nominal Automata

We presented in the introduction a simple program showing a behaviour that is conveniently abstracted as a context-free nominal language. This is often the case when defining static analysis, typically type and effect systems, through which proving program properties, e.g. on resource usage [2]. Nominal context-free languages have therefore both a theoretical and a practical relevance, and received some attention [9,6,5,19]. Below, we extend FSNA with a stack, so obtaining Pushdown Nominal Automata: nominal context-free automata.

Definition 4 (Pushdown Nominal Automata)
A Pushdown Nominal Automata (PSNA) is $A = \langle Q, q_0, \Sigma, \delta, r, F \rangle$ where:

- *Q, q_0, r, F are as in FSNA (Def. 1)*

Table 3. The *Pushreg* operation

Pushreg is an operation taking ζ and S and returning a new stack as follows:

$$Pushreg(z\,\zeta', S) = Pushreg(\zeta', push(\sigma, S)) \qquad \text{where } \sigma = \begin{cases} z & \text{if } z \in \Sigma_s \\ \text{s-top}(N_z) & \text{if } z \in \underline{\mathbf{r}} \end{cases}$$
$$Pushreg(\varepsilon, S) = S$$

- δ *is a relation between triples* (q, σ, Z) *and* (q', Δ, ζ), *with* $(q, \sigma, Z, q', \Delta, \zeta) \in$ δ *written* $q \xrightarrow[\Delta, \zeta]{\sigma, Z} q' \in \delta$

A configuration is a tuple $C = \langle q, w, [N_1, \ldots, N_r], S \rangle$ *where* $q, w, [N_1, \ldots, N_r]$ *are as in FSNA and* S *is a stack with symbols in* Σ.
The configuration $\langle q, w, [N_1, \ldots, N_r], S \rangle$ *is final if* $q \in F, w = \varepsilon$ *and* $S = \_\,$.

Since the label ζ of a transition refers to m-registers, the *push* on the stack S requires an operation (illustrated in Tab. 3) able to also push the s-top of the mentioned m-registers. E.g., $\zeta = a\,3\,b$ causes the string a s-top$(N_3)\,b$ to be pushed.

Definition 5 (Recognizing Step)
Given a PSNA A, the step $\langle q, w, [N_1, \ldots, N_r], S \rangle \;\to\; \langle q', w', [N'_1, \ldots, N'_r], S' \rangle$ *occurs iff* $q \xrightarrow[\Delta, \zeta]{\sigma, Z} q' \in \delta$ *and the following hold*

1. *condition 1 of Def. 2 and* $\sigma = \top \Rightarrow w = top(S)w'$ *and*
2. *condition 2 of Def. 2 and*
3. $\begin{cases} Z = \varepsilon \Rightarrow S' = Pushreg(\zeta, S) \text{ and} \\ Z = i \Rightarrow S' = Pushreg(\zeta, pop(S)) \wedge top(S) = \text{s-top}(N_i) \text{ and} \\ Z = ? \Rightarrow S' = Pushreg(\zeta, pop(S)) \text{ and} \end{cases}$

Finally, the language accepted by R is
$$L(A) = \{w \in \Sigma^* \mid \langle q_0, w, [\_, \ldots, \_], \_\rangle \to^* C_k, \text{ with } C_k \text{ final}\}$$

The definition above extends that for FSNA in handling the stack. In item (1) we add the possibility of checking if the current symbol equals the top of the stack, written $\top$. The top of the stack, say a, is popped if $Z = ?$, as well as if $Z = i$, provided that the s-top of the ith m-register equals a. Finally, if $Z = \varepsilon$ the only action is pushing the string obtained from ζ, as explained above.

Example 6 Fig. 2(a) shows a PSNA accepting $L_p = \{ww^R \mid w \in \Sigma_d^*\}$, and a run accepting *aabbaa* (for brevity, configurations are without strings, the symbols read being in the labels of steps).

As done for FSNA we present a sub-class of PSNA without delete transitions.

Definition 6 (PSNA$_+$). *A PSNA$_+$ is a PSNA with no edges* $q \xrightarrow[i-, \zeta]{\sigma, Z} q'$.

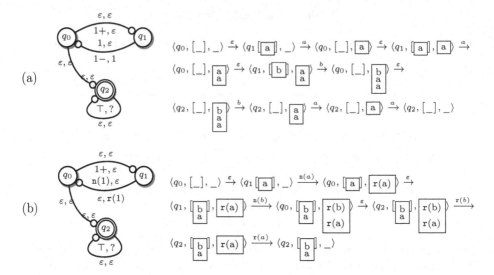

Fig. 2. (a) A PSNA accepting $\{ww^R \mid w \in \Sigma_d^*\}$, and a run on *aabbaa*. (b) A PSNA$_+$ for the data word language of the introduction and a run on $\mathbf{n}(a)\,\mathbf{n}(b)\,\mathbf{r}(b)\,\mathbf{r}(a)$ (n and r stand for **new** and **release**). Strings are omitted in configurations.

Example 7. Consider again the **new-release** (abbreviated n, r) language on data words of the introduction, with all sockets different. The PSNA$_+$ accepting this language is in Fig. 2(b). The labels of transitions, but Δ, contain $\mathbf{n}(u), \mathbf{r}(u), u \in \underline{\mathbf{r}} \cup \Sigma_s$. Fig. 2(b) also shows the run for $\mathbf{n}(a)\,\mathbf{n}(b)\,\mathbf{r}(b)\,\mathbf{r}(a)$; here we omit the strings in configurations and we only mention the sockets in the m-registers. Note that, only keeping the names of the sockets, we get $\cup_{r \in \mathbb{N}} L_r$, for L_r of Ex. 3.

A few properties of closure with respect to language operations follow.

Theorem 3 (Closures)

	$\cup$	$\cap$	$\overline{}$	$\bullet$	$*$
PSNA	✓	×	×	✓	✓
PSNA$_+$	✓	×	×	×	×

We now relate our proposal to *quasi context-free languages* (QCFL) [9] and Usages [2], that are an automata-like and a process calculi-like models for nominal context-free languages, respectively.

Theorem 4 (Comparison)

- $\mathcal{L}(PSNA) \supset \mathcal{L}(QCFL)$ — $\mathcal{L}(PSNA) \supset \mathcal{L}(Usages)$
- $\mathcal{L}(PSNA_+) \nsubseteq \mathcal{L}(QCFL)$ — $\mathcal{L}(PSNA_+) = \mathcal{L}(Usages)$

3   Towards Nominal Model Checking

This section studies the feasibility of model-checking models expressed by nominal context-free languages A against regular nominal properties R, similarly to,

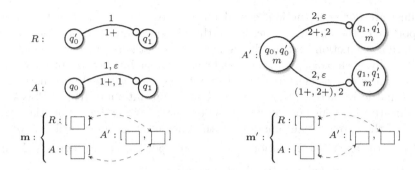

Fig. 3. A' is a portion of the automaton recognizing the intersection of the languages of R and A, the diagrams at the bottom represent the *merges* m, m'. Note that we are using the equivalent version of automata that allows for updating two m-registers at the same time.

e.g., [8,12,2,11]. We give a first positive answer to this problem, which is a main result of this paper.

The standard *automata based model-checking* procedure [24] requires to verify the emptiness of the intersection of the two languages: $A \cap \overline{R} = \emptyset$. A first point seems to arise in our case, because $\overline{R}$ is not necessarily regular, by Thm. 1. Luckily, this is not a problem, since our main concern is now verifying access control policies expressed within the *default-accept* paradigm, where *unwanted* traces are defined, rather than those obeying the required properties [1]. We are therefore left to prove that the intersection above is still a context-free language and that its emptiness is decidable. Indeed, both properties hold for our classes of languages, when their recognizing automata have no transitions that s-pop any m-registers.

Our first theorem states that the emptiness problem is decidable for $PSNA_+$.

Theorem 5. *Given a $PSNA_+$ A, it is decidable whether $L(A) = \emptyset$.*

Proof. (Sketch) The proof relies on a restricted form of the pumping lemma: roughly, there exists a constant n depending on A, s.t. any string $w \in L(A), |w| > n$ can be decomposed into $w = uvxyz$, s.t. also $w = u'x'z'$ belongs to $L(A)$ with u', x', z' obtained from u, x, z by carefully substituting (distinguished) dynamic symbols. By repeatedly applying this kind of pumping lemma, $L(A)$ is non empty if it contains a word w', made of distinguished symbols, and s.t. $|w'| \leq n$.

Our next theorem guarantees that model-checking is feasible in our case: $PSNA_+$ are closed under intersection with $FSNA_+$.

Theorem 6. *Let A be a $PSNA_+$ and R be a $FSNA_+$. Then, $L(A) \cap L(R)$ is a $PSNA_+$ A'.*

Lack of space prevents us from detailing the proof, and we only present below some ingredients of the above sketched construction, ignoring a few mild conditions and assuming all the registers to be active in the initial configuration.

To illustrate our construction, consider the automata A and R in Fig. 3 and the portion of the PSNA$_+$ A' accepting their intersection. A' is obtained by the standard construction that builds the new states as the product of the old ones. Additionally, each pair $\langle q, q' \rangle$ is enriched with a *merge* function m. Intuitively, m describes how the m-registers of the two automata are mapped into those of A' — as a matter of fact, we shall use a variant of our automata (called PSNA$_{+2}$), the transitions of which can update two m-registers at the same time; the equivalence of the extended automata with the ones used so far is easily shown by rendering the extended transition $q \xrightarrow[\Delta_1, \Delta_2, \zeta]{\sigma, Z} q'$ updating the m-registers Δ_1 and Δ_2 with two sequentialized "standard" transitions.

The idea underlying the definition of a merge m is to guarantee the following invariant $\mathcal{I}$ along the runs: if A and R are in configurations $\langle q_0, w, \boxed{\text{y}}, S \rangle$ and $\langle q_0', w, \boxed{\text{x}} \rangle$ then A' will be in configuration $\langle \langle q_0, q_0', m \rangle, w, \boxed{\text{h}}, \boxed{\text{z}}, S' \rangle$ and if two m-registers have the same s-tops then they are merged by m (and vice versa). This is illustrated in the three left-most configurations of Fig. 4: if $x = y = a$ then m maps the two registers to one register of A' (here the second one), and $z = a$. The edges of the automaton are also defined in the standard way. However, the m-registers mentioned in σ, Δ, ζ, Z of A' are those merged by m, provided that R and A agree on both σ and Δ under m.

Consider again Fig. 3. The transition $t : \langle \langle q_0, q_0' \rangle, m \rangle \xrightarrow[2+, 2]{2, \epsilon} \langle \langle q_1, q_1' \rangle, m \rangle$ is present because there are $q_0' \xrightarrow[1+]{1} q_1'$ and $q_0 \xrightarrow[1+, 1]{1, \epsilon} q_1$ and m maps the first m-register of R and that of A to the second of A'. Instead, the state $\langle \langle q_0, q_0' \rangle, m' \rangle$ (omitted in the figure) has no outgoing edges, because the symbols read by R and A are kept apart by m'.

There are transitions that only differ for the merge function in their target state. Not all the possible merges can however be taken, but only those "compatible" with the update Δ, in order to keep the invariant mentioned above. For example, the transition $\langle \langle q_0, q_0' \rangle, m \rangle \xrightarrow[(1+, 2+), 2]{2, \epsilon} \langle \langle q_1, q_1' \rangle, m' \rangle$ permits the recognizing step $C \xrightarrow{a} C'$, where the m-register of R now has got a d, while that of A has got c, and m' keeps them apart. Instead, if both m-registers store the same dynamic symbol c, the merge is still m, and the transition t above enables the step $C \xrightarrow{a} C''$ and guarantees the invariant.

Definition 7 (Merge function). *The function* $m : \{1, 2\} \times \underline{\mathbf{r}} \to \underline{\mathbf{2r}}$ *is a merge iff* $m_1(x) = m(1, x), m_2(x) = m(2, x)$ *are injective. The following are shorthands:*
- *two m-registers* i, j *are merged* $(i \xleftrightarrow{m} j)$ *iff* $m_1(i) = m_2(j)$, *or are taken apart* $(i \xnleftrightarrow{m} j)$, *otherwise;*
- $m(\zeta), m(Z)$ *homomorphically apply* m *to* ζ, Z, *leaving untouched* $\sigma \in \Sigma_s \cup \{?\}$;
- $m[N_1^1, \ldots, N_r^1, N_1^2, \ldots, N_r^2] = [M_1, \ldots, M_{2r}]$, *iff the s-tops of the merged m-registers match and additionally* $\bigcup_{i \in \underline{\mathbf{r}}} \|N_i^1\| \cup \bigcup_{i \in \underline{\mathbf{r}}} \|N_i^2\| = \bigcup_{i \in \underline{\mathbf{2r}}} \|M_i\|$.

A further notion is in order. As shown in Fig. 4, from C one reaches both C' and C'', the states of which however have different merge functions (m' and

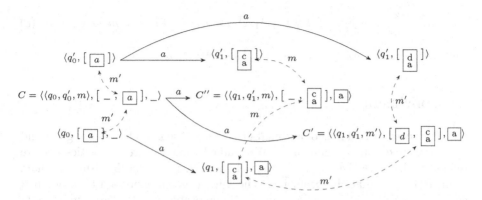

Fig. 4. Two recognizing steps of A' (middle), built from steps of R (top), and steps of A (bottom) (see Fig. 3). The step $C \xrightarrow{a} C'$ simultaneously updates two m-registers.

m, resp.). To account for that, we explicitly represent the changes made by the transition in the registers of the automata to be intersected, call them Δ_1, Δ_2. We then say that a merge m *evolves* into m' ($m \xrightarrow{\Delta_1, \Delta_2} m'$), provided that m' only differs on the updated m-registers Δ_1, Δ_2, possibly merging either of them with other m-registers by m' or taking them apart from some to which they were associated with by m. The registers updated in the intersection automata are actually computed by the suitable function $\overset{m}{m'}(\Delta_1, \Delta_2)$, the definition of which we omit. It guarantees that the invariant $\mathcal{I}$ discussed above, and the compatibility between the actual content of m-registers and m'.

The intersection of PSNA$_+$ with FSNA$_+$ is given as follows:

Definition 8 (Intersection Automaton)
Given the PSNA$_+$ $\langle Q_1, q_0^1, \Sigma, \delta_1, r, F_1 \rangle$ and the FSNA$_+$ $\langle Q_2, q_0^2, \Sigma, \delta_2, r, F_2 \rangle$, their intersection automaton *(of type PSNA$_{+2}$) is $\langle \overline{Q}, \overline{q_0}, \Sigma, \overline{\delta}, 2r, \overline{F} \rangle$, where*

- $\overline{Q} = Q_1 \times Q_2 \times M$, *with* M *set of merge functions*
- $\overline{q_0} = \langle q_0^1, q_0^2, \langle id_{\underline{r}}, id_{\underline{r}} \rangle \rangle$ — $\overline{F} = \{ \langle q_1, q_2, m \rangle \mid q_1 \in F_1, q_2 \in F_2, m \in M \}$
- $\langle q_1, q_2, m \rangle \xrightarrow[\overline{\Delta_1, \Delta_2, \overline{\zeta}}]{\overline{\sigma}, \overline{Z}} \langle q_1', q_2', m' \rangle \in \overline{\delta}$ *iff* $m \xrightarrow{\Delta_1, \Delta_2} m'$ *and*

  $q_1 \xrightarrow[\Delta_1, \zeta]{\sigma_1, Z} q_1' \in \delta_1$ *and* $q_2 \xrightarrow[\Delta_2]{\sigma_2} q_2' \in \delta_2$ *and* $(\sigma_1, \sigma_2 \in \underline{r}$ *or* $\sigma_1, \sigma_2 \in \Sigma_s)$ *and*
  - *if* $\sigma_1, \sigma_2 \in \underline{r}$ *then* $\overline{\sigma} = m_1(\sigma_1) = m_2(\sigma_2)$ *and*
  - *if* $\sigma_1, \sigma_2 \in \Sigma_s$ *then* $\overline{\sigma} = \sigma_1 = \sigma_2$ *and*
  - $(\overline{\Delta_1}, \overline{\Delta_2}) = \overset{m}{m'}(\Delta_1, \Delta_2), \overline{Z} = m_1(Z), \overline{\zeta} = m_1(\zeta)$

  or $q_1 \xrightarrow[\Delta_1, \zeta]{\top, Z} q_1' \in \delta_1$ *and* $q_2 \xrightarrow[\Delta_2]{\sigma_2} q_2' \in \delta_2$ *and* $\sigma_2 \in \underline{r}, \overline{\sigma} = m_2(\sigma_2)$ *and* $\overline{\zeta} = m_1(\zeta)$

  and either $Z = k \in \underline{r}$ *implies* $k \overset{m}{\longleftrightarrow} \sigma_2, \overline{Z} = m_2(\sigma_2), (\overline{\Delta_1}, \overline{\Delta_2}) = \overset{m}{m'}(\Delta_1, \Delta_2),$
  or $Z = ?$ *implies* $\overline{Z} = m_2(\sigma_2), (\overline{\Delta_1}, \overline{\Delta_2}) = \overset{m}{m'}(\Delta_1, \Delta_2)$

$$or \quad q_1 \xrightarrow[\Delta_1\zeta]{\varepsilon, Z} q_1' \in \delta_1 \ and \ \overline{\sigma} = \varepsilon, \ (\overline{\Delta_1}, \overline{\Delta_2}) = \overset{m}{m'}(\Delta_1, \varepsilon), \ \overline{Z} = m_1(Z), \overline{\zeta} = m_1(\zeta)$$

$$or \quad q_2 \xrightarrow[\Delta_2]{\varepsilon} q_2' \in \delta_2 \ \overline{\sigma} = \varepsilon, \ (\overline{\Delta_1}, \overline{\Delta_2}) = \overset{m}{m'}(\Delta_1, \varepsilon), \ \overline{Z} = \varepsilon, \overline{\zeta} = \varepsilon$$

4   Conclusions

We introduced novel kinds of automata that recognise new classes of regular and of context-free nominal languages. We studied their closure properties and we related their expressive power to that of the models in the literature. A main result of ours is that the problem of checking a model expressed by a nominal context-free language against a regular nominal property is decidable, under mild assumptions. Our contribution addresses therefore the shortcoming of standard automata based model-checking approaches in the nominal setting, that only considered regular languages. Ours is a further step towards developing methods for formally verifying computational systems that handle an unbounded number of resources. This is more and more the case nowadays, e.g. XML schemas, web and cloud system, security protocols.

Further investigation is required on the impact that a disposal mechanism has on the feasibility of model-checking. Preliminary results suggest us that intersecting a regular and a context-free language results in a context-free one, when at most one of them has a disposal mechanism.

Future research is needed to fill in the gap between our foundational results and the concrete case studies and prototypal implementation of our abstract model-checking procedure. Another line of research will study a logical characterization of our nominal regular automata, as done e.g. by [7,11].

References

1. Bartoletti, M., Degano, P., Ferrari, G.L.: Planning and verifying service composition. JCS 17(5), 799–837 (2009)
2. Bartoletti, M., Degano, P., Ferrari, G.L., Zunino, R.: Model checking usage policies. In: Kaklamanis, C., Nielson, F. (eds.) TGC 2008. LNCS, vol. 5474, pp. 19–35. Springer, Heidelberg (2009)
3. Benedikt, M., Ley, C., Puppis, G.: Automata vs. logics on data words. In: Dawar, A., Veith, H. (eds.) CSL 2010. LNCS, vol. 6247, pp. 110–124. Springer, Heidelberg (2010)
4. Bojanczyk, M.: Data monoids. In: Dürr, C., Wilke, T. (eds.) STACS 2011, vol. 9, pp. 105–116. Schloss Dagstuhl–Leibniz-Zentrum fuer Informatik (2011)
5. Bojańczyk, M., Klin, B., Lasota, S.: Automata theory in nominal sets (2011), http://www.mimuw.edu.pl/~sl/PAPERS/lics11full.pdf
6. Bojanczyk, M., Klin, B., Lasota, S.: Automata with group actions. LICS, pp. 355–364. IEEE Computer Society, Washington, DC (2011)
7. Bollig, B.: An automaton over data words that captures EMSO logic. In: Katoen, J.-P., König, B. (eds.) CONCUR 2011. LNCS, vol. 6901, pp. 171–186. Springer, Heidelberg (2011)

8. Bollig, B., Cyriac, A., Gastin, P., Narayan Kumar, K.: Model checking languages of data words. In: Birkedal, L. (ed.) FOSSACS 2012. LNCS, vol. 7213, pp. 391–405. Springer, Heidelberg (2012)
9. Cheng, E.Y.C., Kaminski, M.: Context-free languages over infinite alphabets. Acta Inf. 35(3), 245–267 (1998)
10. Degano, P., Ferrari, G.-L., Mezzetti, G.: Nominal automata for resource usage control. In: Moreira, N., Reis, R. (eds.) CIAA 2012. LNCS, vol. 7381, pp. 125–137. Springer, Heidelberg (2012)
11. Demri, S., Lazić, R., Nowak, D.: On the freeze quantifier in constraint ltl: Decidability and complexity. Information and Computation 205(1), 2–24 (2007)
12. Ferrari, G.L., Gnesi, S., Montanari, U., Pistore, M.: A model-checking verification environment for mobile processes. TOSEM 12(4), 440–473 (2003)
13. Gordon, A.D.: Notes on nominal calculi for security and mobility. In: Focardi, R., Gorrieri, R. (eds.) FOSAD 2000. LNCS, vol. 2171, pp. 262–330. Springer, Heidelberg (2001)
14. Grumberg, O., Kupferman, O., Sheinvald, S.: Variable automata over infinite alphabets. In: Dediu, A.-H., Fernau, H., Martín-Vide, C. (eds.) LATA 2010. LNCS, vol. 6031, pp. 561–572. Springer, Heidelberg (2010)
15. Hazel, P.: Pcre: Perl compatible regular expressions (2005), http://www.pcre.org/pcre.txt
16. Kaminski, M., Francez, N.: Finite-memory automata. TCS 134(2), 329–363 (1994)
17. Kurz, A., Suzuki, T., Tuosto, E.: On nominal regular languages with binders. In: Birkedal, L. (ed.) FOSSACS 2012. LNCS, vol. 7213, pp. 255–269. Springer, Heidelberg (2012)
18. Montanari, U., Pistore, M.: π-calculus, structured coalgebras, and minimal hd-automata. In: Nielsen, M., Rovan, B. (eds.) MFCS 2000. LNCS, vol. 1893, pp. 569–578. Springer, Heidelberg (2000)
19. Parys, P.: Higher-order pushdown systems with data. In: Faella, M., Murano, A. (eds.) GandALF. EPTCS, vol. 96, pp. 210–223 (2012)
20. Perrin, D., Pin, J.E.: Infinite words: automata, semigroups, logic and games. Pure and Applied Mathematics, vol. 141. Elsevier (2004)
21. Pitts, A.M.: Nominal sets names and symmetry in computer science: Names and symmetry in computer science. Cambridge Tracts in Theoretical Computer Science, vol. 57. Cambridge University Press
22. Pitts, A.M., Stark, I.D.B.: Observable properties of higher order functions that dynamically create local names, or what's new? In: Borzyszkowski, A.M., Sokolowski, S. (eds.) MFCS 1993. LNCS, vol. 711, pp. 122–141. Springer, Heidelberg (1993)
23. Segoufin, L.: Automata and logics for words and trees over an infinite alphabet. In: Ésik, Z. (ed.) CSL 2006. LNCS, vol. 4207, pp. 41–57. Springer, Heidelberg (2006)
24. Vardi, M.Y., Wolper, P.: An automata-theoretic approach to automatic program verification. In: LICS, pp. 332–344. IEEE Computer Society Press (1986)

Implementation Concepts in Vaucanson 2

Akim Demaille[1], Alexandre Duret-Lutz[1], Sylvain Lombardy[2],
and Jacques Sakarovitch[3]

[1] LRDE, EPITA
{akim,adl}@lrde.epita.fr
[2] LaBRI, Université de Bordeaux
Sylvain.Lombardy@labri.fr
[3] LTCI, CNRS / Télécom-ParisTech
sakarovitch@telecom-paristech.fr

Abstract. VAUCANSON is an open source C++ platform dedicated to
the computation with finite *weighted* automata. It is *generic*: it allows
to write algorithms that apply on a wide set of mathematical objects.
Initiated ten years ago, several shortcomings were discovered along the
years, especially problems related to code complexity and obfuscation as
well as performance issues. This paper presents the concepts underlying
VAUCANSON 2, a complete rewrite of the platform that addresses these
issues.

1 Introduction

VAUCANSON[1] is a free-software[2] *platform* dedicated to the computation of and
with *finite automata*. As a "platform" it is composed of a high-performance
(somewhat unfriendly) low-level C++ library, on top of which more humane
interfaces are provided: a comfortable high-level Application Program Interface (API), flexible formats for Input/Output, easy-to-use command-line tools,
and eventually, a Graphical User Interface (GUI). Here, "automata" is to be
understood in the broadest sense: weighted automata on a free monoid — that
is automata that not only accept, or recognize, *words* but compute for every
word a *multiplicity* which is taken in an arbitrary semiring — and even weighted
automata on *non-free* monoids. As for now, are implemented in VAUCANSON
only the (weighted) automata on (direct) products of free monoids, machines
that are often called *transducers* — automata that realize (weighted) relations
between words.

VAUCANSON was started about ten years ago [7] with two main goals and
a constraint in mind: genericity (offering support for a wide set of automaton
types) and a natural (to mathematicians) programming style, while keeping
performances that compete with similar platforms. Genericity was obtained by
extensive use of C++ overloading and template programming, which are well

---

[1] Work supported by ANR Project 10-INTB-0203 VAUCANSON 2.

[2] http://vaucanson-project.org, http://vaucanson.lrde.epita.fr

S. Konstantinidis (Ed.): CIAA 2013, LNCS 7982, pp. 122–133, 2013.
© Springer-Verlag Berlin Heidelberg 2013

known means to avoid the costly dynamic polymorphism of traditional object-oriented programming. A novel design, dubbed ELEMENT/METAELEMENT, allowed a consistent pattern of implementation of mathematical concepts (such as monoids, semirings, series, rational expressions, automata and transducers). Annual releases have shown that genericity was indeed achieved: for instance TAF-KIT, the command-line tools exposed by VAUCANSON 1, features about 70 commands on 18 different kinds of automata/transducers [9].

However performances did not keep on par. The causes for these poor performances are manifold. For instance "genericity" was sometimes implemented by "generality": instead of a carefully crafted specialized structure, a very general one was used. Most prominently, usual automata, labeled with letters, were implemented as automata whose transitions are labeled by polynomials. The aforementioned ELEMENT/METAELEMENT design also conducted to error-prone, obfuscated code, which eventually impeded the performance of the *developers* themselves.

The VAUCANSON 2 effort aims at keeping the best of VAUCANSON 1 (genericity, rich feature set, easy-to-use shell interface, etc.) while addressing its shortcomings.

This paper presents the design of VAUCANSON 2, which is only partly implemented. Although much remains to be done, enough is already functional so that we can report on our experience in the redesign of VAUCANSON. The remainder of this paper is structured as follows. Section 2 introduces the fundamental types used in VAUCANSON 2, and their implementation is presented in Section 3. In Section 4 we explain how we provide flexible and easy-to-use interfaces on top of an efficient but low-level library. We conclude in Section 5.

2   Types Implementation

2.1   From Typing Automata to Typing Transition

The type of a weighted automaton is classically given by the *monoid* M of its labels and the *semiring* $\mathbb{K}$ of its weights. For instance, if M is a free monoid over a finite alphabet A, the weighted automaton is an automaton over finite words; if M is a product $A^* \times B^*$ of two free monoids (A and B are finite alphabets), then the weighted automaton is actually a transducer. The Boolean semiring $\mathbb{B}$ corresponds to usual NFAs, while $\mathbb{R}$ can be used for probabilistic automata and $\langle \mathbb{Z}, \min, + \rangle$ for distance automata.

Such a weighted automaton realizes an application from M into $\mathbb{K}$. The applications that can be realized this way are called ($\mathbb{K}$-)*rational series*. They are a subset of formal power series over M with coefficients in $\mathbb{K}$, that are usually denoted $\mathbb{K}\langle\langle M \rangle\rangle$, but seen as applications from M into $\mathbb{K}$, they can also be denoted $\mathbb{K}^M$, or, with a notation closer to type theory $M \to \mathbb{K}$.

VAUCANSON 1 follows this characterization, and, on top of its implementation, the type of a weighted automaton is made of both the type of a monoid and the type of a semiring. The pair Monoid/Semiring is called the *context* of the automaton (in VAUCANSON 1 vocabulary).

This definition of types, based on the algebraic characterization, has two main drawbacks. First, each implementation must support the most general automata of a given type. Therefore, any element of M is accepted as a label of a transition; actually, in VAUCANSON 1, labels and weights were handled at the same time, since each transition can be labeled by a polynomial (a linear combination of letters). Second, the application of many algorithms requires the label of the automaton to have a particular form. For instance, the product of two automata (that recognizes the intersection of languages in the case of NFAs) requires the labels of both automata to be letters, or the usual composition algorithm for transducers requires that the transducers are *subnormalized*. In VAUCANSON 1, these prerequisites, crucial for the effective computation, are not enforced statically (i.e., by specific C++ types): they must be checked at runtime.

In the design of VAUCANSON 2 we decided to refine our typing system to fit with the effective requirements of algorithms and implementations. It led to focus on typing the *transitions* rather than the automata themselves. Indeed, even if different automata on words have the same context in the sense of VAUCANSON 1, depending whether they are labeled by letters, or letters and ε, or by words, their implementation and their behavior w.r.t. algorithms may be very different. Hence, in VAUCANSON 2 the type of an automaton is characterized by the type of its transitions, which we name a *context*. Since the type of a transition depends on the type of its label and of its weight, the context of an automaton in VAUCANSON 2 is a pair made from a *LabelSet* and a *WeightSet* (Section 2.3). OPENFST [2] preceded us, and our contexts are alike their "arctypes".

2.2   Bridge between Structures and Values

The implementation of mathematical objects in VAUCANSON 2 follows a pattern, which we name ELEMENT/ELEMENTSET: the C++ type of values is separated from the C++ type of their set of operations. For instance, weights of both semirings $\mathbb{Z}$ and $\langle \mathbb{Z}, \min, + \rangle$ are implemented by `int`, but their operations are defined by the corresponding WeightSets: `z` and `zmin`. Other ELEMENT/ELEMENTSET pairs include `bool/b`, `double/r`, etc., but also the rational expressions and their set: `ratexp<Context>/ratexpset<Context>`. Thanks to this compliance with this ELEMENT/ELEMENTSET duality, rational expressions can be used as weights. The behavior of labels (concatenation, neutral element...) is also provided by a LabelSet.

Whereas VAUCANSON 1's ELEMENT/METAELEMENT is a rich and complex design in itself, our segregation between values and their operations is hardly novel. It is a return to N. Wirth's "Algorithms + Data Structures = Programs" equation [10], phased out by traditional object-oriented programming's "Programs = Objects", and resurrected by generic programming à la STL.

2.3   LabelSet and WeightSet

LabelSet: Different Kinds of Label. The LabelSet contains the information on the type of the labels, that is the implementation of the labels themselves, the implementation of the elements of the monoid (that may be different: a letter is not implemented as a word). It also knows how to multiply (concatenate) labels.

Last, but not least, it provides a type called *Kind* that is used to choose the most appropriate algorithm that has to be applied to the automaton (or rational expression). There are two different kinds: LAW, LAL, and a third sort of automata that is treated as a kind in itself: LAU.

LAW, 'Labels Are Words'. The most general weighted automata have transitions labeled by the elements of a monoid (and weighted by the elements of a semiring). This class is very general, does not correspond to the usual definition of finite automata and is not suitable for many algorithms (determinization, product, evaluation, etc.). However it is the only kind so far that describes automata over direct product of monoids. The LabelSet for the free monoid is called `wordset`.

LAL, 'Labels Are Letters'. This class of automata corresponds to the usual way NFAs, or even WFAs, are defined: labels are restricted to being letters. It also corresponds to the description of automata as *linear representations*. Some algorithms require the automata to be in this class, such as the product (which implements the tensor product of the representations) or the reduction algorithm (which can be applied when the weights belong to a field). The LabelSet implementing this kind is `letterset`.

LAU, 'Labels Are Unit'. This kind corresponds to the case where there is no label, or, equivalently, where every label is the element of a trivial monoid. These automata are therefore weighted graphs (with initial and final values attached to vertices). They prove to be very useful since they allow to unify for instance the state elimination (which computes a rational expression from an automata) and the ε-removal algorithms. Its LabelSet is `unitset`.

WeightSet: Different Values. The WeightSets define the nature of the weights, and the operations that apply. They must provide addition, multiplication and Kleene-star operators. Currently, basic WeightSets include (i) Booleans $\langle \mathbb{B}, \vee, \wedge \rangle$, implemented by the class `b`; (ii) integers, $\langle \mathbb{Z}, +, \times \rangle$ and $\langle \mathbb{Z}, \min, + \rangle$, implemented by `z` and `zmin`; (iii) double precision floats $\langle \mathbb{R}, +, * \rangle$, implemented by `r`.

Every object that offers the same set of methods, i.e., every object that satisfies the *concept* of WeightSet can be used. For instance, rational expressions can be used as weights, even though they do not form a semiring.

2.4   Implementation of Contexts

In VAUCANSON 2, the LabelSets manage the labels. Most aspects of labels are *static*: their type (e.g., `char`, `std::string`, etc.) as well as the operations or services that are available (algebraic operations, iteration on generators, etc.). Others aspects may vary at runtime: an alphabet (a set of letters) is needed to fully define a LabelSet (except for LAU). Therefore, LabelSets are C++ objects (values), not just classes.

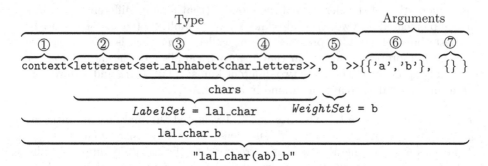

Fig. 1. Anatomy of `lal_char(ab)_b`, the C++11 implementation of $\{a, b\} \to \mathbb{B}$, the context for Boolean automata. *Class template* (named *generics* in C# and Java) generate classes once provided with parameters, in angle brackets. For instance, `set_alphabet` is a class template expecting a single parameter (a class), and `set_alphabet<char_letters>` is class.

Basic WeightSets (`b`, `zmin`, etc.) could be defined only as types, but more complicated weights, like rational expressions (cf. Section 3.2), require runtime values.

Because they depend on runtime values, *contexts* are *objects* (not just classes) which aggregate an instance of a LabelSet and an instance of a WeightSet. Contexts are the cornerstone on top of which the whole VAUCANSON 2 platform types its mathematical entities (words, weights, automata, rational expressions etc.). As such, they could be named "types", but it would lead to too much confusion with C++'s own types. However, these (VAUCANSON 2) "types" have *names* which we write with quotes (e.g., `"lal_char(ab)_b"`), which should not be confused with the C++ objects that implement them, even though C++ names match those of the types, written without quotes (e.g., `lal_char_b`).

The context of Boolean automata (see Fig. 1) will help understanding the C++ implementation of contexts. This context object is named `"lal_char(ab)_b"`. As any object, it is the instantiation of a *Type*, provided with *Arguments*.

We name the C++ Type `lal_char_b`. It is a context ①, which aggregates a LabelSet (`lal_char`) and a WeightSet (`b`). Since we define an LAL context, the LabelSet is `letterset` ②, which must be provided with the type of the letters. The alphabet ③ is a set of letters. We designed alphabets to work on different types of letters, `char`, but also `int`; in Fig. 1 we work with the traditional C++ type for characters, `char`, provided by the `char_letters` class ④. The effective set of these letters is given in ⑥. The WeightSet is `b`, i.e., $\langle \mathbb{B}, \vee, \wedge \rangle$. In this case, `b` needs no argument, so none is provided ⑦. Table 2 gives more examples of contexts.

Table 2. Examples of contexts. `chars` stands for `set_alphabet<char_letters>`, an alphabet whose letters and words are of type `char` and `std::string`.

C++ Type VAUCANSON Static Name	Context C++ Initializers VAUCANSON Dynamic Name
	$\{a, b\} \to \mathbb{B}$
`context<letterset<chars>, b>` `"lal_char_b"`	`{{'a', 'b'}, {}}` `"lal_char(ab)_b"`
	$\{a, b\}^* \to \mathbb{Z}$
`context<wordset<chars>, z>` `"law_char_r"`	`{{'a', 'b'}, {}}` `"law_char(ab)_z"`
	$\{1\} \to \{a, b\} \to \mathbb{Z}$

`context<unitset,`
      `ratexpset<letterset<chars>, z>>`
`"lau_ratexpset<lal_char_z>"`
	`{{}, {'a', 'b'}}`
	
	`"lau_ratexpset<lal_char(ab)_z>"`

	$\{a, b\} \to \{x, y\} \to \mathbb{Z}$

`context<letterset<chars>,`
      `ratexpset<letterset<chars>, z>>`
`"lal_char_ratexpset<lal_char_z>"`
	`{{'a', 'b'}, {'x', 'y'}}`
	
	`"lal_char(ab)_ratexpset<lal_char(xy)_z>"`

3 Object Implementation

3.1 Automata

The design of automata in VAUCANSON 2 is driven by their API, which was designed after the experience of VAUCANSON 1 to allow easy and efficient implementation of algorithms. It offers a number of methods to access both in reading and writing to the automaton content. The automaton is the object that "knows" how to manipulate all sub-entities like states, transitions, labels or weights. Following the ELEMENT/ELEMENTSET principle, these are only values (that may be numbers, pointers or structures) with no methods; they can be handled directly by the automaton or in some cases by the ELEMENTSET objects (the LabelSet or the WeightSet) that can be retrieved from the automaton.

To make the use of the VAUCANSON library as intuitive as possible, all the algebraic information is confined in the context object. An automaton is therefore a class template that depends on the implementation of the automaton, where the template parameter is the context. For instance, the class template `mutable_automaton<Context>` is a linked list based implementation that supports all the API. VAUCANSON 2 provides also some wrappers that can be applied to a totally defined automaton (implementation and context). For instance, the class template `transpose_automaton<Automaton>` wraps an automaton to present a reverse interface, read and write.

The API of `mutable_automata` is totally symmetrical: the forward and backward transition functions are both accessible. Other implementations can be provided that partially implement the API, e.g., `forward_mutable_automaton` (and `backward_mutable_automaton` for symmetry), which does not track incoming transitions, should improve performances in most cases. More wrappers can also be provided, for instance to change the alphabets, to change the weight sets and so on.

Our API is designed to support implementation of automata computed on-the-fly as well [6].

Pre and Post States. Weighted automata feature not only weights on transitions but also on the initial and final states. An implementation keeping these initial and final weights on states is awkward, as we found out in VAUCANSON 1. The data structure is heavy, and the API error-prone.

In VAUCANSON 2, weights on initial states are modeled as weights on (plain) transitions from `pre`, a (unique) invisible "preinitial" state, to every initial state. These transitions are labeled by a fresh label, denoted by $, which can be interpreted as the empty word. As a consequence the initial weights are handled seamlessly, which simplifies significantly the implementation of several algorithms. Besides, since there is a single `pre` state, algorithms that start by loading a queue with the initial states (e.g., determinization) can now push only `pre` and let the loop iterate on the successors.

The final weights are handled in a similar fashion, with `post`, a (unique) invisible "postfinal" state.

3.2 Rational Expressions

Although the concept of "context" emerged to design our implementation of automata, it fits perfectly the same job to define the nature of the corresponding rational expressions. By virtue of the ELEMENT/ELEMENTSET principle, rational expressions are implemented by `ratexp<Context>` and manipulated by `ratexpset<Context>`.

Syntax. VAUCANSON, 1 and 2, supports weighted rational expressions, such as $a + (2b)^* + 3(c^*)$. Because weights can be arbitrarily complex, we delimit them with braces: 'a+({2}b)*+{3}(c*)'.

VAUCANSON 1 provides customizable syntax for rational expressions: one can define which symbols are used for the constants 0 and 1 (denoting the empty language and empty word), and so forth. In the predefined parser of VAUCANSON 2 '\z' and '\e' denote 0 and 1. Because braces are already used for weights they cannot be used for the generalized quantifiers as in POSIX extended notation ($a\{min, max\}$), we denote them $a(*min, max)$. Other parsers can be added to support alternative syntaxes.

Contexts. Like automata, rational expressions support not only various types of weights, but also of labels — for simplicity and by symmetry with automata, we name "labels" the "atoms" of rational expressions. They are therefore parameterized by a context. Classical rational expressions such as '(a+b)*a(a+b)' use letters ('a' and 'b') as atoms, and Booleans as weights ("true" is implicit): their context is "lal_char(ab)_b".

Labels may also be words (LAW) or reduced to 1 (LAU). Weights may be integers (e.g., 'one+{2}(two)+{3}(three)' from "law_char(ehnortw)_z") or even rational expressions: '{{1}x}a+{{2}y}b' is a rational expression from "lal_char(ab)_ratexpset<lal_char(xy)_z>" (see Table 2).

Associativity, Commutativity. We intend to experiment with different implementations of rational expressions, including relying on associatitivity and commutativity. To this end, our `ratexp` structure supports variadic sums and products. It is up to `ratexpset` to decide whether to exploit this feature or not. Currently sums and products are binary, and only straightforward rewritings are performed on rational expressions ("Trivial Identities", [9, Table 2.5]).

4 Dynamic Polymorphism Implementation

4.1 Dynamic I/O Routines

Contexts Input/Output. As its predecessor, VAUCANSON 2 is more than just a C++ library: it is a platform that provides a set of tools to manipulate automata and rational expressions, means to save and restore them, eventually a GUI, etc. I/O of automata and rational expressions is therefore crucial, which requires a means to save, and read, their type. In short, a file containing a Boolean automaton must specify that it is a Boolean automaton. This is achieved by I/O support for contexts.

VAUCANSON 2 can compute a name from a context, and conversely it can build a context from its name. Examples of (dynamic) context names are provided in Table 2; for instance the name for $\{a, b\} \to \mathbb{B}$ is "lal_char(ab)_b". Static names are computed from the (dynamic) names by removing the alphabets: "lal_char_b".

VAUCANSON 1 relies on FSMXML [4] for typed I/O: specific tags define the algebraic context of the stored automaton or rational expression. Context names play a similar role in VAUCANSON 2's I/Os.

Dynamic Automaton/Rational Expression Input. VAUCANSON 2 aims at providing an even larger set of algorithms than VAUCANSON 1, yet in a more flexible way. TAF-KIT 2 consists of a unique binary that replaces the set of binaries of TAF-KIT 1. The following command lines for the computation of the product of two automata and for the construction of the standard (or position) automaton of an expression show the contrast between both interfaces.

```
# Vaucanson 1: use the context–specific tool.
$ vcsn-char-z product a.xml b.xml > c.xml   # product of automata a and b
$ vcsn-char-z  -a abc standard '{2}a+{3}c' > s.xml  # automaton for 2a+3c

# Vaucanson 2: use the generic tool, the context is handled dynamically.
$ vcsn product a.xml b.xml > c.xml
$ vcsn -C 'lal_char(abc)_z' standard -e '{2}a+{3}c' > s.xml
```

In TAF-KIT 1, there is a command for every single context (here, vcsn-char-z), while in TAF-KIT 2 context is just data of a unique command (vcsn). The context must be either carried by the argument, for instance in the FSMXML description of an automaton, or provided with a simple command line option (-C 'lal_char(abc)_z'). In either case, the very profound difference between the static nature of the C++ library and the dynamic one of data must be resolved. This is addressed by our static/dynamic bridge.

4.2 Dynamic Calls to Static Algorithms

Static/Dynamic Polymorphisms. *Polymorphism* is the ability for a single function name to denote several implementations that depend on the type of its argument(s). It provides a means to specialize algorithms: some can be run on automata of special kind, with special type of weights, or special type of labels, others may have different versions according to the type of weights or of labels. For instance, determinization applies to Boolean automata only, product to LAL automata only, etc. The elimination of ε-transitions gives an example of the second kind: any closure algorithm works on Boolean automata, for automata with weights in $\mathbb{Z}$ it boils down to an algorithm that tests for acyclicity, and for automata with weights in $\mathbb{R}$, it is again another algorithm [8].

The *dynamic* polymorphism (i.e., specialization selection at runtime via virtual tables) provides a flexible solution to these problems: roughly, when a method is called on an object, the exact type of this object is examined at run time and the corresponding implementation is then called. It is the essence of object-oriented programming. Unfortunately this elegant mechanism slows down the execution of the program when used in intensive computation loops.

C++ offers another mechanism, *templates*, that allows *static* polymorphism (i.e., specialization selection at compile time). Functions and classes can be parameterized by one or several types (or constants); some specific implementations can also be provided for particular types (see Fig. 1). When a method is called on an object, the type of this object is known at compile time and the method, if needed, is compiled especially for this type. Method invocation is then as efficient as a plain function call. There are some drawbacks to this mechanism. First, every object must be precisely typed: the type of an automaton shows its implementation and is moreover parameterized by the type of its labels and weights, which can themselves be parameterized. Hence, users of the library may handle quite complicated types. For instance the type for a classical Boolean automaton is mutable_automaton<context<letterset<set_alphabet<char_letters >>, b> (see Table 2 for a description of the context part). Second, to compile binaries that offer all the required algorithms for each type of automata, the corresponding functions must be compiled for each of these types. The compilation, as a consequence, is an extremely long process, and the introduction of a new "context" (that is, a new kind of weights, labels or implementation) needs a configuration to specify which algorithms need to be compiled.

This is our experience with VAUCANSON 1; compiling the whole package and the 18 commands of TAF-KIT takes hours (annoying for users, crippling for

```
template <typename Context> void
static(const string& lhs, const string& rhs, const string& word)
{
  using automaton_t = vcsn::mutable_automaton<Context>;
  automaton_t l = vcsn::read_automaton_file<automaton_t>(lhs);
  automaton_t r = vcsn::read_automaton_file<automaton_t>(rhs);
  automaton_t prod = vcsn::product<automaton_t, automaton_t>(l, r);
  typename Context::weight_t w = vcsn::eval<automaton_t>(prod, word);
  prod.context().weightset()->print(std::cout, w);
}

void dynamic(const string& lhs, const string& rhs, const string& word)
{
  using namespace vcsn::dyn;
  automaton l = read_automaton_file(lhs);
  automaton r = read_automaton_file(rhs);
  automaton prod = product(l, r);
  weight w = eval(prod, word);
  print(w, std::cout);
}
```

Fig. 3. Static/Dynamic APIs: evaluation of the word **word** by a product of the automata stored in the files named **lhs** and **rhs**. The **dynamic** routine hides the complexity of templated programming, demonstrated by **static**. It is also more flexible, as it can be invoked with automata of equal algebraic type, but of different C++ types (e.g., a **transpose_automaton** and a **mutable_automaton**).

developers!). Beside, the resulting API was repelling, and very few people dared programming with the library, and preferred to use TAF-KIT.

To address these shortcomings while keeping the good properties of static polymorphism (efficiency and rigorous type checking), VAUCANSON 2 provides a two-level API (see Fig. 3). The low-level API, named "static", is fully typed: the C++ types of the object are precise and heavily templated. When programming at this level, the user is in charge of every detail, including memory management. For sake of efficiency, all the algorithms of the library are written at the static level: dynamic polymorphism and its costs are avoided. The high-level API, named "dynamic", provides the user with the comfort of dynamic polymorphism, but at such a coarse grain that its cost is negligible (to select the appropriate version of an algorithm, not in the implementation of the algorithms themselves). This layer takes in charge details such as the exact type of the objects, and memory management.

Dynamic Access to Template Algorithm. Fig. 4 demonstrates how the dynamic API invokes the low-level one. In the high-level API, an automaton **a1** is handled as a **dyn::automaton**, whatever its exact type. When the user invokes **dyn::determinize(a1)**, several steps must occur. First **a1** is asked for

132   A. Demaille et al.

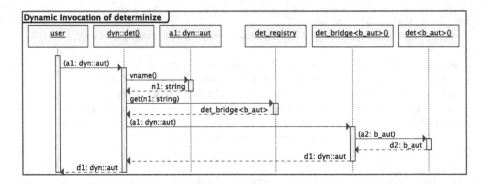

Fig. 4. Dynamic invocation on top of templated implementations. For conciseness, in this diagram `b_aut` denotes the type of Boolean automata, `mutable_automaton<context<letterset<set_alphabet<char_letters>>>, b>`, the "exact" type of the automaton `a2`, contained by the dynamic automaton `a1`.

the name of its exact type (`n1` in Fig. 4). This string is used to query a registry of all the static *bridges* of `determinize`. These bridges all have the same type (`dyn::automaton -> dyn::automaton`), yet they are parameterized by the exact type. Once the specific `det_bridge<b_aut>` is obtained, it is invoked with `a1` as argument. The bridge extracts `a2`, the static automaton, from `a1`, and invokes the low-level `determinize`, whose type is `b_aut -> b_aut`. Its result, the static automaton `d2`, is returned to the bridge, that wraps it in the dynamic automaton `d1`, which is eventually returned to the user.

As far as we know, this two-layer technique to implement dynamic polymorphism on top of static polymorphism is unpublished. However several of its components have already been documented in the literature. For instance, dynamic objects such as `dyn::automaton` implement the EXTERNAL POLYMORPHISM design pattern [3]. The implementation of algorithm registries is very similar to "Object Factories" [1, Chap 8]., an implementation of the Abstract Factories [5] tailored for C++.

5   Conclusion and Future Work

VAUCANSON 2 is a complete rewrite of the VAUCANSON platform. Its whole design relies on objects called "contexts" that behave as a custom typing-system for automata and rational expressions. They are also the cornerstone of the two-level API, which provides an easy-to-use and flexible dynamic interface on top of an efficient static C++ library. Although VAUCANSON 2 is still in its early phase, it already features 30+ different dynamic algorithms (including I/O) and 15+ predefined contexts. The benchmarks show that the new concepts have a clear and positive effect on performances. For instance, as of mid 2013, VAUCANSON 2 is about 2.5 times faster on determinization than its predecessor, and 50% slower than OPENFST.

Much remains to do. Importing the algorithms from VAUCANSON 1 will be a non-straightforward effort, as it will require adjusting to the new API, taking advantage of C++11 features, and binding with dynamic API. It is expected to improve both the readability of the algorithms, and their efficiency.

Adding more contexts should be reasonably simple. We are confident that adding new WeightSets is simple and straightforward, as there are already many very different implementations. Variety in the LabelSets is more challenging: supporting other concepts of generators (integers for instance), and different label kinds (tuples of letters and empty word, for instance, for implementing automata on multiple tapes, a generalization of VAUCANSON 1's support for transducers [7]).

References

1. Alexandrescu, A.: Modern C++ Design: Generic Programming and Design Patterns Applied. Addison-Wesley (2001)
2. Allauzen, C., Riley, M., Schalkwyk, J., Skut, W., Mohri, M.: OpenFst: A general and efficient weighted finite-state transducer library. In: Holub, J., Žďárek, J. (eds.) CIAA 2007. LNCS, vol. 4783, pp. 11–23. Springer, Heidelberg (2007), http://www.openfst.org
3. Cleeland, C., Schmidt, D.C., Harrison, T.: External polymorphism — an object structural pattern for transparently extending C++ concrete data types. In: Proceedings of the 3rd Pattern Languages of Programming Conference (1997)
4. Demaille, A., Duret-Lutz, A., Lesaint, F., Lombardy, S., Sakarovitch, J., Terrones, F.: An XML format proposal for the description of weighted automata, transducers, and regular expressions. In: Piskorski, J., Watson, B.W., Yli-Jyrä, A. (eds.) Post-Proceedings of the Seventh International Workshop on Finite-State Methods and Natural Language Processing (FSMNLP 2008). Frontiers in Artificial Intelligence and Applications, vol. 19, pp. 199–206. IOS Press (2009)
5. Gamma, E., Helm, R., Johnson, R., Vlissides, J.: Design Patterns: Elements of Reusable Object-Oriented Software. Addison-Wesley (1995)
6. Guingne, F., Nicart, F.: Finite state lazy operations in NLP. In: Champarnaud, J.-M., Maurel, D. (eds.) CIAA 2002. LNCS, vol. 2608, pp. 138–147. Springer, Heidelberg (2003)
7. Lombardy, S., Poss, R., Régis-Gianas, Y., Sakarovitch, J.: Introducing Vaucanson. In: Ibarra, O.H., Dang, Z. (eds.) CIAA 2003. LNCS, vol. 2759, pp. 96–107. Springer, Heidelberg (2003)
8. Lombardy, S., Sakarovitch, J.: The validity of weighted automata. Int. J. of Algebra and Computation (2013) (to appear) Available on ArXiV
9. Vaucanson Group. Vaucanson TAF-Kit Documentation, 1.4.1 edn. (September 2011), http://www.vaucanson-project.org
10. Wirth, N.: Algorithms + Data Structures = Programs. Prentice Hall (1976)

A Completion Algorithm
for Lattice Tree Automata

Thomas Genet[1], Tristan Le Gall[2], Axel Legay[1], and Valérie Murat[1]

[1] INRIA/IRISA, Rennes, France
[2] CEA, LIST, Centre de recherche de Saclay, France

Abstract. When dealing with infinite-state systems, Regular Tree Model Checking approaches may have some difficulties to represent infinite sets of data. We propose Lattice Tree Automata, an extended version of tree automata to represent complex data domains and their related operations in an efficient manner. Moreover, we introduce a new completion-based algorithm for computing the possibly infinite set of reachable states in a finite amount of time. This algorithm is independent of the lattice making it possible to seamlessly plug abstract domains into a Regular Tree Model Checking algorithm. As a first instance, we implemented a completion with an interval abstract domain. We provide some experiments showing that this implementation permits to scale up regular tree model-checking of Java programs dealing with integer arithmetics.

1   Introduction

In verification, infinite-state models are often used to avoid assumptions on data structures and architectures, e.g. an artificial bound on the size of a stack or on the value of a variable. At the heart of most of the techniques that have been proposed for exploring infinite state spaces, is a symbolic representation that can finitely represent infinite set of states. In Regular Tree Model Checking (RTMC), states are represented by trees, set of states by tree automata, and behavior of the system by tree transducers [1,8] or rewriting rules [11,16]. Any RTMC approach is equipped with an acceleration algorithm to compute possibly infinite sets of states in a finite amount of time. Among such algorithms, completion by equational abstraction [16] computes successive automata obtained by application of the rewriting rules, and merges intermediary states according to an equivalence relation to enforce the termination of the process.

In [6], the authors proposed an exact translation of the semantics of the Java Virtual Machine to tree automata and rewriting rules. This translation permits to analyze Java programs with Regular Tree Model checkers. One of the major difficulties of this encoding is to capture and handle the two-side infinite dimension that can arise in Java programs. Indeed, in such models, infinite behaviors may be due to unbounded number of calls to method and object creation, or simply because the program is manipulating unbounded data such as integer variables. While multiple infinite behaviors can be over-approximated with completion and equational abstraction [16], their combinations may require the use of artificially large structures.

S. Konstantinidis (Ed.): CIAA 2013, LNCS 7982, pp. 134–145, 2013.

We address this issue by defining *Lattice Tree Automata* (LTA). LTA have special transitions to abstract possibly infinite sets of values by a single element of a lattice. For example, we may abstract a set of integer values by a single interval instead of using an unary or binary encoding of those integers and recognizing the corresponding terms [6]. LTA recognize terms built over such intervals, and the completion algorithm built on LTA will perform each basic arithmetic operation in a single completion step, thanks to abstract interpretation techniques [9].

In this paper, we first define the LTA structure, then we propose a completion algorithm (by equational abstraction) that returns an approximation of the set of reachable states of an infinite-state systems whose behavior is modeled by rewriting rules. Finally, we provide some experimental results on the verification of Java programs using a RTMC environment. More details can be found in [15].

Related Work. [20] defined *lattice automata* to represent sets of words over an infinite alphabet. LTA are an extension of lattice automata to trees. Other models like modal automata [4] or data trees [12,13] consider tree structure with infinite alphabets but do not exploit the lattice structure as we do. Lattice (-valued) automata [19] map words over a finite alphabet to a lattice value, while LTA map trees over an infinite alphabet to $\{0, 1\}$. Similar automata may define fuzzy tree languages [10]. Verification of particular classes of properties of Java programs with interpreted terms can be found in [23].

Many techniques aim at the verification of programs with integer arithmetics. Among them, abstract interpretation [9] computes over-approximations of reachability sets, but requires a complete evaluation of arithmetic expressions. LTA can handle expressions that are only partially evaluated, thus may be useful in interprocedural analysis. There are other ways to deal with arithmetic efficiently in a regular model-checking framework such as [21]. However, we think that LTA provide a way to abstract many different types of data (integers, strings, etc.) by simply plugging the adapted abstract domain (and using its best available implementation) in a RTMC framework. In particular, LTA could be used by other RTMC techniques like [1,8] where such an ability does not exist.

2   Background

Rewriting Systems and Tree Automata. Let $\mathcal{F}$ be a finite set of *functional symbols*, where each symbol is associated with an *arity*, and let $\mathcal{X}$ be a countable set of *variables*. $\mathcal{T}(\mathcal{F}, \mathcal{X})$ denotes the set of *terms* and $\mathcal{T}(\mathcal{F})$ denotes the set of *ground terms* (terms without variables). $Var(t)$ denotes the set of variables of a term t, and $\mathcal{F}^n$, the set of functional symbols of arity n. We denote by $Pos(t)$ the set of positions of a term t, *i.e.* the set of positions of all its subterms, where a position is a world over $\mathbb{N}$ and ε denotes the top-most position. If $p \in Pos(t)$, then $t|_p$ denotes the subterm of t at position p and $t[s]_p$ denotes the term obtained by replacement of the subterm $t|_p$ at position p by the term s. A *Term Rewriting System* (*TRS*) $\mathcal{R}$ is a set of *rewrite rules* $l \rightarrow r$, where $l, r \in \mathcal{T}(\mathcal{F}, \mathcal{X})$, and $Var(l) \supseteq Var(r)$. A rewrite rule $l \rightarrow r$ is *left-linear* if each variable of l occurs only once in l. A *TRS* $\mathcal{R}$ is left-linear if every rewrite rule of $\mathcal{R}$ is left-linear.

We now define Tree Automata that are used to recognize possibly infinite sets of terms. Let $\mathcal{Q}$ be a finite set of symbols of arity 0, called *states*, such that $\mathcal{Q} \cap \mathcal{F} = \emptyset$. The set of *configurations* is $\mathcal{T}(\mathcal{F} \cup \mathcal{Q})$. A *transition* is a rewrite rule $c \to q$, where c is a configuration and q is a state. A transition is *normalized* when $c = f(q_1, \ldots, q_n)$, $f \in \mathcal{F}$ is of arity n, and $q_1, \ldots, q_n \in \mathcal{Q}$. A bottom-up non-deterministic finite tree automaton (tree automaton for short) over the alphabet $\mathcal{F}$ is a tuple $\mathcal{A} = \langle \mathcal{F}, \mathcal{Q}, \mathcal{Q}_F, \Delta \rangle$, where $\mathcal{Q}_F \subseteq \mathcal{Q}$ is the set of final states, Δ is a set of normalized transitions. The transitive and reflexive *rewriting relation* on $\mathcal{T}(\mathcal{F} \cup \mathcal{Q})$ induced by Δ is denoted by $\to_{\mathcal{A}}^*$. The tree language recognized by $\mathcal{A}$ in a state q is $\mathcal{L}(\mathcal{A}, q) = \{t \in \mathcal{T}(\mathcal{F}) \mid t \to_{\mathcal{A}}^* q\}$. We define $\mathcal{L}(\mathcal{A}) = \bigcup_{q \in \mathcal{Q}_F} \mathcal{L}(\mathcal{A}, q)$.

Lattices, Atomic Lattices, Galois Connections. A partially ordered set $(\Lambda, \sqsubseteq)$ is a *lattice* if it admits a *smallest element* $\bot$ and a *greatest element* $\top$, and if any finite set of elements $X \subseteq \Lambda$ admits a *greatest lower bound (glb)* $\sqcap X$ and a *least upper bound (lub)* $\sqcup X$. A lattice is complete if the *glb* and *lub* operators are defined for all possibly infinite subsets of Λ. An element x of a lattice $(\Lambda, \sqsubseteq)$ is an *atom* if it is minimal, *i.e.* $\bot \sqsubset x \wedge \forall y \in \Lambda : \bot \sqsubset y \sqsubseteq x \Rightarrow y = x$. The set of atoms of Λ is denoted by $Atoms(\Lambda)$. A lattice $(\Lambda, \sqsubseteq)$ is atomic if any element $x \in \Lambda$ where $x \neq \bot$ is the least upper bound of atoms, *i.e.* $x = \sqcup\{a \mid a \in Atoms(\Lambda) \wedge a \sqsubseteq x\}$.

Considered two lattices $(C, \sqsubseteq_C)$ (the concrete domain) and $(A, \sqsubseteq_A)$ (the abstract domain), there is a *Galois connection* between the two if there are two monotonic functions $\alpha : C \to A$ and $\gamma : A \to C$ such that : $\forall x \in C, y \in A$, $\alpha(x) \sqsubseteq_A y$ if and only if $x \sqsubseteq_C \gamma(y)$. As an example, sets of integers $(2^{\mathbb{Z}}, \subseteq)$ can be abstracted by the atomic lattice $(I, \sqsubseteq)$ of intervals, whose bounds belong to $\mathbb{Z} \cup \{-\infty, +\infty\}$ and whose atoms are of the form $[x, x]$, for each $x \in \mathbb{Z}$. Any operation op defined on a concrete domain C can be lifted to an operation $op^{\#}$ on the corresponding abstract domain A, thanks to the Galois connection.

3   Lattice Tree Automata

In this section, we first explain how to add elements of a concrete domain into terms, which has been defined in [18]. Then we propose a new type of tree automata recognizing terms with elements of an abstract lattice.

3.1   Interpreted Symbols and Evaluation

In what follows, elements of a possibly infinite concrete domain $\mathcal{D}$ will be represented by a set of *interpreted* symbols $\mathcal{F}_{\bullet}$. The set of symbols is now $\mathcal{F} = \mathcal{F}_{\circ} \cup \mathcal{F}_{\bullet}$, where $\mathcal{F}_{\circ}$ is the set of *passive* (uninterpreted) symbols. The set of *interpreted* symbols $\mathcal{F}_{\bullet}$ is composed of elements of $\mathcal{D}$ (notice that $\mathcal{D} \subseteq \mathcal{F}_{\bullet}^0$), and is also composed of some predefined operations $op : \mathcal{D}^n \to \mathcal{D}$, where $op \in \mathcal{F}_{\bullet}^n$ and $n > 0$. We denote by OP the set of predefined operations, thus we have $\mathcal{F}_{\bullet} = \mathcal{D} \cup OP$. For example, if $\mathcal{D} = \mathbb{Z}$, then $\mathcal{F}_{\bullet}$ can be $\mathbb{Z} \cup \{+, -, *\}$. Passive symbols can be seen as usual non-interpreted functional operators, and interpreted symbols stand for *built-in* operations on the domain $\mathcal{D}$. The set $\mathcal{T}(\mathcal{F}_{\bullet})$ of terms built on $\mathcal{F}_{\bullet}$ (called interpreted terms) can be evaluated by using an eval function $eval : \mathcal{T}(\mathcal{F}_{\bullet}) \to \mathcal{D}$. The purpose of $eval$ is to simplify a term using the built-in operations of the domain $\mathcal{D}$. $eval$ naturally extends to $\mathcal{T}(\mathcal{F})$: (1) $eval(f(t_1, \ldots, t_n)) = f(eval(t_1), \ldots, eval(t_n))$ if $f \in \mathcal{F}_\circ$ or $\exists i = 1 \ldots n : t_i \notin \mathcal{T}(\mathcal{F}_\bullet)$, or (2) the evaluation returns an element of $\mathcal{D}$ if $f(t_1, \ldots, t_n) \in \mathcal{T}(\mathcal{F}_\bullet)$.

We want to define tree automata to recognize sets of interpreted terms. To recognize $\{f(1), f(2), f(3), f(4)\}$, we would like to have tree automata with special transitions to handle sets of integers for instance: $\{1, \ldots, 4\} \to q, f(q) \to q_f$. We propose to generalize this encoding and to define tree automata with some transitions to recognize elements of a lattice (sets of integers are elements of the lattice $(2^{\mathbb{Z}}, \subseteq)$). By considering generic lattices, we can also improve the efficiency of the approach. Since RTMC only requires an over-approximation of the set of reachable states, we have special transitions to recognize elements of a simple, *abstract lattice* $(\Lambda, \sqsubseteq)$ such as the lattice of intervals. Moreover, we assume that this abstract lattice is atomic (cf. Section 2).

Each built-in operation $op \in OP$ defined on $\mathcal{D}$, is also abstracted by $op^\# \in OP^\#$. Since we have that $\mathcal{F}_\bullet = \mathcal{D} \cup OP$, the set of *abstract symbols* is $\mathcal{F}_\bullet^\# = \Lambda \cup OP^\#$. The arity of $op^\#$ is the same as the one of op. Assuming there is a Galois connection between the concrete domain and the abstract one (cf. Section 2), then $op^\# = \alpha \circ op \circ \gamma$ and $eval^\# : \mathcal{T}(\mathcal{F}_\bullet^\#) \mapsto \Lambda$ is the best approximation of $eval$.

Example 1. There is a Galois connection between $(2^{\mathbb{Z}}, \subseteq)$ and the lattice of intervals $(I, \sqsubseteq)$. $eval^\#([2, 3] +^\# [-1, 2]) = [1, 5])$.

3.2   Definition and Semantics

Definition 1 (Lattice tree automaton). *A bottom-up non-deterministic finite tree automaton with lattice (lattice tree automaton for short, LTA) for a given lattice Λ, is a tuple $\mathcal{A} = \langle \mathcal{F} = \mathcal{F}_\circ \cup \mathcal{F}_\bullet^\#, \mathcal{Q}, \mathcal{Q}_F, \Delta \rangle$, where $\mathcal{F}_\circ$ is a set of passive symbols and $\mathcal{F}_\bullet^\# = \Lambda \cup OP^\#$ a set of interpreted symbols, $\mathcal{Q}$ a set of states, $\mathcal{Q}_F \subseteq \mathcal{Q}$ are the final states, and Δ is a set of normalized transitions.*

The set of *lambda transitions*, which recognize elements of the lattice, is defined by $\Delta_\Lambda = \{\lambda \to q \mid \lambda \to q \in \Delta \wedge \lambda \neq \bot \wedge \lambda \in \Lambda\}$. The set of *ground transitions* is formally defined by $\Delta_G = \{f(q_1, \ldots, q_n) \to q \mid f \in \mathcal{F} \wedge f(q_1, \ldots, q_n) \to q \in \Delta \wedge q, q_1, \ldots, q_n \in \mathcal{Q}\}$. *Epsilon transitions* are transitions of the form $q \to q'$ where $q, q' \in \mathcal{Q}$. We extend the partial ordering $\sqsubseteq$ (on Λ) on $\mathcal{T}(\mathcal{F})$:

Definition 2. *Given $s, t \in \mathcal{T}(\mathcal{F})$, $s \sqsubseteq t$ iff :*
(1) $eval(s) \sqsubseteq eval(t)$ (if both s and t belong to $\mathcal{T}(\mathcal{F}_\bullet^\#)$), or (2) $s = f(s_1, \ldots, s_n)$, $t = f(t_1, \ldots, t_n)$, $f \in \mathcal{F}_\circ^n$ and $s_1 \sqsubseteq t_1 \wedge \ldots \wedge s_n \sqsubseteq t_n$.

Example 2. $f(g(a, [1, 5])) \sqsubseteq f(g(a, [0, 8]))$, and $h([0, 4] + [2, 6]) \sqsubseteq h([1, 3] + [1, 9])$.

In what follows we may omit $\#$ on abstract operations when it is clear from the context. We now define the transition relation and recognized language of an LTA. A term t is recognized by an LTA $\mathcal{A}$ if $eval(t)$ can be reduced in $\mathcal{A}$.

Definition 3 ($t_1 \to_\mathcal{A} t_2$ for LTA). *Let $t_1, t_2 \in \mathcal{T}(\mathcal{F} \cup \mathcal{Q})$. $t_1 \to_\mathcal{A} t_2$ iff, for all position $p \in Pos(t_1)$:*

- if $t_1|_p \in \mathcal{T}(\mathcal{F}_\bullet^\#)$, there is a transition $\lambda \to q \in \Delta$ such that $eval(t_1|_p) \sqsubseteq \lambda$ and $t_2 = t_1[q]_p$
- if $t_1|_p = q$ where $q \in \mathcal{Q}$, there is an epsilon-transition $q \to q' \in \Delta$, where $q' \in \mathcal{Q}$ such that $t_2 = t_1[q']_p$
- if $t_1|_p = f(s_1, \dots, s_n)$ where $f \in \mathcal{F}^n$ and $s_1, \dots s_n \in \mathcal{T}(\mathcal{F} \cup \mathcal{Q})$, $\exists s_i' \in \mathcal{T}(\mathcal{F} \cup \mathcal{Q})$ such that $s_i \to_\mathcal{A} s_i'$ and $t_2 = t_1[f(s_1, \dots, s_{i-1}, s_i', s_{i+1}, \dots, s_n)]_p$
- if $t_1|_p = f(q_1, \dots, q_n)$ where $f \in \mathcal{F}^n$ and $q_1, \dots q_n \in \mathcal{Q}$, there is a transition $f(q_1, \dots, q_n) \to q \in \Delta$ such that $t_2 = t_1[q]_p$.

$\to_\mathcal{A}^*$ is the reflexive transitive closure of $\to_\mathcal{A}$. There is a run from t_1 to t_2 if $t_1 \to_\mathcal{A}^* t_2$. If a LTA has a transition $[0, 2] \to q$ then $[0, 0] \to_\mathcal{A}^* q$, $[1, 2] \to_\mathcal{A}^* q$, $\dots$, i.e. all possible unions of atoms $[0, 0], [1, 1], [2, 2]$. The language recognized by a LTA is thus defined over $\mathcal{T}(\mathcal{F}, Atoms(\Lambda))$, where $\mathcal{T}(\mathcal{F}, Atoms(\Lambda))$ is the set of ground terms built over $(\mathcal{F} \setminus \Lambda) \cup Atoms(\Lambda)$.

Definition 4 (Recognized language). *The tree language recognized by $\mathcal{A}$ in a state q is $\mathcal{L}(\mathcal{A}, q) = \{t \in \mathcal{T}(\mathcal{F}, Atoms(\Lambda)) \mid \exists\, t' \text{ such that } t \sqsubseteq t' \text{ and } t' \to_\mathcal{A}^* q\}$. The language recognized by $\mathcal{A}$ is $\mathcal{L}(\mathcal{A}) = \bigcup_{q \in \mathcal{Q}_f} \mathcal{L}(\mathcal{A}, q)$.*

Example 3 (Run, recognized language). Let $\mathcal{A} = \langle \mathcal{F} = \mathcal{F}_\circ \cup \mathcal{F}_\bullet^\#, \mathcal{Q}, \mathcal{Q}_f, \Delta \rangle$ be an LTA where $\Delta = \{[0, 4] \to q_1, f(q_1) \to q_2\}$ and $\mathcal{Q}_f = \{q_2\}$. We have: $f([1, 4]) \to_\mathcal{A}^* q_2$ and $f([0, 1] + [0, 1]) \to_\mathcal{A}^* q_2$, and the recognized language of $\mathcal{A}$ is given by $\mathcal{L}(\mathcal{A}, q_2) = \{f([0, 0]), f([1, 1]), \dots, f([4, 4])\}$.

4   Completion Algorithm

We only present here the completion algorithm on LTA, other operations are detailed in [15]. We are interested in computing the set of reachable states of an infinite state system. We propose to represent states by (built-in) terms and possibly infinite set of states by an LTA. In this section, we assume that the behavior of the system can be represented by conditional term rewriting systems, *i.e.* TRS equipped with conjunction of conditions used to restrain the applicability of the rule. Our conditional TRS, which extends the classical definition of [2], rewrites terms defined on the concrete domain. This makes them independent from the abstract lattice. We first start with the definition of predicates that allows us to express conditions in TRS.

Definition 5 (Predicates). *Let $\mathcal{P}$ be the set of predicates over $\mathcal{D}$. Let ρ be a n-ary predicate of $\mathcal{P}$ such that $\rho : \mathcal{D}^n \mapsto \{true, false\}$. We extend the domain of ρ to $\mathcal{T}(\mathcal{F})^n$ in the following way:*

$$\rho(t_1, \dots, t_n) = \begin{cases} \rho(u_1, \dots, u_n) \text{ if } \forall i = 1 \dots n : t_i \in \mathcal{T}(\mathcal{F}_\bullet) \text{ and } u_i = eval(t_i) \\ false \text{ if } \exists j = 1 \dots n : t_j \notin \mathcal{T}(\mathcal{F}_\bullet) \end{cases}$$

Observe that if one of the predicate parameters cannot be evaluated into a built-in term, then the predicate returns false and the rule is not applied.

Definition 6 (Conditional Term Rewriting System (*CTRS*) on $\mathcal{T}(\mathcal{F}_\circ \cup \mathcal{F}_\bullet, \mathcal{X})$). *In our setting, a Conditional Term Rewriting System $\mathcal{R}$ is a set of rewrite rules $l \to r \Leftarrow c_1 \wedge \ldots \wedge c_n$, where $l \in \mathcal{T}(\mathcal{F}_\circ, \mathcal{X})$, $r \in \mathcal{T}(\mathcal{F}_\circ \cup \mathcal{F}_\bullet, \mathcal{X})$, $l \notin \mathcal{X}$, $Var(l) \supseteq Var(r)$ and $\forall i = 1 \ldots n : c_i = \rho_i(t_1, \ldots, t_m)$ where ρ_i is a m-ary predicate of $\mathcal{P}$ and $\forall j = 1 \ldots m : t_j \in \mathcal{T}(\mathcal{F}_\bullet, \mathcal{X}) \wedge Var(t_j) \subseteq Var(l)$.*

Example 4. Using conditional rewriting rules, the factorial can be encoded by the CTRS: $fact(x) \to 1 \Leftarrow x \geq 0 \wedge x \leq 1$, $fact(x) \to x * fact(x-1) \Leftarrow x \geq 2$.

Let $\mathcal{X}$ a set of variables, $\mathcal{Q}$ a set of states, and $\mathcal{F}$ a set of symbols. A *substitution* σ is a function $\sigma : \mathcal{X} \mapsto \mathcal{Q} \cup \mathcal{T}(\mathcal{F})$ that can be extended to $\mathcal{T}(\mathcal{F}, \mathcal{X})$ in this way: for all $t \in \mathcal{T}(\mathcal{F}, \mathcal{X})$, we define $t\sigma$ as: (1) if $t = f(t_1, \ldots, t_n)$ then $t\sigma = f(t_1\sigma, \ldots, t_n\sigma)$, where $t, t_1, \ldots, t_n \in \mathcal{T}(\mathcal{F}, \mathcal{X}), f \in \mathcal{F}^n$, (2) if $t = x \in \mathcal{X}$ then $t\sigma = \sigma(x)$. Recall that $\mathcal{F} = \mathcal{F}_\circ \cup \mathcal{F}_\bullet$. The *CTRS* $\mathcal{R}$ and the *eval* function induces a rewriting relation $\to_\mathcal{R}$ on $\mathcal{T}(\mathcal{F})$: in the following way: for all $s, t \in \mathcal{T}(\mathcal{F})$, we have $s \to_\mathcal{R} t$ if there exist: (1) a rewrite rule $l \to r \Leftarrow c_1 \wedge \ldots \wedge c_n \in \mathcal{R}$, (2) a position $p \in \mathcal{P}os(s)$, and (3) a substitution $\sigma : \mathcal{X} \mapsto \mathcal{T}(\mathcal{F})$ s.t. $s|_p = l\sigma$, $t = eval(s[r\sigma]_p)$ and $\forall i = 1 \ldots n : c_i\sigma = true$. The reflexive transitive closure of $\to_\mathcal{R}$ is denoted by $\to_\mathcal{R}^*$.

Let $\mathcal{A}$ be an LTA representing the set of initial states, and $\mathcal{R}$ be a *CTRS*. Our objective is to compute another LTA representing (an over-approximation of) the set $\mathcal{R}^*(\mathcal{L}(\mathcal{A})) = \{t \mid \exists t_0 \in \mathcal{L}(\mathcal{A}), t_0 \to_\mathcal{R}^* t\}$. We adopt the completion approach of [16,11], which intends to compute a tree automaton $\mathcal{A}_\mathcal{R}^k$ such that $\mathcal{L}(\mathcal{A}_\mathcal{R}^k) \supseteq \mathcal{R}^*(\mathcal{L}(\mathcal{A}))$ for a left-linear *CTRS* $\mathcal{R}$. The algorithm proceeds by computing the sequence of automata $\mathcal{A}_\mathcal{R}^0, \mathcal{A}_\mathcal{R}^1, \mathcal{A}_\mathcal{R}^2, \ldots$ that represents successive applications of $\mathcal{R}$. Computing $\mathcal{A}_\mathcal{R}^{i+1}$ from $\mathcal{A}_\mathcal{R}^i$ is called a *one-step completion*. In general the sequence of automata may not converge in a finite amount of time. To accelerate the convergence, we perform an abstraction operation that will be described in section 4.3. We now give details on the above constructions, which will be illustrated step by step by a running example.

4.1 Computation of $\mathcal{A}_\mathcal{R}^{i+1}$

In our setting, $\mathcal{A}_\mathcal{R}^{i+1}$ is built from $\mathcal{A}_\mathcal{R}^i$ by using a *completion step* that relies on finding critical pairs. Given a substitution $\sigma : \mathcal{X} \mapsto \mathcal{Q}$ and a rule $l \to r \Leftarrow c_1 \wedge \ldots \wedge c_n \in \mathcal{R}$, a critical pair is a pair $(r\sigma', q)$ where $q \in \mathcal{Q}$ and σ' is the greatest substitution w.r.t $\sqsubseteq$ such that $l\sigma \to_{\mathcal{A}_\mathcal{R}^i}^* q$, $\sigma \sqsupseteq \sigma'$ and $c_1\sigma' \wedge \ldots \wedge c_n\sigma'$. Since $\mathcal{R}$, $\mathcal{A}_\mathcal{R}^i$, $\mathcal{Q}$ are finite, there is only a finite number of such critical pairs. For each critical pair such that $r\sigma' \not\to_{\mathcal{A}_\mathcal{R}^i}^* q$, the algorithm adds two new transitions $r\sigma' \to q'$ and $q' \to q$ to $\mathcal{A}_\mathcal{R}^i$, in order to enrich the language of the previous automaton. To find all critical pairs, in what follows, we use the standard *matching algorithm* introduced and described in [11]. This algorithm $Matching(l, \mathcal{A}, q)$ matches a linear term l with a state q in the automaton $\mathcal{A}$. The solution returned by *Matching* is a set of substitutions $\{\sigma_1, \ldots, \sigma_n\}$ so that $l\sigma_i \to_\mathcal{A}^* q$. However, as our *TRS* relies on conditions, we have to extend this matching algorithm in order to guarantee that each substitution σ_i that is a solution of $l \to r \Leftarrow c_1 \wedge \ldots \wedge c_n$ satisfies $c_1 \wedge \ldots \wedge c_n$.

Example 5. Let $\mathbb{Z}$ be the concrete domain, and intervals on $\mathbb{Z}$ be the lattice, $\mathcal{R} = \{f(x) \to cons(x, f(x+1)) \Leftarrow x \geq 1\}$ be the *CTRS*, $\mathcal{A}_0$ the LTA representing the set of initial configurations, with transitions: $\Delta_0 = \{[0,2] \to q_1, f(q_1) \to q_2\}$. To build $\mathcal{A}_\mathcal{R}^1$ from $\mathcal{A}_0$, we have to find all possible substitutions. The matching algorithm tells that the rewrite rule applies with the substitution $\{x \mapsto q_1\}$. To satisfy the constraint $x \geq 1$, the substitution $\{x \mapsto q_1\}$ with $[0,2] \to q_1$ will be restricted to $\{x \mapsto [1,2]\}$.

Restricting substitutions is done by a solver *Solve* on abstract domains. The output of $Solve(\sigma, \mathcal{A}, c_1 \wedge \ldots \wedge c_n)$ is either a set of substitutions σ' which is a restriction of σ satisfying $c_1 \wedge \ldots \wedge c_n$ or $\emptyset$ if such a restriction does not exist. On the previous example, $Solve(\{x \mapsto q_1\}, \mathcal{A}, x \geq 1) = \{\{x \mapsto [1,2]\}\}$. See [15] for details about the properties the solver needs to have. Such properties are generally fulfilled by usual abstract domains implementations.

Definition 7 (Matching solutions of conditional rewrite rules). *Let $\mathcal{A}$ be a tree automaton, $rl = l \to r \Leftarrow c_1 \wedge \ldots \wedge c_n$ a rewrite rule and q a state of $\mathcal{A}$. The set of all possible substitutions for the rewrite rule rl is $\Omega(\mathcal{A}, rl, q) = \{\sigma' \mid \sigma \in Matching(l, \mathcal{A}, q) \wedge \sigma' \in Solve(\sigma, \mathcal{A}, c_1 \wedge \ldots \wedge c_n) \wedge \nexists \sigma'' : r\sigma' \sqsubseteq r\sigma'' \to_\mathcal{A}^* q\}$.*

Once the set of all possible restricted substitutions σ_i has been obtained, we must add the rules $r\sigma_i \to q'$ and $q' \to q$ in the automaton, where q' is a new state. However, $r\sigma_i \to q'$ is not necessarily a normalized ground transition of the form $f(q_1, \ldots, q_n) \to q$ or a lambda transition of the form $\lambda \to q$, which means it must be normalized first in order to be added to the LTA.

Definition 8 (Normalization). *Let $s \in \mathcal{T}(\mathcal{F} \cup \mathcal{Q})$, $q \in \mathcal{Q}$, $\mathcal{A} = \langle \mathcal{F}, \mathcal{Q}, \mathcal{Q}_f, \Delta \rangle$ an LTA, where $\mathcal{F}_\bullet$ is the set of concrete interpretable symbols used in the CTRS, $\mathcal{F}_\bullet^\#$ the set of abstract symbols used in $\mathcal{A}$, $\mathcal{F} = \mathcal{F}_\bullet^\# \cup \mathcal{F}_\circ$, and $\alpha : \mathcal{F}_\bullet^0 \to \mathcal{F}_\bullet^{\#^0}$ the abstraction function, mapping concrete constants to elements of Λ. A new state is a state of $\mathcal{Q}$ not occurring in Δ. $Norm(s \to q)$ returns the set of normalized transitions deduced from s. $Norm(s \to q)$ is defined by:*

1. *if $s \in \mathcal{F}_\bullet^0$ then $Norm(s \to q) = \{\alpha(s) \to q\}$.*
2. *if $s \in \mathcal{F}_\circ^0 \cup \mathcal{F}_\bullet^{\#^0}$ then $Norm(s \to q) = \{s \to q\}$,*
3. *if $s = f(t_1, \ldots, t_n)$ where $f \in \mathcal{F}_\circ^n \cup \mathcal{F}_\bullet^n$, then $Norm(s \to q) = \{f(q_1', \ldots, q_n') \to q\} \cup Norm(t_1 \to q_1') \cup \ldots \cup Norm(t_n \to q_n')$ where for $i = 1 \ldots n$, q_i' is either:*
   - *the right-hand side of a transition of Δ such that $t_i \to_\Delta^* q_i'$*
   - *or a new state, otherwise.*

Example 6. From Ex.5, we have to add the normalized form of $cons([1,2], f([1,2] + 1)) \to q_2'$ and $q_2' \to q_2$ (where q_2' is a new state) to the set of transitions: 1 has to be abstracted by $[1,1]$ and $f([1,2])$ has to be replaced by a state recognizing this term. So $\Delta_1 = \Delta_0 \cup Norm(cons([1,2], f([1,2]+1)) \to q_2') \cup \{q_2' \to q_2\} = \Delta_0 \cup \{[1,2] \to q_3, [1,1] \to q_{[1,1]}, q_3 + q_{[1,1]} \to q_4, f(q_4) \to q_5, cons(q_3, q_5) \to q_2', q_2' \to q_2\}$, where $q_{[1,1]}, q_3, q_4, q_5$ are new states induced by normalization.

Observe that the normalization algorithm always terminates. We conclude by the formal characterization of the one step completion.

Definition 9 (One step completed automaton $C_\mathcal{R}(\mathcal{A})$). *Let $\mathcal{A} = \langle \mathcal{F}, \mathcal{Q}, \mathcal{Q}_f, \Delta \rangle$ be a tree automaton, $\mathcal{R}$ be a left-linear CTRS. We denote by $C_\mathcal{R}(\mathcal{A})$ the one step completed automaton $C_\mathcal{R}(\mathcal{A}) = \langle \mathcal{F}, \mathcal{Q}', \mathcal{Q}_f, \Delta' \rangle$ where:*

$$\Delta' = \Delta \cup \bigcup_{l \to r \in \mathcal{R},\, q \in \mathcal{Q},\, \sigma \in \Omega(\mathcal{A}, l \to r, q)} Norm(r\sigma \to q') \cup \{q' \to q\}$$

where $\Omega(\mathcal{A}, l \to r, q)$ is the set of all possible substitutions defined in Def.7, $q' \notin \mathcal{Q}$ a new state and $\mathcal{Q}'$ contains all the states of Δ'.

4.2 Evaluation of a Lattice Tree Automaton

Any set of concrete terms that contains the term $1 + 2$ should also contain the term 3. While this property can be true on the initial automaton, it may be broken when performing a completion step.

Example 7. The first completion step described in Ex.6 adds the transition $q_3 + q_{[1,1]} \to q_4$. Since we have that $[1,2] \to q_3$ and $[1,1] \to q_{[1,1]}$, the language recognized by q_4 should also contain the term $[2,3]$.

The objective of the *propag* function is to evaluate the LTA and to add the transition $[2,3] \to q_4$ in the above example.

Definition 10 (*propag*). *Let Δ be the set of transitions of a LTA. Let $f(q_1,\ldots, q_n) \to q \in \Delta$, where $f \in \mathcal{F}_\bullet^n$ is an interpreted symbol and $q, q_1, \ldots, q_n \in \mathcal{Q}$. If there exists $\lambda_1, \ldots, \lambda_n \in \Lambda$ such that $\lambda_1 \to_\Delta^* q_1, \ldots, \lambda_n \to_\Delta^* q_n$, then one step of evaluation of $f(q_1, \ldots, q_n) \to q$ is defined by:*

$$propag(\Delta, f(q_1,\ldots,q_n) \to q) = \begin{cases} \Delta \text{ if } \exists \lambda \to q \in \Delta \wedge eval(f(\lambda_1,\ldots,\lambda_n)) \sqsubseteq \lambda \\ \Delta \cup \{eval(f(\lambda_1,\ldots,\lambda_n)) \to q\}, otherwise. \end{cases}$$

One step of evaluation for Δ is defined by:

$$propag(\Delta) = \bigcup_{\forall f(q_1,\ldots,q_n) \to q \in \Delta \text{ s.t. } f \in \mathcal{F}_\bullet^n} propag(\Delta, f(q_1,\ldots,q_n) \to q)$$

Since *propag* can add new transitions, it must be applied until a fix-point is reached. Then using *propag*, we can extend the *eval* function to sets of transitions and to tree automata in the following way.

Definition 11 (*eval* on transitions and automata). *$\mu X.f(X)$ denotes the least fix-point of a generic function f. We define: $eval(\Delta) = \mu X.propag(X) \cup \Delta$ and $eval(\langle \mathcal{F}, \mathcal{Q}, \mathcal{Q}_f, \Delta \rangle) = \langle \mathcal{F}, \mathcal{Q}, \mathcal{Q}_f, eval(\Delta) \rangle$.*

Example 8. In our example, $eval(\Delta_1) = \Delta_1 \cup \{[2,3] \to q_4\}$.

Theorem 1. $\mathcal{L}(\mathcal{A}) \subseteq \mathcal{L}(eval(\mathcal{A}))$.

4.3 Equational Abstraction

If we perform another completion step on our example, we see that we can apply the rewrite rule with a new substitution mapping x to q_4.

Example 9. Then $Norm(cons(q_4, f(q_4 + 1))) \to q_5'$ and $q_5' \to q_5$ will be added to $eval(\Delta_1)$ to build $\mathcal{A}_\mathcal{R}^2$ from $\mathcal{A}_\mathcal{R}^1$. We have $\Delta_2 = eval(\Delta_1) \cup \{q_4 + q_{[1,1]} \to q_6, f(q_6) \to q_7, cons(q_4, q_7) \to q_5', q_5' \to q_5\}$. If we perform the evaluation step, we have $eval(\Delta_2) = \Delta_2 \cup \{[3,4] \to q_6\}$. We can see that this process is infinite, because it will compute the infinite term $cons([1,2], cons([2,3], cons([3,4], \ldots)))$.

Termination of completion can be enforced using a set E of *approximation equations* as in [22,16]. Depending on the objective, E can either be defined by hand (e.g. [22]), by hand and automatically refined [5], or automatically generated from a static analysis of the *TRS* (e.g. [7]). In our example, the infinite behavior is due to transitions of the form $q_i + q_{[1,1]} \to q_j$. An equation such as $x = x + 1$ is needed to ensure termination of completion. Equations of E will be of the form $u = v \Leftarrow c_1 \wedge \ldots \wedge c_n$, where $u, v \in \mathcal{T}(\mathcal{F}_\circ \cup \mathcal{F}_\bullet, \mathcal{X})$. Let $\sigma : \mathcal{X} \mapsto \mathcal{Q}$ be a substitution s.t. $u\sigma \to_{\mathcal{A}_\mathcal{R}^{i+1}} q$, $v\sigma \to_{\mathcal{A}_\mathcal{R}^{i+1}} q'$ and $q \neq q'$. An over-approximation of $\mathcal{A}_\mathcal{R}^{i+1}$ (denoted by $\mathcal{A}_{\mathcal{R},E}^{i+1}$) can be obtained by merging states q and q', i.e. replacing each occurrence of q' by q in $\mathcal{A}_\mathcal{R}^{i+1}$. Contrary to the completion case, we do not need to restrict the substitutions obtained by the matching algorithm with respect to the constraints of the equation, but simply guarantee that such constraints are satisfiable, i.e., $Solve(\sigma, \mathcal{A}, c_1 \wedge \cdots \wedge c_n) \neq \emptyset$.

For instance, $E = \{x = x + 1 \Leftarrow x > 2\}$ can be used on Ex.9. We have two possible substitutions: $\sigma_1 = \{x \mapsto q_3\}$ and $\sigma_2 = \{x \mapsto q_4\}$. σ_1 is due to the transition $q_3 + q_{[1,1]} \to q_4$. However, since $[1,2] \to q_3$ we have $Solve(\{x \mapsto q_3\}, \mathcal{A}_2, x > 2) = \emptyset$ and thus σ_1 does not satisfy the condition. Substitution σ_2, due to the transition $q_4 + q_{[1,1]} \to q_6$, satisfies the condition because $[2,3] \to q_4$ and $Solve(\{x \mapsto q_4\}, \mathcal{A}_2, x > 2) = \{x \mapsto [3,3]\} \neq \emptyset$. Hence, the equation is applied for σ_2 and results in the merging of q_4 and q_6 according to E.

Theorem 2. *Let $\mathcal{A}$ be an LTA and E a set of equations. We denote by $\leadsto_E^!$ the transformation of $\mathcal{A}$ by merging all equivalent states according to E. If $\mathcal{A} \leadsto_E^! \mathcal{A}'$ then $\mathcal{L}(\mathcal{A}) \subseteq \mathcal{L}(\mathcal{A}')$.*

Widening Step. Any set containing the term $1+2$ should also contain the term 3. However, this can be broken by merging. Merging of states changes transitions of the LTA. So we have to perform an evaluation step after merging by equations.

Example 10. After merging q_4 and q_6, we have $Merge(\Delta_2, q_4, q_6) = eval(\Delta_1) \cup \{q_3 + q_{[1,1]} \to q_4, f(q_4) \to q_5, cons(q_3, q_5) \to q_2', q_2' \to q_2, [2,3] \to q_4, q_4 + q_{[1,1]} \to q_4, f(q_4) \to q_7, cons(q_4, q_7) \to q_5', q_5' \to q_5, [3,4] \to q_4\}$. We have to evaluate the transition $q_4 + q_{[1,1]} \to q_4$. The first iteration will evaluate the term $[3,4] + [1,1]$ which adds the transition $[4,5] \to q_4$. Since a new element is in the state q_4, the second iteration will evaluate the term $[4,5] + [1,1]$ recognized by the transition $q_4 + q_{[1,1]} \to q_4$. Since there will always be a new element of the lattice that will be associated to q_4, the computation of the evaluation will not terminate.

Since $eval$ is defined as a fix-point of *propag*, this computation may not terminate without the application of a widening operator $\nabla_\Lambda : \Lambda \times \Lambda \mapsto \Lambda$. It is a classical way to compute over-approximation of fix-points within the abstract interpretation framework [9].

Example 11. If we apply such a widening operator on our example after 3 iterations (for instance) of the *propag* function, then the transitions: $[2,3] \rightarrow q_4$, $[3,4] \rightarrow q_4$, $[4,5] \rightarrow q_4$ will be replaced by $[2,+\infty[\rightarrow q_4$.

4.4 LTA Completion and Its Soundness

Definition 12 (Automaton completion for LTA). *Let $\mathcal{A}$ be a tree automaton, $\mathcal{R}$ a CTRS and E a set of equations.*

- $\mathcal{A}_{\mathcal{R},E}^0 = \mathcal{A}$,
- *Repeat* $\mathcal{A}_{\mathcal{R},E}^{n+1} = \mathcal{A}'$ *with* $eval(\mathcal{C}_{\mathcal{R}}(\mathcal{A}_{\mathcal{R},E}^n)) \rightsquigarrow_E^! \mathcal{A}''$ *and* $eval(\mathcal{A}'') = \mathcal{A}'$,
- *Until a fixpoint* $\mathcal{A}_{\mathcal{R},E}^* = \mathcal{A}_{\mathcal{R},E}^k = \mathcal{A}_{\mathcal{R},E}^{k+1}$ *(with $k \in \mathbb{N}$) is reached.*

Theorem 3 (Soundness). *Let $\mathcal{R}$ be a left-linear CTRS, $\mathcal{A}$ be a tree automaton and E be a set of linear equations. If completion terminates on $\mathcal{A}_{\mathcal{R},E}^*$ then $\mathcal{L}(\mathcal{A}_{\mathcal{R},E}^*) \supseteq \mathcal{R}^*(\mathcal{L}(\mathcal{A}))$*

Example 12. In our example, thanks to the widening performed at the previous evaluation step, completion adds no more rule to the current automaton and stops. We have a fixed-point which is an over-approximation of the set of reachable states.

5   Experiments

LTA completion has been developed and integrated into Timbuk [14]. For those experiments, we choose to instantiate the generic LTA-completion algorithm with the lattice of integer intervals: TimbukLTA. Experiments are detailed in [15].

We compare the efficiency of LTA completion w.r.t. standard completion on *TRS* produced by Copster [3]. Copster compiles *Java* `.class` files into a *TRS* modeling exactly the semantics of the Java program[1]. We extend Copster to produce either *TRSs* or conditional *TRS* (*CTRS*) as in Section 4. *CTRSs* do not use Peano integers or arithmetic but assume that all integer arithmetic is built-in. On the Java program examples, we prove the same properties using either Timbuk or TimbukLTA and compare their efficiency. We made several experiments on three different Java programs that are detailed in [15]. On the first one, called "Threads", we prove that whatever the scheduling of Java threads may be, the access to a critical section is protected using the **synchronized** Java mechanism. The second one "Euclid" consists of an implementation of integer division in a recursive way using addition and subtraction. In the third one, called "FactoList", there is an unbounded number of integers which are read on the input channel and their factorial values are stored into a singly linked list. In the end, the content of the list is printed to the output stream. Depending on the possible values for integers read on the input stream, we can prove different properties on the integers printed on the output stream.

---

[1] Copster covers basic types, arithmetic, object creation, heap management, field manipulation, virtual method invocation, threads, as well as a subset of the System and String library. Details about this compilation can be found in [6].

Table 1. Performances of standard completion against LTA completion

Examples	Standard completion		LTA completion	
	Compl. steps	Compl. time	Compl. steps	Compl. time
Threads	306	56s	328	280s
Euclid	2019	59s	727	14s
FactoList, input stream=$(3,1,2,0)$	799	17s	538	33s
FactoList, input stream=$(7,5,6,4,1)$	>9465	>2h	1251	250s
FactoList, any input stream of $[-\infty; +\infty]$	467	20s	349	40s
FactoList, any input stream of $[2; +\infty]$	468	21s	430	14s
FactoList, any input stream of $[3; +\infty]$	953	320s	467	15s
FactoList, any input stream of $[4; +\infty]$	>1500	> 2h	641	32s

Table 1 shows that integration of LTA in completion may reduce its efficiency when the *TRS* to verify does not rely on arithmetic ("Threads" example). On the opposite, unlike standard completion, LTA completion scales up when arithmetic is used in the analysis ("Euclid" and "FactoList" example). TimbukLTA and the adapted Copster can be downloaded from their respective pages [14,3].

6   Conclusion and Future Work

We have proposed LTA, a new extension of tree automata for tree regular model checking of infinite-state systems with interpreted terms. One of our main contributions is the development of a new completion algorithm for such automata. A nice property of this adapted algorithm is that it is independent of the lattice: it only has to be atomic and equipped with a solver for the predicates of the *CTRS* [15]. Any lattice fulfilling those requirements can be seamlessly plugged into the regular tree model checking algorithm. We developed TimbukLTA which is the implementation of completion for LTA. We presented a first instance of TimbukLTA where we plugged in an integer interval abstract domain. This simple abstract domain permitted to drastically improve the efficiency of completion for the verification of Java programs dealing with integer arithmetic. The resulting LTA homogeneously combine abstract domains to approximate numerical values with tree automata to approximate structures: thread states, stacks, heaps and objects. Future plans are to integrate in TimbukLTA more abstract domains dealing with other kinds of built-ins: strings, reals, etc. and to define syntactic constraints on equations to guarantee termination of LTA completion like in [17].

References

1. Abdulla, P.A., Jonsson, B., Mahata, P., d'Orso, J.: Regular tree model checking. In: Brinksma, E., Larsen, K.G. (eds.) CAV 2002. LNCS, vol. 2404, p. 555. Springer, Heidelberg (2002)
2. Baader, F., Nipkow, T.: Term Rewriting and All That. Cambridge University Press (1998)

3. Barré, N., Besson, F., Genet, T., Hubert, L., Le Roux, L.: Copster homepage (2009), http://www.irisa.fr/celtique/genet/copster
4. Bauer, S.S., Fahrenberg, U., Juhl, L., Larsen, K.G., Legay, A., Thrane, C.: Quantitative refinement for weighted modal transition systems. In: Murlak, F., Sankowski, P. (eds.) MFCS 2011. LNCS, vol. 6907, pp. 60–71. Springer, Heidelberg (2011)
5. Boichut, Y., Boyer, B., Genet, T., Legay, A.: Equational Abstraction Refinement for Certified Tree Regular Model Checking. In: Aoki, T., Taguchi, K. (eds.) ICFEM 2012. LNCS, vol. 7635, pp. 299–315. Springer, Heidelberg (2012)
6. Boichut, Y., Genet, T., Jensen, T., Le Roux, L.: Rewriting Approximations for Fast Prototyping of Static Analyzers. In: Baader, F. (ed.) RTA 2007. LNCS, vol. 4533, pp. 48–62. Springer, Heidelberg (2007)
7. Boichut, Y., Héam, P.-C., Kouchnarenko, O.: Approximation-based tree regular model-checking. Nord. J. Comput. 14(3), 216–241 (2008)
8. Bouajjani, A., Touili, T.: Extrapolating tree transformations. In: Brinksma, E., Larsen, K.G. (eds.) CAV 2002. LNCS, vol. 2404, p. 539. Springer, Heidelberg (2002)
9. Cousot, P., Cousot, R.: Abstract interpretation: A unified lattice model for static analysis of programs by construction or approximation of fixpoints. In: POPL, pp. 238–252 (1977)
10. Ésik, Z., Liu, G.: Fuzzy tree automata. Fuzzy Sets Syst. 158, 1450–1460 (2007)
11. Feuillade, G., Genet, T., Viet Triem Tong, V.: Reachability Analysis over Term Rewriting Systems. JAR 33(3-4), 341–383 (2004)
12. Figueira, D., Segoufin, L.: Bottom-up automata on data trees and vertical xpath. In: STACS (2011)
13. Genest, B., Muscholl, A., Wu, Z.: Verifying recursive active documents with positive data tree rewriting. In: FSTTCS (2010)
14. Timbuk, T.G.: http://www.irisa.fr/celtique/genet/timbuk/
15. Genet, T., Le Gall, T., Legay, A., Murat, V.: Tree regular model checking for lattice-based automata. Technical Report RT-0424, INRIA (2012), http://hal.inria.fr/hal-00687310
16. Genet, T., Rusu, V.: Equational approximations for tree automata completion. Journal of Symbolic Computation 45(5), 574–597 (2010)
17. Genet, T., Salmon, Y.: Tree Automata Completion for Static Analysis of Functional Programs. Technical report, INRIA (2013), http://hal.archives-ouvertes.fr/hal-00780124/PDF/main.pdf
18. Kaplan, S., Choppy, C.: Abstract rewriting with concrete operations. In: Dershowitz, N. (ed.) RTA 1989. LNCS, vol. 355, pp. 178–186. Springer, Heidelberg (1989)
19. Kupferman, O., Lustig, Y.: Lattice automata. In: Cook, B., Podelski, A. (eds.) VMCAI 2007. LNCS, vol. 4349, pp. 199–213. Springer, Heidelberg (2007)
20. Le Gall, T., Jeannet, B.: Lattice Automata: A Representation for Languages on Infinite Alphabets, and Some Applications to Verification. In: Riis Nielson, H., Filé, G. (eds.) SAS 2007. LNCS, vol. 4634, pp. 52–68. Springer, Heidelberg (2007)
21. Leroux, J.: Structural Presburger digit vector automata. TCS 409(3) (2008)
22. Meseguer, J., Palomino, M., Martí-Oliet, N.: Equational Abstractions. In: Baader, F. (ed.) CADE 2003. LNCS (LNAI), vol. 2741, pp. 2–16. Springer, Heidelberg (2003)
23. Otto, C., Brockschmidt, M., von Essen, C., Giesl, J.: Automated termination analysis of java bytecode by term rewriting. In: RTA. LIPIcs. Dagstuhl (2010)

Approximate Matching between a Context-Free Grammar and a Finite-State Automaton

Yo-Sub Han[1], Sang-Ki Ko[1], and Kai Salomaa[2]

[1] Department of Computer Science, Yonsei University
50, Yonsei-Ro, Seodaemun-Gu, Seoul 120-749, Republic of Korea
{emmous,narame7}@cs.yonsei.ac.kr
[2] School of Computing, Queen's University
Kingston, Ontario K7L 3N6, Canada
ksalomaa@cs.queensu.ca

Abstract. Given a context-free grammar (CFG) and a finite-state automaton (FA), we tackle the problem of computing the most similar pair of strings from two languages. We in particular consider three different gap cost models, linear, affine and concave models, that are crucial for finding a proper alignment between two bio sequences. We design efficient algorithms for computing the edit-distance between a CFG and an FA under these gap cost models. The time complexity of our algorithm for computing the linear or affine gap distance is polynomial and the time complexity for computing the concave gap distance is exponential.

Keywords: approximate matching, edit-distance, context-free grammars, finite-state automata.

1 Introduction

The string matching problem aims to find exact matches of a pattern w from an input text T and the approximate matching problem is to find similar occurrences of w that are within the distance k in T. Many researchers studied the approximate pattern matching problem that allows various types of mismatches [1,6,16,17,19,22]. For example, Aho and Peterson [1], and Lyon [16] introduced an $O(n^2m^3)$ algorithm for the problem of approximately matching a string of length n and a context-free language specified by a grammar of size m. They generalized Earley's algorithm [6] for parsing context-free languages and considered the edit-distance model [15] that has a unit-cost function. Myers [19] considered the variants of the problem under various gap costs such as linear, affine and concave gap costs; these gap cost models are very important to find proper alignment between two bio sequences in practices [20,21]. For the linear and affine gap costs, Myers designed $O(mn^2(n+\log m))$ algorithms and sketched an $O(m^5n^88^m)$ algorithm for the concave gap costs. His algorithm generalizes the Cocke-Younger-Kasami (CYK) algorithm [4,8,12].

The approximate matching problem is based on the edit-distance between two strings, or between a string and a language. This led researchers to examine the edit-distance between two formal languages. Mohri [18] proved that the

S. Konstantinidis (Ed.): CIAA 2013, LNCS 7982, pp. 146–157, 2013.
© Springer-Verlag Berlin Heidelberg 2013

edit-distance between two context-free languages is undecidable and provided a quadratic algorithm for two regular languages. Choffrut and Pighizzini [3] considered the relative edit-distance between languages and defined the reflexivity of binary relations based on the definition. Recently, the authors [9] studied the problem of computing the Levenshtein distance [15] between a context-free language and a regular language given by a pushdown automaton (PDA) P and a finite-state automaton (FA) A, respectively. We constructed an *alignment PDA* that computes all possible alignments between $L(A)$ and $L(P)$, converted the alignment PDA into a CFG and found the optimal alignment from the resulting grammar. The overall runtime is $O((n_1 n_2) \cdot 2^{(m_1 m_2)^2})$, where m_1 is the number of states of A, m_2 is the number of states of P, n_1 is the number of transitions of A and n_2 is the number of transitions of P. We also showed that we can compute the optimal edit-distance value in $O((m_1 m_2)^4 \cdot (n_1 n_2))$ time. Note that the conversion from a PDA of size n into a CFG takes $O(n^3)$ time and the size of the resulting grammar is at most $O(n^3)$ [10]. If a context-free language is given by a CFG instead of a PDA, then we need to construct a PDA for an input CFG before computing the alignment PDA. This motivates us to design algorithms that compute the edit-distance between a CFG and an FA without constructing a PDA, and extend this problem to the approximate matching between a CFG and an FA. In other words, we calculate the minimum edit-distance and the optimal alignment between the most similar pair of strings generated by a CFG and an FA, respectively. We introduce algorithms for computing the various gap distances and the optimal alignments between a CFG and an FA. While the previous research [9,11,14,18] on computing the edit-distance of formal languages rely on variants of the Cartesian product, the proposed algorithms are based on the dynamic programming approach that are generalized from the CYK algorithm. Given an FA of size n and a CFG of size m, our algorithms compute linear and affine gap distances in $O(mn^2(n+\log m))$ time. Furthermore, the worst-case time complexity of our algorithm for computing the concave gap distance is $O(mn^8 8^m)$.

In Section 2, we give a basic notations and terminology used here. We present the definitions for the edit-distance model in Section 3. In Section 4, we introduce a dynamic programming algorithm for computing the edit-distance between a CFG and an FA. The following two sections extend the algorithm to the problems of computing affine and concave gap distance.

2   Preliminaries

Let Σ denote a finite alphabet of characters and Σ^* denote the set of all strings over Σ. The size $|\Sigma|$ of Σ is the number of characters in Σ. A language over Σ is any subset of Σ^*. Given a set X, 2^X denotes the power set of X.

The symbol $\emptyset$ denotes the empty language and the character λ denotes the null string. A finite-state automaton (FA) A is specified by a tuple $(Q, \Sigma, \delta, s, F)$, where Q is a finite set of states, Σ is an input alphabet, $\delta : Q \times \Sigma \to 2^Q$ is a multi-valued transition function, $s \in Q$ is the start state and $F \subseteq Q$ is a set of final states. If F consists of a single state f, we use f instead of $\{f\}$ for simplicity.

For a transition $q \in \delta(p,a)$ in A, we say that p has an *out-transition* and q has an *in-transition*. Furthermore, p is a *source state* of q and q is a *target state* of p. The transition function δ can be extended to a function $Q \times \Sigma^* \to 2^Q$ that reflects sequences of inputs. A string x over Σ is accepted by A if there is a labeled path from s to a state in F such that this path spells out the string x, namely, $\delta(s,x) \cap F \neq \emptyset$. The language $L(A)$ of an FA A is the set of all strings that are spelled out by paths from s to a final state in F.

A context-free grammar (CFG) G is specified by a tuple $G = (V, \Sigma, R, S)$, where V is a set of variables, $R \subseteq V \times (V \cup \Sigma)^*$ is a finite set of productions and $S \in V$ is the start symbol. Let $\alpha A \beta$ be a string over $V \cup \Sigma$ with A a variable and $A \to \gamma$ be a production of G. Then, we say that $\alpha A \beta \Rightarrow \alpha \gamma \beta$. The reflexive, transitive closure of $\Rightarrow$ is $\overset{*}{\Rightarrow}$. Then the context-free language defined by G is $L(G) = \{w \in \Sigma^* \mid S \overset{*}{\Rightarrow} w\}$.

A CFG is in Chomsky normal form (CNF) if all of its production rules are of the form: $A \to BC$ or $A \to a$, where $A, B, C \in V$ and $a \in \Sigma$. Note that every context-free grammar can be converted into the CNF grammar with the size of $O(P^2)$ where P is the size of the original grammar. We consider the *pseudo-CNF* grammars that consist of the rules of the form $A \to BD$ or $A \to B$ or $A \to a$ where $A, B, C \in V$ and $a \in \Sigma$. We can transform every grammar into the pseudo-CNF grammar whose size is still $O(P)$ [19].

For more details on automata theory, we refer the reader to the books [10,23].

3   Edit-Distance

The edit-distance between two strings x and y is the smallest number of operations that transform x to y. People consider different edit operations depending on the applications. We consider three basic operations, insertion, deletion and substitution for simplicity. Given an alphabet Σ, let

$$\Omega = \{(a \to b) \mid a, b \in \Sigma \cup \{\lambda\}\}$$

be a set of edit operations. Namely, Ω is an alphabet of all edit operations for *deletions* $(a \to \lambda)$, *insertions* $(\lambda \to a)$ and *substitutions* $(a \to b)$. We call a string $\omega \in \Omega^*$ an *edit string* [11] or an *alignment* [18].

Let h be the morphism from Ω^* into $\Sigma^* \times \Sigma^*$ defined by setting

$$h((a_1 \to b_1) \cdots (a_n \to b_n)) = (a_1 \cdots a_n, b_1 \cdots b_n).$$

For example, a string $\omega = (a \to \lambda)(b \to b)(\lambda \to c)(c \to c)$ over Ω is an alignment of abc and bcc, and $h(\omega) = (abc, bcc)$. Thus, from an alignment ω of two strings x and y, we can retrieve x and y using h: $h(\omega) = (x,y)$.

Definition 1. *An edit string ω is a sequence of edit-operations transforming a string x into a string y if and only if $h(\omega) = (x,y)$.*

We associate a non-negative edit cost $c(\omega)$ to each edit operation $\omega \in \Omega$ where c is a function $\Omega \to \mathbb{R}_+$. We can extend the function to the cost $c(\omega)$ of an alignment $\omega = \omega_1 \cdots \omega_n$ as follows:

$$c(\omega) = \sum_{i=1}^{n} c(\omega_i).$$

Definition 2. *The edit-distance $d(x, y)$ of two strings x and y over Σ is the minimal cost of an alignment ω between x and y:*

$$d(x, y) = \min\{c(\omega) \mid h(\omega) = (x, y)\}.$$

We say that ω is optimal if $d(x, y) = c(\omega)$.

We can extend the edit-distance definition to languages.

Definition 3. *The edit-distance $d(L, R)$ between two languages $L, R \subseteq \Sigma^*$ is the minimum edit-distance of two strings, one is from L and the other is from R:*

$$d(L, R) = \min\{d(x, y) \mid x \in L \text{ and } y \in R\}.$$

The edit-distance in Definition 3 is the distance between the closest pair of strings from L and R under the considered edit operations. In other words, the most similar pair of strings defines the edit-distance between L and R.

4 Algorithm

We compute the edit-distance between a CFG and an FA. We assume that an input CFG $G = (V, \Sigma, R, S)$ is in pseudo-CNF and an input FA $M = (Q, \Sigma, \delta, s, F)$ has no λ-production. We use a pseudo-CNF (instead of CNF) because an arbitrary grammar can be converted to a pseudo-CNF grammar with only constant increase in size. First, we define $\mathcal{C}(A, q, p)$ to be the minimum edit-distance between one string v derivable from a variable A and a string w that spells out a computation of M from q to p. We can compute the edit-distance between $L(G)$ and $L(M)$ by computing $\mathcal{C}$-values for all $A \in V$ and $q, p \in Q$. We formally define it as follows:

$$\mathcal{C}(A, q, p) = \min\{d(v, w) \mid v \in L(G_A) \text{ and } w \in L(M_{q,p})\},$$

where $G_A = (V, \Sigma, R, A)$ and $M_{q,p} = (Q, \Sigma, \delta, q, \{p\})$. Then, $\min\{\mathcal{C}(S, s, f) \mid f \in F\}$ is the edit-distance between $L(G)$ and $L(M)$. In Theorem 3, we provide a recurrence for computing the $\mathcal{C}$-values. For this purpose we first need to establish some preliminary properties and introduce notation.

First we establish the unsurprising property that among the strings $w \in L(M_{q,p})$ that take state q to state p, the string that minimizes the distance to an individual alphabet symbol uses a computation from q to p that does not repeat any loop. For states q and p, we define $L_{\text{one-cyclic}}(q, p)$ to consist of those strings w such that M has a computation on w from q to p that does not visit any state more than twice.

Lemma 1. *For any states of M, $q, p \in Q$ and $a \in \Sigma$,*

$$d(a, L(M_{q,p})) = d(a, L_{\text{one-cyclic}}(q, p)).$$

Note that, when the cost function is allowed to be arbitrary, a property analogous to Lemma 1 would not hold for strings that correspond to an acyclic computation of M from state q to p. If any string corresponding to an acyclic computation does not contain occurrences of the symbol a and the cost of deleting a is considerably larger than the costs of insertions of any other symbol, it is possible that the distance of a and $L(M_{q,p})$ cannot be minimized by a string that would not repeat any state in the computation from q to p.

Now corresponding to a variable $A \in V$ and states $q, p \in Q$ of M, we define the following sets:

(i) $X(A, q, p) = \{\mathcal{C}(B, q, r) + \mathcal{C}(D, r, p) \mid r \in Q, \ A \to BD \in R\}$.
(ii) $Y(A, q, p) = \{\mathcal{C}(B, q, p) \mid A \to B \in R\}$.
(iii) $Z(A, q, p) = \{d(a, L_{\text{one-cyclic}}(q, p)) \mid A \to a \in R, \ a \in \Sigma\}$.

Theorem 1. *For all $A \in V$ and $q, p \in Q$,*

$$\mathcal{C}(A, q, p) = \min[X(A, q, p) \cup Y(A, q, p) \cup Z(A, q, p)]. \tag{1}$$

Note that Equation (1) in Theorem 1 is an essential recurrence equation for computing $d(L(G), L(M))$ in bottom-up dynamic programming approach. First, we compute $\mathcal{C}(A, q, p)$ where the distance between two states q and p is 0. For convenience, we define the distance between two states q and p in the FA as the minimum number of transitions required to reach p from q and denote it by $\mathfrak{d}(p, q)$. For the basis, we start from when $\mathfrak{d}(p, q)$ is 0, thus, two states are the same. Let us assume that there is a cycle in M from q to q of length n. Since there can be a set of strings L_q accepted through the cycle including the self-loop, we should consider L_q for computing $\mathcal{C}(A, q, q)$. Therefore, we should compute all $\mathcal{C}$-values where the distance between two states is less than n to compute the basis. We denote the $\mathcal{C}$-values not considering the cycles in paths by $\mathcal{C}'$-values to avoid confusion.

Now we consider $\mathcal{C}(A, q, q)$, which is a basis for recursive definition of $\mathcal{C}$-values. First, for a variable $A \in V$ and a state $q \in Q$ of M, we define the following sets:

(i) $X(A, q, q) = \{\mathcal{C}'(B, q, r) + \mathcal{C}'(D, r, q) \mid r \in Q, \ A \to BD \in R\}$.
(ii) $Y(A, q, q) = \{\mathcal{C}'(B, q, q) \mid A \to B \in R\}$.

Then, for all $A \in V$ and $q, p \in Q$, we can establish another recursion for the basis of $\mathcal{C}$-values as follows:

$$\mathcal{C}(A, q, q) = \min[\mathcal{C}'(A, q, q) \cup X(A, q, q) \cup Y(A, q, q)].$$

Now it seems that we are ready to compute $\mathcal{C}$-values. However, we still have a problem to solve the recurrence step. Consider $Y(A, q, p)$ in Equation (1). We need to know $\mathcal{C}(B, q, p)$ to compute $\mathcal{C}(A, q, p)$ that is in the same level of recursion. Similarly, when r is the same state with q or p in the first term $X(A, q, p)$ of the recurrence, we need to compute $\mathcal{C}(B, q, p)$ or $\mathcal{C}(D, q, p)$ to compute $\mathcal{C}(A, q, p)$. This problem also arises when we compute $\mathcal{C}'$-values. These dependencies between the recursive values in the same level prohibit us to compute the next level of recursion. Thus, we define an independent recursive definition for this problem. First, we define the following sets:

(i) $X(A, q, p) = \{\mathcal{C}(B, q, r) + \mathcal{C}(D, r, p) \mid r \in Q, \ A \rightarrow BD \in R\}$.
(ii) $Y(A, q, p) = \{d(a, L_{\text{one-cyclic}}(q, p)) \mid A \rightarrow a \in R, \ a \in \Sigma\}$.

Here, r should not be q or p. Then, $\mathcal{K}$-values are defined as follows:

$$\mathcal{K}(A, q, p) = \min[X(A, q, p) \cup Y(A, q, p)].$$

Note that all $\mathcal{K}$-values can be computed by assuming that all $\mathcal{C}(A, q', p')$ are already computed where $\mathfrak{d}(q', p') < \mathfrak{d}(q, p)$. Now, we can redefine $\mathcal{C}(A, q, p)$ as the minimum of the following four values:

(i) $\mathcal{K}(A, q, p)$.
(ii) $\min_{A \rightarrow B} \mathcal{C}(B, q, p)$.
(iii) $\min_{A \rightarrow BD} \mathcal{C}(B, q, p) + \mathcal{C}(D, p, p)$.
(iv) $\min_{A \rightarrow BD} \mathcal{C}(B, q, q) + \mathcal{C}(D, q, p)$.

We can solve the dependencies between $\mathcal{C}$-values by the construction of a weighted graph, which has a vertex for each variable $A \in V$ and a special source vertex ϕ. Then, we connect ϕ to each vertex for a variable A with an edge whose weight is $\mathcal{K}(A, q, p)$. Also there are the edge of weight 0 from B to A if and only if $A \rightarrow B \in R$ and the edge of weight $\mathcal{C}(D, p, p)$ from B to A if and only if $A \rightarrow BD \in R$ or $A \rightarrow DB \in R$. Then, from the construction, $\mathcal{C}(A, q, p)$ becomes the shortest path from ϕ to A in the graph. Similarly, we can also solve the dependency problem for $\mathcal{C}'$-values. We give an algorithm for computing $d(L(G), L(M))$ in Algorithm 1.

Theorem 2. *Given a CFG $G = (V, \Sigma, R, S)$, an FA $M = (Q, \Sigma, \delta, s, F)$ and a non-negative cost function c, we can compute the edit-distance between $L(G)$ and $L(M)$ in $O(mn^2(n + \log m))$ worst-case time, where $m = |G|$ and $n = |Q|$.*

Lemma 2. *Given a CFG $G = (V, \Sigma, R, S)$, an FA $M = (Q, \Sigma, \delta, s, F)$ and an arbitrary cost function c, we can compute the edit-distance between $L(G)$ and $L(M)$ in $O(mn^2(n + m))$ worst-case time, where $m = |G|$ and $n = |Q|$.*

We can also observe that it is possible to retrieve the optimal alignment by backtracking the optimal path.

Lemma 3. *Given a CFG $G = (V, \Sigma, R, S)$ and an FA $M = (Q, \Sigma, \delta, s, F)$, we can compute the optimal alignment of length k between $L(G)$ and $L(M)$ in $O(mnk)$ worst-case time, where $m = |G|$ and $n = |Q|$.*

Algorithm 1. The algorithm for computing $d(L(G), L(M))$

Input: A CFG $G = (V, \Sigma, R, S)$ and an FA $M = (Q, \Sigma, \delta, s, F)$
1: **for** $q \in Q$ **do**
2:     **for** $d \leftarrow 1$ **to** $|Q| - 1$ **do**
3:        **for** $p \in Q$ **and** $\mathfrak{d}(p, q) = d$ **do**
4:           **for** $A \in V$ **do**
5:              $\mathcal{C}(A, q, p) \leftarrow \mathcal{K}(A, q, p)$
6:           **end for**
7:           $H \leftarrow$ heap of V (ordered by $\mathcal{C}(?, q, p)$)
8:           **while** $H \neq \emptyset$ **do**
9:              $A \leftarrow$ extract_min(H)
10:             **for** $A \in H$ **and** $(A \rightarrow BD \in R$ **or** $A \rightarrow DB \in R)$ **do**
11:                $\mathcal{C}(B, q, p) \leftarrow \min\{\mathcal{C}(A, q, p), \mathcal{C}(B, q, p) + \mathcal{C}(D, p, p)\}$, reheap$(H, A)$
12:             **end for**
13:             **for** $A \in H$ **and** $A \rightarrow B \in R$ **do**
14:                $\mathcal{C}(A, q, p) \leftarrow \min\{\mathcal{C}(A, q, p), \mathcal{C}(B, q, p)\}$, reheap$(H, A)$
15:             **end for**
16:           **end while**
17:        **end for**
18:     **end for**
19: **end for**
20: **return** $\min\{\mathcal{C}(S, s, f) \mid f \in F\}$
Output: $d(L(G), L(M))$

$$S_1 = A\ C\ T\ T\ A\ G\ T\ A\ G\ A\ T\ C\ C \qquad S_1 = A\ C\ T\ T\ A\ G\ T\ A\ G\ A\ T\ C\ C$$
$$\downarrow\downarrow\downarrow\downarrow\downarrow\downarrow\downarrow\downarrow\downarrow\downarrow\downarrow\downarrow\downarrow \qquad\qquad \downarrow\downarrow\downarrow\downarrow\downarrow\downarrow\downarrow\downarrow\downarrow\downarrow\downarrow\downarrow\downarrow$$
$$S_2 = A\ C\ T\ T\ -\ G\ -\ A\ -\ -\ T\ C\ C \qquad S_2 = A\ C\ T\ T\ -\ -\ -\ -\ G\ A\ T\ C\ C$$

$$\text{(a)} \qquad\qquad\qquad\qquad\qquad\qquad \text{(b)}$$

Fig. 1. Two alignment examples that align S_2 to the target sequence S_1. The first alignment contains three short gaps while the second contains one long gap.

5   Affine Gap Distance

The approximate pattern matching problem is often used for the sequence alignment in bioinformatics [20,21]. A biological sequence alignment is a process of arranging the sequences of DNA, RNA or protein, and examining the similarities between the sequences. Consider the two alignments of sequences described in Fig. 1. Both have gaps of length four, which can be defined as deletion or insertion edit operations. However, the second alignment is biologically better since a deletion or insertion of four consecutive elements is more likely to occur than of three separated elements. Therefore, we need to give more penalty to the alignments containing many short gaps than few long gaps. Note that we can consider a sequence of consecutive deletion or insertion operations as a gap. Assume that an alignment ω consists of k consecutive insertions or deletions, in other words, a gap of the length k. Then, the cost of ω is linearly dependent on $|\omega|$. Namely, $c(\omega) = g \cdot |\omega|$ where g is a constant. Instead of using this linear gap penalty function, we can use the *affine gap penalty* function to obtain biologically better alignments. The affine gap penalty function is defined as follows. Here the alphabet Ω of edit operations consists only of deletions ($a \to \lambda$), insertions ($\lambda \to a$) and trivial substitutions ($a \to a$) that do not change the symbol. We denote $\Omega_{\mathrm{del}} = \{(a \to \lambda) \mid a \in \Sigma\}$, $\Omega_{\mathrm{ins}} = \{(\lambda \to a) \mid a \in \Sigma\}$ and $\Omega_{\mathrm{triv}} = \{(a \to a) \mid a \in \Sigma\}$, and thus

$$\Omega = \Omega_{\mathrm{del}} \cup \Omega_{\mathrm{ins}} \cup \Omega_{\mathrm{triv}}.$$

Let $\omega \in \Omega^+$ be a sequence of edit operations. The *(maximal) ID-decomposition of* ω (insertion–deletion decomposition of ω) is the tuple

$$\mathrm{comp}_{\mathrm{ID}}(\omega) = (\omega_1, \omega_2, \ldots, \omega_k)$$

where $\omega_i \in \Omega_{\mathrm{del}}^+ \cup \Omega_{\mathrm{ins}}^+ \cup \Omega_{\mathrm{triv}}^+$, for $i = 1, \ldots, k$, and for any $1 \leq j < k$ the strings ω_j and ω_{j+1} belong to different sets Ω_{del}^+, Ω_{ins}^+ and Ω_{triv}^+.

The ID-decomposition of ω is obtained simply by subdividing ω into maximal substrings each consisting only of insertions, or only of deletions, or only trivial substitutions and thus $\mathrm{comp}_{\mathrm{ID}}(\omega)$ is uniquely defined.

Now for a sequence consisting only of deletions or only of insertions, $\omega \in \Omega_{\mathrm{del}}^+ \cup \Omega_{\mathrm{ins}}^+$, we define the affine gap cost of ω as $c_{\mathrm{affine}}(\omega) = e + g \cdot |\omega|$, where e and g are constants. For a sequence consisting of trivial substitutions, $\omega \in \Omega_{\mathrm{triv}}^+$, we set $c_{\mathrm{affine}}(\omega) = 0$.

Now the *affine gap cost* of an arbitrary sequence of edit operations $\omega \in \Omega^+$, where $\mathrm{comp}_{\mathrm{ID}}(\omega) = (\omega_1, \omega_2, \ldots, \omega_k)$ is defined as

$$c_{\mathrm{affine}}(\omega) = \sum_{i=1}^{k} c_{\mathrm{affine}}(\omega_i).$$

The affine gap cost gives, for a sequence of edit operations, a constant e penalty for each gap opening (consisting of consecutive insertions or consecutive deletions) and additionally a penalty that is linear in the length of the gap. The edit distance based on the affine gap cost function is called the *affine gap distance*.

We introduce an algorithm for computing the affine gap distance between a CFG and an FA. This is an extension of the previous algorithm, yet has the same time complexity. The key difference is that we define four types of C-values as follows:

$$\mathcal{C}_{\rhd}^{\lhd}(A, q, p) =$$
$$\min\{d(x, \lambda) + d(v, w) + d(y, \lambda) \mid A \overset{*}{\Rightarrow} xvy, |x| \lhd 0, |y| \rhd 0 \text{ and } p \in \delta(q, w)\},$$

where $\rhd, \lhd \in \{=, \neq\}$. We illustrate four cases in Fig. 2. The affine gap distance becomes $\min\{\mathcal{C}_{\rhd}^{\lhd}(S, s, f) \mid f \in F \text{ and } \rhd, \lhd \in \{=, \neq\}\}$.

Before introducing the recurrence for $\mathcal{C}_{\rhd}^{\lhd}$-values, corresponding to a variable $A \in V$ and states $q, p \in Q$ of M, we define the following sets:

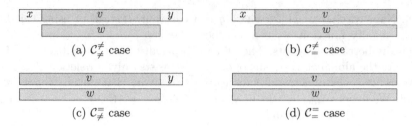

Fig. 2. The pictorial representations of $\mathcal{C}_{\triangleright}^{\triangleleft}$-values where $\triangleright, \triangleleft \in \{=, \neq\}$

(i) $X(A, q, p) = \{\mathcal{C}_{\triangleright}^{\triangleleft}(B, q, r) + \mathcal{C}_{\triangleright}^{\triangleleft}(D, r, p) - (h \text{ if } \trianglerighteq = \trianglelefteq = \text{'}\neq\text{'}) \mid r \in Q, A \to BD \in R\}$.

(ii) $Y(A, q, p) = \{\mathcal{C}_{\triangleright}^{\triangleleft}(B, q, p) \mid A \to B \in R\}$.

(iii) $Z(A, q, p) = \{\mathcal{I}_{\triangleright}^{\triangleleft}(a, w) \mid A \to a \in R, a \in \Sigma, w \in L_{\text{one-cyclic}}(q, p)\}$.

Here, $\mathcal{I}$-values are also defined as follows:

- $\mathcal{I}_{=}^{=}(a, w) = \min_{k \in [1, |w|]}\{c(a, w_k) + (|w| - 1) \cdot g + (h \text{ if } k > 1) + (h \text{ if } k < |w|)\}$,
- $\mathcal{I}_{\neq}^{=}(a, w) = \mathcal{I}_{=}^{\neq}(a, w) = (|w| + 1) \cdot g + 2h$, and
- $\mathcal{I}_{\neq}^{\neq}(a, w) = g + h$.

Now we establish a recursive definition for $\mathcal{C}_{\triangleright}^{\triangleleft}$-values.

Theorem 3. *For all $A \in V$ and $q, p \in Q$,*

$$\mathcal{C}_{\triangleright}^{\triangleleft}(A, q, p) = \min[X(A, q, p) \cup Y(A, q, p) \cup Z(A, q, p)].$$

Note that the time complexity of this algorithm is still $O(mn^2(n + \log m))$, the same as in Theorem 2. Since we consider the four variations of $\mathcal{C}$-values, the time complexity increases to four times the runtime of Theorem 2.

Theorem 4. *Given a CFG $G = (V, \Sigma, R, S)$, an FA $M = (Q, \Sigma, \delta, s, F)$ and a non-negative cost function c, we can compute the affine gap distance between $L(G)$ and $L(M)$ in $O(mn^2(n + \log m))$ worst-case time, where $m = |G|$ and $n = |Q|$.*

Lemma 4. *Given a CFG $G = (V, \Sigma, R, S)$, an FA $M = (Q, \Sigma, \delta, s, F)$ and an arbitrary cost function c, we can compute the affine gap distance between $L(G)$ and $L(M)$ in $O(mn^2(n + m))$ worst-case time, where $m = |G|$ and $n = |Q|$.*

6   Concave Gap Distance

Many researchers consider non-linear gap penalty functions including the affine gap penalty function [13,17,22]. Although the affine gap penalty function prefers few longer gaps to many smaller gaps, the alignment results based on the affine gap penalty function are not practically the best. For example, assume that there are two alignments s_1 and s_2 aligning two sequences. Alignment s_1 contains two gaps whose lengths are 99 and 100, respectively, while s_2 contains just one gap of length 240. Note that the remaining parts of s_1 and s_2 are perfectly matched. By employing the affine gap penalty function with $h = 5$ and $g = 1$, we obtain $c(s_1) = 99 + 100 + 5 \times 2 = 209$ and $c(s_2) = 240 + 5 = 245$. Even though the gap opening penalty is introduced in the affine gap distance, it may not be sufficient to consider some practical cases such as this example. This is why the *concave gap distance* is introduced and replaces other distances considering the linear or affine gap penalties. For a sequence consisting only of deletions or only of insertions, $w \in \Omega_{\text{del}}^+ \cup \Omega_{\text{ins}}^+$, we define the concave gap cost of w as

$$c_{\text{concave}}(w) = e + g \cdot \log |w|,$$

where e and g are constants. For a sequence consisting of trivial substitutions, $w \in \Omega_{\text{triv}}^+$, we set $c_{\text{concave}}(w) = 0$. Now the *concave gap cost* of an arbitrary sequence of edit operations $w \in \Omega^+$, where $\text{comp}_{\text{ID}}(w) = (w_1, w_2, \ldots, w_k)$ is defined as

$$c_{\text{concave}}(w) = \sum_{i=1}^{k} c_{\text{concave}}(w_i).$$

Under this gap penalty function, the shape of the penalty score with respect to the length of the gap is concave in the sense that its forward differences are non-increasing. In other words, $\Delta c(w_1) \geq \Delta c(w_2) \geq \Delta c(w_3) \geq \cdots$ where $\Delta c(w_k) \equiv c(w_{k+1}) - c(w_k)$ and $|w_k| = k$. We define new $\mathcal{C}$-values for computing the concave gap distance as follows:

$$\mathcal{C}(A, q, p, i, j) = \min\{d(x, \lambda) + d(v, w) + d(y, \lambda) \mid A \overset{*}{\Rightarrow} xvy \neq \lambda\},$$

where $|x| = i$, $|y| = j$ and $p \in \delta(q, w)$. Here we use two additional parameters i and j for maintaining the lengths of gaps on both sides. We also define a set $\mathcal{V}(t)$ of variables that can derive strings of length t as follows:

$$\mathcal{V}(t) = \{A \mid A \in V, A \overset{*}{\Rightarrow} w \text{ and } |w| = t\}.$$

We can compute a set $\mathcal{V}(t)$ of variables as follows:

$$\bigcup_{k=1}^{t-1} \{A \mid A \to BD \in \mathcal{V}(k) \times \mathcal{V}(t-k)\} \cup \{A \mid A \to B \in \mathcal{V}(t)\} \cup \{A \mid A \to a\}.$$

Then, before introducing the recurrence for $\mathcal{C}$-values for the concave gap distance, we define the following sets corresponding to a variable $A \in V$ and states q, $p \in Q$ of M:

(i) $X(A, q, p) = \{\mathcal{C}(B, q, r, i, m) + \mathcal{C}(D, r, p, n, j) + g \cdot \log \frac{m+n}{mn} - h \mid r \in Q, A \to BD \in R\}$.

(ii) $Y(A, q, p) = \{\mathcal{C}(B, q, p, i, j - t) \mid A \to BD \in R, 1 \leq t \leq j, D \in \mathcal{V}(t)\}$.

(iii) $Z(A, q, p) = \{\mathcal{C}(D, q, p, i - t, j) \mid A \to BD \in R, 1 \leq t \leq i, B \in \mathcal{V}(t)\}$.

(iv) $U(A, q, p) = \{\mathcal{C}(B, q, p, i, j) \mid A \to B \in R\}$.

(v) $W(A, q, p) = \{\mathcal{I}(a, w, i, j) \mid A \rightarrow a, \; 0 \leq i + j \leq 1, \; w \in L_{\text{one-cyclic}}(q, p)\}$.

Here, $\mathcal{I}$-values are also defined as follows:

- $\mathcal{I}(a, w, 0, 0) = c(a, w_k) + 2h + \log(k-1)(|w| - k)$,
- $\mathcal{I}(a, w, 0, 1) = \mathcal{I}(a, w, 1, 0) = h + \log|w|$.

Now we establish a recurrence for computing the concave gap distance between a CFG and an FA.

Theorem 5. *For all $A \in V$, $q, p \in Q$ and $1 \leq i, j \leq |Q| \cdot 2^{\frac{h}{g}|V|}$,*

$$C_{\triangleright}^{\triangleleft}(A, q, p) = \min[X(A, q, p) \cup Y(A, q, p) \cup Z(A, q, p) \cup U(A, q, p) \cup W(A, q, p)].$$

Based on the recurrence, we can compute the concave gap distance between $L(G)$ and $L(A)$ in exponential runtime.

Theorem 6. *Given a CFG $G = (V, \Sigma, R, S)$, an FA $M = (Q, \Sigma, \delta, s, F)$ and a non-negative cost function c, we can compute the concave gap distance between $L(G)$ and $L(A)$ in $O(mn^8 8^m)$ worst-case time, where $m = |G|$ and $n = |Q|$.*

7   Conclusions

We have considered the problem of approximately matching a context-free language specified by a CFG and a regular language specified by an FA. We have examined three types of gap cost functions that are used for approximate string matching: linear, affine and concave. Based on the dynamic programming approach, we have introduced algorithms for computing the linear, affine and concave gap distance between an FA and a CFG.

Given an FA of size n and a CFG of size m, we have presented algorithms for computing linear and affine gap distances in $O(nm^2(n + \log m))$ time under a non-negative cost function and $O(nm^2(n + m))$ time under an arbitrary cost function. We have also shown that computing the optimal alignment of length k takes $O(nmk)$ time by our algorithm when we consider linear or affine gap distance. Finally, we have proposed an $O(mn^8 8^m)$ time algorithm for computing the concave gap distance.

It will be interesting to see if we can compute the max-min distance between an FA and a CFG, or find a k optimal alignment between an FA and a CFG using a similar approach.

Acknowledgements. We wish to thank the referees for the careful reading of the paper and their constructive suggestions.

Han and Ko were supported by the Basic Science Research Program through NRF funded by MEST (2012R1A1A2044562) and Salomaa was supported by the Natural Sciences and Engineering Research Council of Canada Grant OGP0147224.

References

1. Aho, A., Peterson, T.: A minimum distance error-correcting parser for context-free languages. SIAM Journal on Computing 1(4), 305–312 (1972)
2. Bellman, R.: On a routing problem. Quarterly of Applied Mathematics 16, 87–90 (1958)
3. Choffrut, C., Pighizzini, G.: Distances between languages and reflexivity of relations. Theoretical Computer Science 286(1), 117–138 (2002)
4. Cocke, J.: Programming languages and their compilers: Preliminary notes. Courant Institute of Mathematical Sciences. New York University (1969)
5. Dijkstra, E.: A note on two problems in connexion with graphs. Numerische Mathematik 1, 269–271 (1959)
6. Earley, J.: An efficient context-free parsing algorithm. Communications of the ACM 13(2), 94–102 (1970)
7. Floyd, R.W.: Algorithm 97: Shortest path. Communications of the ACM 5(6), 345–348 (1962)
8. D.H.: Younger. Recognition and parsing of context-free languages in time n^3. Information and Control 10(2), 189–208 (1967)
9. Han, Y.-S., Ko, S.-K., Salomaa, K.: Computing the edit-distance between a regular language and a context-free language. In: Yen, H.-C., Ibarra, O.H. (eds.) DLT 2012. LNCS, vol. 7410, pp. 85–96. Springer, Heidelberg (2012)
10. Hopcroft, J., Ullman, J.: Introduction to Automata Theory, Languages, and Computation, 2nd edn. Addison-Wesley, Reading (1979)
11. Kari, L., Konstantinidis, S.: Descriptional complexity of error/edit systems. Journal of Automata, Languages and Combinatorics 9, 293–309 (2004)
12. Kasami, T.: An efficient recognition and syntax-analysis algorithm for context-free languages. Technical report, Air Force Cambridge Research Lab, Bedford, MA (1965)
13. Knight, J.R., Myers, E.W.: Approximate regular expression pattern matching with concave gap penalties. Algorithmica 14, 67–78 (1995)
14. Konstantinidis, S.: Computing the edit distance of a regular language. Information and Computation 205, 1307–1316 (2007)
15. Levenshtein, V.I.: Binary codes capable of correcting deletions, insertions, and reversals. Soviet Physics Doklady 10(8), 707–710 (1966)
16. Lyon, G.: Syntax-directed least-errors analysis for context-free languages: a practical approach. Communications of the ACM 17(1), 3–14 (1974)
17. Miller, W., Myers, E.W.: Sequence comparison with concave weighting functions. Bulletin of Mathematical Biology 50(2), 97–120 (1988)
18. Mohri, M.: Edit-distance of weighted automata: General definitions and algorithms. International Journal of Foundations of Computer Science 14(6), 957–982 (2003)
19. Myers, G.: Approximately matching context-free languages. Information Processing Letters 54, 85–92 (1995)
20. Needleman, S.B., Wunsch, C.D.: A general method applicable to the search for similarities in the amino acid sequence of two proteins. Journal of Molecular Biology 48(3), 443–453 (1970)
21. Sankoff, D., Kruskal, J.B.: Time Warps, String Edits, and Macromolecules: The Theory and Practice of Sequence Comparison. Addison-Wesley (1983)
22. Waterman, M.S.: Efficient sequence alignment algorithms. Journal of Theoretical Biology 108, 333–337 (1984)
23. Wood, D.: Theory of Computation. Harper & Row (1987)

On Palindromic Sequence Automata
and Applications

Md. Mahbubul Hasan, A.S.M. Sohidull Islam, M. Sohel Rahman,
and Ayon Sen

AℓEDA group
Department of CSE, BUET, Dhaka - 1000, Bangladesh
{mahbub86,sohansayed,msrahman,ayonsn}@cse.buet.ac.bd

Abstract. In this paper, we present a novel weighted finite automata
called PSA (Palindromic Subsequence Automata) that is a compact representation of all the palindromic subsequences of a string. Then we use
PSA to solve the LCPS (Longest Common Palindromic Subsequence)
problem. Our automata based algorithms are efficient both in theory
and in practice.

1 Introduction

A *string* is a sequence of symbols drawn from an alphabet Σ. A *subsequence*
of a string is a sequence that can be derived by deleting zero or more symbols
from it without changing the order of the remaining symbols. A *palindrome* is
a string w such that $w = w^R$, where w^R is the reverse of w; often, w is said to
be a *palindromic string*. For example, $ATTA$ and $CATTAC$ are palindromes.
In the Palindromic Subsequence Problem, all the palindromic subsequences of
a string are to be computed. A *common subsequence* of two strings is a subsequence common to both the strings. Additionally, if the common subsequence is
a palindrome, it is called a *common palindromic subsequence*.

Stringology researchers have been conducting research on different problems
related to palindromes on strings and sequences since long [1,8,15,18,16,14,11].
Palindromes appear frequently in DNA and are widespread in human cancer
cells. Identifying these parts of DNAs could aid in the understanding of genomic
instability [3,19]. Biologists believe that palindromes play an important role
in regulation of gene activity and other cell processes because these are often
observed near promoters, introns and specific untranslated regions. So, finding
palindromic subsequences in any genome sequence is important. Also finding
common palindromes in two genome sequences can be an important criterion
for comparing them, and also to find common relationships between them.

The problem of computing palindromes and variants in a single sequence
has received much attention in the literature. An on-line sequential algorithm
was given by Manacher [15] that finds all initial palindromes in a string. Another algorithm to find long approximate palindromes was given by Porto and
Barbosa [18]. Gusfield gave a linear-time algorithm to find all maximal palindromes in a string [9]. Matsubara et al. in [16] solved the problem of finding

S. Konstantinidis (Ed.): CIAA 2013, LNCS 7982, pp. 158–168, 2013.

all palindromes in SLP (Straight Line Programs)-compressed strings. Additionally, a number of variants of palindromes have also been investigated in the literature [14,11,2]. Very recently, Tomohiro et al. worked on pattern matching problems involving palindromes [12]. Chuang, Lee and Huang [6] proposed an algorithm to solve the palindromic subsequence problem.

1.1 Our Contribution

In this paper we present a weighted finite automata that is a compact representation of all the palindromic subsequences of a string. The space complexity of our approach is better than that of [6]. In particular, we need only $O(n^2)$ space to represent all the palindromic subsequences of the given string while the space complexity of [6] is directly proportional to the total number of palindromes which is exponential in nature. Furthermore, as an interesting application of PSA, we show how we can solve the LCPS problem of two given strings efficiently. The time complexity of our algorithm is $O(\mathcal{R}_1\mathcal{R}_2|\Sigma|)$ where $\mathcal{R}_1$ and $\mathcal{R}_2$ are the number of states of respective automata.

1.2 Roadmap

In Section 2, we present the Palindromic Subsequence Automata. In Section 3, we present another automata to find the LCPS of two strings which is derived from PSA presented in Section 2.

In Section 4, we present extensive experimental results. Finally we briefly conclude in Section 5.

2 PSA: An Automaton to Generate All Palindromic Subsequence of a String

To design this automata we need a string and its reverse. We will find the common subsequence automata of these two strings. We will follow the techniques provided in [10,7]. The resulting automata will be a weighted automata. This automata can be seen as a weighted version of a Common Subsequence Automata (CSA) between a string and its reverse.

Definition 1. *Palindromic Subsequence Automata (PSA): Given a string S and let the reverse of S be S^R. A Palindromic Subsequence Automata (PSA) M accepts all palindromic subsequence of the given string S.*

The PSA M is 6 tuple $(Q, \sum, \delta, \sigma, q_0, F)$, where

- *Q is a finite set of states. Here, Q is a subset of pairs of positions in S and S^R*
- *$\sum$ is an input alphabet*
- *$\delta : Q \times \sum \longrightarrow Q$ is a transition function*
- *$\sigma : Q \times \sum \times Q \longrightarrow K$ assigns a edge cost between pair of states*

- $q_0 \in Q$ is the initial state
- $F \subseteq Q$ is the set of final states. However, in our formulation all the states are valid final states.

Here, Q and δ are defined in the same way as in [10,7]. Each state $q_a \in Q$ is associated with pair of positions (a_i, a_j) where a_i and a_j refers to positions in S and S^R respectively and $S[a_i] = S^R[a_j]$. Let the length of S is n. So we have $1 \leqslant a_i, a_j \leqslant n$ for any state q_a. Now, for any state $q_b \in Q$ and $c \in \Sigma$ we have edge from q_b to state q_a if and only if $S[a_i] = S^R[a_j] = c$, $b_i < a_i$, $b_j < a_j$ and there is no l, k such that $b_i < l < a_i$, $b_j < k < a_j$ and $S[l] = c$ or $S[k] = c$. Now we explain the definition of the function σ as follows. We restrict the edge cost K between 0, 1 and 2. The reason for such values will be clear from the following definition of cost function:

$$\sigma(q_b, c, q_a) = \begin{cases} 2, & \text{if } a_i < n - a_j + 1. \\ 1, & \text{if } a_i = n - a_j + 1. \\ 0, & \text{if } a_i > n - a_j + 1. \end{cases} \qquad (1)$$

Example 1. Let, $S = abacbca$. The PSA of S is given in Fig. 1. Each state in the PSA is an ending state. In Fig. 1 each edge represents a transition. The character above each edge represents the transition character and the number above each edge represents the weight of it. Each state is represented by $[i, j]$, where i and j are indexes of S and S^R. For example, for the edge between states $[1,1]$ and $[2,3]$, the transition character is b and the weight is 2. For simplicity of the figure, the 'error' state is omitted, since the edges from or to this state has weight 0.

A brief discussion on the possible values of K is in order. The reason for the values of K is quite intuitive. We will construct a palindromic string based on accepted strings of PSA. If the accpeted string is w, we find the reverse of w which is w^R and concatenate both to create a palindrome of even length. However for a palindrome having odd length we can not do that. In this case, we have to compute u such that $w = ua$ where $a \in \Sigma$. In this case, we can get a palindrome of the form uau^R. To determine whether an odd-length palindrome is there (in addition to the even-length one), we use the value two. The value of one is used to indicate that only an even-length palindrome of the form ww^R is there. A value of zero indicates that the corresponding character does not take part in the construction of a palindrome.

Now we discuss how PSA works. Assume that we have computed a PSA, M_1 of a given string S_1. Now we want to check whether a string $S = a_1 a_2 ... a_k$ is a palindromic subsequence of S_1. Intuitively, we can always obtain a longest palindromic subsequence of S_1 by first taking the LCS (Longest Common Sequence) L of S_1 and S_1^R and then "reflecting" the first half of the result onto the second half; that is, if L has k characters, then we replace the last $\lfloor \frac{k}{2} \rfloor$ characters of L by the reverse of the first $\lfloor \frac{k}{2} \rfloor$ characters of L to obtain a longest palindromic subsequence of S_1. Obviously, this argument can be extended for any palindromic subsequence by taking any common subsequence of S_1 and S_1^R. Now suppose that $S_h = a_1 a_2 ... a_{\lfloor \frac{k+1}{2} \rfloor}$. It follows from the discussion above that S will be a palindromic subsequence of S_1 if and only if S_h is a common subsequence of both S_1 and S_1^R. So, to decide whether S is a palindromic subsequence of S_1, it is sufficient to check whether S_h is accepted by M_1. Hence, we have the following lemma.

Lemma 1. *Suppose we have constructed M_1 based on the string S_1. Further suppose that $S = a_1 a_2 ... a_k$ is a palindromic string and $S_h = a_1 a_2 ... a_{\lfloor \frac{k+1}{2} \rfloor}$. If S_h is accepted by M_1 then S is a palindromic subsequence of S_1.*□

2.1   Computing All Palindromic Subsequences

To find all the palindromic subsequences, all we need is to traverse M_1 and find the strings accepted by M_1 as follows. Suppose we get a string V_1 reaching a final state q_n. The state before that is q_{n-1} and the transition character is c. Let, $\sigma(q_{n-1}, c, q_n) = k_n$. Clearly, $k_n \in \{0, 1, 2\}$. If $k_n = 2$, we can get two palindromic subsequences. The first string is $V_1 V_1^R$. Suppose $V_1 = Uc$. Then, the second string is UcU^R. If $k_n = 1$ only one palindromic subsequence can be formed which is UcU^R. If $k_n = 0$, we can not get a palindromic subsequence with the current string reaching the final state q_n.

The following example explains the characteristics of PSA. The algorithm for constructing PSA is formally presented in Algorithm 1.

Example 2. In Fig. 1 a path from the starting node to [2,3] is [0,0], [1,1], [2,3]. In this case $w_n = 2$. So, we can form two palindromic subsequences: *abba* and *aba*. But the transition from [1,1] to [3,5] has weight 1. A path from the starting node to [3,5] is [0,0], [1,1], [3,5]. As $w_n = 1$ only one palindromic subsequence, *aaa* can be formed.

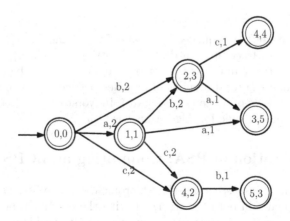

Fig. 1. A Palindromic Subsequence Automata for string "abacbca"

Algorithm 1. PSA Construction

input:S_1 : input string, S_2 : $REVERSE(S_1)$, n : $LENGTH(S_1)$
output: PSA $M = (Q, \sum, \delta, \sigma, q_0, F)$

1: **begin**
2:   $q_0 \leftarrow [0, 0]$
3:   $Q \leftarrow \{[0, 0]\}$
4:   $F \leftarrow \{q_0\}$
5:   $C \leftarrow NEW - QUEUE()$
6:   $ENQUEUE(C, [0, 0])$
7: **while not** EMPTY(C) **do**
8:     $[i_1, j_1] \leftarrow DEQUEUE(C)$
9:     **for all** $s \in \Sigma$ **do**
10:        $i_2 \leftarrow NEXT\_MATCH(S_1, i_1, s)$
11:        $j_3 \leftarrow NEXT\_MATCH(S_2, j_1, s)$
12:        $j_2 \leftarrow n - j_3 + 1$
13:        **if** $i_2 \leqslant j_2$ **then**
14:          $\delta([i_1, j_1], s) \leftarrow [i_2, j_2]$
15:          **if** $[i_2, j_2] \notin Q$ **then**
16:            $ENQUEUE(C, [i_2, j_2])$
17:            $Q \leftarrow Q \cup [i_2, j_2]$
18:            $F \leftarrow Q \cup [i_2, j_2]$
19:          **end if**
20:          **if** $i_2 < j_2$ **then**
21:            $\sigma([i_1, j_1], s, [i_2, j_2]) \leftarrow 2$
22:          **else**
23:            $\sigma([i_1, j_1], s, [i_2, j_2]) \leftarrow 1$
24:          **end if**
25:        **end if**
26:     **end for**
27: **end while**
28: **end**

2.2 Analysis

As PSA is derived from subsequence automata of two strings [10,7] so the running time for Algorithm 1 to compute a PSA of two strings is $O(\mathcal{R}|\Sigma|)$ [10], where $\mathcal{R}$ is the number of states and Σ is the set of characters. It can be easily verified that $\mathcal{R}$ is less than the total number of matches between the two strings (each match does not always produce a valid state). In worst case, $\mathcal{R} = O(n^2)$. Hence the worst case running time for Algorithm 1 is $O(n^2|\Sigma|)$.

3 An Application of PSA: Computing an LCPS

The *longest common subsequence* (LCS) problem for two strings is to find a common subsequence in both the strings, having the maximum possible length. In the *longest common palindromic subsequence* (LCPS) problem, the computed longest common subsequence must also be a palindrome. More formally, given a pair of strings X and Y over an alphabet Σ, the goal of the LCPS problem is to compute an LCS Z such that Z is a palindrome. In what follows, for the sake of convenience we will assume, that X and Y have equal length, n. But our result can be easily extended to handle two strings of different length.

Despite a plethora of work on problems related to palindromes concerning a single sequence, to the best of our knowledge, there has not been any work on the LCPS problem until very recently, when Chowdhury et al. [5,4] introduced two algorithms to solve the LCPS problem with time complexity $O(n^4)$ and $R^2 \log^2 n \log \log n$, respectively. Here, the set of all ordered pairs of matches between the two strings is denoted by $\mathcal{M}$ and $|\mathcal{M}| = R$. Readers are kindly noted regarding the subtle difference of the two parameters R and $\mathcal{R}$ (see Section 2.2).

In this section we discuss how to use PSA to compute an LCPS of two given strings. Our idea is to compute an automata called Common Palindromic Subsequence Automata (CPSA) as defined below.

Definition 2. *Common Palindromic Subsequence Automata (CPSA): Given two strings, a Common Palindromic Subsequence Automata (CPSA) accepts all common palindromic subsequences of the given strings.*

Now in the LCPS problem, given two strings S_1 and S_2 our task is to find the longest common palindromic subsequence (LCPS). Now we have the following lemma.

Lemma 2. *To find the longest common palindromic subsequence (LCPS) of two strings S_1 and S_2 we need to find the Max Length Automata [13] of the intersection of the PSA of S_1 and S_2*

Proof. Since the PSA of S_1 and S_2 can generate all the palindromic sub-sequences of S_1 and S_2, their intersection automata can generate all the palindromic common sub-sequences of S_1 and S_2. So, the Max Length Automata will find the LCPS of S_1 and S_2.□

So we have the following simple algorithm to compute an LCPS.

- Step 1: Find the PSA M_1 of S_1
- Step 2: Find the PSA M_2 of S_2
- Step 3: Find the CPSA M_3 by intersecting M_1 and M_2
- Step 4: Find the Max Length Automaton M_4 of M_3

Notably, in Step 3, we use the algorithm presented in [13,17] with a slight modification. Example 3 shows the construction of a CPSA. The algorithm for constructing CPSA is formally presented in Algorithm 2.

Example 3. Let $S_1 = abba$ and $S_2 = abca$ be two strings. Figures 2 and 3 represent the PSA's of S_1 and S_2 respectively. The CPSA is shown in Fig. 4. From that CPSA we can get four common palindromic subsequences of S_1 and S_2 namely a, b, aa and aba based on the path and weight. The LCPS of S_1 and S_2 is aba.

Algorithm 2. Algorithm for Construction of CPSA

input: PSA $M_1 = (Q^1, \Sigma^1, \delta^1, \sigma^1, q_0^1, F^1), M_2 = (Q^2, \Sigma^2, \delta^2, \sigma^2, q_0^2, F^2)$
output: PSA M = $(Q, \Sigma, \delta, \sigma, q_0, F)$, $\mathcal{L}(M)$ = $\mathcal{L}(M_1)$ $\cap$ $\mathcal{L}(M_2)$

```
1: begin
2:   q₀ ← [q₀¹, q₀²]
3:   Q ← {[q₀¹, q₀²]}
4:   F ← {q₀}
5:   C ← NEW − QUEUE()
6:   ENQUEUE(C, [q₀¹, q₀²])
7:   while not EMPTY(C) do
8:     [q¹, q²] ← DEQUEUE(C)
9:     for all s ∈ Σ do
10:      [p¹, p²] ← [δ¹(q¹, s), δ²(q², s)]
11:      δ([q¹, q²], s) ← [p¹, p²]
12:      if σ(q¹, s, p¹) = σ(q², s, p²) then
13:        σ([q¹, q²], s, [p¹, p²]) ← σ(q¹, s, p¹)
14:      else
15:        σ([q¹, q²], s, [p¹, p²]) ← 1
16:      end if
17:      if [p¹, p²] ∉ Q then
18:        ENQUEUE(C, [p¹, p²]
19:        Q ← Q ∪ {[p¹, p²]}
20:        F ← {[p¹, p²]}
21:      end if
22:    end for
23:  end while
24: end
```

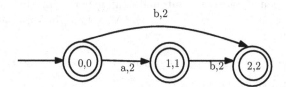

Fig. 2. A PSA for string "abba"

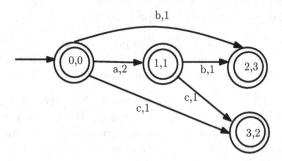

Fig. 3. A PSA for string "abca"

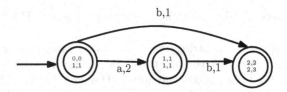

Fig. 4. A CPSA for string "abba" and "abca"

3.1   Analysis

Let the number of states in M_1 and M_2 be $\mathcal{R}_1$ and $\mathcal{R}_2$, respectively. Then the time complexity of constructing M_1 and M_2 will be $O(\mathcal{R}_1|\Sigma|)$ and $O(\mathcal{R}_2|\Sigma|)$, respectively. So, the time complexity of constructing M_3 (presented in Algorithm 2) will be $O(\mathcal{R}_1\mathcal{R}_2|\Sigma|)$ [13]. The time complexity of constructing M_4 will also be $O(\mathcal{R}_1\mathcal{R}_2|\Sigma|)$. We can traverse through the longest path of M_4 and the accepting string will be LCPS of the given two strings. As the length of LCPS is at most n so finding LCPS in M_4 needs only $O(n)$ times. Thus the total complexity of finding LCPS is $O(\mathcal{R}_1\mathcal{R}_2|\Sigma|)$. In worst case the running time is $O(n^4|\Sigma|)$

3.2   Comparison of the Algorithms

As has been mentioned above, to the best of our knowledge the only work in the literature that presents algorithm to compute LCPS is the very recent work of Chowdhury et al. [5,4]. In [5,4] two algorithms were provided. The first algorithm, referred to as CHIR-1 henceforth, runs in $O(n^4)$ time and the second one, referred to as CHIR-2 henceforth, runs in $R^2 \log^2 n \log \log n$ time where R is the number of matches. In this section we compare the running time of our algorithm with CHIR-1 and CHIR-2.

Since $\mathcal{R}_1 = O(n^2)$ and $\mathcal{R}_2 = O(n^2)$, the running time of Algorithm 2 becomes $O(n^4|\Sigma|)$ in the worst case, which is not better than that CHIR-1 presented in [5,4]. But in cases where we have $\mathcal{R}_1 = O(n)$ and $\mathcal{R}_2 = O(n)$, it exhibits very good performance. In such cases the running time reduces to $O(n^2|\Sigma|)$. Even for $\mathcal{R}_1 = \mathcal{R}_2 = O(n^{1.5})$ this algorithm performs better $(O(n^3|\Sigma|))$ than CHIR-1.

Algorithm 2 also performs better than CHIR-2 of [5,4] in some of the cases. Clearly $R \geqslant \mathcal{R}_1$ and $R \geqslant \mathcal{R}_2$ and in most cases $|\Sigma| < \log^2 n \log \log n$. And if Σ is constant then our algorithm always outperforms CHIR-2 of [5,4].

4   Experimental Result

In addition to a theoretical analysis presented above we make an effort to experimentally analyse the performance of PSA and CPSA. Both the running time and space complexity of the algorithms presented here depend on the number of states of PSA and CPSA. So in our experiments we investigate how the number of states increases in practice with the increase of n and Σ for both PSA and CPSA.

In our experiments the strings were generated randomly based on a given character set, Σ. For PSA the length of the string, n varies among 100, 250, 500, 850, 1000, 1200, 1500, 5000 and 10000. The alphabet size $|\Sigma|$ varies among 4, 8, 12, 16, 20, 24 and 28. For each combination of n and $|\Sigma|$, we generate 50 strings and calculate (and report) the average number of states of the PSA. These are presented in Table 1. It can be seen from Table 1 that by increasing n the number of states in PSA increases. However this increase is not really as much as suggested by the theoretical analysis. For example when $n = 10000$ and $|\Sigma| = 4$ the number of states of PSA should be in the range of 100000000 whereas in practice, it is only 7489599. Increasing $|\Sigma|$ decreases the number of states of PSA as is evident in Table 1.

Now, let us focus our attention to our algorithm for LCPS computation. The running time of our algorithm for computing an LCPS of two strings mainly depends on the construction time of the PSA. The construction of PSA in turn depends on the total number of states of PSA. While conducting the experiments the lengths of the two strings were kept equal. The length of the string, n varies among 30, 40, 50, 62, 80, 100, 120, 150, 170, 200. The alphabet size $|\Sigma|$ varies

Table 1. Number of states in PSA

n	$|\Sigma| = 4$	$|\Sigma| = 8$	$|\Sigma| = 12$	$|\Sigma| = 16$	$|\Sigma| = 20$	$|\Sigma| = 24$	$|\Sigma| = 28$
100	715	494	384	322	282	255	232
250	4582	3061	2275	1831	1546	1350	1198
500	18588	12148	8962	7111	5903	5083	4486
850	53707	35150	25723	20353	16776	14356	12551
1000	74401	48645	35478	28008	23167	19720	17261
1200	107539	69964	51211	40257	33194	28248	24669
1500	167600	109669	79886	62637	51792	43929	38378
5000	1873262	1214192	79886	691858	567412	481795	419028
10000	7489599	4862742	3529004	2762079	2263614	1922768	1669577

Table 2. Number of states in CPSA

n	$|\Sigma| = 4$	$|\Sigma| = 8$	$|\Sigma| = 12$	$|\Sigma| = 16$	$|\Sigma| = 20$	$|\Sigma| = 24$	$|\Sigma| = 28$
30	127	61	39	27	24	19	16
40	426	165	88	58	46	38	32
50	1027	385	199	125	92	70	58
62	2604	895	441	276	186	147	106
80	8525	2799	1304	720	482	334	245
100	24634	7786	3299	1705	1122	788	564
120	58311	17239	7137	3786	2350	1574	1212
150	170738	46648	19237	10026	3957	4033	2782
170	277262	84245	33298	17043	9995	6463	4567
200	673053	17239	68762	36851	20198	11937	9730

among 4, 8, 12, 16, 20, 24. For each combination of n and $|\Sigma|$ we generate 50 strings and calculate (and report) the average number of states in the CPSA. The results are given in Table 2. As is evident from Table 2 the number of states in CPSA is far less than n^4, the theoretical worst case bound. It is even less than the square of the number of states in PSA (see Table 1). For example when $n = 200$ and $|\Sigma| = 4$, the number of states of CPSA should be in the range of 200000000 where in practice it is only 673053. So despite a theoretical worst case bound of $O(n^2)$ for PSA and $O(n^4)$ for CPSA, practically the states never reach that limit (and hence the time requirement as well is much low in practice). Increasing $|\Sigma|$ decreases the number of states of CPSA as is evident in Table 2.

5   Conclusion

In this paper, we have introduced and presented PSA, a novel weighted finite automata that is a compact representation of all the palindromic subsequences of a string. The space complexity of our approach is better than that of [6]. Furthermore we use the PSA to solve the LCPS problem. In particular we present CPSA which is the intersection automata of two PSA of two given strings. The time complexity of our algorithm to solve the LCPS problem is $O(\mathcal{R}_1 \mathcal{R}_2 |\Sigma|)$ where $\mathcal{R}_1$ and $\mathcal{R}_2$ are the number of states of respective automata. Our algorithm also performs better than the algorithms presented in [5,4]. We also present experimental results which suggest that both PSA and CPSA perform very well in practice.

References

1. Breslauer, D., Galil, Z.: Finding all periods and initial palindromes of a string in parallel. Algorithmica 14(4), 355–366 (1995)
2. Chen, K.-Y., Hsu, P.-H., Chao, K.-M.: Identifying approximate palindromes in run-length encoded strings. In: Cheong, O., Chwa, K.-Y., Park, K. (eds.) ISAAC 2010, Part II. LNCS, vol. 6507, pp. 339–350. Springer, Heidelberg (2010)
3. Choi, C.Q.: Dna palindromes found in cancer. The Scientist (2005)
4. Chowdhury, S.R., Hasan, M.M., Iqbal, S., Rahman, M.S.: Computing a longest common palindromic subsequence. Fundamneta Informaticae
5. Chowdhury, S.R., Hasan, M. M., Iqbal, S., Rahman, M.S.: Computing a longest common palindromic subsequence. In: Arumugam, S., Smyth, B. (eds.) IWOCA 2012. LNCS, vol. 7643, pp. 219–223. Springer, Heidelberg (2012)
6. Chuang, K., Lee, R., Huang, C.: Finding all palindrome subsequences in a string. In: The 24th Workshop on Combinatorial Mathematics and Computation Theory (2007)
7. Farhana, E., Rahman, M.S.: Doubly-constrained lcs and hybrid-constrained lcs problems revisited. Inf. Process. Lett. 112(13), 562–565 (2012)
8. Galil, Z.: Real-time algorithms for string-matching and palindrome recognition. In: STOC, pp. 161–173 (1976)
9. Gusfield, D.: Algorithms on Strings, Trees, and Sequences - Computer Science and Computational Biology. Cambridge University Press (1997)

10. Hoshino, H., Shinohara, A., Takeda, M., Arikawa, S.: Online construction of sub-sequence automata for multiple texts. In: SPIRE, pp. 146–152 (2000)
11. Hsu, P.-H., Chen, K.-Y., Chao, K.-M.: Finding all approximate gapped palin-dromes. In: Dong, Y., Du, D.-Z., Ibarra, O. (eds.) ISAAC 2009. LNCS, vol. 5878, pp. 1084–1093. Springer, Heidelberg (2009)
12. I., T., Inenaga, S., Takeda, M.: Palindrome pattern matching. In: Giancarlo, R., Manzini, G. (eds.) CPM 2011. LNCS, vol. 6661, pp. 232–245. Springer, Heidelberg (2011)
13. Iliopoulos, C.S., Rahman, M.S., Vorácek, M., Vagner, L.: Finite automata based algorithms on subsequences and supersequences of degenerate strings. J. Discrete Algorithms 8(2), 117–130 (2010)
14. Kolpakov, R., Kucherov, G.: Searching for gapped palindromes. Theor. Comput. Sci. 410(51), 5365–5373 (2009)
15. Manacher, G.K.: A new linear-time "on-line" algorithm for finding the smallest initial palindrome of a string. J. ACM 22(3), 346–351 (1975)
16. Matsubara, W., Inenaga, S., Ishino, A., Shinohara, A., Nakamura, T., Hashimoto, K.: Efficient algorithms to compute compressed longest common substrings and compressed palindromes. Theor. Comput. Sci. 410(8-10), 900–913 (2009)
17. Melicher, B., Holub, J., Muzatko, P.: Language and Translation. Publishing House of CTU (1997)
18. Porto, A.H.L., Barbosa, V.C.: Finding approximate palindromes in strings. Pattern Recognition 35(11), 2581–2591 (2002)
19. Tanaka, H., Tapscott, S.J., Trask, B.J., Yao, M.C.: Short inverted repeats initiate gene amplification through the formation of a large dna palindrome in mammalian cells. National Academy of Science 99(13), 8772–8777 (2002)

LALBLC
A Program Testing the Equivalence of dpda's

Patrick Henry and Géraud Sénizergues

LaBRI and Université de Bordeaux, Talence, France
{Patrick.Henry,ges}@labri.fr

Abstract. We describe the program LALBLC which tests whether two deterministic pushdown automata recognize the same language.

Keywords: Deterministic pushdown automata, deterministic context-free grammars, equivalence problem.

1 Introduction

The so-called "equivalence problem for deterministic pushdown automata" is the following decision problem:

INSTANCE : two dpda A, B; QUESTION : $L(A) = L(B)$?

i.e. do the given automata recognize the *same* language? This problem was shown to be decidable in ([Sén97],[Sén01a, sections 1-9])[1]. Beside crude decidability, the intrinsic complexity of this problem is far from being understood. A progress in this direction has been achieved in [Sti02] by showing that the general problem is *primitive recursive* while subclasses with complexity in P (resp. co-NP) have been discovered in [BCFR06, BG11, BGJ13](resp.[Sén03]). Any further progress in this direction is likely to have some impact on other areas of computer science, as is shown by the numerous applications that were found even before proving decidability of the problem (see [Sén01b] for a survey and [MOW05, CCD13] for more recent connections).

The contribution presented here consists in showing that it is *practically feasable* to solve the equivalence problem for general dpda on non-trivial examples (see section 5). We have implemented, and to some extent refined, the main ideas of [Sén01a]. For every pair (A, B) the program returns, either a proof of $L(A) = L(B)$ (see section 4 for a precise notion of proof) or a terminal word witnessing the fact that $L(A) \neq L(B)$.

The sources of our (Python) program, as well as as additional information, can be uploaded from http://dept-info.labri.u-bordeaux.fr/~ges.

2 Automata, Grammars

We introduce here the notions of automata and grammars that are manipulated by the program.

---

[1] A similar method is exposed within the framework of term root-rewriting in [Jan12].

S. Konstantinidis (Ed.): CIAA 2013, LNCS 7982, pp. 169–180, 2013.

2.1   Deterministic Matricial fa

A *finite automaton* is, as usual, a tuple, $\mathcal{A} = <X, Q, Q_-, Q_+, \delta>$ where X is the input alphabet, Q is the set of states, $Q_- \subset Q$ is the set of initial states, $Q_+ \subset Q$ is the set of terminal states, $\delta \subset Q \times X \times Q$ is the set of transitions, and all of these five sets are finite.

The main object that our program handles is a *matrix* of languages which is defined by some finite automaton with one set of initial states for each line and one set of terminal states for each column.

Let us recall that a language $L \subset X^*$ is a *prefix language* iff, $\forall u, v \in L, u \preceq v \Rightarrow u = v$. The line-vectors of the matrices we are interested in are *prefix vectors* in the following sense:

Definition 1. *A vector* $(L_1, \ldots, L_i, \ldots, L_n) \in \mathcal{P}(X^*)^n$ *is said to be* prefix *iff it fulfills:* $\forall i, j \in [1, n], i \neq j \Rightarrow L_i \cap L_j = \emptyset$ and $\bigcup_{i=1}^{n} L_i$ is a prefix language.

We thus consider the following variant of the notion of d.f.a. which recognizes a *prefix* matrix of languages i.e. where each row-vector is prefix.

Definition 2 (complete deterministic matricial f.a.). *A* finite *complete deterministic matricial finite automaton is a tuple,* $\mathcal{A} = <X, Q, Q_{1,-}, \ldots, Q_{n,-}, Q_{1,+}, \ldots, Q_{m,+}, \delta>$ *where* X *is the input alphabet,* Q *is the set of states,* $Q_{i,-} \subseteq Q$ *is the set of initial states of the i-th line,* $Q_{j,+} \subseteq Q$ *is the set of final states of the j-th column,*
$\forall i \in [1, n], \mathrm{Card}(Q_{i,-}) = 1$
$\forall j, k \in [1, m], j \neq k \Rightarrow Q_{j,+} \cap Q_{k,+} = \emptyset$
$\delta : (Q \times X) \to Q$ *is a total map, which is called the transition map.*
$\forall j \in [1, m], \forall q \in Q_{j,+}, \forall x \in X, \delta(q, x)$ *is not co-accessible from* $\cup_{1 \leq k \leq m} Q_{k,+}$.
All the items of this tuple are assumed to be finite sets.

Such an automaton defines a prefix *matrix of languages* $\mathrm{L}(\mathcal{A}) := (L_{i,j})_{(i,j) \in [1,n] \times [1,m]}$ where $L_{i,j} := \{u \in X^* \mid \exists q \in Q_{i,-}, \delta^*(q, u) \in Q_{j,+}\}$.

The usual theory of recognizable languages, complete deterministic automata and residuals can be adapted to (prefix) matrices of languages, cdmfa's and residuals of matrices.

Implementation. Our module `fautomata` deals with f.automata and their analogues. The class `dr-matrix` implements the notion of cdmfa. The program stores every rational prefix matrix under the form of a *canonical* dcmfa i.e. a minimal dcmfa, in which the states are integers that are completely determined by some depth-first traversal of the minimal automaton. The equality of two rational (prefix) matrices is implemented as an isomorphism-test for the corresponding canonical dcmfa.

2.2   Pushdown Automata and Context-Free Grammars

The notions of *pushdown automaton* and *context-free grammar* are well-known. A pda is said *deterministic* if, informally, on every triple (state, stack-contents, tape-contents), *at most one* transition is applicable. It is called *strict* if it recognizes by empty stack and a finite set of final sates and *normal* if every ϵ-transition is *popping*.

Definition 3. *([Har78, Definition 11.4.1 p.347]) Let* $G = < X, V, P >$ *be a context-free grammar.* G *is said* strict-deterministic *iff there exists an equivalence relation* $\smile$ *over* V *fulfilling the following conditions:*

1- X is a class (mod $\smile$)
2- for every $v, v' \in V, \alpha, \beta, \beta' \in (X \cup V)^*$, if $v \longrightarrow_P \alpha \cdot \beta$ and $v' \longrightarrow_P \alpha \cdot \beta'$ and $v \smile v'$, then either:
2.1- both $\beta, \beta' \neq \epsilon$ and $\beta[1] \smile \beta'[1]$
2.2- or $\beta = \beta' = \epsilon$ and $v = v'$.

(In the above definition, for every word γ, $\gamma[1]$ denotes the first letter of the word γ). Any equivalence $\smile$ satisfying the above condition is said to be a *strict equivalence* for the grammar G. It is known that, given a strict dpda $\mathcal{M}$, one can construct, in polynomial time, an associated grammar $G_\mathcal{M} = < X, V_\mathcal{M}, P_\mathcal{M} >$ which is strict-deterministic and generates the language recognized by $\mathcal{M}$.

Implementation Our module `grammars` deals with dpda and dcf grammars. The translation of a dpda into a dcf grammar is realized by the `autotogram(A)` function; some routine functions around these notions are implemented (test for determinism of a cf grammar, elimination of non-productive non-terminals and reduction in Greibach normal-form, for grammars translating a normal strict dpda), see Figure 1.

```
Non-terminal symbols :
[<q2-A-q4> <q2-A-qb> ][<q4-O-q3> ][<q1-O-q3> ]
[<q2-O-q3> ][<q3p-O-q3> ][<q4-A-q4> ][<q3-O-q3> ]
[<q3-A-q3> ][<q1-A-q3> <q1-A-q5> ][<q3b-O-q3> ]

Terminal symbols : # a b x
Rewriting rules:
 <q1-A-q3>::=a
 <q1-A-q3>::=x<q1-A-q3><q3-A-q3>
 <q1-A-q5>::=b
 <q1-A-q5>::=x<q1-A-q5>
 <q1-O-q3>::=#<q1-A-q3><q3-O-q3>
 <q1-O-q3>::=#<q1-A-q5>
```

```
<q2-A-q4>::=a<q4-A-q4><q4-A-q4>
<q2-A-q4>::=x<q2-A-q4><q4-A-q4>
<q2-A-qb>::=b
<q2-A-qb>::=x<q2-A-qb>
<q2-O-q3>::=#<q2-A-q4><q4-O-q3>
<q2-O-q3>::=#<q2-A-qb>
<q3-A-q3>::=a
<q3-O-q3>::=a<q3p-O-q3>
<q3b-O-q3>::=a
<q3p-O-q3>::=a<q3b-O-q3>
<q4-A-q4>::=a
<q4-O-q3>::=a

Axiom: <q1-O-q3>
```

Fig. 1. A dcf grammar G2 (obtained from some dpda)

3   Algebraic Framework

We recall here the algebraic framework which is the base of our program (see [Sén01a, sections 2,3] for more details).

3.1   Semi-rings and Right-Actions

Semi-ring $\mathbb{B}\langle\langle\ W\ \rangle\rangle$. Let $(B, +, \cdot, 0, 1)$ where $B = \{0, 1\}$ denote the semi-ring of "booleans". Let W be some alphabet. By $(\mathbb{B}\langle\langle\ W\ \rangle\rangle, +, \cdot, \emptyset, \epsilon)$, we denote the semi-ring of *boolean series* over W (which is, up to isomorphism, nothing else than the semi-ring of subsets of W^*: $(\mathcal{P}(W^*), \cup, \cdot, \emptyset, \{\epsilon\})$.

Right-actions over $\mathbb{B}\langle\langle W \rangle\rangle$. We recall the following classical right-action $\bullet$ of the monoid W^* over the semi-ring $\mathbb{B}\langle\langle W \rangle\rangle$: for all $S, S' \in \mathbb{B}\langle\langle W \rangle\rangle, u \in W^*$

$$S \bullet u = S' \Leftrightarrow \forall w \in W^*, (S'_w = S_{u \cdot w}),$$

(i.e. $S \bullet u$ is the *residual* of S by u). Let $(V, \smile)$ be the structured alphabet associated with a strict-deterministic grammar (see paragraph §2.2). We define the right-action $\odot$ over non-terminal words by:

$$\epsilon \odot x = \emptyset. \quad (v \cdot \beta) \odot x = (\sum_{(v,h) \in P} h \bullet x) \cdot \beta,$$

The action is then extended to arbitrary boolean series (on the left) and to arbitrary terminal words (on the right) by:

$$(\sum_{w \in W^*} S_w \cdot w) \odot x := \sum_{w \in W^*} S_w(w \odot x), \quad S \odot \epsilon := S, \quad S \odot wx := (S \odot w) \odot x$$

3.2 Deterministic Matrices

We recall here the notion of *deterministic* series and, more generally, deterministic matrices[2,3]. Let us consider a pair $(W, \smile)$ where W is an alphabet and $\smile$ is an equivalence relation over W. We call $(W, \smile)$ a *structured* alphabet.

Let us denote by $\mathbb{B}_{n,m}\langle\langle W \rangle\rangle$ the set of (n, m)-matrices with entries in the semi-ring $\mathbb{B}\langle\langle W \rangle\rangle$ (the index (m, n) will continue to mean "of dimension (m, n)" for all subsequent subsets of matrices).

Definition 4. *Let* $m \in \mathbb{N}, S \in \mathbb{B}_{1,m}\langle\langle W \rangle\rangle$: $S = (S_1, \cdots, S_m)$. S *is said* left-deterministic *iff either* $\forall i \in [1, m], S_i = \emptyset$ *or* $\exists i_0 \in [1, m], S_{i_0} = \epsilon$ *and* $\forall i \neq i_0, S_i = \emptyset$ *or* $\forall w, w' \in W^*, \forall i, j \in [1, m], (S_i)_w = (S_j)_{w'} = 1 \Rightarrow [\exists A, A' \in W, w_1, w'_1 \in V^*, A \smile A', w = A \cdot w_1 \text{ and } w' = A' \cdot w'_1].$

Both right-actions $\bullet, \odot$ on $\mathbb{B}\langle\langle W \rangle\rangle$ are extended componentwise to $\mathbb{B}_{n,m}\langle\langle W \rangle\rangle$.

Definition 5. *A row-vector* $S \in \mathbb{B}_{1,m}\langle\langle W \rangle\rangle$ *is said* deterministic *iff for every* $u \in W^*$, $S \bullet u$ *is left-deterministic.*
A matrix $S \in \mathbb{B}_{n,m}\langle\langle W \rangle\rangle$ *is said* deterministic *iff for every* $i \in [1, n]$, $S_{i,*}$ *is a deterministic row-vector.*

The classical definition of *rationality* of series in $\mathbb{B}\langle\langle W \rangle\rangle$ is extended componentwise to matrices. Given $A \in \mathbb{B}_{1,m}\langle\langle W \rangle\rangle$ and $1 \leq j_0 \leq m$, we define the vector $\nabla^*_{j_0}(A) :=$ A by:
if $A = (a_1, \ldots, a_j, \ldots, a_m)$ then $A' := (a'_1, \ldots, a'_j, \ldots, a'_m)$ where

$$a'_j := a^*_{j_0} \cdot a_j \text{ if } j \neq j_0 , \quad a'_j := \emptyset \text{ if } j = j_0.$$

---

[2] These series play, for dcf grammars the role that *configurations* play for a dpda.

[3] It extends the notion of (finite) *set of associates* defined in [HHY79, definition 3.2 p. 188].

Note that every deterministic matrix is prefix; it follows that every deterministic ratio-nal matrix is recognized by some cdmfa. We use the acronyms $DB_{n,m}\langle\langle\ W\ \rangle\rangle$ (resp. $DRB_{n,m}\langle\langle\ W\ \rangle\rangle$) for the sets of *Deterministic* (resp. *Deterministic Rational*) matrices. The main closure properties of deterministic rational matrices are summarized below.

Proposition 1. *Let* $S \in DRB_{n,m}\langle\langle\ W\ \rangle\rangle$, $T \in DRB_{m,s}\langle\langle\ W\ \rangle\rangle$, $w \in W^*$, $u \in X^*$, *Then*
$$S \cdot T \in DRB_{n,s}\langle\langle\ W\ \rangle\rangle, S \bullet w \in DRB\langle\langle\ W\ \rangle\rangle, S \odot u \in DRB\langle\langle\ W\ \rangle\rangle$$
If $n = 1, 1 \leq j_0 \leq m$, *then* $\nabla_{j_0}^*(S) \in DRB_{1,m}\langle\langle\ W\ \rangle\rangle$.

These closure properties are effective.

Terminal matrices versus non-terminal matrices. Let us denote by $L : DB\langle\langle\ V\ \rangle\rangle \to DB\langle\langle\ X\ \rangle\rangle$ the map sending every deterministic series S on the language $L(S) := \{u \in X \mid S \odot u = \varepsilon\}$ (i.e. the set of terminal words generated from all non-terminal words of S via the derivation w.r.t. the rules of G). For every integers $n, m \geq 1$, L is extended componentwise as a map $DB_{n,m}\langle\langle\ V\ \rangle\rangle \to DB_{n,m}\langle\langle\ X\ \rangle\rangle$.

Lemma 1. *For every* $S \in DB_{n,m}\langle\langle\ V\ \rangle\rangle$, $T \in DB_{m,s}\langle\langle\ V\ \rangle\rangle$, $u \in X^*$,
$$L(\varepsilon) = \varepsilon, \quad L(S \cdot T) = L(S) \cdot L(T), \quad L(S \odot u) = L(S) \bullet u.$$

Implementation. The module fautomata implements the matricial product · (prod), the right-actions • (bullet), ⊙ (odot) and the operation $\nabla_{j_0}^*$ (nablastar).

3.3   Linear Combinations

Let us call *linear combination* of the series $S_1, \ldots, S_j, \ldots, S_m$ any series of the form $\sum_{1 \leq j \leq m} \alpha_j \cdot S_j$ where $\alpha \in DRB_{1,m}\langle\langle\ V\ \rangle\rangle$. Let $S_1, \ldots, S_j, \ldots, S_m \in DRB\langle\langle\ V\ \rangle\rangle$. We call *dependency* of order 0 between the S_j's, an equality of the form:

$$S_{j_0} = \sum_{1 \leq j \leq m} \gamma_j' \cdot S_j, \tag{1}$$

where $j_0 \in [1, m]$, $\gamma' \in DRB_{1,m}\langle\langle\ V\ \rangle\rangle$ and $\gamma_{j_0}' = \emptyset$. [4] Analogously, we call *depen-dency* of order 1 between the S_j's, an equality of the form (1), but where the symbol "=" is replaced by the symbol "≡". It is clear that the homomorphism L maps every dependency of order 1 between the S_j's onto a dependency of order 0 between the $L(S_j)$.

Canonical coordinates. Let $S, T_1, T_2, \ldots, T_n \in DB\langle\langle\ V\ \rangle\rangle$. We assume that $i \neq j \Rightarrow T_i \neq T_j$. For every $i \in [1, n]$, we define
$\alpha_i := \{u \in V^* \mid S \bullet u = T_i$ and $\forall u' \prec u, \forall j \in [1, n], S \bullet u' \neq T_j\}$ and
$\alpha_{n+1} := \{u \in S \mid \forall u' \preceq u, \forall j \in [1, n], S \bullet u' \neq T_j\}$.

---

[4] This terminology originates in [Mei89].

Lemma 2. *The vector α of canonical coordinates fullfils:*
1- $\alpha \in \mathrm{D\mathbb{B}}_{1,n+1}\langle\langle\, V \,\rangle\rangle$, $S = \sum_{i=1}^{n} \alpha_i \cdot T_i + \alpha_{n+1}$
2- S is a linear combination of the T_i, with a vector of coefficients in $\mathrm{D\mathbb{B}}_{1,n}\langle\langle\, V \,\rangle\rangle$ iff
$\alpha_{n+1} = \emptyset$.

Unifiers. The following notion was implicit in [Sén01a, section 5] and explicited in [Sén05, section 11]. It turns out to be central in our implementation. Let $\alpha, \beta \in \mathrm{D\mathbb{B}}_{1,q}$ $\langle\langle\, X \,\rangle\rangle$. A *unifier* of (α, β) is any matrix $U \in \mathrm{D\mathbb{B}}_{q,q}\langle\langle\, X \,\rangle\rangle$ such that: $\alpha \cdot U = \beta \cdot U$. U is a *Most General* Unifier iff every unifier of (α, β) has the form $U \cdot T$ for some $T \in \mathrm{D\mathbb{B}}_{q,q}$. This notion is lifted to $\alpha, \beta \in \mathrm{DR\mathbb{B}}_{1,q}\langle\langle\, V \,\rangle\rangle$ via the map L.

Theorem 1. *1- Every pair $\alpha, \beta \in \mathrm{D\mathbb{B}}_{1,q}\langle\langle\, V \,\rangle\rangle$ has a MGU (up to $\equiv$)*
2- This MGU is unique, *up to $\equiv$ and up to some right-product by a permutation matrix.*
3- For pairs $\alpha, \beta \in \mathrm{DR\mathbb{B}}_{1,q}\langle\langle\, V \,\rangle\rangle$, the MGU has some representative which belongs to $\mathrm{DR\mathbb{B}}_{q,q}\langle\langle\, V \,\rangle\rangle$ and is computable from α, β.

In other words, the MGU of two algebraic row-vectors defined by det. rational vectors over a s.d. grammar G is itself algebraic and definable by a det. rational-matrix over the grammar G. The MGU of $\alpha, \beta \in \mathrm{DR\mathbb{B}}_{1,q}\langle\langle\, V \,\rangle\rangle$ can be computed along the following algorithm scheme:

$M \leftarrow \mathrm{Id}_q;\; cost \leftarrow 0$
while (not $\alpha \cdot M \equiv \beta \cdot M$) **do**
    find $j \in [1, q]$, $w \in X^*$, prefix-minimal, such that:
    $((\alpha \cdot M) \odot w = \varepsilon_j^q)$ iff $((\beta \cdot M) \odot w \neq \varepsilon_j^q)$
    $\gamma \leftarrow (\alpha \cdot M) \odot w$ (if it is equal to ε_j^q) or $\gamma \leftarrow (\beta \cdot M) \odot w$ (if it is equal to ε_j^q)
    $\gamma \leftarrow \nabla_j^*(\gamma))$
    $D \leftarrow \mathrm{Id}_q;\; D_{j,*} \leftarrow \gamma$ {D is the dependency matrix associated to γ and j}
    $M \leftarrow M \cdot D;\; cost \leftarrow cost + |w|$
end while
return $[M, cost]$

(See on Figure 2 an example of mgu computation, where $q = 4$).
The integer *cost* is useful for a proper use of M leading to an equivalence proof (i.e. for ensuring property (3) of §4.5).

Implementation. The module `fautomata` implements the function `coords` that computes the canonical coordinates of a d.r. series over a finite family of d.r. series.

The module `equations` defines a functional `mgu(f-equiv, f-op, vec1, vec2)`: it computes the MGU of two row-vectors by the above algorithm where `f-equiv` is used for testing the equivalence (or returning a witness) of two row-vectors and `f-op` is the right-action used for computing the dependency γ. The MGU's of order 0 or approximated[5] MGU's of order 1 are obtained by application of this functional.

---

[5] I.e. up to some length for the terminal words.

```
v1: list of states [0,1,2,3]
sets of init states [[0]]
sets of fin states [[1],[3],[],[]]
list of (non-sink) transitions:
( 0 <q1-A-q3> )--> 1
( 0 <q1-A-q5> )--> 3

v2: list of states [0,1,2,3]
sets of init states [[0]]
sets of fin states [[],[],[2],[3]]
list of (non-sink) transitions:
( 0 <q2-A-q4> )--> 2
( 0 <q2-A-qb> )--> 3
```

```
mgu list of states [0,1,2,3,4]
sets of init states [[0],[4],[3],[4]]
sets of fin states [[],[],[3],[4]]
list of (non-sink) transitions:
( 0 <q4-A-q4> )--> 2
( 2 <q4-A-q4> )--> 3

cost_mgu 2
```

Fig. 2. A mgu w.r.t. grammar G2

4   Logics

4.1   The Deduction Relation

We denote by $\mathcal{A}$ the set $\mathrm{DRB}\langle\langle\ V\ \rangle\rangle \times \mathrm{DRB}\langle\langle\ V\ \rangle\rangle$. An element $(S, T) \in \mathcal{A}$ is called an equation while a triple (p, S, T) where $p \in \mathbb{N}$ is called a *weighted* equation. The *divergence* of (S, T), denoted by $\mathrm{Div}(S, T)$, is defined by:

$$\mathrm{Div}(S, T) := \inf\{|u| \mid u \in X^*, (S \odot u = \varepsilon) \Leftrightarrow (S \odot u \neq \varepsilon)\}$$

The map Div is extended to sets of equations by: $\mathrm{Div}(P) := \inf\{\mathrm{Div}(p) \mid p \in P\}$. Let $\mathcal{C}$ be the set of meta-rules described in Figure 3. Let $\mathcal{B}$ be the set of meta-rules obtained by forgetting the first component p (an integer) in every weighted equation (p, S, T) of every meta-rule of $\mathcal{C}$. We define the binary relation $\Vdash_{\mathcal{B}} \subseteq \mathcal{P}(\mathcal{A}) \times \mathcal{A}$, as the set of all the instances of meta-rules of $\mathcal{B}$ where $S, T, T', T'', U \in \mathrm{DB}\langle\langle\ V\ \rangle\rangle, (S_1, S_2), (T_1, T_2), (U_1, U_2) \in \mathrm{DB}_{1,2}\langle\langle\ V\ \rangle\rangle, U_1 \neq \epsilon$. The binary relation $\vdash_{\mathcal{B}}$ over $\mathcal{P}(\mathcal{A})$ is defined by: $\forall P, Q \in \mathcal{P}(\mathcal{A})$

$$P \vdash_{\mathcal{B}} Q \Leftrightarrow (\forall q \in Q - P, \exists P' \subseteq P, \text{ such that } P' \Vdash_{\mathcal{B}} q).$$

The relation $\overset{p}{\vdash}_{\mathcal{B}}$ (for $p \in \mathbb{N}$) and $\overset{*}{\vdash}_{\mathcal{B}}$ are then deduced from $\vdash_{\mathcal{B}}$ as usual (and likewise the binary relations $\Vdash_{\mathcal{C}}, \vdash_{\mathcal{C}}, \overset{p}{\vdash}_{\mathcal{C}}, \overset{*}{\vdash}_{\mathcal{C}}$).

$$
\begin{array}{lll}
(W0) & \emptyset & \Vdash (0, T, T) \\
(W0') & \{(p, S, T)\} & \Vdash (p+1, S, T) \\
(W1) & \{(p, T, T')\} & \Vdash (p, T', T) \\
(W2) & \{(p, T, T'), (p, T', T'')\} & \Vdash (p, T, T'') \\
(W3) & \{(p, S_1, T_1), (p, S_2, T_2)\} & \Vdash (p, S_1 + S_2, T_1 + T_2) \\
(W4) & \{(p, T, T')\} & \Vdash (p, T \cdot U, T' \cdot U) \\
(W5) & \{(p, T, T')\} & \Vdash (p, U \cdot T, U \cdot T') \\
(W6) & \{(p, U_1 \cdot T + U_2, T)\} & \Vdash (p, U_1^* \cdot U_2, T)
\end{array}
$$

Fig. 3. System $\mathcal{C}$

Lemma 3. *: For every* $P, Q \in \mathcal{P}(\mathcal{A})$, $P \overset{*}{\vdash}_\mathcal{B} Q \Rightarrow \mathrm{Div}(P) \leq \mathrm{Div}(Q)$.

4.2   Self-Provable Sets

A subset $P \subseteq \mathcal{A}$ is said *self-provable*[6] iff

$\forall (S,T) \in P, (S = \varepsilon) \Leftrightarrow (T = \varepsilon)$ and $\forall x \in X, P \overset{*}{\vdash}_\mathcal{B} P \odot x$.

Lemma 4. *If* P *is self-provable then,* $\forall (S,T) \in P$, $S \equiv T$.

This follows easily from Lemma 3.

4.3   Comparison-Forest

A *comparison-forest* is, informally speaking, a set of oriented trees labeled by weighted equations such that:
- a distinguished root, the *starting-node*, has a label of the form $(0, S, T)$, where $S, T \in$ DRB$\langle\langle\, V\, \rangle\rangle$
- all other roots, the *unifier-nodes* have labels of the form $(0, u.M, v.M)$ where u, v are det. rat. row-vectors of dimension $(1, d)$ and M is a det. rat. matrix of dimension (d, d)
- non-root nodes have labels of the form (p, U, U') where $U, U' \in$ DRB$\langle\langle\, V\, \rangle\rangle$.
Every node can have the status "open" or "closed". In case it is closed, property (3) of §4.5 is satisfied. Open nodes are leaves.

4.4   Tactics and Strategies

The program maintains, at each step of the computation, a comparison-forest.
The program starts from the comparison-forest consisting of just one node, labeled by $(0, S, T)$. Then it iteratively modifies this c.f. by either:
1- closing an open node and adding new sons (the number of new sons ranges from 0 to the maximum cardinality of some class (modulo $\smile$); at this stage, the sons are open.
2- discovering that an open node is obviously false (e.g (p, ε, S) where $S \neq \varepsilon$); a witness $u \in X^*$ of non-equivalence is thus propagated to the root r above this node

2-a if r is a unifier-node, this unifier is improved and all nodes of the forest that are below some node "using" the unifier are destroyed.

2-b if r is the starting-node, the witness u is thus a *witness of falsity* for the initial equation (S, T). The algorithm stops and returns the witness.
3- discovering that the forest has no open node. The set of equations of the forest is thus a *self-provable set*. The algorithm stops and returns the self-provable set.

The precise sequence of actions of the program will be determined by a *strategy*; in turn, the strategy will call *tactics* that are able to perform, given an open node of the current comparison-forest, one of the above kind of actions.

Tactics. The main tactics already implemented are summarized in Table 1. The four last tactics lean on the notions exposed in Section 3. Note that TCM implements the "triangulation process" described in [Sén01a, section 5].

---

[6] Translation into our framework of the notion of "self-proving set of pairs" from [Cou83, p.162].

Table 1.

Trep	**argument**-node: n, open, labeled by (p, S, T) **context**:n', closed, labeled by (p', S, T) where $p' \le p$. **action**: n is closed, "leaning on" n'.
Teq	**argument**-node: n, open, labeled by (p, T, T) **action**: n is closed.
TA	**argument**-node: n, open, labeled by (p, S, T) **action**: n is closed "leaning on his new sons". $\text{Card}(X)$ sons are created, x-ith son is labeled by $(p + 1, S \odot x, T \odot x)$
TD	**argument**-node: n, open, labeled by $(p, \sum_{j=1}^{d} A_j \cdot S_j, \sum_{j=1}^{n} A_j \cdot T_j)$, where A_j are $\frown$-equivalent non-terminals. **action**: n is closed, "leaning on his new sons". d sons are created, j-ith son is labeled by $(p + 1, S_j, T_j)$
TCM	**argument**-node: n, open **context**: $n_0, n_1, \ldots, n_\ell$ is a path with $n_\ell = n$, n_i is labeled by $E_i = (\boldsymbol{\alpha}_i S, \boldsymbol{\beta}_i S)$ with a weight π_i where $\alpha_i, \beta_i \in \text{DRB}_{1,d}\langle\langle\, V\,\rangle\rangle, S \in \text{DRB}_{d,d}\langle\langle\, V\,\rangle\rangle$, **action**: a subsequence $n_0, n_{i_1}, \ldots, n_{i_r}$ is selected and r series S_j are eliminated as follows (w.l.o.g. we assume the eliminated indices are $1, \ldots, r$) $E_0 \odot w_1 = (S_1, \boldsymbol{\gamma_1} \cdot S), E_{i_1} D_1 \odot w_2 = (S_2, \boldsymbol{\gamma_2} \cdot S), \ldots, E_{i_{r-1}} D_1 \cdots D_{r-1} \odot w_r = (S_r, \boldsymbol{\gamma_r} \cdot S)$ each D_i is the dependency matrix associated to line i and vector $\boldsymbol{\gamma}_i$ Successive indices are chosen in a way that $\pi_j \ge \pi_{j-1} + |w_j| + 1$. the sub-tree strictly beneath n_{i_r} is destroyed. $M := D_1 D_2 \cdots D_r$, n_r is given d new open sons n'_j labeled by: $(\pi_{i_r}, (\boldsymbol{\alpha}_{i_r} \cdot M)_j, (\boldsymbol{\beta}_{i_r} \cdot M)_j)$.
TCJ	**argument**-node: n, open **context**: idem as for TCM. In addition, $\forall i < \ell, \exists u_i \in X^*, (\boldsymbol{\alpha}_i \odot u_i, \boldsymbol{\beta}_i \odot u_i) = (\boldsymbol{\alpha}_{i+1}, \boldsymbol{\beta}_{i+1})$. **action**: a candidate mgu M for the vectors $\boldsymbol{\alpha}_0, \boldsymbol{\beta}_0$ is computed together with its cost c. The smallest index i such that $\pi_i \ge \pi_0 + c + 1$ is selected. The subtree strict. beneath n_i is destroyed. n_i is given d new open sons n'_j labeled by: $(\pi_i, (\boldsymbol{\alpha}_i \cdot M)_j, (\boldsymbol{\beta}_i \cdot M)_j)$
TCR	**argument**-node: n, open, labeled by (p, S, T) **context** : idem as for TCJ. **action**: $M, cost, i$ are computed and subtree is destroyed as in TCJ. A new root n' is created, it is closed , n' is given d new open sons n'_j labeled by $(0, (\boldsymbol{\alpha}_0 \cdot M)_j, (\boldsymbol{\beta}_0 \cdot M)_j)$.
TSUN	**argument**-node: n, open, labeled by $(p, \boldsymbol{\alpha} S, \boldsymbol{\beta} S)$, where $\alpha, \beta \in \text{DRB}_{1,d}\langle\langle\, V\,\rangle\rangle, S \in \text{DRB}_{d,1}\langle\langle\, V\,\rangle\rangle$. and all components of $\boldsymbol{\alpha}, \boldsymbol{\beta}$ are null or have length one. **action**: n is closed. A candidate mgu M for the vectors $\boldsymbol{\alpha}, \boldsymbol{\beta}$ is computed The node n is given d new open sons, labeled by: $(p + 1, S_j, (M \cdot S)_j)$.

Error tactics. The tactics `Terror` is responsible for detecting that an open node is labeled by some trivially false equation. Then it returns "failure".

The tactics `Terror-dyn` also detects that an open node is false and then performs action 2-a or 2-b of subsection 4.4

Strategies. Two kinds of strategies have been developed. They all consist of combinations of the above tactics (or variants).The *static* strategies make only one guess of MGU (for each call to a computation of MGU) and either succeed to confirm this guess by terminating the proof, or discover an error and return "failure" as the global result. The *dynamic strategies* start each computation of mgu by a guess which might be improved by successive discoveries of errors by tactics `Terror-dyn`. Finally they return either a proof of the proposed equivalence or a witness of non-equivalence.

Implementation. The module `proofs` defines a class `proof` that implements the notion of comparison-forest. The functions in charge of managing the equations and MGU's are defined there. The module `tactics` implements the above defined tactics. In general we first defined abstract tactics that depend of functionnal arguments. Concrete tactics are obtained by instanciating these arguments by specific functions which compute MGU's. The module `strategies` defines a functional `make-strategy` `(maxsteps,error-tactics,*tactics)` which, in turn, produces concrete strategies.

4.5   Soundness

Our (meta)-proof that the program is *sound* i.e. that its positive outputs are really self-provable sets, leans on the auxiliary system $\mathcal{C}$ (see Figure 3). Let us use the following notation: for every $\pi, n \in \mathbb{N}, S, S' \in \mathrm{DRB}\langle\langle V \rangle\rangle$,

$$[\pi, S, S', n] = \{(\pi + |u|, S \odot u, S' \odot u) \mid u \in X^{\leq n}\}. \tag{2}$$

All the above tactics T enjoy the following fundamental property: if (π, S, S') is the weighted equation labelling a closed node of the forest t on which tactics T has been applied, then, for every terminal letter $x \in X$

$$\bigcup \{[p, U, U, n] \mid (p, U, U) \in \mathrm{im}(t), p + n \leq \pi\} \overset{*}{\vdash}_{\mathcal{C}} \{(\pi + 1, S \odot x, S' \odot x)\} \tag{3}$$

A comparison-forest is said *closed* when all its nodes are closed.

Theorem 2. *Let t be the closed forest computed by some strategy using only the tactics* Trep, Teq, TA, TD, TCM, TCJ, TCR, TSUN. *Then the set of equations labelling t is a* self-provable *set.*

Sketch of proof: Let us note P the set of weighted equations labelling t and let us consider the following property $\mathcal{Q}(\pi, n, p)$: $\forall S, S' \in \mathrm{DRB}\langle\langle V \rangle\rangle, P \overset{p}{\vdash}_{\mathcal{C}} (\pi, S, S') \Rightarrow P \overset{*}{\vdash}_{\mathcal{C}} [\pi, S, S', n]$.

Following the lines of the induction of [Sén01a, subsec. 10.2, eq (136)], one can prove by lexicographic induction over $(\pi + n, n, p)$ the statement: $\forall(\pi, n, p) \in \mathbb{N}^3$, $\mathcal{Q}(\pi, n, p)$.   □

5   Experiments

Out of 17 strategies already experimented, let us show the behaviour of 5 typical ones over 7 positive examples and 5 negative examples. The selected strategies are characterized by 3 parameters: their algebraic tactics [TCM (*triangulation*) or TCJ (*jump*) or TSUN (*quasi division*)], the *connectedness* property for the forests they produce[7] and their *static* (versus *dynamic*) character (see section 4). The *size* is the sum of the lengths

---

[7] Depending on the fact that they launch a new tree for each new mgu-computation or not.

of the rhs of the grammar. The tests have been run on a computer Intel(R) Xeon(R) CPU X5675 @ 3.07GHz. In each positive example we show the number of nodes of the final proof, the number of tactic calls and the CPU-time (number of seconds or "oot" if $>= 3600$).

pos example	ex0	ex1	ex2	ex3
size	36	51	34	86
$trg, c, stat$	44/44/0.88	75/121/10	99/145/11	oot
jp, c, dyn	44/44/0.79	75/117/4	67/123/8	100/1206/88
jp, nc, dyn	44/44/0.8	60/102/3.5	61/117/7	64/1104/83
$qdiv, nc, stat$	51/51/0.87	54/54/1	54/84/4	25/25/15
$qdiv, nc, dyn$	51/60/1	54/70/1	60/140/7	25/39/0.23

pos example	ex4	ex5	ex6
size	179	253	525
$trg, c, stat$	oot	oot	oot
jp, c, dyn	707/1067/476	oot	oot
jp, nc, dyn	251/467/117	732/4220/1245	oot
$qdiv, nc, stat$	134/134/180	149/149/80	502/502/977
$qdiv, nc, dyn$	132/177/3	149/191/9	489/747/66

In each negative example, we show the length of the witness (for dynamic strategies[8]) and the CPU-time (in s.); we mention the behavior of an exhaustive search, for comparison.

neg example	ex2n	ex4n	ex4nn	ex4nnn	ex6n
size	34	168	175	171	525
$trg, c, stat$	−/2	−/2.7	−/52	−/17	oot
jp, c, dyn	4/2.9	8/3.1	13/69	13/20	19/1609
jp, nc, dyn	4/2.8	8/3.2	11/60	13/20	19/2062
$qdiv, nc, stat$	−/1.3	−/0.6	−/61	−/2.8	−/128
$qdiv, nc, dyn$	4/4	7/15	13/35	13/29	23/170
$ex - srch$	4/0.02	4/0.08	7/2	7/1.3	oot

6   Conclusion and Perspectives

The present program is a prototype where the low-level functions are far from being optimized. Its performance on grammar examples of 20 to 100 rules (and size in [30,500]) seems to show that the equivalence problem for dpda (and the computation of algebraic mgu's) is not out of reach from a practical point of view.

Among our perspectives of development we plan: to improve the core of the program by using rewriting techniques; to devise an example-generation module; to add modules implementing the reductions described in [Sén01b].

The program is open-source and we hope other authors will write their own complementary modules (e.g. the authors of [CCD13] are already implementing their reduction).

---

[8] Recall that the " failure" message sent by static strategies is unconclusive.

Acknowledgements. We thank I. Durand for her continuous advices concerning programming, X. Blanc for his lecture on program-testing and the ANR project " 2010 BLAN 0202 02 FREC" for financial support.

References

[BCFR06] Bastien, C., Czyzowicz, J., Fraczak, W., Rytter, W.: Prime normal form and equivalence of simple grammars. TCS 363(2), 124–134 (2006)

[BG11]   Böhm, S., Göller, S.: Language equivalence of deterministic real-time one-counter automata is NL-complete. In: Murlak, F., Sankowski, P. (eds.) MFCS 2011. LNCS, vol. 6907, pp. 194–205. Springer, Heidelberg (2011)

[BGJ13]   Böhm, S., Göller, S., Jancar, P.: Equivalence of deterministic one-counter automata is NL-complete. CoRR, abs/1301.2181 (2013)

[CCD13]   Chrétien, R., Cortier, V., Delaune, S.: From security protocols to pushdown automata. In: Fomin, F.V., Freivalds, R., Kwiatkowska, M., Peleg, D. (eds.) ICALP 2013, Part II. LNCS, vol. 7966, pp. 137–149. Springer, Heidelberg (2013)

[Cou83]   Courcelle, B.: Fundamental properties of infinite trees. Theoretical Computer Science 25, 95–169 (1983)

[Har78]   Harrison, M.A.: Introduction to Formal Language Theory. Addison-Wesley, Reading (1978)

[HHY79]   Harrison, M.A., Havel, I.M., Yehudai, A.: On equivalence of grammars through transformation trees. TCS 9, 173–205 (1979)

[Jan12]   Jancar, P.: Decidability of dpda language equivalence via first-order grammars. In: LICS, pp. 415–424 (2012)

[Mei89]   Meitus, Y.V.: The equivalence problem for real-time strict deterministic pushdown automata. In: Kibernetika 5, pp. 14–25 (1989) (in Russian, english translation in Cybernetics and Systems analysis)

[MOW05]   Murawski, A.S., Ong, C.-H.L., Walukiewicz, I.: Idealized Algol with ground recursion, and DPDA equivalence. In: Caires, L., Italiano, G.F., Monteiro, L., Palamidessi, C., Yung, M. (eds.) ICALP 2005. LNCS, vol. 3580, pp. 917–929. Springer, Heidelberg (2005)

[Sén97]   Sénizergues, G.: The Equivalence Problem for Deterministic Pushdown Automata is Decidable. In: Degano, P., Gorrieri, R., Marchetti-Spaccamela, A. (eds.) ICALP 1997. LNCS, vol. 1256, pp. 671–681. Springer, Heidelberg (1997)

[Sén01a] Sénizergues, G.: L(A) = L(B)? decidability results from complete formal systems. Theoretical Computer Science 251, 1–166 (2001)

[Sén01b] Sénizergues, G.: Some applications of the decidability of dpda's equivalence. In: Margenstern, M., Rogozhin, Y. (eds.) MCU 2001. LNCS, vol. 2055, pp. 114–132. Springer, Heidelberg (2001)

[Sén03]   Sénizergues, G.: The equivalence problem for t-turn dpda is co-NP. In: Baeten, J.C.M., Lenstra, J.K., Parrow, J., Woeginger, G.J. (eds.) ICALP 2003. LNCS, vol. 2719, pp. 478–489. Springer, Heidelberg (2003)

[Sén05]   Sénizergues, G.: The bisimulation problem for equational graphs of finite outdegree. SIAM J. Comput. 34(5), 1025–1106 (2005) (electronic)

[Sti02]   Stirling, C.: Deciding DPDA equivalence is primitive recursive. In: Widmayer, P., Triguero, F., Morales, R., Hennessy, M., Eidenbenz, S., Conejo, R. (eds.) ICALP 2002. LNCS, vol. 2380, pp. 821–832. Springer, Heidelberg (2002)

Brzozowski's Minimization Algorithm—More Robust than Expected

(Extended Abstract)

Markus Holzer and Sebastian Jakobi

Institut für Informatik, Universität Giessen,
Arndtstr. 2, 35392 Giessen, Germany
{holzer,jakobi}@informatik.uni-giessen.de

Abstract. For a finite automaton, regardless whether it is deterministic or nondeterministic, Brzozowski's minimization algorithm computes the equivalent minimal deterministic finite automaton by applying reversal and power-set construction twice. Although this is an exponential algorithm because of the power-set construction, it performs well in experimental studies compared to efficient $O(n \log n)$ minimization algorithms. Here we show how to slightly enhance Brzozowski's minimization algorithm by some sort of reachability information so that it can be applied to the following automata models: deterministic cover automata, almost equivalent deterministic finite state machines, and k-similar automata.

1 Introduction

The study of the minimization problem for finite automata dates back to the early beginnings of automata theory. This problem is also of practical relevance, because regular languages are used in many applications, and one may like to represent the languages succinctly. While the minimization for nondeterministic automata (NFAs) is computationally intractable [13], it becomes efficiently solvable for deterministic finite automata (DFAs). We refer to [15] for a brief summary of DFA minimization algorithms. While the algorithm with the best running time of $O(n \log n)$ remains difficult to understand, the most elegant one is that of Brzozowski [4], which minimizes an automaton A, regardless whether it is deterministic or nondeterministic, by applying the reversal and power-set construction twice in sequence, i.e., it computes $\mathcal{P}([\mathcal{P}(A^R)]^R)$, to obtain an equivalent minimal DFA—here the superscript R refers to the reversal or dual operation on automata and $\mathcal{P}$ denotes the power-set construction. Although, Brzozowski's minimization technique is exponential due to the power-set construction, it is reported in [15] that it usually outperforms Hopcroft's $O(n \log n)$ minimization algorithm. Further studies conducted in [1] show a more complex scenario when comparing minimization algorithms. Nevertheless, Brzozowski's minimization algorithm is identified as superior to any other minimization technique implemented in [1], when starting with an NFA. Why Brzozowski's minimization algorithm is so efficient is not completely answered. Recently, Brzozowski's minimization algorithm was generalized to Moore machines and weighted finite automata in [3,6].

S. Konstantinidis (Ed.): CIAA 2013, LNCS 7982, pp. 181–192, 2013.

A closer look on Brzozowski's minimization algorithm shows that it does compute state equivalence in the second power-set construction. In fact, in [7] a characterization of the equivalence classes of the automaton A is given w.r.t. the set of states of $\mathcal{P}(A^R)$. This is the starting point of our investigation. We utilize Brzozowski's minimization algorithm to compute other types of equivalences such as, e.g., E-equivalence [10], almost-equivalence [2], similarity [5], and k-similarity [8]. Here two automata A and B are E-equivalent if and only if $L(A) \triangle L(B) \subseteq E$, where $\triangle$ refers to the symmetric difference. To this end we enhance Brzozowski's minimization algorithm with reachability information on the states of the intermediate automaton $B = \mathcal{P}(A^R)$, and/or the resulting automaton $P(B^R)$. These aforementioned equivalence concepts, except for E-equivalence, are related to hyper-minimal DFAs [2], minimal deterministic cover automata [5], and k-minimal DFAs [8]. All these minimal automata can be efficiently computed by state merging algorithms [8,9,11,14] by appropriately applying the previously mentioned state relations. As a first result we show that although we can give a characterization of E-mergeability of states in general, there are even finite sets E where a state merging algorithm *cannot* be applied in order to obtain a DFA that is minimal w.r.t. the considered equivalence. This nicely fits a previously obtained result which shows that minimizing DFAs w.r.t. E-equivalence is already NP-complete [10]. Nevertheless, the previously mentioned enhancement of Brzozowski's minimization algorithm allows us to identify almost-equivalent, similar, and k-similar states in $\mathcal{P}(B^R)$. For instance, two states S and T in $\mathcal{P}(B^R)$ are almost equivalent if and only if every element in $S \triangle T$—these are states in B—belongs to the preamble of B, i.e., to the states of B which are reachable from the initial state of B by a *finite* number of inputs only. Similar characterizations are given for the other mentioned relations, too. Then these characterizations can be used to merge states appropriately in order to obtain minimal machines of a certain type. Like Brzozowski's original minimization algorithm, the proposed algorithms run in exponential time. In fact we expect the performance of our algorithms to be comparable to Brzozowski's minimization algorithm, because the additional information that is needed can be computed easily. This shows that Brzozowski's minimization algorithm is more robust and useful than expected. Experimental verification of the performance of these algorithms has still to be conducted and is subject of further research.

2   Preliminaries

We assume the reader to be familiar with the basic concepts of automata theory [12]. A *multiple entry nondeterministic finite automaton* (NNFA) is a quintuple $A = (Q, \Sigma, \delta, I, F)$, where Q is the finite set of *states*, Σ is the finite set of *input symbols*, $I \subseteq Q$ is the set of *initial states*, $F \subseteq Q$ is the set of *accepting states*, and $\delta \colon Q \times \Sigma \to 2^Q$ is the *transition function*, where 2^Q refers to the power set of Q. The *language accepted* by the finite automaton A is defined as

$$L(A) = \{\, w \in \Sigma^* \mid \exists q_0 \in I \colon \delta(q_0, w) \cap F \neq \emptyset \,\},$$

where the transition function is recursively extended to $\delta \colon Q \times \Sigma^* \to 2^Q$. In case I is a singleton set, i.e., $|I| = 1$, we simply speak of a *nondeterministic finite automaton* (NFA). In this case we simply write $A = (Q, \Sigma, \delta, q_0, F)$ for $A = (Q, \Sigma, \delta, \{q_0\}, F)$. If in addition $|\delta(q, a)| = 1$, for all states $q \in Q$ and letters $a \in \Sigma$, then we say that the automaton is *deterministic* (DFA). In this case we simply write $\delta(q, a) = p$ for $\delta(q, a) = \{p\}$, assuming that δ is a mapping of the form $Q \times \Sigma \to Q$. Two automata A and B are *equivalent*, for short $A \equiv B$, if they accept the same language, which means $L(A) = L(B)$ holds. Further, we need some notation on languages associated with states in automata. For a DFA $A = (Q, \Sigma, \delta, q_0, F)$ and a state $q \in Q$, define $_qA = (Q, \Sigma, \delta, q, F)$, and $A_q = (Q, \Sigma, \delta, q_0, \{q\})$. Thus, the language $L(_qA)$ denotes the set of all words that lead from state q to some accepting state, and $L(A_q)$ is the set of words leading from the initial state of A to state q. The languages $L(_qA)$, and $L(A_q)$ are also known as the *right*, and *left language* of q, respectively. Finally, for a language $E \subseteq \Sigma^*$, the set $Q_A(E) = \{ q \in Q \mid L(A_q) \subseteq E \}$ denotes the set of states of the automaton A, which are only reachable by reading words from E. If there is no danger of confusion we simply write $Q(E)$ instead of $Q_A(E)$.

Next we define two important operations on automata: (i) The *reverse* or *dual* of $A = (Q, \Sigma, \delta, I, F)$ is the (multiple entry) automaton $A^R = (Q, \Sigma, \delta^R, F, I)$, that results from swapping the initial and final states in A, and reversing all its transitions, i.e., $p \in \delta^R(q, a)$ if and only if $q \in \delta(p, a)$, for every $p, q \in Q$ and $a \in \Sigma$. For A^R one can show the nice and useful property $p \in \delta^R(q, w^R)$ if and only if $q \in \delta(p, w)$ by induction on the length of the word w. (ii) For a finite automaton $A = (Q, \Sigma, \delta, I, F)$, its *power-set automaton* or *subset automaton* is the DFA referred to $\mathcal{P}(A) = (Q', \Sigma, \delta', q'_0, F')$ with state set $Q' \subseteq 2^Q$ that consists of only those subsets of Q that are reachable from the (singleton) initial state $q'_0 = I$, final states $F' = \{ P \in Q' \mid P \cap F \neq \emptyset \}$, and the transition function of which is defined as $\delta'(P, a) = \bigcup_{q \in P} \delta(q, a)$, for every state $P \in Q'$ and letter $a \in \Sigma$.

The following result was presented in [4] and is commonly referred to as *Brzozowski's minimization algorithm*.

Theorem 1. *If A is a finite automaton (deterministic or nondeterministic), then $A' = \mathcal{P}([\mathcal{P}(A^R)]^R)$ is the minimal DFA for $L(A)$.*

We illustrate Brzozowski's minimization algorithm on a small example. From the example one observes, that the key to the minimization of A lies in the automaton $\mathcal{P}(A^R)$. In the next section we consider this automaton in more detail.

Example 2. Consider the DFAs A, $B = \mathcal{P}(A^R)$, and $A_{\min} = \mathcal{P}(B^R)$ which are depicted from left to right in Figure 1. In order to show that $A_{\min}$ is minimal, one has to verify that all states are pairwise distinguishable. For instance, the initial state qs of $\mathcal{P}(B^R)$ can be distinguished from state pqr with the help of state $s \in qs \bigtriangleup pqr$ of B and the word $(ba)^R = ab$—the word ba is chosen because it leads from the initial state q in B to state s: since reading ab from state qs leads to the accepting state pq, while reading the same word from pqr leads to

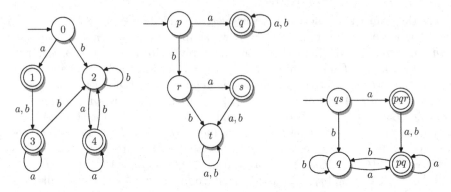

Fig. 1. Left: the DFA A. Middle: the automaton B constructed by reversal and power-set construction from A, i.e., $B = \mathcal{P}(A^R)$. For instance, the initial state of B is $\{1,3,4\}$ simply referred to as p, which has an a-transition to state $\{0,1,2,3,4\} = \bigcup_{x \in \{1,3,4\}} \delta(x,a)$ and a b-transition to $\{1\} = \bigcup_{x \in \{1,3,4\}} \delta(x,b)$. The former state is abbreviated by q and the latter one by r. State s represents the set $\{0\}$, and t stands for the empty set $\emptyset$. Right: the minimal DFA $A_{\min}$ for the language $L(A)$, constructed from B by computing $P(B^R)$. For instance, the state qs is an abbreviation for the subset (state) $\{q,s\}$.

the non-accepting state q. Instead of s, we could also choose p or r from $qs \triangle pqr$ to distinguish these states with the words $\lambda^R = \lambda$ or $b^R = b$, respectively.   □

3   Deciding E-Equivalence of States in $\mathcal{P}(A^R)$

Before we continue our investigations we need some notation, which was introduced in [10], and covers some well-known equivalence concepts. Let $E \subseteq \Sigma^*$ be called the *error language*. Then two languages $L, M \subseteq \Sigma^*$ are *E-equivalent*, denoted by $L \sim_E M$, if and only if their symmetric difference lies in E, i.e., $L \triangle M \subseteq E$. Obviously, the notion of E-equivalence naturally carries over to finite automata and states; two finite automata A and B are E-equivalent, for short $A \sim_E B$, if and only if $L(A) \sim_E L(B)$ is satisfied, and two states p and q of an automaton A are E-equivalent, referred to $p \sim_E q$, if and only if $L(_pA) \sim_E L(_qA)$ holds.

To extend Brzozowski's minimization algorithm, our first goal is to decide for two given states S, T of $\mathcal{P}(A^R)$ and an error language E, whether $S \sim_E T$ holds. The following lemma describes the connection between words accepted from states in $\mathcal{P}(A^R)$, and words leading to states in A. Due to space limitations most of the proofs are omitted.

Lemma 3. *Let $A = (Q, \Sigma, \delta, q_0, F)$ be a DFA and let $B = (Q_B, \Sigma, \delta_B, F, F_B)$ be the power-set automaton of the reverse of A, i.e., $B = \mathcal{P}(A^R)$. Then $L(_sB) = \bigcup_{q \in S} L(A_q)^R$, for all states (sets) $S \in Q_B$, and the union is disjoint. Further, for all states $S, T \in Q_B$, we have $L(_sB) \triangle L(_TB) = \bigcup_{q \in S \triangle T} L(A_q)^R$.*   □

This lemma also holds if the power-set automaton $B = \mathcal{P}(A^R)$ is defined such that $Q_B = 2^Q$, even if not all subsets of Q are reachable in B. From this lemma one can easily conclude the following statement, which can also be used to prove Theorem 1.

Corollary 4. *Let $A = (Q, \Sigma, \delta, q_0, F)$ be a DFA where all states are reachable, that is, $L(A_q) \neq \emptyset$, for every $q \in Q$. Then $\mathcal{P}(A^R)$ is a minimal DFA.*   □

Corollary 4 implies that $\mathcal{P}([\mathcal{P}(A^R)]^R)$ is a minimal DFA, because the "inner" automaton $\mathcal{P}(A^R)$ has no unreachable states. If we leave out the precondition in Corollary 4, that all states in A are reachable, then the resulting DFA is not necessarily minimal, because it may contain distinct states S and T which are not distinguishable. In this case the symmetric difference $L(_S B) \triangle L(_T B) = \bigcup_{q \in S \triangle T} L(A_q)^R$ of the sets of words accepted from states S and T must be empty, which means that all elements $q \in S \triangle T$ satisfy $L(A_q)^R = \emptyset$. In other words, distinct but equivalent states S and T may only differ in elements q which are non-reachable states of A. In this way, we can *syntactically check*, whether or not two states are equivalent: if they differ in an element, which is reachable in A, then they are not equivalent, otherwise they are equivalent. This idea can be adapted also to the general notion of E-equivalence. The next statement is the main lemma of this section.

Lemma 5. *Let A be a DFA with state set Q and input alphabet Σ. Further assume $E \subseteq \Sigma^*$, and let S and T be two states of $B = \mathcal{P}(A^R)$. Then $S \sim_E T$ if and only if $S \triangle T \subseteq Q_A(E^R)$, i.e., if and only if $L(A_q) \subseteq E^R$, for all $q \in S \triangle T$.*   □

Since classical equivalence allows no "errors," two states S and T are equivalent if they are E-equivalent for $E = \emptyset$, i.e., if $S \sim_\emptyset T$. Then Lemma 5 states that S and T are equivalent, if and only if $L(A_q) \subseteq \emptyset^R = \emptyset$, i.e., q is not reachable, for all elements $q \in S \triangle T$. Since Brzozowski's minimization algorithm assumes that the power-set construction does not produce unreachable states, it immediately follows that no two states S, and T in the resulting DFA can be equivalent.

In the classical setting two equivalent states in a DFA can always be merged, and the resulting DFA is still equivalent to the original one. For a DFA $A = (Q, \Sigma, \delta, q_0, F)$, two of its states $p, q \in Q$, and an error language $E \subseteq \Sigma^*$, we say that p is E-*mergeable* to q, if $L(A) \sim_E L(A')$, where A' results from A by *merging* state p to state q, in the following way: $A' = (Q \setminus \{p\}, \Sigma, \delta', q_0', F \setminus \{p\})$, with

$$\delta'(r, a) = \begin{cases} q & \text{if } \delta(r, a) = p, \\ \delta(r, a) & \text{otherwise,} \end{cases} \quad \text{and} \quad q_0' = \begin{cases} q & \text{if } q_0 = p, \\ q_0 & \text{otherwise.} \end{cases}$$

The following result characterizes E-mergeability of states in terms of languages related to these states. We use the following notation: for a DFA $A = (Q, \Sigma, \delta, q_0, F)$, a state $r \in Q$, sets of states $S, T \subseteq Q$, and $L = L(A)$, let $_r L_S^T$ be the set of words that lead from state r to some state $s \in S$, while only reaching states in T in between. If $S = \{s\}$ is a singleton set, we omit the set braces, and write $_qL_s^T$ instead of $_qL_{\{s\}}^T$.

Lemma 6. *In a DFA A, with $L(A) = L$, the state p is E-mergeable to state q if and only if $[_pL_F^Q] \triangle [(_qL_p^{Q\setminus\{p\}})^* \cdot {_qL_F^{Q\setminus\{p\}}}] \subseteq \bigcap_{u \in {_{q_0}L_p^{Q\setminus\{p\}}}} u^{-1}E$.*   □

For E-minimization in general, the approach of merging states in order to obtain an E-minimal automaton does not always work, as the upcoming Theorem 7 shows. Concerning the complexity of the E-minimization problem for DFAs, it is shown in [10] that this problem is NP-complete.

Theorem 7. *There is no algorithm that, for any two given DFAs A and A_E, computes an E-minimal DFA B for the language $L(A)$ by merging states of A, where $E = L(A_E)$. This even holds for a fixed and finite set E.*   □

Nevertheless, for some error sets, minimization can be done by state merging algorithms. This gives rise to the following question: Is there a precise characterization, for which error sets E state-merging algorithms that compute E-minimal DFA representations exist? We have to leave open the answer to this question.

4   Applications

In this section, we present three modifications of Brzozowski's minimization algorithm that allow us to compute hyper-minimal DFAs, minimal deterministic finite cover automata, and k-minimal DFAs. All three automata models share that there is no unique (up to isomorphism) minimal automaton anymore, nevertheless these minimal automata can be computed efficiently by state merging algorithms in $O(n \log n)$ time for hyper-minimization [8,11] and cover-minimization [14], and in time $O(n \log^2 n)$ for k-minimization [8,9]. The herein presented algorithms are comparable in running time to Brzozowski's minimization algorithm, and thus, have an exponential running time, due to the power-set construction.

4.1   A Brzozowski-Like Algorithm for Hyper-Minimizing DFAs

A finite automaton is hyper-minimal if every other automaton with fewer states disagrees on acceptance for an *infinite* number of inputs. In [2] basic properties of hyper-minimization and hyper-minimal DFAs were investigated, and it was shown that a hyper-minimal DFA can be obtained from a given DFA by merging every preamble state p to an almost-equivalent state q, where p is not reachable from q. Here the set of *preamble states*, pre(A), are the states that are reachable from the initial state of A by a *finite* number of inputs only, and the *kernel states*, ker(A), are the states that are reached by an *infinite* number of inputs. Further, two states p and q of A are *almost-equivalent*, $p \sim q$, if $L(_pA) \triangle L(_qA)$ is finite—if A has n states, then p and q are almost-equivalent if and only if they are $\Sigma^{\leq n}$ equivalent [9]. Thus, Lemma 5 implies that two states S and T of $\mathcal{P}(B^R)$,

---

Algorithm 1. Brzozowski-like algorithm for hyper-minimizing finite automata.

Require: a DFA or NFA $A = (Q, \Sigma, \delta, q_0, F)$

 1: construct $B = \mathcal{P}(A^R)$
 2: identify preamble and kernel states of B
 3: construct $A' = \mathcal{P}(B^R)$
 4: identify preamble and kernel states of A'
 5: compute topological order $\prec$ of preamble states of A'
 6: **for all** preamble states S of A' **do**
 7:     **find** $T \neq S$ such that $S \triangle T \subseteq \mathrm{pre}(B)$, and $T \in \ker(A')$ or $S \prec T$
 8:     **if** T exists **then**
 9:         merge S to T
10: **return** A'

---

for some n-state DFA B with state set Q, are almost-equivalent if and only if $S \triangle T \subseteq Q_B(\Sigma^{\leq n}) = \mathrm{pre}(B)$, i.e., if and only if the symmetric difference $S \triangle T$ consists only of preamble states of B. So if there is an element $q \in S \triangle T$, which is a kernel state of B, then S and T are not almost-equivalent, and *vice versa*. We summarize this in the following lemma.

Lemma 8. *Let B be an n-state DFA with state set Q, and let S and T be two states of $\mathcal{P}(B^R)$. Then $S \sim T$ if and only if $S \triangle T \subseteq Q(\Sigma^{\leq n})$, i.e., if and only if all states in $S \triangle T$ are preamble states of B.* □

So, if we use Brzozowski's algorithm on some finite automaton A, we can first build the DFA $B = \mathcal{P}(A^R)$, now mark the preamble and kernel states of B, and then continue by constructing the (minimal) DFA $A' = \mathcal{P}(B^R)$. We can now use Lemma 8 to check which states in A' are almost-equivalent. This gives rise to Algorithm 1 for hyper-minimizing finite automata.

Theorem 9 (Brzozowski-Like Hyper-Minimization). *Given a (deterministic or nondeterministic) finite automaton A, then Algorithm 1 computes a hyper-minimal DFA A' for the language $L(A)$ in exponential time.* □

The identification of preamble and kernel states in the automata B and A' can be done on-the-fly during the power-set construction, e.g., by using Tarjan's algorithm to identify strongly connected components. Also the topological ordering of the preamble states of A' can by computed during the second power-set construction. In order to find an appropriate state T in line 7, one could simply cycle through all states of A' and check whether the desired properties are present. But we do this in a more clever way, by using a hash table H, which gets initialized before the **for all** loop. The entries of H are states of A' (sets of states of B), and they are indexed by their subset of kernel states of B. Since all states in A' that are equivalent to some state S have the same subset of kernel elements from B, namely $K = S \setminus \mathrm{pre}(B)$, the entry $H[K] = S$ is a representative of the almost-equivalence class of state S. The initialization of H, which will be described in a few lines, assures that $H[K]$ is always a state to which other almost-equivalent (preamble) states of A' can be merged: preferably $H[K]$ is a kernel state of A', but if there is no kernel state in the almost-equivalence class of $H[K]$, then $H[K]$ is a preamble state that has a maximal index in the topological order $\prec$ among all states that are almost-equivalent to $H[K]$, i.e., no other almost-equivalent preamble state can be reached from it. The initialization of H can be done as follows: for all kernel states P of A' we construct the set $K = P \setminus \operatorname{pre}(B)$, and if $H[K]$ is not yet defined, then we set $H[K] = P$. Then for all preamble states P of A' we construct the set $K = P \setminus \operatorname{pre}(B)$, and if $H[K]$ is not defined, or if $H[K] = P'$ for some preamble state $P' \prec P$, then we set $H[K] = P$. In this way, the entry $H[K]$ is either a kernel state, or a "maximal" preamble state. Now the search in line 7 for a state T, to which the preamble state S could be merged, can be done as follows. We compute the set of *critical states* $K = S \setminus \operatorname{pre}(B)$ and check, whether the entry $H[K]$ in the hash table H is defined. If $H[K]$ is not defined, or if $H[K] = S$, then no appropriate set T is found, otherwise we have $T = H[K]$.

We illustrate the algorithm with the following example.

Example 10. Consider again the DFAs A, $B = \mathcal{P}(A^R)$, and $A' = \mathcal{P}(B^R)$ from Example 2. The DFA A' is also depicted on the left-hand side of Figure 2. The kernel states in B are q and t, and the preamble states in B are p, r, and s. The kernel states in A' are $\{q\}$ and $\{p,q\}$. If we assume that in the initialization of the hash table H the kernel state $\{q\}$ is processed before the other kernel state $\{p,q\}$, then H has the entry $H[\{q\}] = \{q\}$. Since q is the only element from the two preamble states of A', that is not in $\operatorname{pre}(B)$, no further entry is made in H. The topological order of the preamble of A' would be $\{q,s\} \prec \{p,q,r\}$.

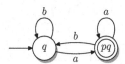

Fig. 2. The hyper-minimal DFA A_{hypermin} for the language $L(A)$

Assume that in the **for all** loop we first process the preamble state $S = \{q,s\}$. We find an appropriate state T by constructing $K = S \setminus \operatorname{pre}(B) = \{q\}$, and obtain $T = H[K] = \{q\}$. Thus, we merge $\{q,s\}$ to state $\{q\}$—note that this has the effect that the new initial state is now $\{q\}$. Now we process the next preamble state $S' = \{p,q,r\}$: again we obtain $K' = S' \setminus \operatorname{pre}(B) = \{q\}$, so we also merge state $\{p,q,r\}$ to state $H[K'] = \{q\}$. Since no further preamble states in A' are left, the algorithm terminates, and we obtain the hyper-minimal DFA, which is depicted on the right-hand side of Figure 2.   □

4.2   A Brzozowski-Like Algorithm for Minimal Cover Automata

Complementary to almost-equivalence, which is $\Sigma^{\le n}$-equivalence, the concept of cover automata can be seen as E-equivalence of automata with $E = \Sigma^{>k}$, for some natural number k, called the cover length. A finite automaton is cover-minimal (w.r.t. number k) if every other automaton with fewer states disagrees on acceptance for at least one input when considering only words up to length k.

Algorithm 2. Brzozowski-like algorithm for finding minimal cover automaton

Require: a DFA or NFA $A = (Q, \Sigma, \delta, q_0, F)$, integer $k \geq 1$

1: construct $B = \mathcal{P}(A^R)$
2: **for all** $i = 0, 1, \ldots, k$ **do**
3:   compute $Q_B(\Sigma^{>i})$
4: construct $A' = \mathcal{P}(B^R)$
5: **for all** states S of A' (in BFS visit order) **do**
6:   let $m = \mathrm{level}_{A'}(S)$
7:   **find** $T \neq S$ such that $S \triangle T \subseteq Q_B(\Sigma^{>k-m})$
8:   **if** T exists **then**
9:     merge S to T
10: **return** A'

The basic properties of cover-minimization and cover-minimal DFAs were studied in [5], where it was shown that a cover-minimal DFA can be obtained by merging similar states (the direction of merging depends on the levels of the states). Here two states p and q of a DFA A with $m = \max(\mathrm{level}_A(p), \mathrm{level}_A(q))$, where $\mathrm{level}_A(p) = \min\{\,|w| \mid w \in L(A_p)\,\}$, are similar (w.r.t. cover length k), denoted by $p \approx_{A,k} q$, if and only if $L(_pA) \cap \Sigma^{\leq k-m} = L(_qA) \cap \Sigma^{\leq k-m}$. If the automaton A and the integer k are clear from the context we omit them and write $\mathrm{level}(p)$ for $\mathrm{level}_A(p)$, and $p \approx q$ for $p \approx_{A,k} q$. One can see that p and q are similar if and only if they are $\Sigma^{>k-m}$-equivalent. If we consider states $S, T \subseteq Q$ in a power-set automaton $\mathcal{P}(B^R)$, where Q is the state set of a DFA B, Lemma 5 implies that S and T are $\Sigma^{>k}$-equivalent if and only if $S \triangle T \subseteq Q(\Sigma^{>k})$. This means that $\mathrm{level}_B(q) > k$, for all elements q in the symmetric difference of S and T. Thus, we obtain the following lemma.

Lemma 11. *Let B be a DFA with state set Q, and let S and T be two states of $A' = \mathcal{P}(B^R)$ with $m = \max(\mathrm{level}_{A'}(S), \mathrm{level}_{A'}(T))$. Further, let k be an integer. Then $S \approx T$ (w.r.t. cover length k) if and only if $S \triangle T \subseteq Q(\Sigma^{>k-m})$, i.e., if and only if $\mathrm{level}_B(p) > k - m$, for all $p \in S \triangle T$.* □

This allows us to use the following algorithm for constructing a cover-minimal DFA, i.e., a $\Sigma^{>k}$-minimal DFA, similar to Brzozowski's minimization algorithm— see Algorithm 2. Given a DFA A and a cover length k, construct the DFA $B = \mathcal{P}(A^R)$, and compute the sets $Q_B(\Sigma^{>i})$, for $0 \leq i \leq k$, by computing the levels of the states in B. Then construct the DFA $A' = \mathcal{P}(B^R)$, and mark each state S in A' with $\mathrm{level}_{A'}(S)$. Finally merge state S to state T, whenever $m = \mathrm{level}_{A'}(S)$, and $S \triangle T \subseteq Q_B(\Sigma^{>k-m})$.

Theorem 12 (Brzozowski-Like Cover-Minimization). *Given a (deterministic or nondeterministic) finite automaton A, and an integer k, then Algorithm 2 computes a cover-minimal DFA A' for the language $L(A)$ in exponential time.* □

The levels of the states of the automata $B = \mathcal{P}(A^R)$ and $A' = \mathcal{P}(B^R)$ can easily be computed on-the-fly, if both power-set constructions are implemented as a breadth-first search. Further, this also allows us to perform the state merging already during the second power-set construction: as soon as we discover a state S on level m, that is $\Sigma^{>k-m}$-equivalent to some previously discovered state T, we can merge S to T. Note that this is different from the hyper-minimization algorithm in Subsection 4.1, where we do not start the merging before the whole state graph of $A' = \mathcal{P}(B^R)$ is discovered.

The search in line 7 for a previously discovered state T, to which a newly discovered state S in $A' = \mathcal{P}(B^R)$ could be merged, can be done by cycling through all states of A'. But again, this can be done more cleverly by using hash tables, similar as in the previous Subsection 4.1. This time we need several hash tables $H_0, H_1, \ldots, H_k$, since we also work with several relations, depending on the level of the currently processed state. An entry $H_i[K] = T$ means that $T \setminus Q_B(\Sigma^{>i}) = K$, so if we have some other state S, with $S \setminus Q_B(\Sigma^{>i}) = K$, then we know that $S \sim_{E'} T$, for $E' = \Sigma^{>i}$. We initialize and use these hash tables as follows. Whenever a state S with $m = \text{level}_{A'}(S)$ is discovered, we compute the set of *critical states* $K = S \setminus Q_B(\Sigma^{>k-m})$ and check, whether there exists an entry $H_{k-m}[K]$ in the hash table H_{k-m} for key K. If $H_{k-m}[K] = T$, then we merge S to T, because we know that $\text{level}_{A'}(S) \geq \text{level}_{A'}(T)$, since we discovered T before S, and we also know that $T \setminus Q(\Sigma^{>k-m}) = K$, so $S \triangle T \subseteq Q_B(\Sigma^{>k-m})$. If $H_{k-m}[K]$ is not defined yet, then we assign $H_i[S \setminus Q(\Sigma^{>i})] = S$, for $0 \leq i \leq k - m$. Compared to Brzozowski's minimization algorithm, the only additional resources we need are the $k + 1$ hash tables $H_0, H_1, \ldots, H_k$.

We illustrate the algorithm with the following example.

Example 13. Consider again the DFAs A, $B = \mathcal{P}(A^R)$, and $A' = \mathcal{P}(B^R)$ from Example 2—the DFA A' is also depicted on the left-hand side of Figure 3. Assume we want to find a minimal cover automaton for A, with cover length $k = 2$, i.e., a minimal cover DFA for $L(A) \cap \Sigma^{\leq 2}$. The sets $Q_B(\Sigma^{>i})$, for $0 \leq i \leq k$ are $Q_B(\Sigma^{>0}) = \{q, r, s, t\}$, $Q_B(\Sigma^{>1}) = \{s, t\}$, and $Q_B(\Sigma^{>2}) = \emptyset$. In the **for all** loop we first process the initial state $S = \{q, s\}$ on level 0. The set of *critical states* is $K = S \setminus Q_B(\Sigma^{>2}) = \{q, s\}$. Since all hash tables are still empty, we do not find an appropriate state T, and so we set $H_i[S \setminus Q_B(\Sigma^{>i})] = S$,

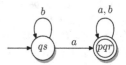

Fig. 3. The minimal cover DFA A_{cover} for the language $L(A) \cap \Sigma^{\leq 2}$

for $0 \leq i \leq 2$, which gives $H_0[\emptyset] = \{q, s\}$, $H_1[\{q\}] = \{q, s\}$, and $H_2[\{q, s\}] = \{q, s\}$. Assume that from the two states in A' on level 1, we first process state $S = \{p, q, r\}$. The *critical states* are $K = S \setminus Q_B(\Sigma^{>1}) = \{p, q, r\}$. Since $H_1[K]$ is not defined, we do not find an appropriate state T and add the entries $H_i[S \setminus Q_B(\Sigma^{>i})] = S$, for $0 \leq i \leq 1$ to the hash tables: $H_0[\{p\}] = \{p, q, r\}$, and $H_1[\{p, q, r\}] = \{p, q, r\}$. Next, we process state $S = \{q\}$ on level 1, where the *critical states* are $K = S \setminus Q_B(\Sigma^{>1}) = \{q\}$. Now we find a matching entry in the hash table $H_{k-m} = H_1$, namely $T = H_1[\{q\}] = \{q, s\}$. This means, that we merge state $S = \{q\}$ to state $T = \{q, s\}$. Since now state $\{q\}$ no longer exists in A', we do not have to explore the state graph "behind" $\{q\}$ any further, but we still process the remaining state $\{p,q\}$, because it gets discovered from state $\{p,q,r\}$. So now we have $S = \{p,q\}$ on level 2, with the set of *critical states* $K = S \setminus Q_B(\Sigma^{>0}) = \{p\}$. In the corresponding hash table H_0 we find the entry $T = H_0[\{p\}] = \{p,q,r\}$, so we merge state $S = \{p,q\}$ to the state $T = \{p,q,r\}$. Since we processed all states in A', the algorithm terminates, and we obtain the minimal cover DFA for $L(A) \cap \Sigma^{\leq 2}$, which is depicted on the right-hand side of Figure 3. □

4.3   A Brzozowski-Like Algorithm for k-Minimization

In this last subsection we discuss k-minimization [8,9], which is closely related to hyper-minimization, and complementary to cover-minimization. A finite automaton is k-minimal if every other automaton with fewer states disagrees on acceptance for at least one input of length at least k. In [9] a simple algorithm for k-minimization of DFAs is presented. Again, it is shown that the minimization can be done by merging states which are k-*similar* to each other, a relation that is complementary to the *similarity relation* used for minimizing cover automata. Two states p and q of a DFA A are k-similar [9], $p \sim_k q$, if $d(p,q) + \min(k, \text{in-level}(p), \text{in-level}(q)) \leq k$. Here the *in-level* of a state is the length of the longest word leading to that state—the in-level is ∞, if the state is a kernel state, and $d(p,q) = \min\{\ell \mid L \cap \Sigma^{\geq \ell} = L' \cap \Sigma^{\geq \ell}\}$, which is one plus the length of the longest word that distinguishes between p and q—and if $p \equiv q$, then $d(p,q) = 0$. One can see that p and q are k-similar if and only if they are $\Sigma^{<k-m}$-equivalent, where $m = \min(k, \text{in-level}(p), \text{in-level}(q))$, thus, Lemma 5 implies the following characterization.

Lemma 14. *Let B be a DFA with state set Q, and let S and T be two states of $A' = \mathcal{P}(B^R)$, with $m = \min(k, \text{in-level}_{A'}(S), \text{in-level}_{A'}(T))$. Further let k be an integer. Then $S \sim_k T$ if and only if $S \triangle T \subseteq Q(\Sigma^{<k-m})$, i.e., if and only if $\text{in-level}_B(p) < k - m$, for all $p \in S \triangle T$.* □

As discussed in [9], the k-minimal DFA can be obtained by merging k-similar states, where the state with the lower in-level has to be merged to the state with the higher in-level. This enables us to use the following modification of Algorithm 2 for k-minimization: (i) instead of $\Sigma^{>i}$, and $\Sigma^{>k-m}$ in lines 3, and 7, we use $\Sigma^{<i}$, and $\Sigma^{<k-m}$, respectively; (ii) instead of level in line 6, we use in-level; and (iii) in the **for all** loop in line 5 we only consider states S of A' with $\text{in-level}_{A'}(S) < k$.

Theorem 15 (Brzozowski-Like k-Minimization). *Given a (deterministic or nondeterministic) finite automaton A, and an integer k, then Algorithm 2 with the above mentioned modifications computes a k-minimal DFA A' for the language $L(A)$ in exponential time.* □

The computation of the in-levels, and of the sets $Q_B(\Sigma^{<i})$ can be done by a depth-first search algorithm on the reverse of the underlying state graph. The search for an appropriate state T to which a state S can be merged can be implemented with the help of hash tables, as in the case of cover automata.

References

1. Almeida, M., Moreira, N., Reis, R.: On the performance of automata minimization algorithms. In: Beckmann, A., Dimitracopoulos, C., Löwe, B. (eds.) Proceedings of the 4th Conference on Computation in Europe: Logic and Theory of Algorithms, pp. 3–14. Technical Report, University of Athens, Athens, Greece (2008)

2. Badr, A., Geffert, V., Shipman, I.: Hyper-minimizing minimized deterministic finite state automata. RAIRO–Informatique Théorique et Applications / Theoretical Informatics and Applications 43(1), 69–94 (2009)

3. Bonchi, F., Bonsangue, M.M., Rutten, J.J.M.M., Silva, A.: Brzozowski's algorithm (co)algebraically. In: Constable, R.L., Silva, A. (eds.) Kozen Festschrift. LNCS, vol. 7230, pp. 12–23. Springer, Heidelberg (2012)

4. Brzozowski, J.A.: Canonical regular expressions and minimal state graphs for definite events. Mathematical Theory of Automata, MRI Symposia Series 12, 529–561 (1962)

5. Câmpeanu, C., Sântean, N., Yu, S.: Minimal cover-automata for finite languages. Theoret. Comput. Sci. 267(1-2), 3–16 (2001)

6. Castiglione, G., Restivo, A., Sciortino, M.: Nondeterministic Moore automata and Brzozowski's minimization algorithm. Theoret. Comput. Sci. 450, 81–91 (2012)

7. Champarnaud, J.M., Khorsi, A., Paranthoën, T.: Split and join for minimization: Brzozowski's algorithm. In: Balík, M., Šimánek, M. (eds.) Proceedings of the Prague Stringology Conference, pp. 96–104. No. DC-2002-03 in Research Report, Czech Technical University, Prague, Czech Republic (2002)

8. Gawrychowski, P., Jeż, A.: Hyper-minimisation made efficient. In: Královič, R., Niwiński, D. (eds.) MFCS 2009. LNCS, vol. 5734, pp. 356–368. Springer, Heidelberg (2009)

9. Gawrychowski, P., Jeż, A., Maletti, A.: On minimising automata with errors. In: Murlak, F., Sankowski, P. (eds.) MFCS 2011. LNCS, vol. 6907, pp. 327–338. Springer, Heidelberg (2011)

10. Holzer, M., Jakobi, S.: From equivalence to almost-equivalence, and beyond—minimizing automata with errors. In: Yen, H.-C., Ibarra, O.H. (eds.) DLT 2012. LNCS, vol. 7410, pp. 190–201. Springer, Heidelberg (2012)

11. Holzer, M., Maletti, A.: An $n \log n$ algorithm for hyper-minimizing a (minimized) deterministic automaton. Theoret. Comput. Sci. 411(38-39), 3404–3413 (2010)

12. Hopcroft, J.E., Ullman, J.D.: Introduction to Automata Theory, Languages and Computation. Addison-Wesley (1979)

13. Jiang, T., Ravikumar, B.: Minimal NFA problems are hard. SIAM J. Comput. 22(6), 1117–1141 (1993)

14. Körner, H.: A time and space efficient algorithm for minimizing cover automata for finite languages. Internat. J. Found. Comput. Sci. 14(6), 1071–1086 (2003)

15. Watson, B.W.: Taxonomies and Toolkits of Regular Language Algorithms. PhD thesis, Eindhoven University of Technology, Department of Mathematics and Computer Science, Den Dolech 2, 5612 AZ Eindhoven, The Netherlands (1995)

Some Decision Problems Concerning NPDAs, Palindromes, and Dyck Languages

Oscar H. Ibarra[1,*] and Bala Ravikumar[2]

[1] Department of Computer Science
University of California, Santa Barbara, CA 93106, USA
ibarra@cs.ucsb.edu
[2] Department of Computer & Engineering Science
Sonoma State University, Rohnert Park, CA 94928 USA
ravi@cs.sonoma.edu

Abstract. We address several types of decision questions related to context-free languages when an NPDA is given as input. First we consider the question of whether the NPDA makes a bounded number of stack reversals (over all accepting inputs) and show that this problem is undecidable even when the NPDA is only 2-ambiguous. We consider the same problem for counter machines (i.e., whether the counter makes a bounded number of reversals) and show that it is also undecidable. On the other hand, we show that the problem is decidable for unambiguous NPDAs even when augmented with reversal-bounded counters. Next, we look at problems of equivalence, containment and disjointness with fixed languages. With the fixed language L_0 being one of the following: $P = \{x \# x^r \mid x \in (0 + 1)^*\}$, $P_u = \{x x^r \mid x \in (0 + 1)^*\}$, D_k = Dyck language with k-type of parentheses, or S_k = two-sided Dyck language with k types of parentheses, we consider problems such as: 'Is $L(M) \cap L_0 = \varnothing$?', 'Is $L(M) \subseteq L_0$?', or 'Is $L(M) = L_0$?', where M is an input NPDA (or a restricted form of it). For example, we show that the problem, 'Is $L(M) \cap P$?', is undecidable when M is a deterministic one-counter acceptor, while the problem 'Is $L(M) \subseteq P$?' is decidable even for NPDAs augmented with reversal-bounded counters. Another result is that the problem 'Is $L(M) \subseteq P_u$?' is decidable in polynomial time for M an NPDA. We also show several other related decidability and undecidability results.

Keywords: Context-free language (CFL), nondeterministic pushdown automaton (NPDA), counter acceptor, 1-reversal counters, palindromes, Dyck language, decidable, undecidable.

1 Introduction

Decision problems for context-free languages have been extensively studied because of their wide-ranging applications such as parsers for programming languages, XML, natural languages and even in biological modeling such as RNA

* Supported in part by NSF Grants CCF-1143892 and CCF-1117708.

S. Konstantinidis (Ed.): CIAA 2013, LNCS 7982, pp. 193–207, 2013.
© Springer-Verlag Berlin Heidelberg 2013

folding patterns [9], [3], etc. In this work, we consider decision problems for various restricted classes of NPDAs. In an earlier paper [13], we considered several questions related to bounded context-free languages and their connection to reversal-bounded NPDAs. Specifically we showed that the number of reversals necessary and sufficient to accept a k-bounded language is $2k - 3$. This study naturally led to decision questions such as, given an NPDA or a nondeterministic counter acceptor (NCA) M, and given a k (resp., for some k) whether every string in $L(M)$ has an accepting computation in which the stack (or counter) makes at most k reversals. We show that this problem is undecidable even when the NPDA is 2-ambiguous. It is also undecidable for an NCA, which is no longer assumed to be finitely ambiguous. On the other hand, we show that the problem is decidable for unambiguous NFAs even when augmented with reversal-bounded counters. Note that our problems are different than previously studied questions (see, e.g., [4]), where it was required that in *all* computation paths, accepting or not, the stack (or counter) makes at most k reversals (or turns in the terminology of [4]).

We also consider the problems of containment, equivalence and disjointness in which one of the languages is fixed and the other one is an unrestricted NPDA (or a restricted NPDA such as a counter machine or an extended model of NPDA augmented with reversal-bounded counters). Some of the results are counter-intuitive. For example, let P (P_u) be the set of palindromes with a marker (with no marker) separating the left and right half of the string. We show that the problem of determining if $L(M)$ is disjoint from P (P_u) for an input M is undecidable even if M is restricted to be a deterministic one-counter acceptor (DCA). However, the containment problem, 'Is $L(M) \subseteq P$?' ('Is $L(M) \subseteq P_u$?'), is decidable. We also show that the question, 'Is $L(M) = P$?', is decidable if M is an *unrestricted* NPDA. We show, based on recent results in [1], that the containment problem, 'Is $L(M) \subseteq P_u$?', is decidable in polynomial time (although the equivalence problem, 'Is $L(M) = P_u$?', remains open). Regarding Dyck (D_k) and two-sided Dyck (S_k) languages: we show that the the equivalence problem, 'Is $L(M) = S_k$?', is undecidable (even for $k = 1$), while the problem, 'Is $L(M) = D_k$?', is open.

NOTATION. We will use the following notation throughout the paper:

1. NPDA = nondeterministic pushdown automaton
2. DPDA = deterministic pushdown automaton
3. NCA = NPDA that uses only one stack symbol in addition to the bottom of the stack, which never altered (thus, the stack is a counter)
4. DCA = deterministic NCA
5. NFA = nondeterministic finite automaton
6. DFA = deterministic finite automaton
7. An NPCM M is an NPDA augmented with multiple 1-reversal counters which are initially set to zero. At each step, every counter can be incremented by 1, decremented by 1, or left unchanged, and can be tested for zero. A zero counter cannot be decremented. M is a 1-reversal machine in that it has the property that once a counter is decremented, it can no longer be incremented.

An NFCM (resp., DFCM) is an NFA (resp., DFA) augmented with multiple 1-reversal counters. For $m \geq 0$, NPCM(m) (resp., NFCM(m), DFCM(m)) is an NPCM (resp., NFCM, DFCM) with m 1-reversal counters.

8. CFG = context-free grammar
9. CFL = context-free language
10. An acceptor is k-ambiguous ($k \geq 1$) if every input can be accepted in at most k distinct computations (note that 1-ambiguous is the same as unambiguous). It is finitely ambiguous if it is k-ambiguous for some k.

We will need the following result from [11]:

Theorem 1. *The following problems are decidable for an NPCM M: (a) Is $L(M) = \varnothing$? (b) Is $L(M)$ infinite?*

2   Undecidability of Reversal Bounds

In this section, we look at the question of whether or not an NPDA (NCA) has a reversal-bounded stack (counter).

An NPDA (NCA) M is k-reversal if for every string accepted by M, there is an accepting computation in which the stack (counter) makes no more than k reversals. M is finite-reversal if it is k-reversal for some k. We will need the following lemma.

Lemma 1. *Let $S = \{a^{i_1}ba^{i_1+i_2} \cdots ba^{i_{n-2}+i_{n-1}}ba^{i_{n-1}} \mid n \geq 2, i_1, \ldots, i_{n-1} \geq 1\}$. Note that $S = S^r$.*

1. S can be accepted by a deterministic counter automaton (DCA).
2. Any NPDA (hence, any NCA) accepting S is not finite-reversal.

Proof. Clearly, S can be accepted by a DCA. Now suppose S can be accepted by an NPDA M that is r-reversal for some r. Let n be such that $2n - 3 > r$ and $R_n = (a^+b)^{n-1}$. R_n is regular and can be accepted by a DFA M'. Construct from M and M' an r-reversal NPDA M'' accepting $L(M) \cap L(M')$. Clearly $L(M'') = \{a^{i_1}ba^{i_1+i_2} \cdots ba^{i_{n-2}+i_{n-1}}ba^{i_{n-1}} \mid i_1, i_2, \ldots, i_{n-1} \geq 1\}$.

Now let $a_1, \ldots, a_n$ be distinct symbols. We construct from M'' another NPDA M''' to accept the language $L_2 = \{a_1^{i_1}a_2^{i_1+i_2} \cdots a_{n-1}^{i_{n-2}+i_{n-1}}a_n^{i_{n-1}} \mid i_1, i_2, \ldots, i_{n-1} \geq 1\}$, which makes no more than r reversals. ¿From the proof of Theorem 5 in [16], L_2 cannot be accepted by any NPDA in less than $2n-3$ reversals. But, $2n-3 > r$, a contradiction. □

Theorem 2. *The following problems are undecidable for a 2-ambiguous NPDA M:*

1. Given M and $k \geq 1$, is M k-reversal?
2. Given M, is M k-reversal for some $k \geq 1$?
3. Given M does there exist an NPDA M' (which need not be finitely ambiguous) that is k-reversal for a given $k \geq 1$ (resp., k-reversal for some $k \geq 1$) such that $L(M') = L(M)$?

Proof. We will use the undecidability of the halting problem for Turing machines. Let T be an arbitrary Turing machine. The (unique) halting computation of T on blank tape, if it exists, can be described by a sequence of instantaneous descriptions $H(T) = I_1 \# I_2 \# \ldots \# I_m$, where I_1 is the initial instantaneous description of T, I_m is a halting instantaneous description of T, and I_{j+1} follows from I_j in one step for $j = 1, 2, \ldots, m-1$. Let Σ be set of symbols that can occur in $H(T)$ and a, b, c be new symbols. Define $L = L_1 \cup L_2$ where

$L_1 = \{xcycz \mid x, y \in \Sigma^+, z \in (a+b)^+, y \neq x^r\}$ and
$L_2 = \{xcx^r cz \mid x = H(T), z \in S\}$

where S is the language in Lemma 1. Note that $L_2 = \emptyset$ if and only if the TM does not halt on blank tape.

We can design a 2-ambiguous NPDA M that accepts L as follows. M non-deterministically verifies one of the following two processes when given input w, which we may assume has the form $xcycz$, where $x, y \in \Sigma^+$ and $z \in (a+b)^+$ (since this format can be checked by the finite control):

(a) M checks and accepts w if it is in L_1. Clearly, M can do this deterministically by making only one stack reversal.
(b) M assumes that in w, $y = x^r$ (Note that if, in fact, $y \neq x^r$, w would be accepted in process (a).) M checks deterministically that $x = H(T)$ using only one stack reversal by reading the input segment before the second c, i.e., the segment $xcx^r = I_1 \# I_2 \# \cdots \# I_{m-1} \# I_m c I_m^r I_{m-1}^r \# \cdots \# I_2^r \# I_1^r$. This checking is done as follows: M reads the string $I_1 \# I_2 \# \cdots \# I_{m-1} \# I_m$ and stores it in the stack, and pops I_m. Then M reads $c I_m^r I_{m-1}^r \# \cdots \# I_2^r \# I_1^r$ while popping the stack and checks that I_{j+1} is the valid successor of I_j for $1 \leq j \leq m-1$ (i.e., $I_1 \# I_2 \# \cdots \# I_m = H(T)$). If $x = H(T)$, M then checks and accepts if $z \in S$. Since S can be accepted by a DCA (by Lemma 1), M can do this deterministically. If $x \neq H(T)$, M rejects.

Since processes (a) and (b) are deterministic, M is 2-ambiguous. Clearly, if T does not halt on blank tape, then M only makes one reversal on the stack. On the other hand, if T halts on blank tape, then there is a unique halting computation $x_0 = H(T)$ of T. Then for any $z \in (a+b)^+$, $x_0 c x_0^r cz$ will not be accepted via process (a). However, for any $z \in S$, $x_0 c x_0^r cz$ will be accepted via process (b). We claim that $L(M)$ cannot be accepted by *any* finite-reversal NPDA. For suppose $L(M)$ can be accepted by some k-reversal NPDA M'. Then we can construct from M' another NPDA M'' accepting S as follows: M'' on input z in $(a+b)^+$ simulates the computation of M' on $x_0 c x_0^r cz$ but simulates movements of the input head on the initial segment $x_0 c x_0^r c$ in the finite control. Then S can be accepted by a k-reversal NPDA, a contradiction by Lemma 1.

Items (1), (2), and (3) of the theorem follow from the above discussion. □

We have a similar result for NCAs.

Theorem 3. *The following are undecidable for an NCA M (which is no longer assumed to be finitely ambiguous):*

1. *Given M and $k \geq 1$, is M k-reversal?*

2. *Given M, is M k-reversal for some $k \geq 1$?*
3. *Given M does there exist an NCA M' that is k-reversal for a given $k \geq 1$ (resp., k-reversal for some $k \geq 1$) such that $L(M') = L(M)$?*

Proof. Again, we will use the undecidability of the halting problem for Turing machines on blank tape. Using the notation in the proof of the previous theorem, let

$$L = \{xcz \mid x \in \Sigma^+, x \neq H(T), z \in (a+b)^+\} \cup \{xcz \mid x \in \Sigma^+, z \in S\}$$

where S is the language in Lemma 1. L can be accepted by an NCA M as follows: On input xcz, M nondeterministically verifies one of the following two possibilities:

1. $x \neq H(T)$: either x is not well-formed, or I_1 is not the initial ID, or I_m is not a halting ID, or I_{j+1} does not follow from I_j in one step for some j. The first three conditions can be checked with a DFA. The last condition can be checked by guessing j and guessing the location where I_j and I_{j+1} do not agree using one reversal of the counter. Since the number of candidates for j depends on m, M is not finitely ambiguous.
2. z is in S.
   Since S can be accepted by a DCA (by Lemma 1), M can easily check this deterministically.

If T does not halt on blank tape, then any input that is accepted via a computation of type (2) above is also accepted by a computation of type (1), and hence M is 1-reversal.

On the other hand, if T halts on blank tape, then there is a unique halting computation x_0 of T. Then for any $z \in (a+b)^+$, x_0cz will not be accepted via a computation of type (1). However, for any $z \in S$, x_0xz will be accepted by a computation of type (2). Hence, M will be unbounded-reversal. It follows that M is finite-reversal iff it is 1-reversal iff T halts on blank tape. We also note that in the case when M is not finite-reversal, there is no NCA M' that is finite-reversal that will accept $L(M)$. Otherwise, we can construct from M' (and x_0), another NCA M'' that will accept S, a contradiction by Lemma 1.   □

Remark: In the statement of Theorem 2, we assumed that $k \geq 1$, because the proofs only worked for this case. When $k = 0$, we have:

1. It is decidable, given an NPDA (hence, also an NCA) M (which may be of unbounded ambiguity), whether it is 0-reversal. This is because we can construct an NFA M' which, on any input, simulates M and accepts if M accepts and during the computation, the stack does not reverse. Obviously, $L(M') \subseteq L(M)$. Hence, M is 0-reversal iff $L(M) \subseteq L(M')$ iff $L(M) \cap \overline{L(M')} = \varnothing$, which is decidable.
2. It is undecidable, given an NPDA (resp., NCA) M, whether there exists another NPDA (resp., NCA) M' that is 0-reversal and $L(M') = L(M)$.

Note that for such an M', if it exists, $L(M)$ is regular. The claim then follows from the fact that it is undecidable, given an NCA (hence, also an NPDA), whther the language it accepts is regular [5].

In contrast to Theorems 2 and 3 parts (1) and (2), for 1-ambiguous NPCM (i.e., NPDA with 1-reversal counters):

Theorem 4. *It is decidable, given a 1-ambiguous NPCM M, whether M is finite-reversal (resp., k-reversal for a given $k \geq 0$).*

Proof. Given M, we construct another NPCM M' accepting $L = \{d^m \mid m \geq 0, M$ makes at least m stack reversals on some input $\}$. L can be accepted by an M' which, on unary input $x = d^m$, simulates M by guessing some input y to M symbol-by-symbol, and every time the stack makes a reversal, the input head of M' moves one cell to the right. When x is exhausted, M' continues the simulation (further guessing the symbols in y) and accepts if M accepts. Clearly, M is not finite-reversal iff L is infinite, which is decidable by Theorem 1.

When k is given, M' operates as above and checks that the input d^m has length $k + 1$. Then M is not k-reversal iff $L(M') \neq \varnothing$, which is decidable by Theorem 1.   □

3   Palindromes

In this section, we study the disjointness, containment and equivalence problems for NPDA's when one of the languages is fixed: the set of marked palindromes or the set of unmarked palindromes (these languages are fundamental CFLs). Let Σ be an alphabet with at least two symbols, and $\#$ be a symbol not in Σ. Define:

$P = \{x \# x^r \mid x \in \Sigma^*\}$
$P_u = \{x x^r \mid x \in \Sigma^*\}$

Thus, P is the set of palindromes with a center marker, and P_u is the set of unmarked palindromes. Clearly, P (resp., P_u) can be accepted by a 1-reversal DPDA (NPDA).

Theorem 5. *The following problems are undecidable:*

1. *Given a DCA M, is $L(M) \cap P = \varnothing$?*
2. *Given a DCA M, is $L(M) \cap P_u = \varnothing$?*

Proof. We first prove part (1). The proof uses the undecidability of the halting problem for 2-counter machines. A close look at the proof in [17] of the undecidability of the halting problem for 2-counter machines, where initially one counter has value d_1 and the other counter is zero, reveals that the counters behave in a regular pattern. The 2-counter machine operates in phases in the following way. Let c_1 and c_2 be its counters. The machine's operation can be divided into phases, where each phase starts with one of the counters equal to some positive integer d_i and the other counter equal to 0. During the phase, the positive counter decreases, while the other counter increases. The phase ends with the first counter having value 0 and the other counter having value d_{i+1}. Then in the next phase the modes of the counters are interchanged. Thus, a sequence of configurations corresponding to the phases will be of the form:

$$(q_1, d_1, 0), (q_2, 0, d_2), (q_3, d_3, 0), (q_4, 0, d_4), (q_5, d_5, 0), (q_6, 0, d_6), \ldots$$

where the q_i's are states, with q_1 the initial state, and $d_1, d_2, d_3, \ldots$ are positive integers. Note that the second component of the configuration refers to the value of c_1, while the third component refers to the value of c_2. We assume, w.l.o.g., that $d_1 = 1$.

Let T be a 2-counter machine. We assume that if T halts, it does so in a unique state q_h. Let T's state set be Q, and $1, \#$ be new symbols.

In what follows, α is any sequence of the form $I_1 I_2 \cdots I_{2m}$ (thus we assume that the length is even), where $I_i = 1^k q$ for some $k \geq 1$ and $q \in Q$, represents a possible configuration of T at the beginning of phase i, where q is the state and k is the value of counter c_1 (resp., c_2) if i is odd (resp., even).

Define L_{odd} to be the set of all strings α such that

1. $\alpha = I_1 I_2 \cdots I_{2m}$;
2. $m \geq 1$;
3. $I_1 = 1^{d_1} q_1$, where $d_1 = 1$ and q_1 is the initial state;
4. $I_{2m} = 1^v q_h$ for some positive integer v;
5. for odd j, $1 \leq j \leq 2m-1$, $I_j \Rightarrow I_{j+1}$, i.e., if T begins in configuration I_j, then after one phase, T is in configuration I_{j+1};

Define L_{even} analogously except that the condition "$I_j \Rightarrow I_{j+1}$" now applies to *even* values of j, $2 \leq j \leq 2m-2$.

Define the following languages:
$$L = \{x \# y^r \mid x \in L_{odd}, y \in L_{even}\}$$
$$P = \{w \# w^r \mid w \in (Q \cup \{1\})^*\}$$

We can construct an NCA M accepting L as follows. Given input z, M's finite control can check if z is well formed. So assume that $z = x \# y^r$. M checks that x is in L_{odd} deterministically by simulating the 2-counter machine T as follows: M reads $x = I_1 I_2 \cdots I_{2m} = 1^{d_1} q_1 1^{d_2} q_2 \cdots 1^{d_{2m}} q_{2m}$ and verifies that $d_1 = 1$, q_1 is the initial state, q_{2m} is the halting state q_h, and for odd j, $1 \leq j \leq 2m-1$, $I_j \Rightarrow I_{j+1}$. To check that $I_j \Rightarrow I_{j+1}$, M reads the segment $1^{d_j} q_j$ and stores 1^{d_j} in its counter (call it c) and remembers the state q_j in its finite control. This represents the configuration of T when one of its two counters, say c_1, has value d_j, the other counter, say c_2, has value 0, and its state is q_j. Then, starting in state q_j, M simulates the computation of T by decrementing c (which is simulating counter c_1 of T) and reading the input segment $1^{d_{j+1}}$ until c becomes zero and at which time, the input head of M should be on q_{j+1}. Thus, the process has just verified that counter c_2 of T has value $1^{d_{J+1}}$, counter c_1 has value 0, and the state is q_{j+1}.

Similarly, M checks that y is in L_{even}, i.e., for even j, $2 \leq j \leq 2m - 2$, $I_j \Rightarrow I_{j+1}$. Now $y^r = I_{2m}^r I_{2m-1}^r \cdots I_2^r I_1^r = q_{2m} 1^{d_{2m}} q_{2m-1} 1^{d_{2m-1}} \cdots q_2 1^{d_2} q_1^{d_1}$. So to check that $I_j \Rightarrow I_{j+1}$, M reads the segment $q_{j+1} 1^{d_{j+1}}$ and stores $1^{d_{j+1}}$ in its counter c and remembers q_{j+1} in its finite control. Then, M verifies that $I_j \Rightarrow I_{j+1}$ by reading the segment $q_j 1^{d_j}$ while decrementing counter c.

Clearly, $L \cap P = L(M) \cap P = \varnothing$ iff T halts, which is undecidable.

The proof for part (2) is similar. The languages L and P now become:

$$L = \{x\#\#y^r \mid x \in L_{odd}, y \in L_{even}\}$$
$$P_u = \{ww^r \mid w \in (Q \cup \{1, \#\})^*\}$$

So now, $\#$ is no longer a center marker. The rest the proof is the same as above. □

The above result shows that it is undecidable, given a 1-reversal DPDA M_1 and a DCA M_2, whether $L(M_1) \cap L(M_2) = \varnothing$. It is also known that disjointness of $L(M_1)$ and $L(M_2)$ is undecidable when both machines are DCAs [11]. On the other hand, disjointness is decidable when M_1 is an NPCM and M_2 is an NFCM, since the intersection language can be accepted by an NPCM, and emptiness for NPCMs is decidable by Theorem 1. Hence it is decidable, given an NFCM M, whether $L(M) \cap P = \varnothing$ (resp. $L(M) \cap P_u = \varnothing$).

Proposition 1. *It is decidable, given an NPCM M and a language L whose complement, $\overline{L}$, can be accepted by an NFCM, whether $L(M) \subseteq L$.*

Proof. Let $L' = L(M) \cap \overline{L}$. Then $L(M) \subseteq L$ if and only if $L' = \varnothing$. Clearly, L' can be accepted by an NPCM, and the emptiness problem for NPCMs is decidable by Theorem 1. □

Corollary 1. *It is decidable, given an NPCM M, whether $L(M) \subseteq P$ (resp., whether $L(M) \subseteq P_u$).*

Proof. This follows from Proposition 1, since $\overline{P}$ (resp., $\overline{P_u}$) can be accepted by an NFCM(1). □

The converse of the above corollary is not true. In fact, we have:

Proposition 2. *It is undecidable whether $P \subseteq L(M)$ for M a DCA, DPDA, or NPDA. The result also holds for P_u.*

Proof. . It is sufficient to prove the result for M a DCA. Since DCA languages are closed under complementation, let M' be a DCA accepting $\overline{L(M)}$. Then $P \subseteq L(M)$ if and only if $P \cap L(M) = \varnothing$. The claim follows from Theorem 5. Similar argument works for P_u. □

Clearly, the question, Is $P = L(M)$?, is decidable for M a DCA (or NCA), because the answer is always no, since P cannot be accepted by any NCA. It is also decidable when M is a DPDA, since P can be accepted by a DPDA and equivalence of DPDAs is decidable [18,19]. It is also decidable when M is an NPDA, as the next theorem shows.

Theorem 6. *It is decidable, given a CFG G, whether $L(G) = P$.*

Proof. First we check if $L(G) \subseteq P$. This is decidable by Corollary 1.

Now we show how to test if $P \subseteq L(G)$ as follows. Let $G = < \Sigma \cup \{\#\}, T, R, S >$, T is the set of non-terminals, S is the start symbol and R is the set of rules. Also assume that G is in Chomsky Normal Form. (This is a valid assumption since ε is not in $L(G)$.) Assume that all the non-terminals are useful in that for each $A \in T$, there is a string $w \in \Sigma^*$ such that $S \Rightarrow^* \alpha A \beta \Rightarrow^* w$ for some $w \in \Sigma^*$. In the following, lower case letters (e.g. w) denote terminal strings.

Define a non-terminal $X \in T$ to be *self-embedding* if $X \Rightarrow^* \alpha X \beta$ for some α and β, (at least one of which is non-null) or $X \Rightarrow^* \alpha Y \beta$ where Y is self-embedding. We note that for every self-embedding variable A, if $A \Rightarrow^* w$, then w must include exactly one $\#$. (Clearly it can't include more than one; if it includes less than one, then, we can use the following pumping argument: We have $S \Rightarrow^* \alpha' A \beta' \Rightarrow^* \alpha' \alpha^n A \beta^n \beta' \Rightarrow^* \alpha' \alpha^n w \beta^n \beta'$. Since w does not contain $\#$, it must be in one of the strings α', α, β or β'. But this leads to a contradiction by choosing two different values of n.

It follows that if X is a non-embedded variable, then the language $L(X) = \{x \mid X \Rightarrow^* x\}$ is finite. Now we describe an NFA M that accepts the language $\{x \mid x \# y \in L(G)$ for some $y\}$. The NFA remembers one non-terminal in the finite control during the simulation of the PDA. Initially, this symbol will be S. If the current symbol it holds in finite-control is X, then M guesses a rule with X on the left-hand side. Since G is in CNF, the rule will be either $X \to a$ for some $a \in \Sigma \cup \{\#\}$ or $X \to BC$. In the former case, if $a = \#$, M will accept, else it will reject the input. In the latter case, at most one of the two variables B or C will be self-embedding. (They both can't be self-embedding as shown in the above paragraph.) Suppose B is self-embedding. Then, it will simply switch X to B on an ε move and continue the simulation. (Note that in this case it is discarding C completely since the part of the string generated by C is to the right of the $\#$ symbol and hence is irrelevant.) If C is the self-embedding variable, then $L(B)$ is finite, so M guesses one of the strings in $L(B)$ and matches this string with a prefix of the string on the input tape, and then remembers C in finite control. This completes the description of one step of the simulation. M repeats this step until it halts by accepting or rejecting as described above. It is easy to see that $L(M) = \{x \mid x \# y \in L(G)$ for some $y\}$.

Next, we show that if a CFG G is such that $L(G) \subseteq P$, then, $L(M) = \{x \mid x \# y \in L(G)$ for some $y\} = \Sigma^*$ if and only if $P \subseteq L(G)$.

Let x be an arbitrary string in Σ^*. Then, x is in $L(M)$. This means, there is a y such that $x \# y$ is in $L(G)$. By the hypothesis, y must be x^r (else G would generate a non-palindrome). Thus, $x \# x^r$ is in $L(G)$. The converse is obvious.

Thus, the problem of testing if $P \subseteq L(G)$ reduces to checking if $L(M) = \Sigma^*$ which is clearly decidable. $\qquad \square$

Theorem 4.9 of Hunt and Rosenkrantz [10] which states that equivalence testing for an input CFG G to a fixed, unbounded CFL L_0 is PSPACE-hard. It is easy to see that P is unbounded (e.g. using the equivalence between sparseness and boundedness of CFL's [12]). Thus the problem is PSPACE-hard. We believe that this problem is PSPACE-complete. We also do not know if the question $L(G) \subseteq P$? is decidable in polynomial time.

Now consider the language P_u. We know from Corollary 1 that it is decidable, given an NPCM M, whether $LM) \subseteq P_u$. Interestingly, for the case when M is an NPDA, containment in P_u is decidable in polynomial time, as shown in the next theorem. To show this result, we will use a theorem due to [1]. We first introduce some definitions from [1].

Let Σ be a finite alphabet. A *semi-Thue system* (or *string rewriting system*) is a pair of generators and relators $< \Sigma; R >$ where R is a finite subset of $\Sigma^* \times \Sigma^*$. The pair $< \Sigma; R >$ is a *monoid presentation*. It defines a monoid $M(R)$ which is the quotient of Σ^* by the congruence $\sim_R$ generated by the relators R. The canonical morphism ϕ_R maps Σ^* onto $M(R)$ by assigning to each word its congruence class. A pair (u, v) in R is denoted by $u = v$. A monoid is said to be cancellative if $xy = xz$ implies $y = z$ and $yx = zx$ implies $y = z$.

With a total order defined on the strings in Σ^*, we can define an ordered version $\rightarrow_R$ of relation R where $u \rightarrow_R v$ if $(u, v) \in R$ and $v < u$.

A string rewriting system R is said to be complete if it has the property of finite termination and confluence [2]. For such a string rewriting system, each string x is equivalent to a unique reduced string denoted by $Red_R(x)$. It is easy to check that $x \sim_R y$ if and only if $Red_R(x) = Red_R(y)$.

A *straight-line program* (SLP) is a restricted context-free grammar $G = < \Sigma, S, P, N >$ such that for every $X \in N$, there exists exactly one production of the form $X \rightarrow \alpha$ is in P for $\alpha \in (\Sigma \cup N)^*$, and there exists a linear order $<$ on the set of nonterminals N such that $X < Y$ whenever there exists a production of the form $(X, \alpha) \in P$ for $\alpha \in (N \cup \Sigma)^* Y (N \cup \Sigma)^*$.

The following decision problems (with respect to a fixed, complete cancellative rewriting system R) were introduced in [1]:

1. *Compressed-equality problem*: Takes as input two SLP's S_1 and S_2 and returns 'yes' ('no') if $\phi_R(s_1) = \phi_R(s_2)$ where s_1 (s_2) is the unique string generated by S_1 (S_2).
2. *Containment problem*: Takes as input a CFG G (in Chomsky Normal Form) and outputs 'yes' ('no') if $L(G) \subseteq \phi_R^{-1}(1)$.

The following result was shown in [1].

Theorem 7. *Containment problem is polynomial-time reducible to compressed-equality problem.*

Using the above result, we now show the following:

Theorem 8. *It is decidable in polynomial time, given a CFG G, whether $L(G) \subseteq P_u$.*

Proof. Let Σ be the alphabet over which P_u is defined. We note that the rewriting system $R = aa \rightarrow \varepsilon$ for each $a \in \Sigma$ defines the language P_u in the sense that a string $w \in P_u$ if and only if $\phi_R(w) = \varepsilon$. Clearly R is finitely terminating since every rewriting step reduces the length of the string by 2.

To show that R is confluent, suppose $x \to_R u$, $x \to_R v$. We need to show that there is a w such that $u \to w$ and $v \to w$. Let S_1 and S_2 be the sets of pairs of positions that were removed in x to generate u and v respectively. Thus, for example, the rewrite $aabbabaa \to bbabaa$ corresponds to $(1,2)$. If $S_1 = S_2$, then u and v are the same so we can choose w to be u ($= v$) and we are done. Otherwise, define a set $S_3 = \{p \mid p \in S_1 \text{ or } S_2 \text{ but not both }\}$. It can be checked that applying S_3 to u and v will result in a common string w that is the result of applying $S_1 \cup S_2$ to the original string x. This proves that R a confluent rewriting system.

Next we show that R is cancellative for which we need to show that if (xy, xz) (or (yx, zx)) is in R, (y, z) is in R. Suppose (xy, xz) in R. This means xy has been rewritten as xz. This implies that during the rewriting process, x has not been affected by the rewriting. Thus it is clear that (y, z) is in R since we can still apply exactly the rewriting rules on y to derive z.

Thus from Theorem 7, the containment problem $L(G) \subseteq P_u$ reduces to compressed-equality problem. To complete the proof, it is enough to show that compressed equality problem is solvable for the rewriting system R, which we do in the following. We can show this using the result of [15] that the compressed equality problem is solvable in polynomial time for 2-homogeneous N-free rewriting systems. A rewriting system is 2-homogeneous if the left-side of all the rewrite rules in R have length 2 and this is clearly true for R. A rewriting system is N-free if ac, ad, bc in the left-side of some rules of R, then bd should be on the left-side of some rule. This condition is vacuously satisfied for R. This completes the proof. □

Our results give rise to the following interesting situations (where G is a CFG):

1. $L(G) \subseteq P$? is decidable but not known to be in polynomial time; $P \subseteq L(G)$? is undecidable; $L(G) = P$? is decidable.
2. $L(G) \subseteq P_u$? is decidable in polynomial time; $P_u \subseteq L(G)$? is undecidable; $L(G) = P_u$? is open.

4   Dyck and Two-Sided Dyck Languages

Here, we look at the question, $L(M) \subseteq L_0$? and $L(M) = L_0$, where L_0 is either the Dyck language D_k or the 2-sided Dyck language S_k, and M is an NPCM, NFCM, or DFCM. (i.e., NPDA, NFA, DFA augmented with 1- reversal counters).

Let $k \geq 1$, and $\Sigma_k = \{a_1, \ldots, a_k, b_1, \ldots, b_k\}$. Let D_k be the Dyck language over the alphabet Σ_k. Thus, D_k is generated by the CFG:

$$S \to SS$$
$$S \to a_i S b_i \text{ for } 1 \leq i \leq k$$
$$S \to \varepsilon$$

Clearly, D_k can be accepted by a DPDA. It is known that $L(G) \subseteq D_k$? is decidable when G is a CFG [6]. In fact it was shown in [1] that the problem is decidable in polynomial time.

Lemma 2. $\overline{D}_1$ *can be accepted by a machine* M *in* $NFCM(2)$.

Proof. We first note that a string w is in $\overline{D}_1$ if and only if one of the following conditions hold: (i) the number of occurrences of b_1 is larger than the number of occurrences of a_1 in some prefix of w; (ii) the number of occurrences of a_1 and b_1 in w are not equal [8]. We can construct an NFCM with two 1-reversal counters as follows. M on input w in Σ_1^* guesses one of the two processes below to execute:

1. M accepts w if (i) holds. M does this by scanning w while keeping track of the number of a_1's (resp., the number of b_1's) in a counter C_1 (resp., counter C_2). At some point, nondeterministically chosen, M checks if $C_2 > C_1$ (by decrementing the counters). If so, M accepts w.
2. M accepts w if (ii) holds. Again, M uses two 1-reversal counters to accomplish this, as in item (1).

It is straightforward to verify that $L(M) = \overline{D}_1$.   □

Theorem 9. *It is decidable, given an NPCM(m)* M*, whether* $L(M) \subseteq D_1$.

Proof. Clearly, $L(M) \subseteq D_1$ if and only if $L(M) \cap \overline{D_1} = \varnothing$. Since, by Lemma 2, $\overline{D}_1$ can be accepted by an $NFCM(2)$, we can construct an $NPCM(m+2)$ M' to accept $L(M) \cap \overline{D_1}$. The result follows since emptiness is decidable for NPCMs.   □

Notation. NLOG denotes the class of languages accepted by nondeterministic TMs in *log n* space. PTIME denotes the class of languages accepted by deterministic TMs in polynomial time.

Corollary 2. *For fixed* $m \geq 0$*, deciding, given an NFCM(m)* M*, whether* $L(M) \subseteq D_1$ *is in* $NLOG$ *(hence, in* $PTIME$*).*

Proof. As in the proof of Theorem 9, we can construct an NFCM$(m+2)$ to accept $L(M) \cap \overline{D_k}$. The result follows, since emptiness of NFCMs with a fixed number of 1-reversal counters is in $NLOG$ [7].   □

It is an interesting open question whether the inclusion problem, $D_1 \subseteq L(M)$?, is decidable for M an NPCM. We do not even know if it holds when M is a DPDA. When M is a DFCM, the inclusion problem, $D_k \subseteq L(M)$?, is decidable (for any $k \geq 1$) as shown below.

Theorem 10. *It is decidable, given a DFCM(m)* M*, whether* $D_k \subseteq L(M)$.

Proof. $D_k \subseteq L(M)$ if and only if $D_k \cap \overline{L(M)} = \varnothing$. Since M is deterministic, $\overline{L(M)}$ can also be accepted by a DFCM(m). Hence $D_k \cap \overline{L(M)}$ can be accepted by a DPCM$(m+2)$ (since D_k can be accepted by a DPDA), and therefore its emptiness is decidable.   □

Remark. We conjecture that Lemma 2 can be generalized as follows: there is a fixed constant c such that for any $k \geq 1$, $\overline{D_k}$ can be accepted by a machine M in $NFCM(c)$. Then Theorem 9 and Corollary 2 would also generalize for D_k.

Now consider the two-sided Dyck language S_k over an alphabet $\Sigma = \{a_1, ..., a_k, b_1, ..., b_k\}$ which is defined by a CFG:

$$S \rightarrow SS$$
$$S \rightarrow a_i S b_i \text{ for } 1 \leq i \leq k$$
$$S \rightarrow b_i S a_i \text{ for } 1 \leq i \leq k$$
$$S \rightarrow \varepsilon$$

We consider the containment, disjointness and equivalence problems of CFL to S_k. We use the following result from [1]:

Theorem 11. *The problem, Is $L(G) \subseteq S_k$?, given input CFG G (in CNF) is in PTIME.*

Our next result settles the containment problem in the reverse direction and the equivalence problem.

Theorem 12. *The problems, Is $S_k \subseteq L(G)$? and Is $L(G) = S_k$?, given input CFG G are undecidable even for $k = 1$.*

Proof. In [10] it was shown that the problem, Is $L(G) = L_0$?, for a fixed CFL L_0 is undecidable if L_0 contains an unbounded regular subset. It is easy to show that S_1 over alphabet $\{0, 1\}$ is the set of strings with equal number of 0's and 1's. Since $(01 + 10)^*$ is an unbounded regular subset of S_1, the undecidability of equivalence is immediate. Now suppose, the containment, Is $S_1 \subseteq L(G)$?, is decidable. This, combined with the decidability result of the previous theorem would lead to an algorithm for equivalence problem which we just showed to be undecidable.   □

Finally, we summarize the following results for the case of S_1.

Theorem 13. *Let S_1 be over alphabet $\{0, 1\}$.*

1. *It is decidable, given an NPCM M, whether $L(M) \subseteq S_1$.*
2. *It is decidable, given a DPCM M, whether $S_1 \subseteq L(M)$ and whether $S_1 = L(M)$.*
3. *For a fixed m, deciding for a given NFCM(m) M, whether $L(M) \subseteq S_1$ is in NLOG (hence, in PTIME).*
4. *For a fixed m, deciding for a given DFCM(m) M, whether $S_1 \subseteq L(M)$ and whether $S_1 = L(M)$ are in NLOG (hence, in PTIME).*

Proof. For part (1), as noted above, S_1 is the set of strings with equal number of 0's and 1's. Hence S_1 and $\overline{S_1}$ can be accepted by DFCM(2)s. It follows that $L(M) \cap \overline{S_1}$ can be accepted by an NPCM, and its emptiness is decidable.

For part (2), it is sufficient to show that $S_1 \subseteq L(M)$ is decidable. Since M is deterministic, $\overline{L}(M)$ can be accepted by a DPCM [14]. It follows that $S_1 \cap \overline{L}(M)$ can be accepted by a DPCM and its emptiness is decidable.

For part (3), as in part (1), $L(M) \cap \overline{S}_1$ can be accepted by an NFCM($m+2$) and its emptiness is in NLOG.

For part (4), as in part (2), since M is deterministic, $S_1 \cap \overline{L}(M)$ can be accepted by a DFCM($m+2$), and its emptiness is in NLOG.   □

5   Conclusion

We conclude with some open problems:

1. Characterize the class of CFLs L for which the decision problem $Disjoint_L$ (for this problem L is fixed and is not part of the input) is decidable: Input is a DCA (or NCA or NPDA) M and output is 'yes' ('no') if $L(M) \cap L = \varnothing$. L clearly has to be an unbounded language for this problem to be undecidable.
2. Characterize the class of CFLs L for which the decision problem $Equivalence_L$ (for this problem L is fixed and is not part of the input) is decidable: Input is a DCA (or NCA or NPDA) M and output is 'yes' ('no') if $L(M) = L$. Specifically, is the equivalence problem "$L(G) = P_u$?" decidable? Recall the result of Hunt and Rosenkrantz [10]: If a CFL L_0 has an unbounded regular subset, testing if $L(G) = L_0$ is undecidable. In terms of the type of regular subsets, the two languages P and P_u are quite different: the former does not have an infinite regular subset - this readily follows from the pumping lemma for regular languages [9]. The latter has infinite regular subsets (e.g. 0*). However, it does not have unbounded regular subsets. Thus, the result of Hunt and Rosenkrantz is not helpful in resolving this problem.
3. Two problems left open related to P are: if the input is a NCA M, and $L(M) = \overline{P}$, is the equivalence problem decidable? We have shown that containment $L(M) \subseteq \overline{P}$ is undecidable. (In fact, our result shows that this problem is undecidable even if M is a DCA.) But the containment in the other direction and the equivalence problem are open. Also the complexity of the equivalence problem "Is $L(G) = P$?" remains open. We showed that the problem is PSPACE-hard, but we do not know if it is in PSPACE.

References

1. Bertoni, A., Choffrut, C., Radicioni, R.: The inclusion problem of context-free languages: some tractable cases. In: Diekert, V., Nowotka, D. (eds.) DLT 2009. LNCS, vol. 5583, pp. 103–112. Springer, Heidelberg (2009)
2. Book, R., Otto, F.: String Rewriting Systems. Springer, New York (1993)
3. Durbin, R.: Biological Sequence Analysis: Probabilistic Models of Proteins and Nucleic Acids. Cambridge University Press (1998)
4. Ginsburg, S., Spanier, E.: Finite-turn pushdown automata. SIAM J. on Control 4, 429–453 (1966)
5. Greibach, S.A.: An infinite hierarchy of context-free languages. J. Assoc. Comput. Mach. 16, 91–106 (1969)
6. Greibach, S.A., Friedman, E.P.: Superdeterministic PDAs: a subcase with a decidable inclusion problem. J. Assoc. Comput. Mach. 27, 675–700 (1980)

7. Gurari, E., Ibarra, O.H.: The complexity of decision problems for finite-turn multicounter machines. J. Comput. System Sci. 22, 220–229 (1981)
8. Harrison, M.A.: Introduction to Formal Languages. Addison-Wesley (1978)
9. Hopcroft, J.E., Motwani, R., Ullman, J.D.: Introduction to Automata Theory, Languages and Computation, 3rd edn. Addison-Wesley (2006)
10. Hunt, H.B., Rosenkrantz, D.J.: Computational parallels between the regular and context-free languages. In: Symp. on Theory of Computing, pp. 64–74 (1974)
11. Ibarra, O.H.: Reversal-bounded multicounter machines and their decision problems. J. Assoc. Comput. Mach. 25, 116–133 (1978)
12. Ibarra, O.H., Ravikumar, B.: On sparseness and ambiguity for acceptors and transducers. In: Monien, B., Vidal-Naquet, G. (eds.) STACS 1986. LNCS, vol. 210, pp. 171–179. Springer, Heidelberg (1986)
13. Ibarra, O.H., Ravikumar, B.: On bounded languages and reversal-bounded automata. In: Dediu, A.-H., Martín-Vide, C., Truthe, B. (eds.) LATA 2013. LNCS, vol. 7810, pp. 359–370. Springer, Heidelberg (2013)
14. Ibarra, O.H., Yen, H.-C.: On the containment and equivalence problems for two-way transducers. TCS 429, 155–163 (2012)
15. Lohrey, M.: Word problems and membership problems on compressed words. SIAM Journal on Computing 35(5), 1210–1240 (2006)
16. Malcher, A., Pighizzini, G.: Descriptive complexity of bounded context-free languages. In: Harju, T., Karhumäki, J., Lepistö, A. (eds.) DLT 2007. LNCS, vol. 4588, pp. 312–323. Springer, Heidelberg (2007)
17. Minsky, M.: Recursive unsolvability of Post's problem of Tag and other topics in the theory of Turing machines. Ann. of Math. (74), 437–455 (1961)
18. Sénizergues, G.: Complete formal systems for equivalence problems. Theoret. Comput. Sci. 231(1), 309–334 (2000)
19. Stirling, C.: Deciding DPDA equivalence is primitive recursive. In: Widmayer, P., Triguero, F., Morales, R., Hennessy, M., Eidenbenz, S., Conejo, R. (eds.) ICALP 2002. LNCS, vol. 2380, pp. 821–832. Springer, Heidelberg (2002)

On the Boundary of Regular Languages

Jozef Jirásek [1,*] and Galina Jirásková [2,**]

[1] Institute of Computer Science, Faculty of Science, P.J. Šafárik University
Jesenná 5, 040 01 Košice, Slovakia
jozef.jirasek@upjs.sk
[2] Mathematical Institute, Slovak Academy of Sciences
Grešákova 6, 040 01 Košice, Slovakia
jiraskov@saske.sk

Abstract. We prove that the tight bound on the state complexity of the boundary of regular languages, defined as $\mathrm{bd}(L) = L^* \cap (\overline{L})^*$, is $2^{2n-2} + 2^{2n-3} + 2^{n-2} + 2 - 2 \cdot 3^{n-2} - n$. Our witness languages are described over a five-letter alphabet. For a four-letter alphabet, the lower bound is smaller by just one, and we conjecture that the upper bound cannot be met in the quaternary case.

1   Introduction

The famous Kuratowski's "14-theorem" states that, in a topological space, at most 14 sets can be produced by applying the operations of closure and complement to a given set [2,4]. In analogy with this theorem, Brzozowski et al. [1] proved that there is only a finite number of distinct languages that arise from the operations of Kleene (or positive) closure and complement performed in any order and any number of times. Every such language can be expressed, up to inclusion of the empty string, as one of the following five languages and their complements: $L, L^+, (\overline{L})^+, (\overline{L^+})^+, ((\overline{L})^+)^+$, where $\overline{L}$ and L^+ denote the complement and positive closure of L, respectively.

If the *state complexity* of a regular language L, that is, the number of states of the minimal deterministic finite automaton for L, is n, then the state complexity of $\overline{L}$ is also n, and the state complexity of L^+ and $(\overline{L})^+$ is $3/4 \cdot 2^n - 1$ [5,12]. The state complexity of $(\overline{L^+})^+$ could potentially be double-exponential [8], however, as shown in [3], it is only $2^{\Theta(n \log n)}$.

Brzozowski, Grant, and Shallit in [1] also studied the concepts of "open" and "closed" sets. A language L is said to be Kleene-closed if $L = L^*$, where L^* is the Kleene closure of L. A languages is Kleene-open if its complement is Kleene-closed. The same notions can be defined for positive closure. These are natural analogues of the concepts with the same names from point-set topology, and in [1], the authors found many natural analogues of the classical theorems.

In point-set topology, the concept of the "boundary" of a set S is studied [8]. The boundary of S, $\mathrm{bd}(S)$, is defined as the intersection between the closures

---

* Research supported by grants VEGA 1/0479/12 and APVV-0035-10.
** Research supported by grants VEGA 2/0183/11 and APVV-0035-10.

S. Konstantinidis (Ed.): CIAA 2013, LNCS 7982, pp. 208–219, 2013.
© Springer-Verlag Berlin Heidelberg 2013

of S and $\overline{S}$. This is a natural concept which corresponds well to the geometrical notion of boundary. For example, when S is the unit disk, that is, the set of points $\{(x,y) \mid x^2 + y^2 \leq 1\}$ in the plane, the boundary $\mathrm{bd}(S)$ is just the circle $\{(x,y) \mid x^2 + y^2 = 1\}$.

The boundary of a language is defined as $\mathrm{bd}(L) = L^* \cap (\overline{L})^*$, respectively, as $L^+ \cap (\overline{L})^+$ for positive closure [1,8,9]. In this paper, we study the state complexity of the boundary of regular languages in the case of Kleene closure. To simplify the exposition, we will write everything in an exponent notation, using c to represent complement, thus L^{c*} stands for $(\overline{L})^*$, and so $\mathrm{bd}(L) = L^* \cap L^{c*}$.

We show that if a language L over an alphabet Σ is accepted by an n-state deterministic finite automaton (DFA), then the boundary $\mathrm{bd}(L)$ is accepted by a DFA of at most $2^{2n-2} + 2^{2n-3} + 2^{n-2} + 2 - 2 \cdot 3^{n-2} - n$ states. We also prove that this bound is tight in case when the alphabet Σ has at least five symbols. For a four-letter alphabet, the lower bound is smaller by just one. Our calculations show that four symbols are not enough to meet the upper bound, and we strongly conjecture that a five letter alphabet used for defining our witnesses is optimal.

2 Upper Bound: Construction of DFAs for Boundary

Let us start with the construction of a DFA for the boundary of a regular language L defined by $\mathrm{bd}(L) = L^* \cap L^{c*}$. Without loss of generality, we may assume that the empty string is in L. Let a language L be accepted by a DFA $A = (Q, \Sigma, \cdot, s, F)$, where $|Q| = n$, $s \in F$, $|F| = k$, and $\cdot$ is the transition function extended to the domain $2^Q \times \Sigma^*$ in a natural way. Let $F^c = Q \setminus F$.

Construct an NFA N for the language L^* from the DFA A by adding an ε-transition from each state in F to the initial state s. Next, construct an NFA N' for the language L^{c*} from the DFA A as follows. First, interchange the sets of final and non-final states to get a DFA for L^c. Then add a new initial and final state q_0 going to state s by ε. Finally, add an ε-transition from each state in F^c to the state s. Fig. 1 illustrates the construction of NFAs N and N'.

Let D and D' be the DFAs obtained from NFAs N and N', respectively, by the subset construction [6,10]. Then the language $L^* \cap L^{c*}$ is accepted by the cross-product automaton $D \times D'$, the states of which are pairs of subsets of Q. The initial state of the cross-product automaton is $(\{s\}, \{q_0, s\})$, and a state (S,T) is final if S is a final state in D and T is a final state in D'.

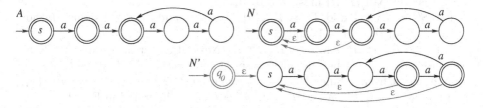

Fig. 1. A DFA A of a language L and the NFAs N and N' for L^* and L^{c*}

Since the initial state contains s in both components, we can make the following observation.

Proposition 1. *If (S,T) is a reachable state of the cross-product automaton, then S and T have a non-empty intersection.*   □

Denote the transition functions of the NFAs N and N' (extended to subsets of Q, see [10]) by $\circ$ and $\bullet$, respectively. Then in the cross-product automaton, a state (S,T) goes to the state $(S \circ a, T \bullet a)$ by any symbol a in Σ. Notice that

$$S \circ a = \begin{cases} S \cdot a, & \text{if } S \cap F = \varnothing, \\ S \cdot a \cup \{s\}, & \text{otherwise,} \end{cases} \qquad T \bullet a = \begin{cases} T \cdot a, & \text{if } T \cap F^c = \varnothing, \\ T \cdot a \cup \{s\}, & \text{otherwise;} \end{cases}$$

recall that $\cdot$ is the (deterministic) transition function of the DFA A. It follows that in the DFA D, a set S may go a set of (at most by one) larger cardinality by any symbol a only if S contains a final state of A. In the DFA D', a set going by a to a larger set, must contain a rejecting state. This gives the following results.

Proposition 2. *If (S,T) is a reachable state of the cross-product automaton, then at least one of the sets S and T contains the initial state s of the DFA A.*

Proof. By Proposition 1, we have $S \cap T \neq \varnothing$. Let $q \in S \cap T$. If $q \in F$, then $s \in S$, otherwise $s \in T$.   □

Proposition 3. *Let $S, T \subseteq Q$ and $a \in \Sigma$. Then*

(i) $|S \circ a| \leq |S| + 1$ and if $|S \circ a| = |S| + 1$ then $S \cap F \neq \varnothing$ and $s \in S \circ a$;
(ii) $|T \bullet a| \leq |T| + 1$ and if $|T \bullet a| = |T| + 1$ then $T \cap F^c \neq \varnothing$ and $s \in T \bullet a$.   □

We use the above mentioned observations to show that in the cross-product automaton, we cannot reach certain pairs (S,T) having only s in their intersection. Recall that $n = |Q|$, $k = |F|$, and $s \in F$. In what follows we assume that $n \geq 3$ and $k \geq 2$.

Lemma 1. *Let $\mathcal{N} = \{(S,T) \subseteq Q \times Q \mid S \cap T = \{s\}, |T| = k, |S| = n - k + 1\}$. No pair in $\mathcal{N}$ is reachable in the cross-product automaton $D \times D'$.*

Proof. Since $n \geq 3$ and $k \geq 2$, no pair in $\mathcal{N}$ can be reached from the initial state $(\{s\}, \{q_0, s\})$. Assume that a pair (S,T) in $\mathcal{N}$ is reached from a pair (P,R) in $Q \times Q$ with $P \cap R \neq \varnothing$ by a symbol a, that is, $(S,T) = (P \circ a, R \bullet a)$. Let us show that the pair (P,R) must also be in $\mathcal{N}$.

Let $S' = S \setminus \{s\}$ and $T' = T \setminus \{s\}$. Then S' and T' are disjoint, do not contain s, and $|T'| = k - 1$, and $|S'| = n - k$.

Let $P' = P \setminus (P \cap R)$ and $R' = R \setminus (P \cap R)$. Notice that for each state q in $P \cap R$, we must have $q \cdot a = s$ because otherwise the intersection $S \cap T$ would contain a state different from s. Since $P \cap R \neq \varnothing$, we have $(P \cap R) \circ a = (P \cap R) \bullet a = \{s\}$, and therefore $S' \subseteq P' \circ a$ and $T' \subseteq R' \bullet a$, which is illustrated in Fig. 2.

By Proposition 3, since S' and T' do not contain s, we must have $|S'| \leq |P'|$ and $|T'| \leq |R'|$. Thus $|P'| \geq n - k$ and $|R'| \geq k - 1$, so $|P'| + |R'| \geq n - 1$. Since

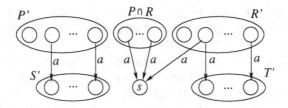

Fig. 2. Unreachable states; $S' \cap T' = \varnothing$, $|S'| = n - k$ and $|T'| = k - 1$

P' and R' are disjoint subsets of Q with $|Q| = n$, it follows that $|P \cap R| = 1$. Hence $|P| = |P'| + |P \cap R| = n - k + 1$, and $|R| = |R'| + |P \cap R| = k$.

Next, since for the state (P, R) we have $|P| = n - k + 1$, where $k = |F|$, the set P must contain a state from F. Therefore it also contains the state s. Since $|R| = k$, then either $R = F$ or R contains a state from F^c. In both cases, the state s is in R. Thus $P \cap R = \{s\}$ which means that the state (P, R) is in $\mathcal{N}$. The proof is complete. □

Now we are able to prove an upper bound on the state complexity of the boundary of L. Recall that the *state complexity* of a regular language L, sc(L), is the number of states of the minimal DFA recognizing the language L.

The following lemma provides an upper bound that depends on the number of final states in the minimal DFA for L. Then we show that this upper bound is maximal if the minimal DFA has two final states. In the end of this section, we discuss the case when the initial state is a unique final state.

Lemma 2. *Let $n \geq 3$ and $2 \leq k \leq n-1$. Let L be a regular language with $\varepsilon \in L$ and* sc(L) $= n$. *Let the minimal DFA for L have k final states. Then*

$$\mathrm{sc}(L^* \cap L^{c*}) \leq 4^{n-1} - \binom{n-1}{k-1} + 2^{n-k}\,2^{n-1} - 3^{n-k}\,2^{k-1} + 2^{k-1}\,2^{n-1} - 3^{k-1}\,2^{n-k} + 1.$$

Proof. Let L be accepted by a minimal DFA $A = (Q, \Sigma, \cdot, s, F)$ with $|F| = k$. Since $\varepsilon \in L$, the initial state s is in F. Let $F^c = Q \setminus F$.

Construct the NFAs N and N' and the DFAs D and D' as described above, and consider the cross-product automaton $D \times D'$ for the language $L^* \cap L^{c*}$. Let us count the number of reachable pairs in the cross-product automaton.

By Propositions 1 and 2, the sets S and T have a non-empty intersection, and at least one of them contains the initial state s of the DFA A. We now count the number of reachable pairs (S, T) in $Q \times Q$ such that (i) $s \notin S$ and $s \in T$, (ii) $s \in S$ and $s \notin T$, and (iii) $s \in S$ and $s \in T$.

(i) If $s \notin S$ and $s \in T$, then S must be a subset of F^c and T is a subset of Q containing s. The number of all such pairs is $2^{n-k}2^{n-1}$. However, the subsets S and T must have a non-empty intersection, so we need to subtract all the pairs with S and T disjoint. The number of such pairs is $3^{n-k}2^{k-1}$ since every function $f \colon F^c \to \{1, 2, 3\}$ may be viewed as a code of two disjoint subsets of F^c: $S = \{i \mid f(i) = 1\}$ and $T = \{i \mid f(i) = 2\}$. Some other elements of T may be chosen arbitrarily in the set $F \setminus \{s\}$. Thus in this case, we get $2^{n-k}2^{n-1} - 3^{n-k}2^{k-1}$ pairs.

(ii) The case of $s \in S$ and $s \notin T$ is symmetric. The set T must be a subset of F not containing s, and after subtracting the disjoint subsets, we get $2^{k-1}2^{n-1} - 3^{k-1}2^{n-k}$ pairs in this case.

(iii) Now consider the case of $s \in S$ and $s \in T$. By Lemma 1, the pairs in $\{(S,T) \in Q \times Q \mid S \cap T = \{s\}, |T| = k, |S| = n - k + 1\}$ are unreachable. Hence in this case, we have to subtract $\binom{n-1}{k-1}$ from $2^{n-1}2^{n-1}$.

By counting the number of pairs in all three cases, and by adding the initial state $(\{s\}, \{q_0, s\})$, we get the expression in the statement of the lemma. $\quad\square$

Straightforward calculations show that the upper bound given by Lemma 2 is maximal if the minimal DFA for the language L has 2 or $n - 1$ final states.

Lemma 3. *Let $n \geq 3$ and $2 \leq k \leq n - 1$. The value of the function*

$$f(k) = 4^{n-1} - \binom{n-1}{k-1} + 2^{n-k} \cdot 2^{n-1} - 3^{n-k}2^{k-1} + 2^{k-1} \cdot 2^{n-1} - 3^{k-1}2^{n-k} + 1$$

is maximal if $k = 2$ or $k = n - 1$, and the maximum of the function $f(k)$ is $2^{2n-2} + 2^{2n-3} + 2^{n-2} + 2 - 2 \cdot 3^{n-2} - n$. $\quad\square$

Next, we consider the case when a language is accepted by a DFA, in which the initial state is the sole final state.

Lemma 4. *Let $n \geq 3$ and let L be accepted by an n-state DFA, in which only the initial state is final. Then $\mathrm{sc}(L^* \cap L^{c*}) \leq (n + 1) \cdot 2^{n-2}$.*

Proof. If L is accepted by a DFA, in which the initial state is a unique final state, then $L = L^*$. Thus, the DFA D for the language L^* is the same as the DFA A. Therefore the language $L^* \cap L^{c*}$ is accepted by the cross-product automaton $A \times D'$, where D' is the DFA for L^{c*} described in the beginning of this section.

Let s be the initial and the unique final state of A. All the other states of A are non-final. Therefore in every reachable state (S, T) of the cross-product automaton, the set S is equal to a set $\{q\}$, where q is a state of A.

Next, if $(\{q\}, T)$ is a reachable state of the cross-product automaton, then $T \cap \{q\} \neq \varnothing$, and moreover, if $q \neq s$, then the initial state s must be in T. This gives 2^{n-1} states $(\{s\}, T)$, and 2^{n-2} states $(\{q\}, T)$ for every non-final q. The lemma follows. $\quad\square$

Since $(n+1) \cdot 2^{n-2} < 2^{2n-2} + 2^{2n-3} + 2^{n-2} + 2 - 2 \cdot 3^{n-2} - n$ for $n \geq 2$, we get the following upper bound on the state complexity of boundary of regular languages.

Theorem 1 (Boundary: Upper Bound). *Let $n \geq 3$ and let L be a language with $\mathrm{sc}(L) = n$. Then $\mathrm{sc}(L^* \cap L^{c*}) \leq 2^{2n-2} + 2^{2n-3} + 2^{n-2} + 2 - 2 \cdot 3^{n-2} - n$.* $\quad\square$

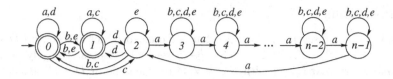

Fig. 3. The DFA A over $\{a, b, c, d, e\}$ accepting the witness language L

3   Matching Lower Bound

In this section, we show that the upper bound given by Theorem 1 is tight. Our witness languages will be defined over a five-letter alphabet. However, the fifth symbol will be used only to prove the reachability of one particular pair. As a consequence, a lower bound for a four-letter alphabet is just by one smaller. Our calculations show that the upper bound cannot be met in the quaternary case.

Let L be the language accepted by the DFA $A = (Q, \{a, b, c, d, e\}, \cdot, 0, \{0, 1\})$ in Fig. 3 with the state set $Q = \{0, 1, \ldots, n-1\}$, $n \geq 4$, and the transitions defined as follows. By a, states 0 and 1 go to themselves, state $n-1$ goes to 2, and every other state i goes to $i+1$. By b, the states 0 and 1 are interchanged, state 2 goes to state 0, and every other state goes to itself. The inputs c, d, and e interchange the states 0 and 2, 1 and 2, and 0 and 1, respectively.

Construct an NFA N for the language L^* from the DFA A by adding an ε-transition from state 1 to state 0 as shown in Fig. 4 (up). Next, construct an NFA N' for the language L^{c*} from the DFA A first by exchanging the final and non-final states, and then by adding a new initial and final state q_0 going to state 0 by ε, and by adding ε-transitions from states $2, 3, \ldots, n-1$ to state 0. Fig. 4 (down) illustrates the construction of N'.

Apply the subset construction to the ε-NFAs N and N' [10] to get the DFAs D and D' for the languages L^* and L^{c*}, respectively. In what follows we consider the cross-product automaton $D \times D'$ for the language $L^* \cap L^{c*}$.

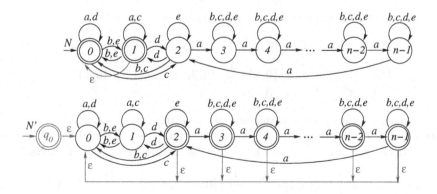

Fig. 4. The NFAs N and N' for L^* and L^{c*}, respectively

3.1 Reachability

The aim of this subsection is to show that the cross-product automaton $D \times D'$ has $2^{2n-2} + 2^{2n-3} + 2^{n-2} + 2 - 2 \cdot 3^{n-2} - n$ reachable states. We only need to show that every pair (S, T) with $S \cap T \neq \varnothing$ is reachable in $D \times D'$, except for the pairs with $S \cap T = \{0\}$ and $|T| = 2$ and $|S| = n - 1$.

We start with the reachability of some special sets, and then we prove the reachability of the pairs (S, T) with $S \cap T = \{0\}$. Notice that we need the symbol e only to get the reachability of the pair $(Q, \{0\})$.

Lemma 5. *Let $k, \ell \geq 0$ and $k + \ell \leq n - 1$. Moreover, if $\ell = 1$ let $k \leq n - 3$. Then the pair*

$$\bigl(\{0, \ell + 1, \ell + 2, \ldots, \ell + k\}, \{0, 1, 2, \ldots, \ell\} \bigr)$$

is reachable in the cross-product automaton.

Proof. Consider three cases.

(i) Let $\ell = 0$. Let us show by induction on k that every pair $(\{0, 1, 2, \ldots, k\}, \{0\})$ is reachable in the cross-product automaton. The claim holds if $k = 0$ or $k = 1$ since we have

$$(\{0\}, \{q_0, 0\}) \xrightarrow{a} (\{0\}, \{0\}) \xrightarrow{b} (\{0, 1\}, \{1\}) \xrightarrow{b} (\{0, 1\}, \{0\}).$$

Assume that $2 \leq k \leq n - 2$, and that $\bigl(\{0, 1, 2, \ldots, k-1\}, \{0\} \bigr)$ is reachable. Since

$$(\{0, 1, 2, \ldots, k - 1\}, \{0\}) \xrightarrow{a} (\{0, 1, 3, 4 \ldots, k\}, \{0\}) \xrightarrow{d}$$

$$(\{0, 2, 3, 4, \ldots, k\}, \{0\}) \xrightarrow{a} (\{0, 3, 4, 5, \ldots, k + 1\}, \{0\}) \xrightarrow{bb}$$

$$(\{0, 1, 3, 4, 5, \ldots, k + 1\}, \{0\}) \xrightarrow{a^{n-3}} (\{0, 1, 2, 3, 4, \ldots, k\}, \{0\}),$$

the pair $(\{0, 1, 2, \ldots, k\}, \{0\})$ is reachable as well.

Finally, if $k = n - 1$, then $(\{0, 1, 2, \ldots, n - 1\}, \{0\})$ is reached from the pair $(\{0, 1, 2, \ldots, n - 2\}, \{0\})$ by ade. Notice that this is *the only place* in the proof of reachability and distinguishability where we use the symbol e.

(ii) Let $\ell = 1$. By the assumption of the lemma, we have $k \leq n - 3$. As shown in case (i), the set $(\{0, 1, 2, \ldots, k\}, \{0\})$ is reachable, and

$$(\{0, 1, 2, 3, \ldots, k\}, \{0\}) \xrightarrow{ada} (\{0, 3, 4, \ldots, k + 2\}, \{0\}) \xrightarrow{cc}$$

$$(\{0, 3, 4, \ldots, k + 2\}, \{0, 2\}) \xrightarrow{d} (\{0, 3, 4, \ldots, k + 2\}, \{0, 1\}) \xrightarrow{a^{n-3}}$$

$$(\{0, 2, 3, \ldots, k + 1\}, \{0, 1\}).$$

Thus the lemma holds in the case of $\ell = 1$ and $k \leq n - 3$.

(iii) Let $2 \leq \ell \leq n - 1$ and $k + \ell \leq n - 1$. Then $k \leq n - 3$, and therefore the pair $(\{0, 2, 3, \ldots, k + 1\}, \{0, 1\})$ is reachable as shown in case (ii). Next

$$(\{0, 2, 3, \ldots, k + 1\}, \{0, 1\}) \xrightarrow{(acc)^{\ell - 1}} (\{0, \ell + 1, \ell + 2, \ldots, \ell + k\}, \{0, 1, \ldots, \ell\}),$$

which completes the proof. $\qquad\square$

Lemma 6. *Let S, T be subsets of Q with $S \cap T = \{0\}$. Moreover, if $|T| = 2$ let $|S| \leq n - 2$. Then the pair (S, T) is reachable in the cross-product automaton.*

Proof. Let $S = \{0, s_1, s_2, \ldots, s_k\}$ and $T = \{0, t_1, t_2, \ldots, t_\ell\}$ be subsets of Q with $\{s_1, s_2, \ldots, s_k\} \cap \{t_1, t_2, \ldots, t_\ell\} = \varnothing$. By the assumption of the lemma, if $\ell = 1$, then $k \leq n - 3$. By Lemma 5, the pair $\big(\{0, \ell+1, \ell+2, \ldots, \ell+k\}, \{0, 1, 2, \ldots, \ell\} \big)$ is reachable in the cross-product automaton $D \times D'$. Next, notice that in the DFA A, the string ad performs circular shift $(1, 2, \ldots, n-1)$, while the input d swaps the states 1 and 2. Recall that swap and circular shift generate the whole symmetric group. Therefore for every permutation π of $\{1, 2, \ldots, n-1\}$, there is a string w_π in $\{ad, d\}^*$ such that in the DFA A, each state i in $\{1, 2, \ldots, n-1\}$ goes to the state $\pi(i)$ by w_π. Moreover, by both a and d, the state 0 goes to itself in the DFA A. Thus in the cross-product automaton, the state

$$\big(\{0, \ell+1, \ell+2, \ldots, \ell+k\}, \{0, 1, 2, \ldots, \ell\} \big)$$

goes to the state

$$\big(\{0, \pi(\ell+1), \pi(\ell+2), \ldots, \pi(\ell+k)\}, \{0, \pi(1), \pi(2), \ldots, \pi(\ell)\} \big)$$

by the string w_π. Now, by considering a permutation $\hat{\pi}$ such that $\hat{\pi}(i) = t_i$ for $i = 1, 2, \ldots, \ell$, and $\hat{\pi}(\ell + i) = s_i$ for $i = 1, 2, \ldots, k$, we get the reachability of (S, T). Note that $(Q, \{0\})$ is the only pair that needs symbol e to be reached. $\square$

Next, we prove the reachability of pairs (S, T) such that $S \cap T$ contains 0 and at least one more state.

Lemma 7. *Let S and T be subsets of Q such that $\{0\} \subsetneq S \cap T$. Then the pair (S, T) is reachable in the cross-product automaton.*

Proof. Let $\{0\} \subsetneq S \cap T$. Let $S' = (S \setminus (S \cap T)) \cup \{0\}$ and $T' = (T \setminus (S \cap T)) \cup \{0\}$. Then the subsets S' and T' satisfy the conditions in Lemma 6 since if $|T'| = 2$, then $|S'| \leq n - 2$ because S and T have a non-zero state in their intersection. Therefore the pair (S', T') is reachable by a string over $\{a, b, c, d\}$. Now it is enough to prove that if $0 \in S \cap T$ and $i \notin S \cup T$, then the pair $(\{i\} \cup S, \{i\} \cup T)$ is reached from the pair (S, T).

First, let $i = 2$. If $1 \in S$, then the pair (S, T) goes to $\big(\{2\} \cup S, \{2\} \cup T \big)$ by input c, otherwise by the string bd.

Now, let $i \geq 3$. Then we can rotate the sets S, T using transitions by a to some sets S', T' that do not contain the state 2. Namely, we use the string a^{n-i}, by which the sets S, T go to some sets S', T', respectively, such that S' and T' do not contain the state 2. Moreover, these sets S', T' go back to the sets S, T by a^{i-2}; notice that there is a self-loop under a in the states 0 and 1 in the DFA A. As shown in the previous case, the pair $(\{2\} \cup S', \{2\} \cup T')$ is reached from the pair (S', T') by a string x in $\{c, bd\}$. Hence

$$(S, T) \xrightarrow{a^{n-i}} (S', T') \xrightarrow{x} (\{2\} \cup S', \{2\} \cup T') \xrightarrow{a^{i-2}} (\{i\} \cup S, \{i\} \cup T).$$

Finally, let $i = 1$. Then by d, the sets S, T go to some sets S', T', respectively, such that S', T' do not contain state 2, and go back to S, T by d. Similarly as in the previous case we get the reachability of $(\{1\} \cup S, \{1\} \cup T)$. $\square$

The next two lemmata prove the reachability of all pairs that contain 0 only in one of their components.

Lemma 8. *Let S be a subset of Q containing the states 0 and 1. Then the pair $(S, \{1\})$ is reachable in the cross-product automaton.*

Proof. First, let $2 \notin S$. By Lemma 6, the pair $(S, \{0\})$ is reachable by a string over $\{a, b, c, d\}$, and it goes to $(S, \{1\})$ by b.

If $2 \in S$, then $(S \setminus \{2\}, \{1\})$, which is reachable as shown in the former case, goes to $(S, \{1\})$ by c.   □

Lemma 9. *Let $S \subseteq \{2, 3, \ldots, n-1\}$ and $T \subseteq Q$ be subsets such that $0 \in T$ and $S \cap T \neq \varnothing$. Then the pair (S, T) is reachable in the cross-product automaton.*

Proof. Let $i \in S \cap T$, thus $i \geq 2$. Let $S' = S \setminus \{i\}$ and $T' = T \setminus \{i\}$. Since S' and T' do not contain i, by the string a^{n-i}, the sets S', T' go to some sets S'', T'' that do not contain state 2, and go back to S', T', respectively, by a^{i-2}. By Lemmata 6 and 7, the pair $(\{0\} \cup S'', T'')$ is reachable by a string over $\{a, b, c, d\}$, and

$$(\{0\} \cup S'', T'') \xrightarrow{c} (\{2\} \cup S'', \{2\} \cup T'') \xrightarrow{a^{i-2}} (\{i\} \cup S', \{i\} \cup T') = (S, T).$$

This proves the lemma.   □

As a consequence of Lemmata 5-9, we get the following result.

Corollary 1. *Let L be the language accepted by the DFA over $\{a, b, c, d, e\}$ shown in Fig. 3. Then the cross-product automaton $D \times D'$ for the language $L^* \cap L^{c*}$ has $2^{2n-2} + 2^{2n-3} + 2^{n-2} + 2 - 2 \cdot 3^{n-2} - n$ reachable states. Moreover, all the states, except for $(Q, \{0\})$, can be reached via strings in $\{a, b, c, d\}^*$.*

3.2 Distinguishability

The idea of the proof of distinguishability of the states in the cross-product automaton of $D \times D'$ for the language $L^* \cap L^{c*}$ is the following. We show that for every state q of the DFA A, there exist strings u_q and v_q such that in the NFA N for the language L^*, the string u_q is accepted only from the state q, and the string v_q is accepted from each of its states; while in the NFA N' for the language L^{c*}, the string u_q is accepted from each of its states, and the string v_q is accepted only from from the state q.

This is enough to prove distinguishability since if (S, T) and (S', T') are two distinct states of the cross-product automaton $D \times D'$, then either $S \neq S'$ or $T \neq T'$. In the first case, without loss of generality, there is a state q with $q \in S$ and $q \notin S'$. Then the string u_q is accepted in D from S and rejected from S'. Moreover, in D', the string u_q is accepted from T, and therefore, this string is accepted by the cross-product automaton $D \times D'$ from (S, T), but rejected from (S', T'). The second case is symmetric, and v_q distinguishes (S, T) and (S', T').

Assume that $n \geq 4$, and let us start with the following technical result.

Lemma 10. *Let $\circ$ and $\bullet$ denote the (nondeterministic) transition functions of the NFAs N and N' for the languages L^* and L^{c*}, respectively. For every state q in Q, there exists a string w_q in $\{a, b, c, d\}^*$ such that*

$$q \circ w_q = \{2\}, \quad q \bullet w_q = \{0, 1, 2\},$$

and for every state p in Q with $p \neq q$,

$$p \circ w_q = p \bullet w_q = \{0, 1\}.$$

Proof. First, let $y = (ab)^{n-3}a$. Then $2 \circ y = \{2\}$ and $2 \bullet y = \{0, 1, 2\}$. However, if $r \neq 2$, then $r \circ y = r \bullet y = \{0, 1\}$.

Now, for $q = 0, 1, \ldots, n - 1$, define the string x_q by $x_0 = c$, $x_1 = d$, and $x_q = a^{n-q}$ if $q \geq 2$. Then $q \circ x_q = \{2\}$ and $q \bullet x_q = \{0, 2\}$.

Finally, let $w_q = x_q y$ for $q = 0, 1, \ldots, n - 1$. Then we have

$$q \circ w_q = q \circ x_q y = \{2\} \circ (ab)^{n-3}a = \{n-1\} \circ a = \{2\},$$
$$q \bullet w_q = q \bullet x_q y = \{0, 2\} \bullet (ab)^{n-3}a = \{0, 1, n-1\} \bullet a = \{0, 1, 2\},$$

which proves the first part of the lemma.

Next, let $p \neq q$. Then p goes by x_q to a set not containing state 2 in both NFAs N and N'. Therefore $p \circ w_q = p \circ x_q y = \{0, 1\} = p \bullet x_q y = p \bullet w_q$. $\square$

Now we are ready to define the strings u_q and v_q as described above.

Lemma 11. *For every state q in Q, there exist strings u_q and v_q in $\{a, b, c, d\}^*$ such that*

(i) *in the NFA N, the string u_q is accepted only from the state q, while the string v_q is accepted from each of its states;*
(ii) *in the NFA N', the string u_q is accepted from each of its states, while the string v_q is accepted only from the state q.*

Proof. Let w_q be the string defined by Lemma 10. Let

$$u_q = w_q\, dac, \quad v_q = w_q\, ddab.$$

Let $p \neq q$. By Lemma 10, in the NFA N (with final states 0 and 1) we have

$$q \circ u_q = q \circ w_q\, dac = \{2\} \circ dac = \{0, 1, 2\},$$
$$p \circ u_q = p \circ w_q\, dac = \{0, 1\} \circ dac = \{2, 3\},$$
$$q \circ v_q = q \circ w_q\, ddab = \{2\} \circ ddab = \{0, 1, 3\},$$
$$p \circ v_q = p \circ w_q\, ddab = \{0, 1\} \circ ddab = \{0, 1\}.$$

It follows that in N, the string u_q is accepted only from q, while the string v_q is accepted from each state.

By the same lemma, in the NFA N' (with final states $2, 3, \ldots, n-1$) we have

$$q \bullet u_q = q \bullet w_q\, dac = \{0, 1, 2\} \bullet dac = \{0, 1, 2, 3\},$$
$$p \bullet u_q = p \bullet w_q\, dac = \{0, 1\} \bullet dac = \{0, 2, 3\},$$
$$q \bullet v_q = q \bullet w_q\, ddab = \{0, 1, 2\} \bullet ddab = \{0, 1, 3\},$$
$$p \bullet v_q = p \bullet w_q\, ddab = \{0, 1\} \bullet ddab = \{0, 1\}.$$

Therefore in N', the string u_q is accepted from each state, while the string v_q is accepted only from the state q. $\square$

The last lemma proves the distinguishability of the states in the cross-product automaton.

Lemma 12. *Let L be the language accepted by the DFA A shown in Fig. 3. All the reachable states of the cross-product automaton $D \times D'$ for the language $L^* \cap L^{c*}$ are pairwise distinguishable by strings in $\{a, b, c, d\}^*$.*

Proof. Let S, S', T, T' be subsets of Q with $(S, T) \neq (S', T')$. Then either $S \neq S'$ or $T \neq T'$. In the former case, without loss of generality, there is a state q such that $q \in S$ and $q \notin S'$. Let u_q be the string given by Lemma 11. Then u_q is accepted by the NFA N only from the state q. It follows that the DFA D, obtained from N by the subset construction, accepts the string u_q from the subset S, and rejects from the subset S'. Moreover, the string u_q is accepted from each state in the NFA N'. Hence the DFA D', obtained from N' by the subset construction, accepts u_q from the subset T. Therefore, the cross-product automaton $D \times D'$ accepts the string u_q from (S, T) and rejects from (S', T'). The latter case is symmetric; now the sets T and T' differ in a state q, and the string v_q distinguishes states (S, T) and (S', T') of the cross-product automaton.

Finally, we need to show that the initial and final state $(\{0\}, \{q_0, 0\})$ can be distinguished from any other final state. Notice that the initial state $(\{0\}, \{q_0, 0\})$ goes to the non-final state $(\{0\}, \{0\})$ by a. Let us show that a is accepted from any other final state of the cross-product automaton. To this aim let (S, T) be a final state of $D \times D'$. Then $0 \in S$ and $T \cap \{2, 3, \ldots, n-1\} \neq \varnothing$. In the DFA A, the state 0 goes to itself by a, while every state in $\{2, 3, \ldots, n-1\}$ goes to a state in $\{2, 3, \ldots, n-1\}$ by a. This means that the final state (S, T) goes by a to a state (S', T') with $0 \in S'$ and $T' \cap \{2, 3, \ldots, n-1\} \neq \varnothing$. It follows that (S', T') is final, so a is accepted from (S, T). This completes the proof.   □

Corollary 1 and Lemma 12 give the following lower bound.

Theorem 2 (Boundary: Lower Bound). *Let L be the language accepted by the DFA A in Fig. 3. Then $\mathrm{sc}(L^* \cap L^{c*}) = 2^{2n-2} + 2^{2n-3} + 2^{n-2} + 2 - 2 \cdot 3^{n-2} - n$. Moreover, if K is the language accepted by the DFA A restricted to $\{a, b, c, d\}$, then $\mathrm{sc}(K^* \cap K^{c*}) = \mathrm{sc}(L^* \cap L^{c*}) - 1$.*   □

Since the lower bound in Theorem 2 matches our upper bound in Theorem 1, we have the exact value of the state complexity of boundary of regular languages over an alphabet of at least five letters. Moreover, a lower bound for quaternary languages is smaller by just one.

Theorem 3 (State Complexity of Boundary). *Let $n \geq 4$ and let L be a language over an alphabet Σ with $\mathrm{sc}(L) = n$. Then*

$$\mathrm{sc}(L^* \cap L^{c*}) \leq 2^{2n-2} + 2^{2n-3} + 2^{n-2} + 2 - 2 \cdot 3^{n-2} - n,$$

and the bound is tight if $|\Sigma| \geq 5$. Moreover, there is a quaternary language K such that $\mathrm{sc}(K^* \cap K^{c*}) = 2^{2n-2} + 2^{2n-3} + 2^{n-2} + 2 - 2 \cdot 3^{n-2} - n - 1$.   □

4   Conclusions

We studied the state complexity of the boundary of regular languages defined as $\mathrm{bd}(L) = L^* \cap (\overline{L})^*$, where $\overline{L}$ is the complement of L and L^* is the Kleene star of L. We showed that if a language L is accepted by an n-state deterministic finite automaton, then the boundary $\mathrm{bd}(L)$ is accepted by a deterministic finite automaton of at most $2^{2n-2} + 2^{2n-3} + 2^{n-2} + 2 - 2 \cdot 3^{n-2} - n$ states.

We also proved that this upper bound is tight by describing a language over a five-letter alphabet meeting this upper bound for boundary. For a four-letter alphabet, we showed that the lower bound is smaller by just one.

We also did some calculations. If $n = 3$ or $n = 4$, then the upper bound can be met by a four-letter automaton, however, if $n = 5$ then a four-letter alphabet is not enough to define the worst case example. We strongly conjecture that the five-letter alphabet used to define our witness languages is optimal.

The calculations in the binary case show that the upper bound in the case, when only the initial state is final, is met by binary languages. We would guess $\Omega(2^{2n})$ lower bound in the binary case with more final states. All these open problems are of great interest to us.

To conclude the paper, let us notice that the unary case is easy since the string a must be either in L or in $\overline{L}$. Therefore one of L^* or $(\overline{L})^*$ is equal to a^*, and the boundary is equal to the other. The state complexity of star operation in the unary case is known to be $(n - 1)^2 + 1$ [12].

References

1. Brzozowski, J.A., Grant, E., Shallit, J.: Closures in formal languages and Kuratowski's theorem. Int. J. Found. Comput. Sci. 22, 301–321 (2011)
2. Fife, J.H.: The Kuratowski closure-complement problem. Math. Mag. 64, 180–182 (1991)
3. Jirásková, G., Shallit, J.: The state complexity of star-complement-star. In: Yen, H.-C., Ibarra, O.H. (eds.) DLT 2012. LNCS, vol. 7410, pp. 380–391. Springer, Heidelberg (2012)
4. Kuratowski, C.: Sur l'opération $\overline{A}$ de l'analysis situs. Fund. Math. 3, 182–199 (1922)
5. Maslov, A.N.: Estimates of the number of states of finite automata. Soviet Math. Dokl. 11, 1373–1375 (1970)
6. Rabin, M., Scott, D.: Finite automata and their decision problems. IBM Res. Develop. 3, 114–129 (1959)
7. Salomaa, A., Salomaa, K., Yu, S.: State complexity of combined operations. Theoret. Comput. Sci. 383, 140–152 (2007)
8. Shallit, J.: Open problems in automata theory and formal languages, https://cs.uwaterloo.ca/~shallit/Talks/open10r.pdf
9. Shallit, J.: The state complexity of $(\overline{L^*})^*$ and $L^* \cap (\overline{L})^*$. Personal Communication (2010)
10. Sipser, M.: Introduction to the theory of computation. PWS Publishing Company, Boston (1997)
11. Yu, S.: Regular languages. In: Rozenberg, G., Salomaa, A. (eds.) Handbook of Formal Languages, vol. I, ch. 2, pp. 41–110. Springer, Heidelberg (1997)
12. Yu, S., Zhuang, Q., Salomaa, K.: The state complexity of some basic operations on regular languages. Theoret. Comput. Sci. 125, 315–328 (1994)

On the Length of Homing Sequences for Nondeterministic Finite State Machines[*]

Natalia Kushik[1,2] and Nina Yevtushenko[1]

[1] Tomsk State University, Tomsk, Russia
[2] Telecom SudParis, Evry, France
ngkushik@gmail.com, ninayevtushenko@yahoo.com

Abstract. Given a reduced deterministic finite state machine, there always exists a homing sequence of length polynomial with respect to the number of states of the machine. For nondeterministic reduced finite state machines, a homing sequence may not exist, and moreover, if it exists, its length can be exponential. We show that the problem of deriving a homing sequence cannot be reduced to deriving a synchronizing word for underlying automata and should be studied independently. We also propose a novel class of $(n-1)$-input finite state machines with n states whose shortest homing sequence is of length $2^{n-1} - 1$.

1 Introduction

Finite state machines are widely used in many application areas, such as analysis and synthesis of digital circuits, telecommunication protocols, software testing and verification etc. A finite state machine (FSM) is a state transition model with finite non-empty sets of inputs, outputs, states, and transitions. An FSM moves from state to state producing an output when an input is applied, and thus FSMs are widely used for modeling reactive systems where inputs are used for representing queries while outputs are used for representing responses. When a system has no reset there is a problem of determining its initial state and usually this is done using so-called homing or synchronizing experiments [11]. In a homing experiment, a sequence of inputs is applied to an FSM under experiment, output responses are observed and the conclusion is drawn what is the system state after the experiment, while a synchronizing experiment relies on one and the same final state for all output responses to a corresponding input sequence. For a deterministic reduced FSM a homing sequence always exists, and if the FSM has n states, then its shortest homing sequence has length at most $n(n-1)/2$ and this upper bound is reachable [6,1]. Nowadays the behavior of many systems is described by nondeterministic FSMs. Nondeterminism occurs due to various reasons such as performance, flexibility, limited controllability, and abstraction [9]. Synchronizing sequences for (nondeterministic) FMSs are usually derived for the underlying automaton obtained by deleting outputs at each transition of an FSM under experiment. It has

---

[*] Supported by TEMPLAN grant #8.4055.2011 sponsored by the Russian Ministry of Education and Research.

S. Konstantinidis (Ed.): CIAA 2013, LNCS 7982, pp. 220–231, 2013.

been shown [2] that a nondeterministic synchronizing automaton with n states can be synchronized by a word of length $2^n - n - 1$ and this bound is tight. A deterministic automaton can be derived for an observable FSM by considering the Cartesian product of the input and output alphabets as the set of actions. In this case, a synchronizing automaton has a synchronizing word [10,5] of polynomial length, but as an example in Section 2 shows, this word is not necessarily a synchronizing/homing word for the initial FSM. We also show that a homing word can exist when the underlying automaton without outputs is not synchronizing and thus, the existing results on the derivation of a synchronizing word cannot be directly used for deriving homing words for nondeterministic FSMs.

In [7] an algorithm for deriving a homing sequence for a nondeterministic FSM has been proposed and a tight lower bound on the length of a shortest homing sequence for a nondeterministic FSM with n states has been shown to be of the order 2^{n^2}. However, this exponential bound is only shown to be reachable for nondeterministic FSMs where the number of inputs is exponential w.r.t. the number of states, and as usual, there is a challenge to establish the reachability of this bound for FSMs with minimal number of inputs. For synchronizing words the reachability problem can be often solved for 2- and 3-letter automata, see, for example, [8]. Here we show that for homing words, there exists a class of $(n-1)$-input FSMs with n states which have a shortest homing sequence of length $2^{n-1} - 1$. As the maximal length of a homing sequence for NFSM with n states and m initial states is at most $2^{\binom{n}{2}} - 2^{\binom{n}{2} - \binom{m}{2}}$, we conclude that for an $(n-1)$-input FSM with n states the length of a shortest homing word belongs to the segment $[2^{n-1} - 1, 2^{\binom{n}{2}} - 2^{\binom{n}{2} - \binom{m}{2}}]$.

The rest of the paper is organized as follows. Preliminaries are given in Section 2. Section 3 includes a description of a class of $(n-1)$-input FSMs with n states which have a shortest homing sequence of length $2^{n-1} - 1$. Section 4 concludes the paper.

2 Preliminaries

A *finite state machine* (*FSM*), or simply a *machine* throughout this paper is a complete observable, possibly nondeterministic FSM, i.e., a 5-tuple $\mathbf{S} = \langle S, I, O, h_S, S' \rangle$, where S is a finite nonempty set of states with a nonempty subset S' of initial states; I and O are finite input and output alphabets; and $h_S \subseteq S \times I \times O \times S$ is a *behavior* (*transition*) *relation*. For each pair $(s, i) \in S \times I$ there exists $(o, s') \in O \times S$ such that $(s, i, o, s') \in h_S$ (FSM is *complete*) and for each triple $(s, i, o) \in S \times I \times O$ there exists at most one state $s' \in S$ such that $(s, i, o, s') \in h_S$ (FSM is *observable*). An FSM is *nondeterministic* (NFSM), if for some pair $(s, i) \in S \times I$ there exist several pairs $(o, s') \in O \times S$ such that $(s, i, o, s') \in h_S$. We further refer to an FSM with m inputs as an *m-input* FSM.

The notion of an FSM is very close to the notion of an automaton that does not support output responses, i.e., automaton transitions are labeled by actions that are not divided into inputs and outputs. One may propose several ways how to derive an automaton that corresponds to an FSM under experiment, while the

most popular ways are as follows: 1) to eliminate outputs at each transition, and
2) to consider the set $I \times O$ as the set of actions of the automaton. In the former
case, a nondeterministic FSM has a nondeterministic underlying automaton; in
the latter case, the underlying automaton is deterministic if and only if the FSM
is observable. As synchronizing experiments do not rely on output responses of a
system under experiment, almost all the results on the synthesis of synchronizing
words have been obtained for automata. In particular, it has been shown that
the length of a synchronizing word for a deterministic automaton with n states
is bounded by a polynomial of degree three [10,5]. However, as we further show,
the known results on synchronizing automaton cannot be applied to deriving
homing sequences even for observable FSMs.

As usual, the behavior relation is extended to input and output sequences.
Given an FSM $S = \langle S, I, O, h_S, S' \rangle$, states $s, s' \in S$, an input sequence $\alpha =
i_1 i_2 \ldots i_k \in I^*$, and an output sequence $\beta = o_1 o_2 \ldots o_k \in O^*$, there is a transition
under input sequence α with the output sequence β if there exist states $s_1 =
s, s_2, \ldots, s_k, s_{k+1} = s'$ such that $(s_j, i_j, o_j, s_{j+1}) \in h_S$, $j \in \{1, \ldots, k\}$. In this
case, the property that the input/output sequence α/β can *take* (or simply
takes) the FSM S from the state s to the state s' is expressed by the function:
$next\_state(s, \alpha/\beta) = s'$. By $out(s, \alpha)$ we denote the set of all output sequences
(responses) that the FSM S can produce at the state s in response to the input
sequence α. Any pair α/β with $\beta \in out(s, \alpha)$ is an *input/output (I/O) sequence*
(or a *trace*) at the state s. The FSM is *connected* if for any two states s and
s', there exists an I/O sequence that can take the FSM from the state s to the
state s'.

When deriving a homing sequence for an FSM we are interested in pairs of
FSM states. A *pair* of states is an unordered state pattern of length 2 denoted
as $\overline{s_p, s_q}$ with $s_q, s_q \in S$; if $s_p = s_q$ then the pair is a singleton $\overline{s_p, s_p}$. Given
an input/output pair i/o and a pair $\overline{s_p, s_q}$, an *i/o-successor* of $\overline{s_p, s_q}$ is the pair
of i/o-successors of states s_p and s_q (if such successors exist for both states s_p
and s_q). If s_k is an i/o-successor of both states s_p and s_q then the i/o-successor
of $\overline{s_p, s_q}$ is the singleton $\overline{s_k, s_k}$. Given an input i, the i-successor of $\overline{s_p, s_q}$ is the
set of i/o-successors of $\overline{s_p, s_q}$ for all possible outputs $o \in O$. This set is empty
if for each $o \in O$ the pair $\overline{s_p, s_q}$ has no i/o-successor, i.e., for each $o \in O$, the
i/o-successor of $\overline{s_p, s_q}$ exists at most for one state s_p or s_q.

Given an FSM $S = \langle S, I, O, h_S, S' \rangle$, two states $s_1, s_2 \in S$ are *equivalent* if
for each input sequence $\alpha \in I^*$ we have $out(s_1, \alpha) = out(s_2, \alpha)$. The FSM S is
reduced if its states are pair-wise non-equivalent. States s_1, s_2 of S are *separable*
if there exists an input sequence $\alpha \in I^*$ such that $out(s_1, \alpha) \cap out(s_2, \alpha) = \varnothing$; in
this case, α is a *separating* sequence of states s_1 and s_2. If there exists an input
sequence α that separates every two distinct states of the set S', then the FSM
S is separable and α is a separating sequence for the FSM S.

A sequence $\alpha \in I^*$ is a *homing sequence (HS)*[1] for an observable FSM $S =
\langle S, I, O, h_S, S' \rangle$ if for each set $\{s_1, s_2\} \subseteq S'$, it holds that

$$\forall \beta \in out(s_1, \alpha) \cap out(s_2, \alpha) \ [next\_state(s_1, \alpha/\beta) = next\_state(s_2, \alpha/\beta)].$$

[1] In [7], such a sequence is called a *preset homing sequence*.

In [7] we have shown that not every nondeterministic reduced FSM has a homing sequence. Similarly to synchronizing automata which have a synchronizing sequence [13], we call an FSM *homing* if the FSM has a homing sequence. By definition of a homing sequence, the following statements hold.

Proposition 1. *If an FSM* **S** *has a separating sequence, then the FSM is homing; moreover, every separating sequence is a homing sequence for* **S**.

Proposition 2. *Let* $\mathbf{S} = \langle S, I, O, h_S, S' \rangle$ *be an FSM such that*

$$\forall \alpha \in I^* [out(s_1, \alpha) \cap out(s_2, \alpha) \neq \varnothing \ \rightarrow \ \exists \beta \in out(s_1, \alpha) \cap out(s_2, \alpha)$$
$$(next\_state(s_1, \alpha/\beta) \neq next\_state(s_2, \alpha/\beta))].$$

The FSM **S** *is homing if and only if it has a separating sequence; moreover, every homing sequence is a separating sequence for* **S**.

However, there exist FSMs which are not separable but have a synchronizing sequence and, thus, have a homing sequence.

A sequence $\alpha \in I^*$ is a *synchronizing sequence* for an FSM $\mathbf{S} = \langle S, I, O, h_S, S' \rangle$ if $\forall s_j \in S \ (|next\_state(s_j, \alpha)| = 1)$ and $\forall s_j, s_k \in S' \ (next\_state(s_j, \alpha) = next\_state(s_k, \alpha))$. This means that α takes **S** to the same state $next\_state(s_j, \alpha)$ for each state $s_j \in S'$ for all possible output responses that are not analyzed when performing the experiment.

In other words, a homing sequence allows one to detect the current state of an FSM by observing the output response to this sequence while a synchronizing sequence takes an FSM from every initial state to the same final state independently of output responses. As mentioned above, synchronizing sequences and their lengths are well studied for automata but as we show below, the known results cannot be directly applied for generating homing sequences for nondeterministic FSMs.

Example 1. Consider the nondeterministic 2-input FSM **S** in Fig. 1.1. By direct

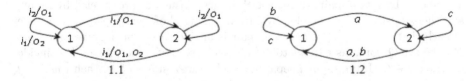

Fig. 1. The FSM **S** and its underlying automaton **A**

inspection, one can verify that the FSM has no homing sequence. Derive the corresponding underlying automaton **A** for the FSM **S** with the set $I \times O$ of actions (Fig. 1.2). In Fig. 1.2, the action a corresponds to the input/output pair i_1/o_1, the action b corresponds to the input/output pair i_1/o_2 while the action c denotes the input/output pair i_2/o_1. The automaton $\mathbf{A}$ in Fig. 1.2 is synchronizing, and moreover, the length of the shortest synchronizing word equals one, since $\mathbf{A}$ is synchronized by the letter b. Thus, the problem of deriving a homing experiment for a nondeterministic FSM cannot be reduced to the problem of deriving a synchronizing experiment for the corresponding underlying automaton where the set of actions is the Cartesian product of the input and output alphabets. On the other hand, consider the FSM in Fig. 2 that is homed be the input i_2. If we observe the output response o_1, then the current state of

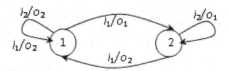

Fig. 2. The homing FSM $\mathbf{S}$ without synchronizing words

$\mathbf{S}$ is 2. When the output response is o_2, the current state is 1. Therefore, for this FSM no synchronizing sequence exists since after applying i_1 (or i_2) at the states 1 or 2 we reach different states with different outputs. Thus the problem of deriving a homing sequence for a nondeterministic FSM cannot be reduced to the problem of deriving a synchronizing sequence. The two problems should be studied independently.

An algorithm for deriving a shortest HS for possibly nondeterministic FSM has been proposed in [7], and we briefly sketch it for establishing the maximal length of a homing sequence for an NFSM.

Procedure 1 for deriving a shortest HS for an FSM
Input: $\mathbf{S} = \langle S, I, O, h_S, S' \rangle$
Output: A shortest HS for $\mathbf{S}$ or the message " the FSM $\mathbf{S}$ is not homing"

Derive a truncated successor tree for the FSM $\mathbf{S}$. The root of the tree is labeled with the set of the pairs $\overline{s_p, s_q}$, where $s_p, s_q \in S'$, $s_p \neq s_q$; the nodes of the tree are labeled by sets of pairs of the set S. Edges of the tree are labeled by inputs and there exists an edge labeled by i from a node P of level j, $j \geq 0$, to a node Q such that a pair $\overline{s_p, s_q} \in Q$ if this pair is an i/o-successor of some pair from P. The set Q contains a singleton if i/o-successors of some pair of P coincide for some $o \in O$. If the input i separates each pair of states of P, then the set Q is empty.

Given a node P at the level k, $k > 0$, the node is *terminal* if one of the following conditions holds.

Rule-1: P is the empty set.
Rule-2: P contains a set R without singletons that labels a node at a level j, $j < k$.
Rule-3: P has only singletons.

If the successor tree has no nodes labeled with a set of singletons or with the empty set, i.e., is not truncated using Rules 1 or 3 then
Return the message "FSM **S** is not homing".
Otherwise,
    Determine a path with minimal length from the root to a node labeled with a set of singletons or with the empty set;
    Return HS as the input sequence α that labels the selected path.
End

Proposition 3. *Given an FSM* $S = \langle S, I, O, h_S, S' \rangle$ *with* $|S| = n$, $|S'| = m$, *the length of a shortest homing sequence is at most* $2^{\binom{n}{2}} - 2^{\binom{n}{2} - \binom{m}{2}}$.

Proof. The length of an HS for an FSM with n states and m initial states is bounded by the number of sets of state pairs that do not include pairs of initial states (Rule-2). The number of all sets of state pairs which are not singletons equals $2^{\binom{n}{2}}$ while the number of sets of state pairs including pairs of initial states equals $2^{\binom{n}{2} - \binom{m}{2}}$. Thus, if an HS exists, then its maximal length is at most the difference $2^{\binom{n}{2}} - 2^{\binom{n}{2} - \binom{m}{2}}$.   □

Thus, according to the results in [7], given a homing FSM with n states and m initial states, the length L of a shortest HS satisfies $2^{n^2/4} \leq L \leq 2^{\binom{n}{2}} - 2^{\binom{n}{2} - \binom{m}{2}}$.
    However, the reachability of the lower bound is shown only for $2^{n^2/4}$-input FSMs with n states, and thus, it is a challenge to assess the number of inputs when the length of a shortest HS is still exponential. In this paper, we propose a new class of $(n - 1)$-input homing FSMs with n states for which a shortest homing sequence has length $2^{n-1} - 1$.

3 Deriving an $(n - 1)$-Input FSM with a Shortest Homing Sequence of Exponential Length

When deriving a class of $(n - 1)$-input FSMs with the state set $\{0, 1, \ldots, n - 1\}$ and an exponential lower bound on the length of a shortest homing sequence, we consider FSMs where a truncated successor tree has a path labeled with a sequence of all subsets of the state set but the next subset in the chain depends not on the previous subset (as in [12]) but rather on the least integer in the current subset. Such a chain of subsets of length $2^{n-1} - 1$ can be derived using a special linear order that is called a deducibility relation.
    We define a linear order (*deducibility* relation) over the set of all nonempty subsets of the set $Z_n = \{0, 1, \ldots, n-1\}$ based on corresponding Boolean vectors of length n. Given a subset $P = \{p_1, p_2, \ldots, p_t\}$, $0 \leq p_1 < p_2 < \cdots < p_t$, the corresponding Boolean vector has 1's in the positions p_1, p_2, $\ldots$, p_t (counting from right to left) and 0's in all other positions. For example, the vector $\mathbf{v} = (1\ 1\ 0\ 1\ 0)$ corresponds to the subset $Q = \{1, 3, 4\} \subseteq Z_5$, and the corresponding integer $B(Q) = 26$. A subset $Q \subseteq Z_n$ is *directly deduced* from the nonempty subset P (written: $P \preceq^d Q$) if $B(Q) = B(P) - 1$. In other words, the subset

$Q \subseteq Z_n$ is directly deduced from the nonempty subset $P = \{p_1, p_2, \ldots, p_t\}$, $0 \le p_1 < p_2 < \cdots < p_t < n$, if one of the following conditions holds[2].

1. If $p_1 = 0$, then $Q = P - \{0\} = \{p_2, \ldots, p_t\}$.
2. If $0 \notin P$, then $Q = \{0, 1, \ldots, p_1 - 1, p_2, \ldots, p_t\}$.

A subset $Q \subseteq Z_n$ is *deduced* from a subset $P \subseteq Zn$ (written: $P \preceq Q$) if there exists a sequence of sets $P = R_1, R_2, \ldots, R_l = Q$ such that each set R_j, $j = 2, \ldots, l$, is directly deduced from the set R_{j-1}. In this case, the sequence of sets $R_1, R_2, \ldots, R_l$ is called a *deduction chain* for the subset Q from the subset P. For instance, the set $\{1, 3, 4\} \subseteq Z_5$ is directly deduced from the set $\{0, 1, 3, 4\} \subseteq Z_5$ while the set $\{3, 4\}$ is deduced from set $\{1, 3, 4\}$ via a sequence $\{0, 3, 4\}$, $\{3, 4\}$, i.e., $\{0, 1, 3, 4\} \preceq^d \{1, 3, 4\}$ and $\{1, 3, 4\} \preceq \{3, 4\}$.

By definition, the deduction relation is a linear order and a subset $Q \subseteq Z_n$ is deduced from a nonempty subset $P \subseteq Z_n$ if and only if $B(Q) = B(P) - k$ for some $0 < k < B(P)$. Correspondingly, when starting with the set Z_n of the initial states each subset can be deduced from Z_n.

When deriving a class of $(n-1)$-input FSMs with n states with the exponential lower bound on the length of a shortest homing sequence, we consider FSMs where the sequence of subsets directly deducible from each other labels a path of a truncated successor tree, i.e., this tree should have a path labeled with the chain of sets

$$\{0, 1, \ldots, n-1\} \preceq^d \{1, 2, \ldots, n-1\} \preceq^d \{0, 2, \ldots, n-1\} \preceq^d$$
$$\{2, 3, \ldots, n-1\} \preceq^d \cdots \preceq^d \{0, n-1\} \preceq^d \{n-1\}. \quad (1)$$

Given a subset $P = \{p_1, p_2, \ldots, p_t\}$, $0 \le p_1 < p_2 < \cdots < p_t = n-1$, an input i_{p_1} is responsible for deriving a subset that is directly deduced from P. Outputs at the transition under input i_{p_1} should truncate branches which are labeled with a sequence headed with the input i_{p_1} for all subsets where the least integer is not p_1. Therefore, at each level of the successor tree a single edge is useful when deriving a homing sequence and this edge is labeled by the input i_{p_1}. This edge bridges a node labeled by all the pairs of the set $P = \{p_1, \ldots, p_t\}$ with a node labeled by pairs of the set $\{0, 1, \ldots, p_1 - 1, p_1 + 1, p_2, \ldots, p_t\}$ if $p_1 > 0$ or with a node labeled by pairs of the set $\{p_2, \ldots, p_t\}$ if $p_1 = 0$. All other inputs take the FSM to a node labeled with pairs of the set Q such that P is deduced from Q or Q contains P. Below we show how an FSM $\mathbf{S_n}$ of this class can be derived.

Consider NFSM $\mathbf{S_n}$, $n > 1$, with the set $S = \{0, 1, 2, \ldots, n-1\}$ of states, the set $I = \{i_0, i_1, \ldots, i_{n-2}\}$ of inputs, and the set $O = \{(i, j) | i, j \in \{0, \ldots, n-1\}$ and $i < j\}$ of outputs. We define the transition relation of the FSM $\mathbf{S_n}$ in the following way. Given the input i_0, there is a single transition at the state 0 under this input, namely, the transition to the state $(n-1)$ with the output $(0, n-1)$. Moreover, at each state j, $0 < j \le n-1$, there is a loop labeled with the I/O pair $i_0/(0, n-1)$. Thus, given a node labeled by all the pairs of the set $P = \{0, p_2, \ldots, p_t\}$, $p_t = n-1$, an outgoing transition at this node labeled

---

[2] Such a chain of subsets is proposed in [4].

with the input i_0 takes the FSM to the node labeled with all the pairs of the set $\{p_2, \ldots, p_t\}$ that is directly deduced from P.

Given the input i_k, $0 < k < n - 1$, we define transitions under this input keeping the two following points in mind. On the one hand, for a node labeled by the set of all pairs of the set $P = \{k, p_2, \ldots, p_t\}$, $0 < k < p_2 < \cdots < p_t = n-1$, the outgoing transition at this node labeled with the input i_k should take the FSM to the node labeled with all the pairs of the set $\{0, \ldots, k-1, p_2, \ldots, p_t\}$ that is directly deduced from P. Therefore, given a state k, the FSM $\mathbf{S_n}$ moves from the state k to a state $q \in \{0, 1, \ldots, k-1\}$ under input i_k producing outputs (q, l) where $l \in \{0, 1, \ldots, k-1, k+1 \ldots, n-1\}$ and $l > q$, i.e., there exist transitions $(k, i_k, (q, l), q)$ and those transitions are shown below. On the other hand, for a node labeled by all the pairs of the set $P = \{p_1, p_2, \ldots, p_t\}$, $p_1 \neq k$, the outgoing transition at this node labeled with the input i_k should take the FSM to an unpromising node, i.e., to a node labeled with pairs of the set Q where Q contains P or P is deduced from Q. Correspondingly, at each state p where $p > k$ the FSM $\mathbf{S_n}$ has loops labeled with $i_k/(l, p)$ for all $l < p$ and with $i_k/(p, l)$ for all $p < l < n$. Moreover, at the state p there are transitions to each state $q \in \{1, 2, \ldots, k-1\}$ labeled with $i_k/(l, q)$ for all $l < q$. At each state $p < k$ the FSM $\mathbf{S_n}$ has transitions to the state k under the input i_k with the output (q, l) for $q, l \neq k$ and $0 \leq q < l \leq n - 1$.

The FSM $\mathbf{S_4}$ constructed by using the above rules is shown below (Fig. 3). Here we notice that there are many ways how to derive an FSM with n states

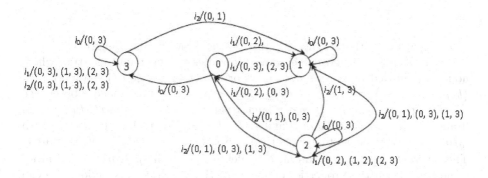

Fig. 3. The NFSM $\mathbf{S_4}$

such that a truncated successor tree in Procedure 2.1 has only one path that is labeled by a shortest HS. In this paper, we just show that the $(n-1)$-input FSM $\mathbf{S_n}$ constructed by the above rules has a shortest homing sequence of length $2^{n-1} - 1$.

For the FSM $\mathbf{S_n}$ we use Procedure 1 in order to derive a shortest HS. For FSM $\mathbf{S_4}$, the corresponding truncated successor tree is represented in Fig. 4 and by direct inspection, one can verify that the sequence $i_0 i_1 i_0 i_2 i_0 i_1 i_0$ is a shortest homing sequence for NFSM $\mathbf{S_4}$. The sequence has seven inputs, i.e., its length is $2^3 - 1 = 2^{n-1} - 1$. We note that in Fig. 4 we do not show singletons of the sets which label nodes of the truncated successor tree.

Fig. 4. The truncated tree for $\mathbf{S_4}$ returned by Procedure 1

We now prove that a shortest HS for the FSM $\mathbf{S_n}$ has length $2^{n-1} - 1$. For this purpose, we establish some properties of the truncated tree where nodes are labeled by sets of state pairs or by the empty set. Given a set $U = \{k, p_1, p_2, \ldots, p_t\}$, $0 \le k < p_1 < p_2 < \cdots < p_t \le n-1$, we denote by $\tilde{U}_2$ the set of all pairs $\overline{i,j}$, $i, j \in U$, $i < j$, and a pair $\overline{i,j}$ belongs to $next\_state(\tilde{U}_2, i_q)$ if and only if there exist $o \in O$ and states $b, c \in U$ such that the state i is an i_q/o-successor of the state b while the state j is an i_q/o-successor of the state c. The set $next\_state(\tilde{U}_2, i_q)$ can have singletons, i.e., pairs $\overline{p,p}$, $p \in S$. We prove some lemmas on the properties of the FSM $\mathbf{S_n}$. By direct inspection, one can verify that for $n > 3$, this FSM is complete, connected and observable.

Lemma 1. *Given* $U = \{k, p_1, p_2, \ldots, p_t\}$, $0 \le k < p_1 < p_2 < \cdots < p_t$, $p_t = n-1$, *the set* $next\_state(\tilde{U}_2, i_k)$ *up to singletons coincides with the set* $\tilde{N}_2$, *where* N *is directly deduced from* U, *i.e.,* $U \preceq^d N$.

Proof. If $k = 0$, then the statement holds since at each state $1, 2, \ldots, n-1$, the FSM $\mathbf{S_n}$ has a loop labeled by an I/O pair $i_0/(0, n-1)$ while at the state 0 there is a transition to the state $(n-1)$ labeled by the same I/O pair. Correspondingly, $next\_state(\tilde{U}_2, i_0) = \{\overline{i,j}, i < j, j = p_1, p_2, \ldots, p_t\} \cup \{\overline{n-1, n-1}\}$, i.e., $next\_state(\tilde{U}_2, i_0) = \tilde{N}_2 \cup \{\overline{n-1, n-1}\}$ where $N = \{p_1, p_2, \ldots, p_t\}$, i.e., $U \preceq^d N$.

If $U = \{k, p_1, p_2, \ldots, p_t\}$, $0 < k < p_1 < p_2 < \cdots < p_t$, $p_t = n - 1$, then by the definition, $next\_state(\tilde{U}_2, i_k)$ has all pairs $\overline{i, j}$, which are i_k/o-successors of the pairs of the set $\tilde{U}_2$ for some output o. Two cases are possible.

Case 1. If $\overline{i, j}$ is in the i_k-successor of a pair $\overline{k, p_z}$ with $z \in \{1, 2, \ldots, t\}$, then by construction, under the input i_k the FSM $\mathbf{S_n}$ moves from the state k to a state $q \in \{0, 1, \ldots, k-1\}$ outputting (q, l) for $l \in \{0, 1, \ldots, k-1, k+1, \ldots, n-1\}$ and $l > q$. At the state $p_z > k$ there are loops labeled by the input/output pair $i_k/(l, p_z)$ for all $l < p_z$, and loops labeled by $i_k/(p_z, l)$ for $p_z < l < n$. Moreover, at the state p_z there are transitions to a state $q \in \{1, 2, \ldots, k-1\}$ labeled by the I/O pair $i_k/(l, q)$ for $l < q$. Therefore, the set of i_k/o-successors of the pairs of the set $\tilde{U}_2$ contains each pair $\overline{q, p_z}$ for $q \in \{0, 1, \ldots, k-1\}$, $z = 1, \ldots, t$, and each pair of the set $\{\overline{0, 1}, \ldots, \overline{0, k-1}, \overline{1, 2}, \ldots, \overline{k-2, k-1}\}$.

Case 2. Let a pair $\overline{i, j}$ be an element of the i_k-successor of the pair $\overline{p_r, p_z}$, $k < p_r < p_z$, $1 < r < z < t$. Each i_k/o-successor of the pair $\overline{p_r, p_z}$, $k < p_r < p_z$, is the pair $\overline{p_r, p_z}$, and this pair is obtained according to outputs (l, p_z) when $l = p_r$ (or (p_r, l) when $l = p_z$). Moreover, an i_k/o-successor can be a singleton $\overline{q, q}, q \in \{1, 2, \ldots, k-1\}$ according to an I/O pair $i_k/(l, q)$ for $l < q$.

Thus, the set $next\_state(\tilde{U}_2, i_k)$ up to singletons coincides with the set $\tilde{N}_2$ where $N = \{0, 1, \ldots, k-1, p_1, p_2, \ldots, p_t\}$, i.e., $U \preceq^d N$. $\square$

Lemma 2. *Given* $U = \{k, p_1, p_2, \ldots, p_t\}$, $p_t = n - 1$, $k < p_1 < p_2 < \cdots < p_t$, $k > 0$, *the union of* i_j-*successors of all pairs of* $\tilde{U}_2$ *contains* $\tilde{U}_2$ *for* $0 < j < k$.

Proof. In the conditions of the lemma, $p > j$ for each $p \in U$. By construction of $\mathbf{S_n}$, at each state p for $p > j$ there are loops labeled by the input/output pair $i_j/(l, p)$ for all $l < p$, and loops labeled by the input/output pair $i_j/(p, l)$ for $p < l < n$. Moreover, at the state p there are transitions to a state $q \in \{0, 1, \ldots, j-1\}$ labeled by $i_j/(l, q)$ for $l < q$. Thus, similar to the proof of Lemma 1, each pair of the i_j-successor of each pair of the set $\tilde{U}_2$ is a pair $\overline{k, p_z}$, $z \in \{1, 2, \ldots, t\}$, or is a singleton $\overline{q, q}$, $q \in 1, 2, \ldots, j - 1$. Correspondingly,

$$next\_state(\tilde{U}_2, i_k) = \tilde{U}_2 \cup \{\overline{1, 1}, \overline{2, 2}, \ldots, \overline{k-1, k-1}\},$$

i.e. the set $next\_state(\tilde{U}_2, i_k)$ contains $\tilde{U}_2$. $\square$

Lemma 3. *Given* $U = \{k, p_1, p_2, \ldots, p_t\}$, $k < p_1 < p_2 < \cdots < p_t$, $p_t = n - 1$, *for each* $n > j > k$ *it holds that* $next\_state(\tilde{U}_2, i_j)$ *contains* $\tilde{N}_2$ *where* U *is deduced from* N, *i.e.,* $N \preceq U$.

Proof. Given $n > j > k$, two cases are possible.

Case 1. If $j \in U$, then similarly to the proofs of Lemmas 1 and 2, it can be shown that the set $next\_state(\tilde{U}_2, i_j)$ contains all the pairs $\overline{p_r, p_z}$ with $p_r, p_z \in U$, $j < p_r < p_z$. Moreover, at the state k there are transitions to the state j with the outputs (q, l), $q, l \neq j$, $0 \leq q < l < n$. Therefore, the set $next\_state(\tilde{U}_2, i_j)$ contains the pairs $\overline{j, p_z}$, with $p_z \in U, j < p_z$. At the state j there are transitions to a state $q \in \{0, 1, \ldots, j-1\}$ with outputs (q, l), $q, l \neq j$, $0 \leq q < l < n$ and at the state $q_t = n - 1$ there are transitions to a state $p \in \{1, 2, \ldots, j-1\}$ with outputs (l, p), $0 \leq l < p < j$, i.e., the set $next\_state(\tilde{U}_2, i_j)$ contains the pairs $\overline{q, p_z}$, $0 \leq q < j < p_z$ and thus, $next\_state(\tilde{U}_2, i_j)$ contains $\tilde{U}_2$.

Case 2. If $j \notin U$, then $U = \{k, p_1, \ldots, p_t\}$ for $k < j < p_1$ or $U = \{k, p_1, \ldots, p_{a-1}, p_a, \ldots, p_t\}$ where $p_{a-1} < j < p_a$. Consider a subset $U' = \{k, p_a, \ldots, p_t\}$ of the set U such that $k < j < p_a < \cdots < p_t$. If $k < j < p_1$ then $U' = U$ and $a = 1$. Similarly to the proofs of Lemmas 1 and 2, it can be shown that the set $next\_state(\tilde{U}_2', i_j)$ contains the pairs $\overline{p_r, p_z}$ with $p_r, p_z \in U$, $a \leq r < z$. Moreover, $k < j$, and for each state $p < j$, the FSM $\mathbf{S_n}$ has transitions to the state j under the input i_j with an output (q, l) for $q, l \neq j$, $0 \leq q < l \leq n - 1$. Therefore, the set $next\_state(\tilde{U}_2', i_j)$ contain all pairs $\overline{j, p_z}$, with $p_z \in U$, due to the I/O pair $i_j/(p_z, n - 1)$.

Correspondingly, the set $next\_state(\tilde{U}_2', i_j)$ contains the subset $\tilde{N}_2$ for $N = \{j, p_a, \ldots, p_t\}$, and $\sum_{i=1}^{t}(2^{p_i} + 2^j)$ is the integer corresponding to the Boolean vector for the subset N. The integer corresponding to the Boolean vector for the subset U equals $\sum_{i=1}^{t} 2^{p_i} + 2^k$ and the difference $B(N) - B(U) = \sum_{i=1}^{t} 2^{p_i} + 2^j - \sum_{i=1}^{t} 2^{p_i} - 2^k$ equals $2^j - (2^k + \sum_{i=1}^{a-1} 2^{p_i})$. If $a = 1$, then $\sum_{i=1}^{a-1} 2^{p_i} = 0$, and the difference $B(N) - B(U) = 2^j - 2^k > 0$. If $a > 1$, then $2^k + \sum_{i=1}^{a-1} 2^{p_i} < 2^j$, since in the worst case the set U contains all the integers of the interval $[0; j - 1]$ ($j > k$ and $j > p_{a-1}$), and thus, the difference $B(N) - B(U) > 0$. For this reason, $B(N) > B(U)$ and the subset U can be deduced from the subset N.   □

Theorem 1. *The length of a shortest homing sequence for NFSM $\mathbf{S_n}$, $n > 3$, is equal to $2^{n-1} - 1$.*

Proof. The homing sequence arises as a sequence that labels a path in the truncated successor tree from the root to a node labeled by the empty set or by a set of singletons. The root of the tree is labeled by the set $\{\overline{i, j} : 0 \leq i < j \leq n - 1\}$. Consider a node of the tree labeled by $\tilde{U}_2$, $U = \{k, p_1, p_2, \ldots, p_t\}$, $k < p_1 < p_2 < \cdots < p_t$, $p_t = n - 1$. All the nodes at the next level labeled with $next\_state(\tilde{U}_2, i_j)$, $j \neq k$, are terminal due to Rule-2 (Lemmas 2 and 3). For the input i_k, the node at the next level is labeled with $next\_state(\tilde{U}_2, i_k) = \tilde{N}_2$ where $U \preceq^d N$. Thus, the shortest input sequence that labels a path from the root to the set of singletons is the sequence that labels the chain (1). In [4], it is shown that this chain has each subset that contains the state $(n - 1)$. The input sequence that labels this path has length $2^{n-1} - 1$.   □

4   Conclusion and Future Work

In this paper, we have shown that the problem of deriving homing sequences for nondeterministic FSMs cannot be reduced to the problem of deriving synchronizing words for the corresponding automata. We also show that an exponential lower bound for the minimum length of a homing sequence is reachable for $(n-1)$-input FSMs with n states. In fact, we have shown that for each $n > 3$ there exists a class of $(n - 1)$-input FSMs with n states with a shortest homing sequence of length $2^{n-1} - 1$. However, there still is a gap between lower and upper bounds on the length of a shortest HS. One of directions of possible future research is to reduce this gap. If there is no HS for a given nondeterministic FSM, an adaptive homing experiment may still exist sometimes. Adaptive homing experiments and their complexity are out of the scope of this paper and we only mention that a preliminary study shows that the complexity of adaptive homing experiments, i.e., length of a longest adaptive input/output sequence, also seems to be exponential w.r.t. the number of states of the FSM under experiment. More research is needed in order to accurately evaluate the complexity of homing experiments for nondeterministic FSMs.

Acknowledgments. The authors are also thankful to the anonymous referees for several useful remarks.

References

1. Agibalov, G., Oranov, A.: Lectures on Automata Theory. Tomsk (1984) (in Russian)
2. Burkhard, H.D.: Zum Längenproblem homogener Experimente an determinierten und nicht-deterministischen Automaten. Elektronische Informationsverarbeitung und Kybernetik 12(6), 301–306 (1976)
3. Gill, A.: Introduction to the Theory of Finite State Machines. McGraw-Hill (1962)
4. Hwang, I., Yevtushenko, N., Cavalli, A.R.: Tight bound on the length of distinguishing sequences for non-observable nondeterministic Finite-State Machines with a polynomial number of inputs and outputs. Inf. Process. Lett. 112(7), 298–301 (2012)
5. Klyachko, A.A., Rystsov, I.K., Spivak, M.A.: An extremal combinatorial problem associated with the bound on the length of a synchronizing word in an automaton. Cybernetics 23, 165–171 (1987)
6. Kohavi, Z.: Switching and Finite Automata Theory. McGraw–Hill (1978)
7. Kushik, N., El-Fakih, K., Yevtushenko, N.: Preset and adaptive homing experiments for nondeterministic finite state machines. In: Bouchou-Markhoff, B., Caron, P., Champarnaud, J.-M., Maurel, D. (eds.) CIAA 2011. LNCS, vol. 6807, pp. 215–224. Springer, Heidelberg (2011)
8. Martugin, P.V.: Lower bounds for the length of the shortest carefully synchronizing words for two- and three-letter partial automata. Diskretn. Anal. Issled. Oper. 15(4), 44–56 (2008) (in Russian)
9. Milner, R.: Communication and Concurrency. Prentice-Hall (1989)
10. Pin, J.-E.: On two combinatorial problems arising from automata theory. Ann. Discrete Math. 17, 535–548 (1983)
11. Sandberg, S.: Homing and synchronization sequences. In: Broy, M., Jonsson, B., Katoen, J.-P., Leucker, M., Pretschner, A. (eds.) Model-Based Testing of Reactive Systems. LNCS, vol. 3472, pp. 5–33. Springer, Heidelberg (2005)
12. Spitsyna, N., El-Fakih, K., Yevtushenko, N.: Studying the separability relation between finite state machines. Software Testing, Verification and Reliability 17(4), 227–241 (2007) (in Russian)
13. Volkov, M.V.: Synchronizing automata and the Černý conjecture. In: Martín-Vide, C., Otto, F., Fernau, H. (eds.) LATA 2008. LNCS, vol. 5196, pp. 11–27. Springer, Heidelberg (2008)

Input-Driven Queue Automata: Finite Turns, Decidability, and Closure Properties*

Martin Kutrib[1], Andreas Malcher[1], Carlo Mereghetti[2],
Beatrice Palano[2], and Matthias Wendlandt[1]

[1] Institut für Informatik, Universität Giessen, Arndtstr. 2, 35392 Giessen, Germany
{kutrib,malcher,matthias.wendlandt}@informatik.uni-giessen.de
[2] Dip. Informatica, Univ. degli Studi di Milano, v. Comelico 39, 20135 Milano, Italy
{mereghetti,palano}@di.unimi.it

Abstract. We introduce and study the model of input-driven queue automata. On such devices, the input letters uniquely determine the operations on the memory store which is organized as a queue. In particular, we consider the case where only a finite number of turns on the queue is allowed. The resulting language families share with regular languages many desirable properties. We show that emptiness and several other problems are decidable. Furthermore, we investigate closure under Boolean operations. The existence of an infinite and tight hierarchy depending on the number of turns is also proved.

Keywords: Input-driven automata, queue automata, finite turns, decidability questions, closure properties.

1   Introduction

Finite automata have been widely investigated from many theoretical as well as practical points of view. On the one hand, they possess many nice properties such as equivalence of nondeterministic and deterministic models, existence of minimization algorithms, closure under many operations, and decidable questions such as emptiness, inclusion, or equivalence (see, e.g., [13]). On the other hand, their computational power is quite low since only regular languages are accepted. It is therefore natural to consider extensions of the model featuring additional storage media such as pushdown tapes [6], stack tapes [8], queue tapes [5], or Turing tapes. In general, such extensions lead to a broader family of accepted languages, but also to a weaker manageability of the models since certain closure properties do not longer hold, minimization algorithms do not exist, and formerly decidable questions become undecidable. Thus, there is an obvious interest in extensions which enlarge the language family, but keep as many of the 'good' properties as possible.

* Partially supported by CRUI/DAAD under the project "Programma Vigoni: Descriptional Complexity of Non-Classical Computational Models", and MIUR under the project "PRIN: Automi e Linguaggi Formali: Aspetti Matematici e Applicativi".

S. Konstantinidis (Ed.): CIAA 2013, LNCS 7982, pp. 232–243, 2013.

One such extension is represented by input-driven automata. Basically, for such devices the operations on the storage medium are dictated by the input symbols. The first references date back to [4,16], where input-driven pushdown automata (PDA) are introduced as classical PDA in which the input symbols define whether a push operation, a pop operation, or no operation on the pushdown store has to be performed. The first results show that input-driven PDA can accept context-free languages in logarithmic space, and that nondeterministic and deterministic models are equivalent. Later, it has been shown in [7] that their membership problem belongs to the parallel complexity class NC^1. The investigation of input-driven PDA has been renewed in [1,2], where such devices are called visibly PDA (VPDA) or nested word automata. The main result is that input-driven PDA describe a language family lying properly in between the regular and the deterministic context-free languages, and sharing with regular languages many desirable features such as many closure properties and decidable questions. Moreover, an exponential upper bound is given for the size increase when removing nondeterminism in input-driven PDA. Extensions have also been studied, e.g., with respect to multiple pushdown stores in [17], or more general auxiliary storages in [14]. Recently, the computational power of input-driven stack automata has been investigated in [3].

In this paper, we consider the storage medium of a *queue*, and investigate deterministic input-driven queue automata (DVQA), as well as general deterministic queue automata (DQA). Note that the letter V in DVQA stands for 'visibly' as in VPDA and means a shorthand notation for input-driven DQA. DQA have been considered in [5] where, among others, it is proved that their computational power equals that of Turing machines. Thus, to focus on a more 'manageable' device, we investigate here also the restricted type of *finite-turn* queue automata, in which the number of alternations between enqueuing and dequeuing is bounded by some fixed finite number. The property of finite turns has been already introduced and studied for PDA in [9,10]. In this model, the number of alternations between increasing and decreasing the height of the pushdown store is bounded by some fixed finite number. The main results are a proper turn hierarchy for PDA and the decidability of the finite-turn property for an arbitrary PDA. On the other hand, decidability problems which are undecidable for PDA remain undecidable for finite-turn PDA. In [15], the finite turn-paradigm is considered for two-way PDA. Precisely, the descriptional power of finite-turn PDA processing bounded languages with a constant number of input head reversals is investigated, as well as some related decidability questions. The formal definition of finite-turn queue automata and examples are given in Section 2. The main result of Section 3 is the existence of an infinite proper hierarchy on the number of turns for DQA and DVQA. In fact, for every $k \geq 0$, we show that $k+1$ turns are more powerful than k turns. One motivation to study input-driven automata is to obtain models with attractive closure properties and decidability questions. Here, we consider both automata with compatible signatures – which roughly speaking means that they agree on the actions induced by input symbols – and automata with incompatible signatures. Concerning closure properties, in Section 4 we show that DVQA

234 of M. Kutrib et al.

are closed under complementation, whereas finite-turn DVQA are not. Moreover, DVQA as well as finite-turn DVQA are closed under union and intersection in case of compatible signatures, whereas non-closure is obtained for incompatible signatures. Concerning decidability questions, in Section 5 we show that emptiness, finiteness, and universality are decidable for finite-turn DQA which are not necessarily input-driven. For finite-turn DVQA with compatible signatures, we show in addition that inclusion and equivalence are decidable. On the other hand, we prove the undecidability of inclusion in case of incompatible signatures, and the undecidability of all questions in case of DVQA with unbounded turns. This shows that it is essential for DVQA to restrict the number of turns in order to get a manageable model. It may be worth noticing that positive decidability results for finite-turn DVQA are achieved although the model is not closed under complementation.

2   Preliminaries and Definitions

We write Σ^* for the set of all words over the finite alphabet Σ. The empty word is denoted by λ, and we set $\Sigma^+ = \Sigma^* \setminus \{\lambda\}$. The reversal of a word w is denoted by w^R, and for the length of w we write $|w|$. We use $\subseteq$ for inclusions and $\subset$ for strict inclusions. The complement of a language $L \subseteq \Sigma^*$ is defined as $\overline{L} = \Sigma^* \setminus L$.

A classical deterministic queue automaton is called input-driven if the next input symbol defines the next action on the queue, that is: entering a symbol at the end of the queue, removing a symbol from the front of the queue, or changing the internal state without modifying the queue content. To this end, we assume the input alphabet Σ is partitioned into the sets Σ_e, Σ_r, and Σ_i, that control the actions enter, remove, state change only (internal). A formal definition reads as:

Definition 1. *A deterministic input-driven queue automaton (DVQA) is a system* $M = \langle Q, \Sigma, \Gamma, q_0, F, \perp, \delta_e, \delta_r, \delta_i \rangle$, *where*

1. Q *is the finite set of* internal states,
2. Σ *is the finite set of* input symbols *consisting of the disjoint union of sets* Σ_e, Σ_r, *and* Σ_i,
3. Γ *is the finite set of* queue symbols,
4. $q_0 \in Q$ *is the* initial state,
5. $F \subseteq Q$ *is the set of* accepting states,
6. $\perp \notin \Gamma$ *is the* empty-queue symbol,
7. δ_e *is the partial transition function mapping* $Q \times \Sigma_e \times (\Gamma \cup \{\perp\})$ *to* $Q \times \Gamma$,
8. δ_r *is the partial transition function mapping* $Q \times \Sigma_r \times (\Gamma \cup \{\perp\})$ *to* Q,
9. δ_i *is the partial transition function mapping* $Q \times \Sigma_i \times (\Gamma \cup \{\perp\})$ *to* Q.

A *configuration* of a DVQA $M = \langle Q, \Sigma, \Gamma, q_0, F, \perp, \delta_e, \delta_r, \delta_i \rangle$ is a triple (q, w, s), where $q \in Q$ is the current state, $w \in \Sigma^*$ is the unread part of the input, and $s \in \Gamma^*$ denotes the current queue content, where the leftmost symbol is at the front. The *initial configuration* for an input string w is set to (q_0, w, λ). During the course of its computation, M runs through a sequence of configurations. One step from a configuration to its successor configuration is denoted by $\vdash$. Let $a \in \Sigma$, $w \in \Sigma^*$, $z, z' \in \Gamma$, and $s \in \Gamma^*$. We set

1. $(q, aw, zs) \vdash (q', w, zsz')$, if $a \in \Sigma_e$ and $(q', z') \in \delta_e(q, a, z)$,
2. $(q, aw, \lambda) \vdash (q', w, z')$, if $a \in \Sigma_e$ and $(q', z') \in \delta_e(q, a, \perp)$,
3. $(q, aw, zs) \vdash (q', w, s)$, if $a \in \Sigma_r$ and $q' \in \delta_r(q, a, z)$,
4. $(q, aw, \lambda) \vdash (q', w, \lambda)$, if $a \in \Sigma_r$ and $q' \in \delta_r(q, a, \perp)$,
5. $(q, aw, zs) \vdash (q', w, zs)$, if $a \in \Sigma_i$ and $q' \in \delta_i(q, a, z)$,
6. $(q, aw, \lambda) \vdash (q', w, \lambda)$, if $a \in \Sigma_i$ and $q' \in \delta_i(q, a, \perp)$.

So, whenever the queue is empty, the successor configuration is computed by the transition functions with the special empty-queue symbol $\perp$. As usual, we define the reflexive and transitive closure of $\vdash$ by $\vdash^*$. The language accepted by the DVQA M is the set $L(M)$ of words for which there exists some computation beginning in the initial configuration and ending in a configuration in which the whole input is read and an accepting state is entered. Formally:

$$L(M) = \{\, w \in \Sigma^* \mid (q_0, w, \lambda) \vdash^* (q, \lambda, s) \text{ with } q \in F, s \in \Gamma^* \,\}.$$

The difference between a DVQA and a classical deterministic queue automaton (DQA) is that the latter makes no distinction on the types of the input symbols, and may perform λ-moves. However, in all cases, there must not be more than one choice of action for any possible configuration. So, the transition function is defined to be a (partial) mapping from $Q \times (\Sigma \cup \{\lambda\}) \times (\Gamma \cup \{\perp\})$ to $Q \times (\Gamma \cup \{\text{remove}, \text{internal}\})$, where it is understood that remove means to remove the symbol from the front of the queue, internal means to let the content of the queue unchanged, and a symbol of Γ means to enter this symbol at the tail of the queue. In general, the family of all languages accepted by an automaton of some type X will be denoted by $\mathscr{L}(X)$.

For a computation of a queue automaton, a turn is a phase in which the length of the queue first increases and then decreases. Formally, a sequence of at least three configurations $(q_1, w_1, s_1) \vdash (q_2, w_2, s_2) \vdash \cdots \vdash (q_m, w_m, s_m)$ is a turn if $|s_1| < |s_2| = \cdots = |s_{m-1}| > |s_m|$. For any given $k \geq 0$, a k-turn computation is any computation containing exactly k turns. A DVQA performing at most k turns in any computation is called k-turn DVQA and will be denoted by DVQA_k.

In order to clarify our notions, we continue with an example.

Example 2. Let $h_p : \{a, b\}^* \to \{a', b'\}^*$ be the homomorphism that is defined by $h_p(a) = a'$, $h_p(b) = b'$. The language $\{\, wh_p(w)\# \mid w \in \{a, b\}^* \,\}$ is accepted by the DVQA_1 $M = \langle Q, \Sigma, \Gamma, q_0, F, \perp, \delta_e, \delta_r, \delta_i, \rangle$, where $Q = \{q_0, q_1, q_2\}$, $F = \{q_2\}$, $\Sigma_i = \{\#\}$, $\Sigma_e = \{a, b\}$, $\Sigma_r = \{a', b'\}$, $\Gamma = \{A, B\}$, and the transition functions are as follows:

(1) $\delta_i(q_0, \#, \perp) = q_2$,
(2) $\delta_i(q_1, \#, \perp) = q_2$,

(3) $\delta_e(q_0, a, Z) = (q_0, A)$ for $Z \in \{A, B, \perp\}$,
(4) $\delta_e(q_0, b, Z) = (q_0, B)$ for $Z \in \{A, B, \perp\}$,

(5) $\delta_r(q_0, a', A) = q_1$,
(6) $\delta_r(q_0, b', B) = q_1$,
(7) $\delta_r(q_1, a', A) = q_1$,
(8) $\delta_r(q_1, b', B) = q_1$.

Consider an accepting computation of M. Basically, the transitions of δ_e enqueue the prefix w. Subsequently, the infix $h_p(w)$ has to be matched with the contents of the queue. When the first primed symbol appears in the input, it is matched by the transitions δ_r. The first application switches from state q_0 to q_1 which is kept during this part of the computation. The transition function is not defined for q_1 in connection with symbol a or b, and it is defined for q_1 and the symbol # only if the queue is empty. Exactly one application of this transition leads M into the sole accepting state q_2, for which no transition is defined. So, the input is accepted if and only if the prefix w matches the infix $h_p(w)$ followed by a single #. Clearly, M performs at most one turn in every computation.   □

3   Turn Hierarchy

In [11,12], extensions of pushdown automata called flip-pushdown automata are investigated. Basically, a flip-pushdown automaton is an ordinary pushdown automaton with the additional ability to flip its pushdown during the computation.

Definition 3. *A deterministic flip-pushdown automaton (DFPDA) is a system* $M = \langle Q, \Sigma, \Gamma, q_0, F, \lhd, \delta, \Delta \rangle$, *where Q is the finite set of internal states, Σ is the finite set of input symbols, Γ is the finite set of pushdown symbols, $q_0 \in Q$ is the initial state, $F \subseteq Q$ is the set of accepting states, $\lhd \in \Gamma$ is a particular pushdown symbol, called the bottom-of-pushdown symbol, which initially appears on the pushdown store, δ is the partial transition function mapping $Q \times (\Sigma \cup \{\lambda\}) \times \Gamma$ to $Q \times \Gamma^*$, and Δ is the partial flip function mapping from Q to Q, so that, for all $q \in Q$, and $Z \in \Gamma$, if $\Delta(q)$ is not empty, then $\delta(q, a, Z)$ is empty, for all $a \in \Sigma \cup \{\lambda\}$, and if $\delta(q, \lambda, Z)$ is not empty, then $\delta(q, a, Z)$ is empty, for all $a \in \Sigma$.*

Here, it is understood that the transition function is such that the symbol $\lhd$ appears at most once, that is, at the bottom of the pushdown only. Moreover, an application of δ is an 'ordinary' transition as for pushdown automata, and an application of Δ flips the content of the pushdown up to the bottom-of-pushdown symbol. So, for $\gamma \in \Gamma^*$, the pushdown content $\gamma\lhd$ becomes $\gamma^R\lhd$. Technical details can be found in [11,12]. However, it is worth mentioning that a DFPDA has never more than one choice of action for any possible configuration. A DFPDA performing at most k flips in any computation is called k-flip DFPDA and will be denoted by $DFPDA_k$.

Now, we turn to simulate DVQA by DFPDA where, as would seem to be natural, the number of flips of the DFPDA depends on the number of turns of the DVQA. The goal is to derive results from known properties of DFPDA. The direct simulation is straightforward:

Lemma 4. *Let $k \geq 1$ be a constant and M be a k-turn DQA. Then, an equivalent $2k$-flip DFPDA can effectively be constructed.*

Proof. The idea of the construction is to use one end of the pushdown store as the front and the other end as the tail of the queue. So, whenever the queue automaton performs a turn, that is, changes from increasing to decreasing mode, the flip-pushdown automaton flips the front end of the pushdown store to the top. Similarly, whenever the queue automaton changes from decreasing to increasing mode, the tail end has to be flipped to the top. So, for any turn of the queue, one flip is necessary, plus one flip to change from one turn to the next, plus possibly another one to change from the decreasing phase following the last turn to a final increasing phase. Altogether, this makes $2k$ flips.   □

An essential technique for flip-pushdown automata is the 'flip-pushdown input-reversal' technique, which has been developed and proved in [11]. It allows to simulate flipping the pushdown by reversing the (remaining) input, and reads as follows:

Theorem 5. *Let $k \geq 0$ be a constant. A language L is accepted by a (nondeterministic) flip-pushdown automaton $M_1 = \langle Q, \Sigma, \Gamma, q_0, F, \lhd, \delta, \Delta \rangle$ by final state with at most $k + 1$ pushdown reversals if and only if the language*

$$L_R = \{ wv^R \mid (q_0, w, \lhd) \vdash^*_{M_1} (q_1, \lambda, \gamma \lhd) \text{ with at most } k \text{ reversals}, q_2 \in \Delta(q_1),$$
$$\text{and } (q_2, v, \gamma^R \lhd) \vdash^*_{M_1} (q_3, \lambda, q_4) \text{ without any reversal}, q_4 \in F \}$$

is accepted by a (nondeterministic) flip-pushdown automaton M_2 by final state with at most k pushdown reversals.

In particular, given a DFPDA$_k$ M, the k-fold application of Theorem 5 yields a context-free language L that is letter equivalent to $L(M)$. An immediate consequence of this theorem and Lemma 4 is that every language accepted by a DQA with a constant number of turns obeys a semilinear Parikh mapping.

Corollary 6. *Let $k \geq 0$ be a constant and M be a k-turn DQA. Then $L(M)$ is semilinear. In particular, if $L(M)$ is a unary language then it is regular.*

To prove a tight turn hierarchy for DVQA, the straightforward simulation of Lemma 4 is too weak. So, let $h_p : \{a, b\}^* \to \{a', b'\}^*$ be the homomorphism defined as $h_p(a) = a'$, $h_p(b) = b'$. For all $j \geq 0$, we define the sets

$$C_j = \{ \#w\#h_p(w) \mid w \in \{a, b\}^* \}^j \cdot \#$$

and, for all $k \geq 0$, the language $L_k = \bigcup_{j=0}^k C_j$.

Lemma 7. *Let $k \geq 1$ be a constant. Then, the language L_k is accepted by a DVQA$_k$.*

However, the language L_k cannot be accepted by any queue automaton with less than k turns.

Lemma 8. *Let $k \geq 1$ be a constant. Then, the language L_k is not accepted by any DQA$_{k-1}$.*

Proof. Assume, by contradiction, that L_k is accepted by some DQA_{k-1} $M_k = \langle Q, \Sigma, \Gamma, q_0, F, \bot, \delta \rangle$. We consider words from $C_k \subset L_k$.

If there exists at least one prefix $\bar{w}_1 = \#w_1\#h_p(w_1)$ so that the computation $(q_0, \bar{w}_1 v, \lambda) \vdash^+ (q, v, \gamma)$, for $v \in \Sigma^*$, contains at least one turn, we fix this prefix and consider next the language L_{k-1}. This language is accepted by a DQA_{k-2} M_{k-1} as follows. Initially, M_{k-1} enqueues γ by extra states and λ-moves, and changes to state q. Then, it simulates M_k on the remaining input. Since M_k performs at most $k-2$ turns on the remaining input and the initialization takes no additional turn, M_{k-1} is in fact a DQA_{k-2}. Further, all words v leading M_k to acceptance belong to L_{k-1}. So, the language accepted by M_{k-1} is L_{k-1}. Now, let M_{k-i} be some DQA_{k-i-1} accepting the language L_{k-i}. We iterate this construction if there exists at least one prefix such that M_{k-i} performs at least one turn on it.

Finally, we distinguish two cases. The first case is that we can iterate the construction until $i = k-1$, that is, we have a DQA_0 accepting L_1. Since 0-turn DQA accept only regular languages and L_1 is not regular, this is a contradiction. The second case is that M_{k-i} does not perform a turn on all prefixes of the form $\#w_1\#h_p(w_1)$. Since M_{k-i} accepts L_{k-i}, it accepts all words from $C_0 \cup C_1$. Simply by counting the number of symbols #, from M_{k-i} one can construct a DQA M' accepting exactly $(C_0 \cup C_1) = L_1$. In particular, M' blocks after reading the third #. So, M' accepts L_1 without any turn, which is again a contradiction. □

Lemma 7 and Lemma 8 prove the following two proper turn hierarchies for general queue automata as well as input-driven queue automata.

Theorem 9. *For all $k \geq 1$, the family of languages accepted by deterministic k-turn (input-driven) queue automata is properly included in the family of languages accepted by $(k+1)$-turn (input-driven) queue automata.*

Though here we do not consider nondeterministic queue automata, it is worth mentioning that L_k is even not accepted by any *nondeterministic* $(k-1)$-turn queue automaton. Thus, we obtain the proper turn hierarchy also for nondeterministic queue automata.

4   Closure Properties

This section is devoted to investigating the closure properties of language families defined by deterministic finite-turn input-driven queue automata. For pushdown automata, strong closure properties have been derived in [1] *provided that* all automata involved share the same partition of the input alphabet into enter, remove, and internal symbols. Here we distinguish this important special case from the general one. For easier writing, we call the partition of an input alphabet a *signature*, and say that two signatures $\Sigma = \Sigma_e \cup \Sigma_r \cup \Sigma_i$ and $\Sigma' = \Sigma'_e \cup \Sigma'_r \cup \Sigma'_i$ are *compatible* if $\bigcup_{j \in \{e,r,i\}}(\Sigma_j \setminus \Sigma'_j) \cap \Sigma' = \emptyset$ and $\bigcup_{j \in \{e,r,i\}}(\Sigma'_j \setminus \Sigma_j) \cap \Sigma = \emptyset$.

Lemma 10. *Let $k \geq 0$ be a constant. The language family $\mathscr{L}(DVQA_k)$ is closed under intersection with regular languages. Moreover, it is closed under union and intersection if the signatures are compatible.*

On the contrary, due to turn restrictions, $DVQA_k$ are not closed under complementation despite their determinism:

Lemma 11. *Let $k \geq 1$ be a constant. The language family $\mathscr{L}(DVQA_k)$ is not closed under complementation.*

Proof. The language L_k of the previous section witnesses the assertion. Lemma 7 shows that L_k is accepted by some $DVQA_k$. In contrast to the assertion, assume there is a $DVQA_k$ M accepting $\overline{L_k}$. We consider the words of the form $\#a^m\#a'^n\#$ that belong to the complement, that is, with $m \neq n$. For m large enough, M will run into a cycle, say of length $p \geq 1$, while processing a^m. Clearly, M has to accept $\#a^m\#a'^{m+p}\#$. If $a \in \Sigma_r \cup \Sigma_i$, then running through the cycle leaves the queue unaffected. So one more cycle cannot lead to rejection, which implies that $\#a^{m+p}\#a'^{m+p}\#$ is accepted as well, though it belongs to L_k. So, we must conclude that $a \in \Sigma_e$. If $a' \in \Sigma_e \cup \Sigma_i$, we argue similarly: While processing a'^m, M runs through cycles in which the queue is either unaffected or its length grows with m. In any case, the final step processing the sole $\#$ at the end cannot distinguish the number of cycles passed through. Therefore, the acceptance of $\#a^{m+p}\#a'^m\#$ implies the acceptance of $\#a^{m+p}\#a'^{m+p}\#$. So, we must have $a' \in \Sigma_r$.

Finally, we consider the input $(aa')^{k+1} \in \overline{L_k}$ that must be accepted by M. However, since $a \in \Sigma_e$ and $a' \in \Sigma_r$, we have that M has to perform one turn on each aa' pair. Thus, it performs at least $k+1$ turns, a contradiction. □

The proof of the previous lemma implies the following stronger result:

Corollary 12. *Let $k \geq 1$ be a constant. The complement of the language L_k is not accepted by any k'-turn DVQA, for any constant $k' \geq 0$.*

Another non-closure result for finite-turn devices is contained in

Lemma 13. *Let $k \geq 1$ be a constant. The language family $\mathscr{L}(DVQA_k)$ is not closed under union.*

Proof. Let us consider the languages $L_1 = \{\, a^m b^m c^n \# \mid m, n \geq 0 \,\}$ and $L_2 = \{\, a^m b^n c^n \# \mid m, n \geq 0 \,\}$. Clearly, we have that $L_1, L_2 \in \mathscr{L}(DVQA_1)$. On the other hand, we have that $L_1 \cup L_2 \notin \mathscr{L}(DVQA)$. To prove this, assume by contradiction that $L_1 \cup L_2$ is accepted by a DVQA A. For such A, suppose that $b \in \Sigma_e \cup \Sigma_i$. Then, by a pigeonhole argument on the accepting computation on $a^m b^m \#$ for sufficiently large m, we would get that A should accept all the strings of the form $a^m b^{m+kp}\#$, for some $p \geq 1$ and any $k \geq 0$. Clearly, this cannot happen, so we must have $b \in \Sigma_r$. On the other hand, a similar reasoning on an input of the form $b^m c^m \#$ would lead to $b \in \Sigma_e$ as well, which clearly contradicts the fact that A is input-driven. □

In contrast to the finite-turn case in Lemma 11, the general model of DVQA turns out to be closed under complementation, as well as under union and intersection for compatible signatures, but not closed under union and intersection for incompatible signatures:

Lemma 14. *The language family $\mathscr{L}(DVQA)$ is closed under intersection with regular languages and complementation. Moreover, it is closed under union and intersection if the signatures are compatible, and not closed under union and intersection if the signatures are incompatible.*

5   Decidability Problems

We recall (see, e.g., [13]) that a decidability problem is *semidecidable* (resp., *decidable*) if and only if the set of all instances for which the answer is "yes" is recursively enumerable (resp., recursive).

Let us begin by showing some decidability results for finite-turn devices.

Theorem 15. *Let $k \geq 0$ be a constant and M be a DQA_k. Then, the emptiness and finiteness of M is decidable.*

Proof. As in the discussion after Theorem 5, by applying Lemma 4 and, repeatedly, Theorem 5 to a given DQA_k M, we obtain a context-free language that is letter equivalent to $L(M)$. Since emptiness and finiteness can be decided for context-free languages [13], both questions can be decided for DQA_k as well.   □

Theorem 16. *Let $k \geq 0$ be a constant and M be a DQA_k. Then, the equivalence with regular sets and, in particular, universality is decidable for M.*

Proof. Let R be a regular set. Then, testing $L(M) = R$ is equivalent to test $L(M) \cap \overline{R} = \emptyset$ and $R \cap \overline{L(M)} = \emptyset$. Since R is regular, $\overline{R}$ is regular as well. By reasoning as in Lemma 10, we have that $\mathscr{L}(DQA_k)$ is closed under intersection with regular sets. Hence, $L(M) \cap \overline{R}$ is accepted by some DQA_k whose emptiness can be tested by Theorem 15. For testing $R \cap \overline{L(M)} = \emptyset$, we first construct a DQA_k M' accepting $\overline{L(M)}$. M' simulates M while counting in an additional component the number of turns executed so far. M' accepts an input as soon as it would require more than k turns. Additionally, M' accepts if the input is rejected by M performing a number of turns which is less than or equal to k. All other inputs are rejected. Clearly, M' is a DQA_k accepting $\overline{L(M)}$. Due to the closure under intersection with regular sets, we obtain again that $R \cap \overline{L(M)}$ is accepted by some DQA_k whose emptiness can again be tested by Theorem 15. We get equivalence if and only if both tests are positive.   □

Theorem 17. *Let $k \geq 0$ be a constant and M, M' be $DVQA_k$ with compatible signatures. Then, inclusion and equivalence of M and M' is decidable.*

Proof. We first show the decidability of the inclusion $L(M) \subseteq L(M')$. The decidability of the inclusion $L(M') \subseteq L(M)$ can be shown analogously and implies the decidability of equivalence. The inclusion $L(M) \subseteq L(M')$ is equivalent to $L(M) \cap \overline{L(M')} = \emptyset$. Since $\mathscr{L}(DVQA_k)$ is not closed under complementation due to Lemma 11 but $\mathscr{L}(DVQA)$ is due to Lemma 14, we obtain that $\overline{L(M')}$ is accepted by some DVQA M'' possibly performing more than k turns but having the same signature as M'. Since $\mathscr{L}(DVQA)$ is closed under intersection with compatible signatures by Lemma 14, we obtain a DVQA N which accepts $L(M) \cap \overline{L(M')}$ by simulating M with at most k turns in its first component and simulating M'' in its second component. Since M' and M'' have compatible signatures, we can observe the following. If M' (resp., M'') accepts or rejects an input with at most k turns, then M'' (resp., M') performs at most k turns as well. Any computation requiring more than k turns is rejected by M and hence by N. Thus, N is a $DVQA_k$ and its emptiness can be tested by Theorem 15. We conclude that the inclusion $L(M) \subseteq L(M')$ is decidable. $\qquad\qquad\qquad\square$

We now turn to undecidable problems. First, the undecidability of the emptiness problem for DVQA is shown by reducing from the emptiness problem for deterministic linearly space bounded one-tape, one-head Turing machines, so-called linear bounded automata (LBA). It is well known that emptiness for LBA is not semidecidable. See, e.g., [13] where also the notion of *valid computations* is introduced. These are, basically, histories of LBA computations which are encoded into single words. We may assume that LBA get their inputs in between two endmarkers, can halt only after an odd number of moves, accept by halting, and make at least three moves.

Let Q be the state set of some LBA M, where q_0 is the initial state, $T \cap Q = \emptyset$ is the tape alphabet, and $\Sigma \subset T$ is the input alphabet. Then a configuration of M can be written as a string of the form T^*QT^* such that $t_1 t_2 \cdots t_i q t_{i+1} \cdots t_n$ is used to express that M is in the state q, scanning tape symbol t_{i+1}, and the string $t_1 t_2 \cdots t_n \in T^*$ is the tape inscription.

Let T' and T'' be copies of T and Q' and Q'' be copies of Q. Furthermore, let $S = T \cup Q \cup \{\triangleright, \triangleleft\}$, $S' = T' \cup Q' \cup \{\overline{\triangleright}, \overline{\triangleleft}\}$, and $S'' = T'' \cup Q'' \cup \{\overline{\triangleright}, \overline{\triangleleft}\}$. Then, consider two mappings $f_1 : S \to S \cdot S'$ and $f_2 : S \to S \cdot S''$ such that $f_1(a) = a\bar{a}$ and $f_2(a) = a\bar{\bar{a}}$, for all $a \in S$. The set VALC(M) of valid computations is now defined to be the set of words of the form $f_2(w_0)\$f_1(w_1)\$f_1(w_2)\$ \cdots \$f_1(w_{2m+1})\$$, where $\$ \notin S \cup S' \cup S''$, $w_i \in T^*QT^*$ are configurations of M, w_0 is an initial configuration of the form $q_0\Sigma^*$, w_{2m+1} is a halting (hence accepting) configuration, and w_{i+1} is the successor configuration of w_i.

Lemma 18. *Let M be an LBA. Then a DVQA accepting VALC(M) can effectively be constructed.*

At this point, we are ready to show non-semidecidability results.

Theorem 19. *Emptiness, finiteness, infiniteness, universality, inclusion, equivalence, regularity, and context-freeness are not semidecidable for DVQA.*

Proof. Let us show exemplarily that emptiness is not semidecidable. Let M be an LBA. According to Lemma 18, we can effectively construct a DVQA M' accepting VALC(M). Clearly, $L(M') = \text{VALC}(M)$ is empty if and only if $L(M)$ is empty. Since emptiness is not semidecidable for LBA, we obtain our claim. $\qquad\square$

Next, we want to show that inclusion is not semidecidable for two $DVQA_k$ with incompatible signatures. To this end, we consider the following variant of the set of valid computations. Let $\hat{S} = \hat{T} \cup \hat{Q} \cup \{\hat{\triangleright}, \hat{\triangleleft}\}$. We consider mappings

$g_1 : S \to S'$, $g_2 : S \to S''$, and $g_3 : S \to \hat{S}$ such that $g_1(a) = \overline{a}$, $g_2(a) = \overline{\overline{a}}$, and $g_3(a) = \hat{a}$, for all $a \in S$. The modified set $\mathrm{VALC}'(M)$ of valid computations is then defined to be the set of words of the form

$$g_2(w_0)\$w_2\$ \cdots \$w_{2m}\#g_1(w_1)\$g_1(w_3)\$ \cdots \$g_1(w_{2m-1})\$g_3(w_{2m+1})\$,$$

where $\$, \# \notin S \cup S' \cup S'' \cup \hat{S}$, $w_i \in T^*QT^*$ are configurations of M, w_0 is an initial configuration of the form $q_0 \Sigma^*$, w_{2m+1} is an accepting configuration, and w_{i+1} is the successor configuration of w_i.

Lemma 20. *Let M be an LBA. Then, $DVQA_1$ M_1 and M_2 can effectively be constructed such that $VALC'(M) = L(M_1) \cap L(M_2)$.*

From Lemma 20, we obtain

Theorem 21. *Let $k \geq 1$ be a constant and M, M' be two $DVQA_k$ with incompatible signatures. Then, the inclusion $L(M) \subseteq L(M')$ is not semidecidable.*

Proof. Consider the DVQA_1 M_1, M_2 from Lemma 20 such that $L(M_1) \cap L(M_2) = \mathrm{VALC}'(M)$ for some LBA M. Let R be the regular language of all words correctly formatted with respect to $\mathrm{VALC}'(M)$, and let M' be a DVQA_1 accepting all words in R which do not belong to $L(M_2)$. To M', the same construction as in the proof of Lemma 20 can be applied. The only difference is that we have to accept only if an error in some configuration is encountered. The correct formatting of the input can be checked in the state set. Obviously, M_1 and M' have incompatible signatures.

Let us now assume, by contradiction, that inclusion is semidecidable. Then, we know that the inclusion $L(M_1) \subseteq L(M') = \overline{L(M_2)} \cap R$ is semidecidable. This latter inclusion holds if and only if $L(M_1) \subseteq (\overline{L(M_2)} \cap R) \cup \overline{R}$, since every $w \in L(M_1)$ is correctly formatted and, hence, is not in $\overline{R}$. Thus, we know that $L(M_1) \subseteq \overline{L(M_2)}$ is semidecidable, which is equivalent to semidecide $L(M_1) \cap L(M_2) = \emptyset$. This implies that the emptiness of $\mathrm{VALC}'(M)$, and hence of the LBA M, is semidecidable. This is a contradiction. □

Another consequence of Lemma 20 is

Lemma 22. *Let $k \geq 1$ be a constant. The language family $\mathscr{L}(DVQA_k)$ is not closed under intersection.*

Proof. Consider the languages $L(M_1)$ and $L(M_2)$ from Lemma 20 for a given LBA M. If $\mathscr{L}(\mathrm{DVQA}_k)$ was closed under intersection, then $L(M_1) \cap L(M_2) = \mathrm{VALC}'(M)$ could be accepted by some DVQA_k. This together with Theorem 15 would lead to decidability of LBA emptiness, a contradiction. □

We conclude this section with the result that there is no algorithm which either tests whether a given DVQA is finite-turn or tests whether a given DVQA is k-turn for some fixed $k \geq 0$.

Theorem 23. *Let M be a DVQA. It is not semidecidable whether M is finite-turn. It is undecidable whether M is k-turn, for some fixed $k \geq 0$.*

Acknowledgements. The authors wish to thank the anonymous referees for their comments.

References

1. Alur, R., Madhusudan, P.: Visibly pushdown languages. In: STOC 2004, pp. 202–211. ACM (2004)
2. Alur, R., Madhusudan, P.: Adding nesting structure to words. J. ACM 56 (2009)
3. Bensch, S., Holzer, M., Kutrib, M., Malcher, A.: Input-driven stack automata. In: Baeten, J.C.M., Ball, T., de Boer, F.S. (eds.) TCS 2012. LNCS, vol. 7604, pp. 28–42. Springer, Heidelberg (2012)
4. von Braunmühl, B., Verbeek, R.: Input-driven languages are recognized in $\log n$ space. In: Karpinski, M. (ed.) FCT 1983. LNCS, vol. 158, pp. 40–51. Springer, Heidelberg (1983)
5. Cherubini, A., Citrini, C., Crespi-Reghizzi, S., Mandrioli, D.: QRT FIFO automata, breadth-first grammars and their relations. Th. Comp. Sci. 85, 171–203 (1991)
6. Chomsky, N.: Formal properties of grammars. In: Handbook of Mathematical Psychology, vol. 2, pp. 323–418. Wiley, New York (1963)
7. Dymond, P.W.: Input-driven languages are in $\log n$ depth. Inform. Process. Lett. 26, 247–250 (1988)
8. Ginsburg, S., Greibach, S.A., Harrison, M.A.: One-way stack automata. J. ACM 14(2), 389–418 (1967)
9. Ginsburg, S., Spanier, E.H.: Finite-turn pushdown automata. SIAM J. Comput. 4(3), 429–453 (1966)
10. Greibach, S.A.: An infinite hierarchy of context-free languages. J. ACM 16, 91–106 (1969)
11. Holzer, M., Kutrib, M.: Flip-pushdown automata: $k + 1$ pushdown reversals are better than k. In: Baeten, J.C.M., Lenstra, J.K., Parrow, J., Woeginger, G.J. (eds.) ICALP 2003. LNCS, vol. 2719, pp. 490–501. Springer, Heidelberg (2003)
12. Holzer, M., Kutrib, M.: Flip-pushdown automata: Nondeterminism is better than determinism. In: Ésik, Z., Fülöp, Z. (eds.) DLT 2003. LNCS, vol. 2710, pp. 361–372. Springer, Heidelberg (2003)
13. Hopcroft, J.E., Ullman, J.D.: Introduction to Automata Theory, Languages, and Computation. Addison-Wesley, Reading (1979)
14. Madhusudan, P., Parlato, G.: The tree width of auxiliary storage. In: Principles of Programming Languages (POPL 2011), pp. 283–294. ACM (2011)
15. Malcher, M., Mereghetti, C., Palano, B.: Descriptional complexity of two-way pushdown automata with restricted head reversals. Th. Comp. Sci. 449, 119–133 (2012)
16. Mehlhorn, K.: Pebbling mountain ranges and its application of DCFL-recongnition. In: de Bakker, J.W., van Leeuwen, J. (eds.) ICALP 1980. LNCS, vol. 85, pp. 422–435. Springer, Heidelberg (1980)
17. Torre, S.L., Madhusudan, P., Parlato, G.: A robust class of context-sensitive languages. In: Logic Comp. Sci. (LICS 2007), pp. 161–170. IEEE Comp. Soc. (2007)

Hyper-optimization
for Deterministic Tree Automata

Andreas Maletti*

Institute for Natural Language Processing
Universität Stuttgart, Pfaffenwaldring 5b, 70569 Stuttgart, Germany
andreas.maletti@ims.uni-stuttgart.de

Abstract. A recent minimization technique, called hyper-minimization, permits reductions of language representations beyond the limits imposed by classical semantics-preserving minimization. Naturally, the semantics is not preserved by hyper-minimization; rather the reduced representation, which is called hyper-minimal, can accept a language that has a finite symmetric difference to the language of the original representation. It was demonstrated that hyper-minimization for (bottom-up) deterministic tree automata (DTAs), which represent the recognizable tree languages, can be achieved in time $\mathcal{O}(m \cdot \log n)$, where m is the size of the DTA and n is the number of its states. In this contribution, this result is complemented by two results on the quantity of the errors. It is shown that optimal hyper-minimization for DTAs (i.e., computing a hyper-minimal DTA that commits the least number of errors of all hyper-minimal DTAs) can be achieved in time $\mathcal{O}(m \cdot n)$. In the same time bound also the number of errors of any hyper-minimal DTA can be computed.

1 Introduction

In many application areas, large finite-state models are approximated automatically from data. Classical examples in the area of natural language processing include the estimation of tree automata [6,7] for parsing [18] and weighted finite-state automata [19] for speech recognition [16]. To keep the size of those models under control, minimization is used whenever possible and efficient. Unfortunately, computing an equivalent minimal nondeterministic (unweighted) finite-state automaton [21] is PSPACE-complete [8] and thus inefficient; this remains true even if the input automaton is deterministic. However, given a deterministic finite-state automaton (DFA) the computation of an equivalent minimal DFA is very efficient [12]. Consequently, we restrict our focus to deterministic finite-state devices. Exactly, the same situation exhibits itself for tree automata [17,3], which are the finite-state models used in this contribution. We note that (bottom-up) deterministic tree automata are as expressive as (nondeterministic) tree automata (albeit the deterministic device might require exponentially more states as in the string case), which recognize exactly the regular tree languages.

* The author was financially supported by the German Research Foundation (DFG) grant MA 4959 / 1-1.

S. Konstantinidis (Ed.): CIAA 2013, LNCS 7982, pp. 244–255, 2013.

In several applications it is beneficial to reduce the size even further at the expense of errors. In hyper-minimization [2] we simply allow any finite number of errors; i.e., the obtained representation might recognize a language that has a finite symmetric difference to the language recognized by the original representation. While this error profile is rather simplistic, it allows a convenient theoretical treatment [2], efficient minimization algorithms [1,4,11,14], and sometimes finitely many errors are even absolutely inconsequential [20]. Moreover, more refined error profiles often yield NP-hard minimization problems [5] and thus inefficient minimization procedures. Recently, an efficient hyper-minimization algorithm [13] for (bottom-up) deterministic tree automata (DTAs) was developed. It runs in time $\mathcal{O}(m \cdot \log n)$, where m is the size of the input DTA and n is the number of its states. Thus, it is asymptotically as efficient as the fastest classical minimization algorithms [9] for DTAs.

The existing hyper-minimization algorithm for DTAs is purely qualitative in the sense that it guarantees that the resulting hyper-minimal DTA (a DTA M is hyper-minimal if there exists no DTA with fewer states[1] that recognizes a tree language with a finite difference to the tree language recognized by M) commits only finitely many errors, but provides no (non-trivial) bound on this number of errors. Since there are (in general) many (non-isomorphic) hyper-minimal DTAs for a given tree language, returning simply any hyper-minimal DTA is short-sighted. In this contribution, we perform a more quantitative analysis in the spirit of [15]. We develop a hyper-minimization algorithm that returns a hyper-minimal DTA (i.e., it has as many states as the DTA returned by the existing algorithm of [13]) that commits the least number of errors among all hyper-minimal DTAs. To this end, we first characterize all hyper-minimal DTAs for a given tree language. For DFAs the structural differences between hyper-minimal DFAs for the same language were characterized in [2, Thms. 3.8 and 3.9]. Despite the additional complications encountered in DTAs hyper-minimization, we faithfully generalize the results for DFAs to DTAs. Thus, any two hyper-minimal DTAs for a given tree language permit a bijection between their states such that the distinction into preamble (i.e., those states that can only be reached by finitely many trees) and non-preamble (or kernel) states is preserved. Moreover, the DTAs behave equivalently on their preambles except for their acceptance decisions and isomorphically on their kernels. Finally, the strange condition on the initial state in [2, Thms. 3.8 and 3.9] disappears completely for DTAs.

With the help of this characterization we can now easily compare different hyper-minimal DTAs provided that we can compute the number of errors that they commit. Thus, we derive a method to compute the number of errors caused by each relevant decision (finality decision for preamble states and transition targets for transitions from the preamble into the kernel). For DFAs the same approach was used in [15], but our method is slightly more complicated because we have to avoid counting errors several times (because an error tree can contain multiple positions at which a switch from preamble to kernel states happens

---

[1] Since we consider only deterministic devices, we can as well use the number of transitions as a size measure.

when processing it by the DTA). We solve this problem by attributing the error
tree to the left-most such transition. It turns out that this change can easily
be incorporated, so that our approach closely resembles the approach of [15].
Overall, we obtain an algorithm that, given a DTA M and a hyper-minimal
DTA N that recognizes a finitely different tree language, can compute the number
of errors committed by N in time $\mathcal{O}(m \cdot n)$, where m is the size of M and n is
the number of states of M. In addition, we can also compute an optimal hyper-
minimal DTA N' in time $\mathcal{O}(m \cdot n)$, which is a hyper-minimal DTA that commits the
least number of errors among all hyper-minimal DTAs that recognize a finitely
different tree language. Of course, we can also compute the exact number of
errors committed by this optimal DTA.

2 Preliminaries

The set $\mathbb{N}$ consists of all nonnegative integers and $[k] = \{i \in \mathbb{N} \mid 1 \leq i \leq k\}$ for
all $k \in \mathbb{N}$. The symmetric difference $S \ominus T$ of sets S and T is $(S-T) \cup (T-S)$. A
binary relation $\cong \subseteq S \times S$ is an equivalence relation if it is reflexive, symmetric,
and transitive. Given such an equivalence relation $\cong$, the equivalence class $[s]$ of
$s \in S$ is $\{s' \in S \mid s \cong s'\}$. If S is finite, then we write $|S|$ for its cardinality.

An alphabet Σ is a finite set, and a ranked alphabet (Σ, rk) consists of an
alphabet Σ and a mapping $\text{rk}: \Sigma \to \mathbb{N}$ that assigns a rank to each symbol of Σ.
The set of all symbols of rank $k \in \mathbb{N}$ is $\Sigma_k = \text{rk}^{-1}(k)$. We typically denote (Σ, rk)
by just Σ, and we let $\Sigma(T) = \{\sigma(t_1, \ldots, t_k) \mid \sigma \in \Sigma_k, t_1, \ldots, t_k \in T\}$ for every
set T. The set $T_\Sigma(Q)$ of Σ-trees with states Q is the smallest set T such that
$Q \cup \Sigma(T) \subseteq T$. We write T_Σ for $T_\Sigma(\emptyset)$. The mapping height $\text{ht}(t): T_\Sigma(Q) \to \mathbb{N}$ is
defined by $\text{ht}(q) = 0$ for all $q \in Q$ and $\text{ht}(\sigma(t_1, \ldots, t_k)) = 1 + \max\{\text{ht}(t_i) \mid i \in [k]\}$
for all $\sigma \in \Sigma_k$ and $t_1, \ldots, t_k \in T_\Sigma(Q)$. The subset $C_\Sigma(Q) \subseteq T_{\Sigma \cup \{\square\}}(Q)$ of
contexts contains all trees in which the special nullary symbol $\square$ occurs exactly
once. Again, we write C_Σ for $C_\Sigma(\emptyset)$. For all $c \in C_\Sigma(Q)$ and $t \in T_{\Sigma \cup \{\square\}}(Q)$, we
write $c[t]$ for the tree obtained from c by replacing $\square$ by t. The tree t is a subtree
of $c[t]$ for all contexts $c \in C_\Sigma(Q)$.

A (total bottom-up) deterministic (finite-state) tree automaton (DTA) [6,7]
is a tuple $M = (Q, \Sigma, \delta, F)$ where Q is the finite, nonempty set of states, Σ is
the ranked alphabet of input symbols, $\delta: \Sigma(Q) \to Q$ is the transition mapping,
and $F \subseteq Q$ is the set of final states. The transition mapping δ extends to
$\delta: T_\Sigma(Q) \to Q$ by $\delta(q) = q$ for all $q \in Q$ and

$$\delta(\sigma(t_1, \ldots, t_k)) = \delta(\sigma(\delta(t_1), \ldots, \delta(t_k)))$$

for all $\sigma \in \Sigma_k$ and $t_1, \ldots, t_k \in T_\Sigma(Q)$. We let $L(M)^q_{q'} = \{c \in C_\Sigma \mid \delta(c[q']) = q\}$
for all $q, q' \in Q$. Moreover, $L(M)_{q'} = \bigcup_{q \in F} L(M)^q_{q'}$ contains all (stateless) contexts that take q' into a final state, and $L(M)^q = \delta^{-1}(q) \cap T_\Sigma$ contains all (stateless) trees that are recognized in the state q. A state $q \in Q$ is a kernel (resp.,
preamble) state [2] if $L(M)^q$ is infinite (resp., finite). The sets Ker_M and Pre_M
contain all kernel and preamble states, respectively. The DTA M recognizes the

tree language $L(M) = \bigcup_{q \in F} L(M)^q$, and all DTAs that recognize the same tree language are equivalent. A DTA is minimal if there exists no equivalent DTA with strictly fewer states. We can compute a minimal DTA that is equivalent to M using an adaptation [9] of HOPCROFT's algorithm [12], which runs in time $\mathcal{O}(|M| \cdot \log n)$ where $|M| = \sum_{k \in \mathbb{N}} k \cdot |\Sigma_k| \cdot n^k$ is the size of M and $n = |Q|$.

3   Structural Characterization of Hyper-minimal DTAs

In this section we develop a characterization that points out the differences between different *hyper-minimal* DTAs for the same tree language. It will tell us which alternatives to consider when we search for an optimal hyper-minimal DTA, which commits the least number of errors. However, before we start, we recall the basic notions (e.g., *almost equivalence* and *hyper-minimality*).

Throughout the paper, let $M = (Q, \Sigma, \delta, F)$ and $N = (P, \Sigma, \mu, G)$ be minimal DTAs. Since we ultimately want to compare two DTAs, we introduce all basic notions for M and N. However, we often use them in the particular case that $M = N$. Two states $q \in Q$ and $p \in P$ are *almost equivalent* [13, Def. 1], written $q \sim p$ or $p \sim q$, if $E_{q,p} = L(M)_q \ominus L(N)_p$ is finite. We also say that q and p *disagree* on each element of $E_{q,p}$. If $E_{q,p} = \emptyset$, then q and p are *equivalent*, which is written $q \equiv p$ or $p \equiv q$. It is well-known [3, Sect. 1.5] that minimal DTAs do not have different, but equivalent states. Correspondingly, the DTAs M and N are *almost equivalent*, also written $M \sim N$, if $E = L(M) \ominus L(N)$ is finite.

Lemma 1 (see [13, Lm. 4]). *If $M \sim N$, then $\delta(t) \sim \mu(t)$ and $\delta(t') \equiv \mu(t')$ if $\mathrm{ht}(t') > |Q \times P|$ for all $t, t' \in T_\Sigma$.[2]*

Proof. The property $\delta(t) \sim \mu(t)$ is proven in [13, Lm. 4]. For the other property, we consider the product DTA $M \times N = (Q \times P, \Sigma, \delta \times \mu, F \times G)$, where

$$(\delta \times \mu)\big(\sigma(\langle q_1, p_1 \rangle, \ldots, \langle q_k, p_k \rangle)\big) = \big\langle \delta(\sigma(q_1, \ldots, q_k)), \mu(\sigma(p_1, \ldots, p_k)) \big\rangle$$

for all $\sigma \in \Sigma_k$ and $\langle q_1, p_1 \rangle, \ldots, \langle q_k, p_k \rangle \in Q \times P$. Clearly, $(\delta \times \mu)(t) = \langle \delta(t), \mu(t) \rangle$ for all $t \in T_\Sigma$. If $\mathrm{ht}(t) > |Q \times P|$, then $(\delta \times \mu)(t)$ is a kernel state of $M \times N$ because the tree t can be pumped [6,7]. For the sake of a contradiction, suppose that $\delta(t) \not\equiv \mu(t)$; i.e., there exists $c \in E_{\delta(t), \mu(t)}$. Since $\langle \delta(t), \mu(t) \rangle$ is a kernel state of $M \times N$, there exist infinitely many $u \in T_\Sigma$ such that $\langle \delta(u), \mu(u) \rangle = \langle \delta(t), \mu(t) \rangle$. However, for each such tree u we have $c[u] \in E$, which contradicts $M \sim N$.   □

The previous lemma shows that almost equivalent DTAs are in almost equivalent states after processing the same (stateless) tree. If the tree is tall, then they are even in equivalent states. Before we proceed with the comparison of almost equivalent DTAs, we recall another notion and a related result. A DTA is *hyper-minimal* if all almost equivalent DTAs have at least as many states.

Theorem 2 ([13, Thm. 7]). *A minimal DTA is hyper-minimal if and only if all pairs of different, but almost equivalent states consist only of kernel states.*

---

[2] If $M = N$, then $\mathrm{ht}(t') > |Q|$ is actually sufficient.

Now we can investigate how almost equivalent hyper-minimal DTAs differ. We extend each mapping $h\colon Q \to P$ to a mapping $h\colon T_\Sigma(Q) \to T_\Sigma(P)$ by $h(\sigma(t_1,\ldots,t_k)) = \sigma(h(t_1),\ldots,h(t_k))$ for every $\sigma \in \Sigma_k$ and $t_1,\ldots,t_k \in T_\Sigma(Q)$. Such a mapping $h\colon Q \to P$ is a *transition homomorphism*[3] if $h(\delta(s)) = \mu(h(s))$ for every $s \in \Sigma(Q)$. Moreover, h is a DTA *homomorphism* if additionally $h(q) \in G$ if and only if $q \in F$. As usual, a bijective homomorphism is called *isomorphism*. Next, we show that two almost equivalent hyper-minimal DTAs have DTA-isomorphic kernels and transition-isomorphic preambles.

Theorem 3. *If $M \sim N$ and both M and N are hyper-minimal, then there exists a bijection $h\colon Q \to P$ such that*

1. *h is bijective on $\mathrm{Ker}_M \times \mathrm{Ker}_N$,*
2. *$h(q) \in G$ if and only if $q \in F$ for all $q \in \mathrm{Ker}_M$, and*
3. *$h(\delta(s)) = \mu(h(s))$ for every $s \in \Sigma(Q) - \{s \in \Sigma(\mathrm{Pre}_M) \mid \delta(s) \in \mathrm{Ker}_M\}$.*

Proof. Clearly, $|Q| = |P|$ since M and N are both hyper-minimal. For every $q \in Q$ select $t_q \in L(M)^q$ such that $\mathrm{ht}(t_q) > |Q \times P|$ whenever $q \in \mathrm{Ker}_M$.[4] We let $h\colon Q \to P$ be such that $h(q) = \mu(t_q)$ for every $q \in Q$, which immediately proves that $h(q) \in \mathrm{Ker}_N$ for all $q \in \mathrm{Ker}_M$ because $\mathrm{ht}(t_q) > |Q \times P|$. Moreover, for each $q \in \mathrm{Ker}_M$ the facts $M \sim N$ and $\mathrm{ht}(t_q) > |Q \times P|$ imply $q = \delta(t_q) \equiv \mu(t_q) = h(q)$ by Lemma 1. Thus, h is injective on $\mathrm{Ker}_M \times \mathrm{Ker}_N$ because $h(q_1) \equiv q_1 \not\equiv q_2 \equiv h(q_2)$ for all different $q_1, q_2 \in \mathrm{Ker}_M$.[5] Finally, for every $p \in \mathrm{Ker}_N$, select $u_p \in L(N)^p$ such that $\mathrm{ht}(u_p) > |Q \times P|$. Clearly, $\delta(u_p) \in \mathrm{Ker}_M$ and by Lemma 1 we obtain $p = \mu(u_p) \equiv \delta(u_p) \equiv h(\delta(u_p))$. Since N is minimal, we can conclude that $p = h(\delta(u_p))$, which shows that h is surjective on $\mathrm{Ker}_M \times \mathrm{Ker}_N$, thereby proving the first item.

Recall from the previous paragraph that $q \equiv h(q)$ for every $q \in \mathrm{Ker}_M$. Thus, $h(q) \in G$ if and only if $q \in F$, which proves the second item. For the third objective, let $s = \sigma(q_1,\ldots,q_k) \in \Sigma(Q)$. Then

$$\delta(s) = \delta(\sigma(t_{q_1},\ldots,t_{q_k})) \overset{\dagger}{\sim} \mu(\sigma(t_{q_1},\ldots,t_{q_k})) = \mu(\sigma(\mu(t_{q_1}),\ldots,\mu(t_{q_k})))$$
$$= \mu(\sigma(h(q_1),\ldots,h(q_k))) = \mu(h(s)) \ ,$$

where the step marked $\dagger$ is due to Lemma 1. In the following, assume that $s \notin \Sigma(\mathrm{Pre}_M)$, which yields that there exists $i \in [k]$ such that $q_i \in \mathrm{Ker}_M$. Consequently, $\mathrm{ht}(\sigma(t_{q_1},\ldots,t_{q_k})) > |Q \times P|$ by the selection of t_{q_i}, which can be used in Lemma 1 to show that the step marked $\dagger$ is actually equivalence ($\equiv$). We obtain that $\delta(s) \equiv \mu(h(s))$ and $\delta(s) \in \mathrm{Ker}_M$. Thus, $h(\delta(s)) \equiv \delta(s) \equiv \mu(h(s))$ by the argument in the previous paragraph. Since N is minimal, we can conclude that $h(\delta(s)) = \mu(h(s))$ as desired.

Before we prove the missing case, in which $s \in \Sigma(\mathrm{Pre}_M)$ with $\delta(s) \in \mathrm{Pre}_M$, we prove that h is bijective on $\mathrm{Pre}_M \times \mathrm{Pre}_N$, which automatically also proves

---

[3] Or a homomorphism between the Σ-algebras [6,7] associated with M and N.

[4] Such trees exist because each state is reachable (by hyper-minimality) and $L(M)^q$ is infinite for each kernel state q.

[5] We have $q_1 \not\equiv q_2$ because M is minimal.

that $h\colon Q \to P$ is a bijection. Since h is bijective on $\mathrm{Ker}_M \times \mathrm{Ker}_N$ by the proven first item, which yields $|\mathrm{Ker}_M| = |\mathrm{Ker}_N|$, and additionally $|Q| = |P|$, we obtain that $|\mathrm{Pre}_M| = |\mathrm{Pre}_N|$. Suppose that $h(q) \in \mathrm{Ker}_N$ for some $q \in \mathrm{Pre}_M$. Then $q \sim h(q) = \mu(u_{h(q)}) \sim \delta(u_{h(q)})$ with $\mathrm{ht}(u_{h(q)}) > |Q \times P|$ because $h(q) \in \mathrm{Ker}_N$, where the almost equivalences are due to Lemma 1. Clearly, $\delta(u_{h(q)})$ is a kernel state of M, which yields that $q \neq \delta(u_{h(q)})$. Together with $q \sim \delta(u_{h(q)})$ and $q \in \mathrm{Pre}_M$, these facts contradict the hyper-minimality of M by Theorem 2. It remains to prove that h is injective on $\mathrm{Pre}_M \times \mathrm{Pre}_N$, which due to $|\mathrm{Pre}_M| = |\mathrm{Pre}_N|$ also proves that h is surjective. For the sake of a contradiction, let $q_1, q_2 \in \mathrm{Pre}_M$ be such that $q_1 \neq q_2$ but $h(q_1) = h(q_2)$. Using Lemma 1 we obtain $q_1 \sim h(q_1) = h(q_2) \sim q_2$, which together with $q_1 \neq q_2$ contradicts the hyper-minimality of M by Theorem 2. Consequently, h is bijective on $\mathrm{Pre}_M \times \mathrm{Pre}_N$.

Now we return to the final missing objective, which requires us to show that $h(\delta(s)) = \mu(h(s))$ if $\delta(s) \in \mathrm{Pre}_M$. Recall that $\delta(s) \sim \mu(h(s))$ for all $s \in \Sigma(Q)$. Moreover, if $\delta(s) \in \mathrm{Pre}(M)$, then $h(\delta(s)) \in \mathrm{Pre}_N$ by the results of the previous paragraph and additionally $h(\delta(s)) \sim \delta(s) \sim \mu(h(s))$ by Lemma 1. Consequently, we have a preamble state $h(\delta(s))$ of N that is almost equivalent to $\mu(h(s))$. Since N is hyper-minimal, we have $h(\delta(s)) = \mu(h(s))$ by Theorem 2. $\qquad\square$

The previous theorem states that two almost equivalent hyper-minimal DTAs are indeed very similar. They have a bijection between their states that preserves the distinction between preamble and kernel states. Moreover, via this bijection the two DTAs behave equally (besides acceptance) on the preamble states and isomorphically on the kernel states. Thus, two such DTAs can only differ in two aspects, which mirror the corresponding aspects for deterministic finite-state string automata [2]:

1. the finality (i.e., whether the state is final or not) of preamble states, and
2. transitions from exclusively preamble states to a kernel state.

4 Computing the Number of Errors

Now we can compute the number of errors made by a particular hyper-minimal DTA N that is almost equivalent to the reference DTA M. In addition, we show how to obtain a hyper-minimal DTA that commits the least number of errors among all almost equivalent hyper-minimal DTAs. More precisely, let N be hyper-minimal and almost equivalent to M, which itself is not necessarily hyper-minimal. Recall that $E = L(M) \ominus L(N)$ is the set of error trees. We partition E into $(E_p)_{p \in P}$, where $E_p = L(N)^p \cap E$ for every $p \in P$. In other words, we associate each error tree $t \in E$ with the state $\mu(t)$. In the following development, we distinguish errors associated to preamble and kernel states. Theorem 3 shows that the preamble-kernel error distinction is stable among all almost equivalent hyper-minimal DTAs.[6] Finally, [13, Sect. 4] shows how to obtain one hyper-minimal DTA N' that is almost equivalent to M. Roughly speaking, we identify

---

[6] An error associated to a preamble state of N can only be associated to a preamble state of another almost equivalent hyper-minimal DTA N'. Naturally, the error can be avoided in N', but the same error cannot be associated to a kernel state of N'.

the almost equivalence $\sim$ on M and then merge each preamble state that is almost equivalent to another state into this state. For every two different states $q, q' \in Q$, the DTA $\text{merge}(M, q \to q')$ is $(Q - \{q\}, \Sigma, \delta', F - \{q\})$ where $\delta'(s) = q'$ if $\delta(s) = q$ and $\delta'(s) = \delta(s)$ otherwise for every $s \in \Sigma(Q - \{q\})$. We start with the errors E_p associated to a preamble state $p \in \text{Pre}_N$. Since the preambles of N and N' are transition-isomorphic by Theorem 3, we can essentially compute with N' and only need to remember that the preamble states of N and N' can differ in finality.

Lemma 4 (see [13]). *Let* $N' = \text{merge}(M, q \to q')$ *for some* $q \sim q'$ *with* $[q] \subseteq \text{Pre}_M$. *Then* $L(N')^{q'} = L(M)^q \cup L(M)^{q'}$. *If* N'' *is the DTA returned by [13] and* $B \in \{[q] \mid q \in Q\}$ *is such that* $B \subseteq \text{Pre}_M$, *Then* $L(N'')^{q_B} = \bigcup_{q \in B} L(M)^q$.

By Theorem 3 there exists a mapping $h \colon P' \to P$ such that N' and N are transition-isomorphic on their preambles via h. Together with Lemma 4 we thus have

$$L(N)^{h(q_B)} = L(N')^{q_B} = \bigcup_{q \in B} L(M)^q$$

for every $B \in \{[q] \mid q \in Q\}$ with $B \subseteq \text{Pre}_M$.[7] Next, we demonstrate how to compute $a_q = |L(M)^q|$ for each state $q \in \text{Pre}_M$.

Proposition 5. *For every* $q \in \text{Pre}_M$

$$a_q = \sum_{\sigma(q_1,\ldots,q_k) \in \delta^{-1}(q) \cap \Sigma(Q)} \left(\prod_{i=1}^{k} a_{q_i} \right) .$$

It is clear that the equations in Proposition 5 yield a recursive algorithm that runs in time $\mathcal{O}(|M|)$, if we do not recompute already computed values. With the help of Lemma 4 and Proposition 5, we can now compute the number of errors made due to the finality of preamble states. For every $q_B \in \text{Pre}_{N'}$, we know that its block $B \subseteq \text{Pre}_M$ consists of exclusively preamble states. Consequently, Lemma 4 can be applied to compute the number of errors associated to q_B.

Theorem 6. *For every* $p \in \text{Pre}_N$,

$$|E_p| = \begin{cases} \sum_{q \in [\delta(u_p)] - F} a_q & \text{if } p \in G \\ \sum_{q \in [\delta(u_p)] \cap F} a_q & \text{otherwise,} \end{cases}$$

where $u_p \in L(N)^p$ *is arbitrary.*

Proof. The result follows from Theorem 3 and Lemma 4. ☐

Since $L(N)^p$ and $L(N)^{p'}$ are disjoint if $p \neq p'$, the total number of errors associated to preamble states is $\sum_{p \in \text{Pre}_N} |E_p|$. To obtain the minimal number of errors, we select the finality of p such that E_p is minimal (see Algorithm 1).[8]

---

[7] The union is actually disjoint as $(L(M)^q)_{q \in Q}$ is a partition of T_Σ.

[8] Clearly, the DTA remains hyper-minimal and almost equivalent as only a finite number of errors is introduced by making p final or non-final.

Finally, we need to compute the number $\sum_{p \in \mathrm{Ker}_N} |E_p|$ of errors associated to kernel states. Recall that N' is the hyper-minimal DTA returned by [13]. Theorem 3 shows that the preambles of N and N' are transition-isomorphic and the kernels are DTA-isomorphic, but the transitions from exclusively preamble to kernel states are not covered in this characterization. As in the string case, we thus try to attribute errors to these preamble-to-kernel transitions because we know what happens before (transition-isomorphic on preamble) and what happens afterwards (DTA-isomorphic on the kernel). However, in the tree case this is complicated by the fact that such transitions can be taken several times in a single error tree as the next example demonstrates.

Example 7. Consider the DTA $M = (Q, \Sigma, \{q_\alpha\}, \delta)$ such that $Q = \{q_\alpha, q_\beta, q_\sigma\}$, $\Sigma = \{\alpha^{(0)}, \beta^{(0)}, \sigma^{(2)}\}$, and

$$\delta(\alpha) = q_\alpha \qquad \delta(\beta) = q_\beta \qquad \delta(\sigma(q_\beta, q_\beta)) = q_\sigma \qquad \delta(\sigma(q, q')) = q_\alpha$$

for all $(q, q') \in Q^2 - \{(q_\beta, q_\beta)\}$. Obviously, $L(M) = T_\Sigma - \{\beta, \sigma(\beta, \beta)\}$. An almost equivalent hyper-minimal DTA is $N = (\{\top\}, \Sigma, \{\top\}, \mu)$, where μ is such that $\mu^{-1}(\top) = \Sigma(\{\top\})$. Since $L(N) = T_\Sigma$, we have that $E = \{\beta, \sigma(\beta, \beta)\}$. However, when N processes the error tree $\sigma(\beta, \beta)$, then it will take two transitions (both times $\delta(\beta) = \top$) that switch from exclusively preamble states (no states in this case as α is nullary) to the kernel state $\top$.

We solve this problem by selecting the left-most occurrence of such a transition and disregarding all other occurrences to avoid counting duplicates. To this end, we first need to introduce positions. Let Δ be a ranked alphabet and $t \in T_\Delta(Q)$. The set $\mathrm{pos}(t) \subseteq \mathbb{N}^*$ of *positions in* t is defined by $\mathrm{pos}(q) = \{\varepsilon\}$ for every $q \in Q$ and $\mathrm{pos}(\sigma(t_1, \ldots, t_k)) = \{\varepsilon\} \cup \{iw \mid i \in [k], w \in \mathrm{pos}(t_i)\}$ for all $\sigma \in \Delta_k$ and $t_1, \ldots, t_k \in T_\Delta(Q)$. For every $w \in \mathrm{pos}(t)$, we write $t|_w$ for the subtree of t that is rooted in position w. A position $w_1 \in \mathrm{pos}(t)$ is *to the left of* another position $w_2 \in \mathrm{pos}(t)$, written $w_1 \sqsubset w_2$, if $w_1 \preceq w_2$ and $w_1 \not\leq w_2$, where $\preceq$ and $\leq$ are the lexicographic and prefix order on $\mathbb{N}^*$, respectively. In a context $c \in C_\Sigma(Q)$ the unique position of $\square$ is denoted by $\mathrm{pos}_\square(c)$. Now we can define the set LC of *left-most contexts*, which have no subtree to the left of (the occurrence of) $\square$ that is recognized (by M) in a state that is almost equivalent to a kernel state, as follows:

$$\mathrm{LC} = \{c \in C_\Sigma \mid \forall w \in \mathrm{pos}(c) \colon w \sqsubset \mathrm{pos}_\square(c) \text{ implies } [\delta(c|_w)] \subseteq \mathrm{Pre}_M\} \ .$$

With the help of the set LC we can now make the error attribution more formal. We already know that each remaining error tree has a special transition that switches from exclusively preamble states to a kernel state. Moreover, we will now prove that every such error tree decomposes uniquely into a context of LC, which encodes the part of the tree that is processed after a special transition, and a tree that uses a special transition at the root. Due to the definition of LC, we know that the decomposition selects exactly the left-most occurrence of a special transition.

.

Lemma 8. *Every error tree $t \in E_p$ with $p \in \mathrm{Ker}_N$ decomposes uniquely via $t = c[u]$ into a left-most context $c \in \mathrm{LC}$ and $u \in T_\Sigma$ such that $\mu(u) \in \mathrm{Ker}_N$, but $\mu(u|_w) \in \mathrm{Pre}_N$ for all $w \in \mathrm{pos}(u)$ with $w \neq \varepsilon$.*

Proof. Since $\mu(t) \in \mathrm{Ker}_N$, there must exist positions $w \in \mathrm{pos}(t)$ such that $\mu(t|_w) \in \mathrm{Ker}_N$ but $\mu(t|_{wv}) \in \mathrm{Pre}_N$ for all $wv \in \mathrm{pos}(t)$ with $v \neq \varepsilon$. Let w be the left-most such position (i.e., a minimal such position with respect to $\sqsubseteq$). It remains to show that $t[\square]_w \in \mathrm{LC}$, where $t[\square]_w$ denotes the context obtained from t by replacing the subtree rooted at w by $\square$. By the selection of w, all positions $v \sqsubset w$ are such that $\mu(t|_v) \in \mathrm{Pre}_N$. Thus, $[\delta(t|_v)] \subseteq \mathrm{Pre}_M$ by Theorem 3 and the earlier discussion, which proves that $t[\square]_w \in \mathrm{LC}$. Thus, we obtain the suitable decomposition $t = c[t|_w]$ with $c = t[\square]_w$. The uniqueness is also easy to show as each other suitable position w' obeys $w \sqsubset w'$ and $\mu(t|_w) \in \mathrm{Ker}_N$, which by Lemma 1 yields that $\delta(t|_w) \sim q$ for some $q \in \mathrm{Ker}_M$. Consequently, $t[\square]_{w'} \notin \mathrm{LC}$ for all other suitable positions $w' \neq w$.   □

The decomposition $t = c[u]$ already hints at the next steps. We can compute $\delta(u)$ and $\mu(u)$, for which we know that $\delta(u) \sim \mu(u)$ by Lemma 1. The error is then made between those two states, so $c \in E_{\delta(u),\mu(u)} = L(M)_{\delta(u)} \ominus L(N)_{\mu(u)}$ is an error context of LC. To make the computation even simpler, we observe that $\mu(u) \in \mathrm{Ker}_N$, which with the help of Theorem 3 yields that there exists a state $q \in Q$ such that $q \equiv \mu(u)$. Consequently, it is sufficient to compute $E_{\delta(u),q}$ for all $\delta(u) \sim q$. In fact, for all $q \sim q'$ we know that $E_{q,q'}$ is finite, but we need the exact cardinality $d(q,q')$ of the subset $E_{q,q'} \cap \mathrm{LC}$. More exactly, for all $q \sim q'$, let $d(q,q') = |E_{q,q'} \cap \mathrm{LC}|$ and

$$C_M = C_\Sigma(Q) \cap \Sigma(Q \cup \{\square\}) \quad \text{and} \quad \overline{C}_M = C_M \cap C_\Sigma(\mathrm{Pre}_M) \ ,$$

of which the elements are called *transition* and *preamble transition contexts*, respectively. To compute d, we adjust the straightforward counting procedure [15].

Lemma 9. *For all $q \sim q'$ we have $d(q,q) = 0$ and*

$$d(q,q') = \left(\sum_{\substack{c \in \overline{C}_M \\ c = \sigma(q_1,\ldots,q_i,\square,q_{i+1},\ldots,q_k) \\ [q_1],\ldots,[q_i] \subseteq \mathrm{Pre}_M}} a_{q_1} \cdot \ldots \cdot a_{q_k} \cdot d\big(\delta(c[q]), \delta(c[q'])\big) \right) + \begin{cases} 1 & \text{if } q \in F \text{ xor } q' \in F \\ 0 & \text{otherwise.} \end{cases}$$

Proof. The first equation is trivial and the second equation straightforwardly formalizes $|E_{q,q'}|$, but only counts the error contexts of LC. More precisely, the final summand checks whether $\square \in \mathrm{LC}$ is in the difference $E_{q,q'}$. Every other difference context $c'' = c'[\bar{c}] \in E_{q,q'}$ consists of (i) a context $\bar{c}$ obtained from a transition context $c = \sigma(q_1,\ldots,q_i,\square,q_{i+1},\ldots,q_k)$ of $\overline{C}_M$ by replacing the states $q_1,\ldots,q_k \in Q$ by $t_1 \in L(M)^{q_1},\ldots,t_k \in L(M)^{q_k}$, respectively, which yields the factors $a_{q_1},\ldots,a_{q_k}$, and (ii) an error context c' for the states $\delta(c[q])$ and $\delta(c[q'])$, which yields the factor $d(\delta(c[q]), \delta(c[q']))$. We can immediately restrict ourselves to preamble transition contexts because $\delta(c[q]) = \delta(c[q'])$ by [13, Prop. 18], which yields that $d(\delta(c[q]), \delta(c[q'])) = 0$, for all $c \in C_M - \overline{C}_M$. Moreover, if the states $q_1, \ldots, q_i$ to the left of $\square$ are not such that $[q_1], \ldots, [q_i] \subseteq \mathrm{Pre}_M$, then the context c'' is not in LC and thus discarded. $\square$

Since we now have a recursive procedure to compute d, let us quickly analyse its time complexity. The analysis is based on the idea that entries in d are never recomputed once they have been computed once.

Corollary 10 (of Prop. 5 and Lm. 9). *For all $q \sim q'$ we can compute $d(q, q')$ in time $\mathcal{O}(m \cdot n)$ where $m = |M|$ and $n = |Q|$.*

Proof. We can trivially compute all a_q with $q \in \mathrm{Pre}_M$ in time $\mathcal{O}(m)$ as already mentioned, and we can compute each entry in d in time $\mathcal{O}(\frac{m}{n})$ without the time needed to compute the recursive calls because there are $\sum_{k \geq 1} k \cdot |\Sigma_k| \cdot |Q|^{k-1}$ transition contexts.[9] Since there are at most n^2 entries in d, we obtain the stated time-bound. $\square$

Thus, we can now identify and count the errors caused in kernel states of M. To this end, we look at all the transitions that switch from exclusively preamble to kernel states and compute the number of errors induced by this transition. Let $s = \sigma(p_1, \ldots, p_k) \in \Sigma(\mathrm{Pre}_N)$ and $\mu(s) \in \mathrm{Ker}_N$ be such a transition. The set E_s of errors caused by this transition s is

$$E_s = E \cap \{c[\sigma(t_1, \ldots, t_k)] \mid c \in L(N)_{\mu(s)} \cap \mathrm{LC}, \forall i \in [k] : t_i \in L(N)^{p_i}\} \ ,$$

which contains all errors that use the transition s as the left-most special transition.

Lemma 11. *For every $s = \sigma(p_1, \ldots, p_k) \in \Sigma(\mathrm{Pre}_N)$ with $\mu(s) \in \mathrm{Ker}_N$*

$$|E_s| = e_{s,q} = \sum_{q_1 \in [\delta(u_{p_1})], \ldots, q_k \in [\delta(u_{p_k})]} a_{q_1} \cdot \ldots \cdot a_{q_k} \cdot d\big(\delta(\sigma(q_1, \ldots, q_k)), q\big) \ ,$$

where $u_p \in L(N)^p$ for every $p \in \mathrm{Pre}_N$ and $q \in \mathrm{Ker}_M$ is such that $q \equiv \mu(s)$.

Proof. Let $N' = (P', \Sigma, \mu', G')$ be the DTA returned by [13], and for every $i \in [k]$ let $p'_i = \mu'(u_{p_i})$. Then $L(N)^{p_i} = L(N')^{p'_i} = \bigcup_{q_i \in [\delta(u_{p_i})]} L(M)^{q_i}$ by Theorem 3 and Lemma 4. Moreover, $L(N)_{\mu(s)} = L(M)_q$ by assumption. Together with these statements, the equation is a straightforward implementation of the definition of E_s. $\square$

By Lemma 8 the sets E_s for suitable $s \in \Sigma(\mathrm{Pre}_N)$ are pairwise disjoint, so the errors just add up. In addition, any state $p \in P$ such that $p \sim \mu(s)$ is a valid transition target, so to optimize the errors, we can simply select the transition target $p \in P$ with $p \sim \mu(s)$ that minimizes the number of caused errors. In summary, this yields our main theorem (and Algorithm 1).

---

[9] This actually needs another trick. Given a transition $\sigma(q_1, \ldots, q_k) \in \Sigma(Q)$ we obtain k transition contexts $c_1, \ldots, c_k$ by replacing in turn each state $q_1, \ldots, q_k$ by $\square$. To avoid a multiplication effort of $\mathcal{O}(k)$ we once compute $a_{q_1} \cdot \ldots \cdot a_{q_k}$ and then compute the value for the context c_i by dividing this product by the value a_{q_i}.

Algorithm 1. Optimal choice of preamble-to-kernel transitions.

Require: a minimal DTA M, its almost equivalence $\sim\,\subseteq Q \times Q$, and
an almost equivalent hyper-minimal DTA N
Return: an almost equivalent hyper-minimal DTA N minimizing $|L(M) \ominus L(N)|$

select $u_p \in L(N)^p$ for all $p \in \mathrm{Pre}_N$
2: $G \leftarrow \{p \in P \mid \sum_{q \in [\delta(u_p)]-F} a_q < \sum_{q \in [\delta(u_p)] \cap F} a_q\}$
  for all $s \in \Sigma(\mathrm{Pre}_N)$ with $\mu(s) \in \mathrm{Ker}_N$ **do**
4:   select $p \in P$ such that $p \equiv \arg\min_q \big(e'_{s,q} \mid q \in \mathrm{Ker}_M, q \sim \mu(s)\big)$
   $\mu(s) \leftarrow p$ // reroute transition
6: **return** N

Theorem 12. *Let $m = |M|$ and $n = |Q|$. For every hyper-minimal DTA N that is almost equivalent to M we can determine $|L(M) \ominus L(N)|$ in time $\mathcal{O}(m \cdot n)$. Moreover, we can compute a hyper-minimal DTA N' that minimizes the number $|L(M) \ominus L(N')|$ of errors in time $\mathcal{O}(m \cdot n)$.*

Future Work

Recently, [10] showed results in the string case for other regular languages of allowed differences. These should translate trivially to tree automata. The difference in the number of errors between the optimal dta and the worst dta can be exponential, so the optimization can avoid a large number of errors. A practical evaluation for dta remains future work, but a simple experiment was already conducted in [15, Sect. 6] for the string case. A reviewer suggested to consider the sum of the error tree sizes instead of the simple count of error trees, but the optimization of that criterion seems closely related to bin packing already in the acyclic case (i.e., the case where the automaton has no kernel states), but the details should still be worked out. In addition, the reviewer suggested to consider those languages $L \subseteq T_\Sigma$, for which the minimal dta is hyper-minimal and optimal and in addition no other dta of strictly smaller size recognizes a tree language that is almost equivalent to L. Clearly, such tree languages exist (e.g., T_Σ), but the author is unaware of the particular properties of those languages.

References

1. Badr, A.: Hyper-minimization in $O(n^2)$. Int. J. Found. Comput. Sci. 20(4), 735–746 (2009)
2. Badr, A., Geffert, V., Shipman, I.: Hyper-minimizing minimized deterministic finite state automata. RAIRO Theor. Inf. Appl. 43(1), 69–94 (2009)
3. Comon, H., Dauchet, M., Gilleron, R., Löding, C., Jacquemard, F., Lugiez, D., Tison, S., Tommasi, M.: Tree automata: Techniques and applications (2007), http://tata.gforge.inria.fr/ (release October 12)

4. Gawrychowski, P., Jeż, A.: Hyper-minimisation made efficient. In: Královič, R., Niwiński, D. (eds.) MFCS 2009. LNCS, vol. 5734, pp. 356–368. Springer, Heidelberg (2009)
5. Gawrychowski, P., Jeż, A., Maletti, A.: On minimising automata with errors. In: Murlak, F., Sankowski, P. (eds.) MFCS 2011. LNCS, vol. 6907, pp. 327–338. Springer, Heidelberg (2011)
6. Gécseg, F., Steinby, M.: Tree Automata. Akadémiai Kiadó, Budapest (1984)
7. Gécseg, F., Steinby, M.: Tree languages. In: Rozenberg, G., Salomaa, A. (eds.) Handbook of Formal Languages, vol. 3, ch. 3, pp. 1–68. Springer (1997)
8. Gramlich, G., Schnitger, G.: Minimizing nfa's and regular expressions. J. Comput. System Sci. 73(6), 908–923 (2007)
9. Högberg, J., Maletti, A., May, J.: Backward and forward bisimulation minimization of tree automata. Theoret. Comput. Sci. 410(37), 3539–3552 (2009)
10. Holzer, M., Jakobi, S.: From equivalence to almost-equivalence, and beyond— minimizing automata with errors. In: Yen, H.-C., Ibarra, O.H. (eds.) DLT 2012. LNCS, vol. 7410, pp. 190–201. Springer, Heidelberg (2012)
11. Holzer, M., Maletti, A.: An $n \log n$ algorithm for hyper-minimizing a (minimized) deterministic automaton. Theoret. Comput. Sci. 411(38-39), 3404–3413 (2010)
12. Hopcroft, J.E.: An $n \log n$ algorithm for minimizing states in a finite automaton. In: Kohavi, Z., Paz, A. (eds.) Theory of Machines and Computations, pp. 189–196. Academic Press (1971)
13. Jeż, A., Maletti, A.: Hyper-minimization for deterministic tree automata. In: Moreira, N., Reis, R. (eds.) CIAA 2012. LNCS, vol. 7381, pp. 217–228. Springer, Heidelberg (2012)
14. Maletti, A., Quernheim, D.: Hyper-minimisation of deterministic weighted finite automata over semifields. In: Proc. 13th Int. Conf. Automata and Formal Languages, pp. 285–299. Nyíregyháza College (2011)
15. Maletti, A., Quernheim, D.: Optimal hyper-minimization. Int. J. Found. Comput. Sci. 22(8), 1877–1891 (2011)
16. Mohri, M.: Finite-state transducers in language and speech processing. Comput. Linguist. 23(2), 269–311 (1997)
17. Nivat, M., Podelski, A.: Tree Automata and Languages. Studies in Computer Science and Artificial Intelligence. North-Holland (1992)
18. Petrov, S., Barrett, L., Thibaux, R., Klein, D.: Learning accurate, compact, and interpretable tree annotation. In: Proc. 44th Ann. Meeting of the ACL, pp. 433–440. Association for Computational Linguistics (2006)
19. Sakarovitch, J.: Rational and recognisable power series. In: Droste, M., Kuich, W., Vogler, H. (eds.) Handbook of Weighted Automata. EATCS Monographs on Theoretical Computer Science, ch. IV, pp. 105–174. Springer (2009)
20. Schewe, S.: Beyond hyper-minimisation — minimising DBAs and DPAs is NP-complete. In: Proc. 30th Int. Conf. Foundations of Software Technology and Theoretical Computer Science. LIPIcs, vol. 8, pp. 400–411. Schloss Dagstuhl (2010)
21. Yu, S.: Regular languages. In: Rozenberg, G., Salomaa, A. (eds.) Handbook of Formal Languages, vol. 1, ch. 2, pp. 41–110. Springer (1997)

Lambda-Confluence Is Undecidable
for Clearing Restarting Automata

František Mráz[1,*] and Friedrich Otto[2,**]

[1] Charles University, Faculty of Mathematics and Physics
Department of Computer Science
Malostranské nám. 25, 118 00 Praha 1, Czech Republic
mraz@ksvi.ms.mff.cuni.cz
[2] Fachbereich Elektrotechnik/Informatik
Universität Kassel, 34109 Kassel, Germany
otto@theory.informatik.uni-kassel.de

Abstract. Clearing restarting automata are based on contextual rewriting. A word w is accepted by an automaton of this type if there is a computation that reduces the word w to the empty word λ by a finite sequence of rewritings. Accordingly, the word problem for a clearing restarting automaton can be solved nondeterministically in quadratic time. If, however, the contextual rewritings happen to be λ-confluent, that is, confluent on the congruence class of the empty word, then the word problem can be solved deterministically in linear time. Here we show that, unfortunately, λ-confluence is not even recursively enumerable for clearing restarting automata. This follows from the fact that λ-confluence is not recursively enumerable for finite factor-erasing string-rewriting systems.

Keywords: Clearing restarting automaton, limited context restarting automaton, factor-erasing string-rewriting system, λ-confluence.

1 Introduction

Restarting automata were introduced in [8] to model the technique of *analysis by reduction*, which is used in linguistics to analyze sentences of natural languages with free word order. Interestingly, a restarting automaton is not only useful for accepting a language, but it also enables error localization in rejected words (see, e.g., [7]). Despite these nice properties, restarting automata are rarely used in practice. One reason for this is certainly the fact that it is quite a complex task to design a restarting automaton for a given language. Accordingly, methods have

* F. Mráz was partially supported by the Grant Agency of the Czech Republic under the project P103/10/0783.
** This paper was written while F. Otto was visiting at the Department of Computer Science of Charles University at Prague. He gratefully acknowledges the hospitality of the Faculty of Mathematics and Physics.

S. Konstantinidis (Ed.): CIAA 2013, LNCS 7982, pp. 256–267, 2013.

been studied for learning a restarting automaton from positive (and negative) examples of sentences and/or reductions (see, e.g., [1,2,5,6,11]).

Specifically, Černo and Mráz introduced a restricted type of restarting automaton, the so-called clearing restarting automaton, in [6], which was later extended to the limited context restarting automaton by Basovník and Mráz in [1,2]. A *clearing restarting automaton* M is defined through a finite set $I(M)$ of instructions of the form $(x \mid z \to \lambda \mid y)$. Based on the local context x and y, this instruction erases the factor z from the tape contents ¢w\$ of M, where ¢ and \$ are the left and right delimiters of the tape. Now M repeatedly applies its instructions, and it accepts if and when its tape contents has been reduced to the word ¢\$. For clearing restarting automata a simple learning algorithm exists [5,6], but on the other hand, they are quite limited in their expressive power. In fact, while these automata accept all regular languages and even some languages that are not context-free, they do not even accept all context-free languages (see [6]).

A *limited context restarting automaton* (lc-R-automaton, for short) M is defined through a finite set of instructions of the form $(x \mid z \to t \mid y)$, where $|z| > |t|$. Based on the local context x and y, this instruction replaces an occurrence of the factor z of the tape contents by the word t. Several different types of lc-R-automata have been defined in [2] and in [17] based on the form of the admissible contexts x and y and the form of the word t. In an lc-R-automaton of type $\mathcal{R}_1$, we have $|t| \leq 1$ for each instruction, and for an lc-R-automaton of type $\mathcal{R}_2$, we require in addition that $x \in \{$¢$, \lambda\}$ and $y \in \{$\$$, \lambda\}$ for all instructions, that is, the left (right) context of each instruction is either the left (right) delimiter, or it is empty. Obviously, the lc-R-automaton of type $\mathcal{R}_1$ is a proper extension of the clearing restarting automaton, while those of type $\mathcal{R}_2$ are incomparable to clearing restarting automata.

To test whether a word w belongs to the language $L(M)$ accepted by a given lc-R-automaton M, one has to check whether w can be reduced to the empty word λ by a sequence of applications of the instructions of M. As each instruction is length-reducing, such a sequence is bounded in length by $|w|$, but as there could be several instructions that are applicable to the same word, or there could be several places at which a given instruction can be applied, all such sequences must be checked. Accordingly, the membership problem for $L(M)$ is decidable nondeterministically in time $O(n^2)$. The situation would be much better if it was known that each and every sequence of applications of instructions of M reduces w to λ, if w does indeed belong to the language $L(M)$. In this case we could concentrate on leftmost sequences of reductions, and accordingly, membership in $L(M)$ would be decidable deterministically in time $O(n)$.

With an lc-R-automaton M, we can associate a finite string-rewriting system $S(M) = \{ (xzy \to xty) \mid (x \mid z \to t \mid y) \in I(M) \} \cup \{($¢\$$ \to \lambda)\}$. Obviously, for all input words w, $w \in L(M)$ if and only if ¢w\$ $\Rightarrow^*_{S(M)} \lambda$ holds, where $\Rightarrow^*_{S(M)}$ denotes the reduction relation induced by $S(M)$ (see below). Now the lc-R-automaton M is called *confluent*, if the string-rewriting system $S(M)$ is confluent. In [17] the expressive power of the various types of confluent lc-R-automata has been investigated, but confluent lc-R-automata of type $\mathcal{R}_1$ ($\mathcal{R}_2$) are much less expressive than the non-confluent lc-R-automata of the same type.

However, for solving the membership problem for the language $L(M)$ in linear time, confluence is actually not needed. In fact, it would suffice that the string-rewriting system $S(M)$ is λ-*confluent*, which means that all words that are congruent to λ reduce to λ. And indeed, we will see below that for lc-R-automata of type $\mathcal{R}_2$, λ-confluence is decidable.

If M is a clearing restarting automaton, then each rule of the string-rewriting system $S(M)$ simply erases a non-empty factor of its left-hand side. Thus, in this case $S(M)$ is a *factor-erasing* string-rewriting system. As our technical main result we will show that it is undecidable in general whether a given finite factor-erasing string-rewriting system is λ-confluent. In fact, we will see that this problem is not even recursively enumerable (r.e.). It will follow that it is undecidable (in fact, not r.e.) in general whether a given clearing restarting automaton is λ-confluent.

Our main result improves upon the result established in [15], which states that λ-confluence is undecidable for finite length-reducing string-rewriting systems. Observe that factor-erasing string-rewriting systems can be seen as a generalization of *special* string-rewriting systems, for which λ-confluence is known to be decidable in polynomial time [16].

This paper is structured as follows. In the next section we introduce the necessary notation and notions on string-rewriting systems. Then we present the clearing restarting automaton and the limited context restarting automaton in short, and we state our results on λ-confluence for these types of automata. In Section 4 we derive our main undecidability result, and in the final section we present some open problems for further work.

2   String-Rewriting Systems

A (finite) *string-rewriting system* S on an alphabet Σ consists of (finitely many) pairs of strings from Σ^*, called *rewrite rules*, which are written as $(\ell \to r)$. By dom(S) we denote the set dom(S) = $\{ \ell \mid \exists r \in \Sigma^* : (\ell \to r) \in S \}$ of left-hand sides of rules of S. The *reduction relation* $\Rightarrow_S^*$ on Σ^* that is induced by S is the reflexive and transitive closure of the *single-step reduction relation* $\Rightarrow_S = \{ (u\ell v, urv) \mid (\ell \to r) \in S, u, v \in \Sigma^* \}$. For a string $u \in \Sigma^*$, if there exists a string v such that $u \Rightarrow_S v$ holds, then u is called *reducible* mod S. If such a string v does not exist, then u is called *irreducible* mod S. By IRR(S) we denote the set of all irreducible strings mod S. As IRR(S) = $\Sigma^* \smallsetminus (\Sigma^* \cdot \mathrm{dom}(S) \cdot \Sigma^*)$, we see that IRR($S$) is a regular language, if S is finite. By $\Leftrightarrow_S^*$ we denote the *Thue congruence* on Σ^* that is induced by S. It is the smallest equivalence relation on Σ^* containing the single-step reduction relation $\Rightarrow_S$. For each word $w \in \Sigma^*$, $[w]_S = \{ u \in \Sigma^* \mid u \Leftrightarrow_S^* w \}$ is the *congruence class* of w.

Here we are interested in certain restricted types of string-rewriting systems. A string-rewriting system S is called

- *length-reducing*, if $|\ell| > |r|$ for each rule $(\ell \to r) \in S$,
- *monadic*, if it is length-reducing and $|r| \leq 1$ for each rule $(\ell \to r) \in S$,
- *special*, if it is length-reducing and $|r| = 0$ for each rule $(\ell \to r) \in S$,
- *erasing*, if, for each rule $(\ell \to r) \in S$, r is a proper scattered subword of ℓ, that is, r is obtained from ℓ by erasing one or more letters of ℓ,
- *factor-erasing*, if, for each rule $(\ell \to r) \in S$, there exists a factorization $\ell = uxv$ such that $|x| \geq 1$ and $r = uv$,
- *confluent*, if, for all $u, v \in \Sigma^*$, $u \leftrightarrow_S^* v$ implies that there exists some $z \in \Sigma^*$ such that $u \Rightarrow_S^* z$ and $v \Rightarrow_S^* z$ hold,
- *λ-confluent*, if it is confluent on the congruence class of the empty word λ, that is, for all $w \in \Sigma^*$, if $w \leftrightarrow_S^* \lambda$, then $w \Rightarrow_S^* \lambda$, and
- *convergent*, if it is confluent and *terminating*, that is, S does not admit any infinite sequence of reductions of the form $u \Rightarrow_S u_1 \Rightarrow_S u_2 \Rightarrow_S \ldots$.

Obviously, each length-reducing system is terminating. Hence, a length-reducing and confluent system is convergent. For a convergent system S, the set $\mathsf{IRR}(S)$ of irreducible strings is a complete set of unique representatives for the Thue congruence $\leftrightarrow_S^*$ (see, e.g., [4]).

For each pair of rewriting rules $\ell_1 \to r_1$ and $\ell_2 \to r_2$, the set of *critical pairs* is defined as the set $\{ (xr_1, r_2y) \mid x, y \in \Sigma^*, x\ell_1 = \ell_2y \text{ and } |x| < |\ell_2| \} \cup \{ (r_1, xr_2y) \mid x, y \in \Sigma^*, \ell_1 = x\ell_2y \}$. We say that a critical pair (p, q) *resolves* if p and q have a common descendant mod S. Finally, for checking that a terminating system S is confluent, it suffices to determine all *critical pairs* (p, q) of S and to check whether they all *resolve mod S* (see, e.g., [4]). In particular, it is known that confluence is decidable in polynomial time for finite length-reducing string-rewriting systems [10].

The problem of checking λ-confluence is much more difficult. It is shown in [15] that λ-confluence is decidable in double exponential time for finite monadic string-rewriting systems, but that it is undecidable in general for finite length-reducing string-rewriting systems. On the other hand, λ-confluence is decidable in polynomial time for finite special string-rewriting systems [16].

3   Clearing and Limited Context Restarting Automata

Let k be a positive integer. A *k-context rewriting system* is a triple $C = (\Sigma, \Gamma, I)$, where Σ is a finite input alphabet, Γ is a finite working alphabet containing Σ but not the special symbols ¢ and \$, called sentinels, and I is a finite set of instructions of the form $(x \mid z \to t \mid y)$, where x is called a *left context*, $x \in \Gamma^k \cup ¢ \cdot \Gamma^{\leq k-1}$, y is called a *right context*, $y \in \Gamma^k \cup \Gamma^{\leq k-1} \cdot \$$, and $z \to t$ is called a *rule*, where $z, t \in \Gamma^*$. A word $w = uzv$ can be rewritten into utv, denoted as $uzv \to_C utv$, if and only if there exists an instruction $i = (x \mid z \to t \mid y) \in I$ such that x is a suffix of ¢ $\cdot u$ and y is a prefix of $v \cdot \$$. The *reduction language* associated with C is defined as $L(C) = \{ w \in \Sigma^* \mid w \to_C^* \lambda \}$, where $\to_C^*$ denotes the reflexive and transitive closure of $\to_C$.

A *k-clearing restarting automaton* is a k-context rewriting system $M = (\Sigma, \Sigma, I)$ such that, for each instruction $i = (x \mid z \to t \mid y) \in I$, we have $z \in \Sigma^+$

and $t = \lambda$. A *clearing restarting automaton* is a k-clearing restarting automaton for some $k \geq 0$.

A *limited context restarting automaton*, an lc-R-automaton for short, is a k-context rewriting system $M = (\Sigma, \Gamma, I)$ for some $k \geq 1$, where I contains *instructions* of the form $(x \,|\, z \to t \,|\, y)$ such that $|z| > |t|$.

We consider two restricted types of lc-R-automata. We say that an lc-R-automaton $M = (\Sigma, \Gamma, I)$ is of type

- $\mathcal{R}_1$, if all instructions of I are of the form $(x \,|\, z \to t \,|\, y)$, where $t \in \Gamma \cup \{\lambda\}$, and $z \in \Gamma^+$ such that $|z| > |t|$;
- $\mathcal{R}_2$, if all instructions of I are of the form $(x \,|\, z \to t \,|\, y)$, where $t \in \Gamma \cup \{\lambda\}$, $x \in \{\lambda, \mathdollar\!\!c\}$, $y \in \{\lambda, \$\}$, and $z \in \Gamma^+$ such that $|z| > |t|$.

Obviously, clearing restarting automata are a subclass of the lc-R-automata of type $\mathcal{R}_1$.

An lc-R-automaton M accepts exactly the set of input words which can be reduced to λ. Obviously, λ is in $L(M)$ for each lc-R-automaton M. Further, an lc-R-automaton M is called *λ-confluent* if the reduction relation $\to_M$ is confluent on λ, that is, if $w \to_M^* \lambda$ and $w \to_M^* w'$, then also $w' \to_M^* \lambda$.

With a finite factor-erasing string-rewriting system S on some alphabet Γ, we can associate a clearing restarting automaton $M_S = (\Gamma, \Gamma, I_S)$ by taking

$$I_S = \{ (u \,|\, x \to \lambda \,|\, v) \mid (uxv \to uv) \in S \}.$$

Now S is λ-confluent iff, for all $w \in \Gamma^*$, $w \leftrightarrow_S^* \lambda$ implies that $w \Rightarrow_S^* \lambda$, which in turn holds iff, for all $w \in \Gamma^*$, $\mathdollar\!\!c w\$ \leftrightarrow_S^* \mathdollar\!\!c \$$ implies that $\mathdollar\!\!c w\$ \Rightarrow_S^* \mathdollar\!\!c \$$, which in turn is equivalent to saying that, for all $w, w' \in \Gamma^*$, $w \to_{M_S}^* \lambda$ and $w \to_{M_S}^* w'$ imply that $w' \to_{M_S}^* \lambda$. Hence, S is λ-confluent iff M_S is λ-confluent.

Below we will show that λ-confluence is not even recursively enumerable for finite factor-erasing string-rewriting systems (Theorem 4). Thus, we immediately obtain the following results.

Theorem 1.

(a) *The problem of deciding λ-confluence for clearing restarting automata is not recursively enumerable.*

(b) *The problem of deciding λ-confluence for lc-R-automata of type $\mathcal{R}_1$ is not recursively enumerable.*

Let $M = (\Sigma, \Gamma, I)$ be an lc-R-automaton of type $\mathcal{R}_2$. Then the string-rewriting system

$$R(M) = \{ (xzy \to xty) \mid (x \,|\, z \to t \,|\, y) \in I \}$$

can be split into four disjoint subsystems:

(a) $R_{bif} = \{ (\mathdollar\!\!c z\$ \to \mathdollar\!\!c t\$) \mid (\mathdollar\!\!c \,|\, z \to t \,|\, \$) \in I \}$, the *bifix rules* of $R(M)$,

(b) $R_{pre} = \{ (\mathdollar\!\!c z \to \mathdollar\!\!c t) \mid (\mathdollar\!\!c \,|\, z \to t \,|\, \lambda) \in I \}$,   the *prefix rules* of $R(M)$,

(c) $R_{suf} = \{ (z\$ \to t\$) \mid (\lambda \,|\, z \to t \,|\, \$) \in I \}$,   the *suffix rules* of $R(M)$,

(d) $R_{inf} = \{ (z \to t) \mid (\lambda \,|\, z \to t \,|\, \lambda) \in I \}$,   the *infix rules* of $R(M)$.

Here we can assume without loss of generality that $|t| = 1$ for all rules in R_{inf}, as each instruction of the form $(\lambda \,|\, z \to \lambda \,|\, \lambda)$ can be replaced by the set of instructions $\{\, (\lambda \,|\, Az \to A \,|\, \lambda) \mid A \in \Gamma \,\} \cup \{(\mathfrak{c} \,|\, z \to \lambda \,|\, \lambda)\}$. Then it can be shown that the lc-R-automaton M is λ-confluent, if and only if the string-rewriting system $R(M)$ is confluent on the congruence class of $\mathfrak{c}\$$.

As for each instruction $(x \,|\, z \to t \,|\, y) \in I$, $|z| > |t|$ and $|t| \le 1$, we see that R_{inf} is a monadic string-rewriting system, and R_{pre} (R_{suf} and R_{bif}) is a monadic string-rewriting system with added left (right, left and right) sentinel. In [15] it is shown that it is decidable whether a finite monadic string-rewriting system is confluent on a given congruence class. This is done as follows. First, it is shown that a finite length-reducing string-rewriting system S is confluent on $[w]_S$, where $w \in \mathsf{IRR}(S)$, if and only if $L_p(w) = L_q(w)$ holds for each unresolved critical pair (p, q) of S. Here $L_z(w)$ is the language $L_z(w) = \{\, x \# y \mid x, y \in \mathsf{IRR}(S),\ xzy \Rightarrow^*_{S,L} w \,\}$, where $\Rightarrow_{S,L}$ denotes the *leftmost* reduction mod S. Secondly, it is shown that, if S is monadic, then from S, z and w, a deterministic one-turn pushdown automaton $A(z, w)$ can be constructed that accepts the language $L_z(w)$. As the set of unresolved critical pairs can be computed in polynomial time from S, and as it is decidable (in double exponential time) whether two deterministic one-turn pushdown automata accept the same language [3,18], this shows that confluence on $[w]_S$ is decidable for a finite monadic string-rewriting system S. On close inspection it turns out that this very algorithm extends to string-rewriting systems of the form of the system $R(M)$ above, which yields the following.

Theorem 2. *It is decidable in double exponential time whether a given lc-R-automaton of type $\mathcal{R}_2$ is λ-confluent.*

4   Undecidability of λ-Confluence

Here we are interested in the problem of deciding λ-confluence for finite factor-erasing string-rewriting systems. These systems can be seen as a generalization of special string-rewriting systems, as in each reduction step a non-empty factor is simply erased, but in contrast to the situation for a special string-rewriting system, the place of this factor is restricted by a two-sided context-condition for a factor-erasing string-rewriting system. As we will see below, this restriction is already sufficient to turn λ-confluence into a property that is not even recursively enumerable.

We will prove this result by an extension of the corresponding result for finite length-reducing string-rewriting systems. Therefore, we first outline the proof of the latter result from [15] in short.

Let Σ be a finite alphabet, let $L \subseteq \Sigma^*$ be a language that is recursively enumerable, but non-recursive, and let $M = (Q, \Sigma, b, q_0, q_n, \delta)$ be a single-tape Turing machine that accepts the language L. Here $Q = \{q_0, q_1, \ldots, q_n\}$ is the set of states of M, Σ is the input alphabet, $b \notin \Sigma$ is the blank symbol, $q_0 \in Q$ is the initial state, $q_n \in Q$ is the unique halting state, and $\delta : ((Q \setminus \{q_n\}) \times (\Sigma \cup \{b\})) \to (Q \times \Sigma \times \{\text{right}, \text{left}\})$ is the transition function. A configuration of M is a word of the form $uqav$ for $u \in \Sigma^*$, $q \in Q$, $a \in \Sigma_b$, and $v \in \Sigma^*$, where $\Sigma_b = \Sigma \cup \{b\}$, the initial configuration on input $w \in \Sigma^*$ is $q_0 w$, and a final (or accepting) configuration has the form $u q_n a v$. W.l.o.g. we may assume that for each final configuration $u q_n a v$, $|v| \geq 1$. Hence, $w \in L$ if and only if $q_0 w \vdash_M^* u q_n a v$ holds for some $u \in \Sigma^*$, $a \in \Sigma$, and $v \in \Sigma^+$. Let $\Pi = \Sigma \cup \{\triangleright, \triangleleft\}$, where $\triangleleft$ and $\triangleright$ are two new symbols that will be used as endmarkers. Further, let $Q_1 = Q_p \cup Q_s$, where $Q_p = \{p_0, p_1, \ldots, p_n\}$ and $Q_s = \{s_0, s_1, \ldots, s_n\}$ are disjoint copies of the set Q of states of M, and let

$$D = \{\, \langle a p_i \rangle, \langle s_i a \rangle \mid a \in \Pi, i \in \{0, 1, \ldots, n-1\} \,\} \cup \{\langle A \rangle, \langle B \rangle\}$$

be a set of additional letters called *dummy symbols*.

Following [12,13] a finite length-reducing string-rewriting system $R(M)$ can now be constructed that simulates the computations of M in reverse order.

Let CONFIG $= \triangleright \cdot (\Sigma \cup D)^* \cdot Q_1 \cdot (\Sigma \cup D)^* \cdot \triangleleft$, and let HALTING $= \triangleright \cdot (\Sigma \cup D)^* \cdot \{p_n, s_n\} \cdot (\Sigma \cup D)^+ \cdot \triangleleft$. The elements of CONFIG can be interpreted as descriptions of possible configurations of M interspersed with occurrences of dummy symbols, and the elements of HALTING correspond to possible halting configurations of M. Now let $\Gamma = \Pi \cup Q_1 \cup D \cup \{\langle C \rangle\}$, where $\langle C \rangle$ is another new symbol, and let $R(M)$ be the finite length-reducing string-rewriting system on Γ that is defined as follows, where $a \in \Sigma$, $c \in \Sigma$, $q_i \in Q$, $p_i \in Q_p$, $s_j \in Q_s$, and $\langle d \rangle \in D$:

$$
\begin{aligned}
\langle a p_i \rangle c s_j &\to a p_i, && \text{if } \delta(q_i, a) = (q_j, c, \text{right}), \\
\langle s_i a \rangle c s_j &\to s_i a, && \text{if } \delta(q_i, a) = (q_j, c, \text{right}), \\
p_j c \langle a p_i \rangle &\to a p_i, && \text{if } \delta(q_i, a) = (q_j, c, \text{left}), \\
p_j c \langle s_i a \rangle &\to s_i a, && \text{if } \delta(q_i, a) = (q_j, c, \text{left}), \\
\triangleright \langle \triangleright p_i \rangle c s_j &\to \triangleright p_i, && \text{if } \delta(q_i, b) = (q_j, c, \text{right}), \\
\langle s_i \triangleleft \rangle c s_j \triangleleft &\to s_i \triangleleft, && \text{if } \delta(q_i, b) = (q_j, c, \text{right}), \\
\triangleright p_j c \langle \triangleright p_i \rangle &\to \triangleright p_i, && \text{if } \delta(q_i, b) = (q_j, c, \text{left}), \\
p_j c \langle s_i \triangleleft \rangle \triangleleft &\to s_i \triangleleft, && \text{if } \delta(q_i, b) = (q_j, c, \text{left}), \\
p_i \langle d \rangle \langle A \rangle &\to \langle d \rangle p_i, && \langle B \rangle \langle d \rangle s_i \to s_i \langle d \rangle, \\
\langle B \rangle p_i s_j &\to \langle C \rangle, && p_i s_j \langle A \rangle \to \langle C \rangle, \\
\langle C \rangle \langle A \rangle &\to \langle C \rangle, && \langle B \rangle \langle C \rangle \to \langle C \rangle.
\end{aligned}
$$

The string-rewriting system $R(M)$ is even confluent, and it satisfies the following properties (see, e.g., [12,14]):

(1) $|\ell| \leq 4$ and $|r| \leq 2$ for all rules $(\ell \to r) \in R(M)$,
(2) $\triangleright s_0 \cdot \Sigma^* \cdot \triangleleft \subseteq \text{IRR}(R(M))$,
(3) if $w \in$ CONFIG and $w \Rightarrow_{R(M)}^* z$, then $z \in$ CONFIG,
(4) if $z \in$ CONFIG and $w \Rightarrow_{R(M)}^* z$, then $w \in$ CONFIG,
(5) for all $x \in \Sigma^*$, $x \in L$ if and only if $\exists w \in$ HALTING $: w \Rightarrow_{R(M)}^* \triangleright s_0 x \triangleleft$.

Let $\bot$ be another new symbol, let $\Delta = \Gamma \cup \{\bot\}$, and let

$$R_L = R(M) \cup \{\, (p_n a \to \bot), (s_n a \to \bot) \mid a \in \Sigma \cup D \,\}.$$

Then R_L is a finite length-reducing string-rewriting system on Δ, and it is easily seen that R_L is *not* confluent, as none of the critical pairs that result from overlapping one of the new rules with a rule of $R(M)$ resolves.

For $x \in \Sigma^*$, if $x \in L$, then there exist words $z_1, z_2 \in (\Sigma \cup D)^*$, $z_2 \neq \lambda$, and a symbol $r \in \{p_n, s_n\}$ such that $\triangleright z_1 r z_2 \triangleleft \Rightarrow^*_{R(M)} \triangleright s_0 x \triangleleft$ holds (see (5)). Since $\triangleright z_1 r z_2 \triangleleft \Rightarrow_{R_L} \triangleright z_1 \bot z_3 \triangleleft$ for some $z_3 \in (\Sigma \cup D)^*$, and since $\triangleright s_0 x \triangleleft$ and $\triangleright z_1 \bot z_3 \triangleleft$ are both irreducible mod R_L, we see that R_L is *not* confluent on the congruence class $[\triangleright s_0 x \triangleleft]_{R_L}$. If, however, $x \notin L$, then it can be shown that R_L is confluent on $[\triangleright s_0 x \triangleleft]_{R_L}$ (see [15]). Thus, R_L is confluent on the congruence class $[\triangleright s_0 x \triangleleft]_{R_L}$ if and only if $x \notin L$.

Finally, for $x \in \Sigma^*$, we take the string-rewriting systems

$$R_L(x) = R_L \cup \{(\triangleright s_0 x \triangleleft \to \lambda)\}.$$

Then $R_L(x)$ is a finite length-reducing string-rewriting system, which satisfies the following equivalences:

$$R_L(x) \text{ is } \lambda\text{-confluent iff } R_L(x) \text{ is confluent on } [\lambda]_{R_L(x)}$$
$$\text{iff } R_L \text{ is confluent on } [\triangleright s_0 x \triangleleft]_{R_L}$$
$$\text{iff } x \notin L.$$

From the choice of the language L it thus follows that the following problem is not recursively enumerable:

INSTANCE: A word $x \in \Sigma^*$.
QUESTION: Is the finite length-reducing string-rewriting system $R_L(x)$ λ-confluent?

Next we extend the result above to finite erasing string-rewriting systems, and then we extend it to finite factor-erasing string-rewriting systems. For the former step we make use of a variant of an encoding introduced in [9].

We introduce a linear ordering on the alphabet $\hat{\Delta} = \Delta \setminus \{\triangleleft, \triangleright\}$. Accordingly, we can write $\hat{\Delta}$ as $\hat{\Delta} = \{d_1, d_2, \ldots, d_m\}$. In addition, we introduce a linear ordering on the rules of the string-rewriting system R_L, that is, R_L can be written as $R_L = \{(\ell_i \to r_i) \mid i = 1, \ldots, s\}$. Let $\Omega = \{\triangleright, \triangleleft, c, d, 0, 1\}$ be a new alphabet, and let $\psi : \Delta^* \to \Omega^*$ be the morphism that is defined as follows, where $\hat{s} = s + 1$ and $\hat{m} = m + 1$:

$$\triangleright \mapsto \triangleright, \quad \triangleleft \mapsto \triangleleft, \quad d_i \mapsto c^{\hat{s}} 1^{\hat{m}-i} 0^i (c^{\hat{s}} d 1^{\hat{m}} 0^{\hat{m}})^2, \; 1 \leq i \leq m.$$

Obviously, ψ is an encoding, that is, an injective mapping. In addition, it has the following nice property.

Lemma 1. [9] *For all $u, v \in \hat{\Delta}^*$, if $|v| < |u| \leq 3$, then $\psi(v)$ is a scattered subword of $\psi(u)$.*

Proof. Let $u = d_{i_1} d_{i_2} d_{i_3}$ and $v = d_{j_1} d_{j_2}$. Then the encoded words $\psi(u)$ and $\psi(v)$ look as follows:

$$\psi(u) = c^{\hat{s}} 1^{\hat{m}-i_1} 0^{i_1} (c^{\hat{s}} d 1^{\hat{m}} 0^{\hat{m}})^2 c^{\hat{s}} 1^{\hat{m}-i_2} 0^{i_2} (c^{\hat{s}} d 1^{\hat{m}} 0^{\hat{m}})^2 c^{\hat{s}} 1^{\hat{m}-i_3} 0^{i_3} (c^{\hat{s}} d 1^{\hat{m}} 0^{\hat{m}})^2,$$
$$\psi(v) = c^{\hat{s}} 1^{\hat{m}-j_1} 0^{j_1} (c^{\hat{s}} d 1^{\hat{m}} 0^{\hat{m}})^2 c^{\hat{s}} 1^{\hat{m}-j_2} 0^{j_2} (c^{\hat{s}} d 1^{\hat{m}} 0^{\hat{m}})^2.$$

Now $\psi(u)$ can be factored as follows:

$$\psi(u) = \underline{c^{\hat{s}}1^{\hat{m}-i_1}0^{i_1}} \cdot c^{\hat{s}} \cdot \underline{d1^{j_1}} \cdot 1^{\hat{m}-j_1}0^{j_1} \cdot \underline{0^{\hat{m}-j_1}} \cdot (c^{\hat{s}}d1^{\hat{m}}0^{\hat{m}}) \cdot \underline{c^{\hat{s}}1^{\hat{m}-i_2}0^{i_2}} \cdot$$
$$(c^{\hat{s}}d1^{\hat{m}}0^{\hat{m}})c^{\hat{s}} \cdot \underline{d1^{j_2}} \cdot 1^{\hat{m}-j_2}0^{j_2} \cdot \underline{0^{\hat{m}-j_2}c^{\hat{s}}1^{\hat{m}-i_3}0^{i_3}} \cdot (c^{\hat{s}}d1^{\hat{m}}0^{\hat{m}})^2,$$

and by erasing the six underlined factors, $\psi(v)$ is obtained from $\psi(u)$. The case that $|v| = 1$ and $|u| \in \{2,3\}$ is dealt with analogously.   □

Let $\hat{R}_L$ be the string-rewriting system on Ω that is defined as

$$\hat{R}_L = \{ (\psi(\ell) \to \psi(r)) \mid (\ell \to r) \in R_L \},$$

and for $x \in \Sigma^*$, let $\hat{R}_L(x)$ be the string-rewriting system

$$\hat{R}_L(x) = \{ (\psi(\ell) \to \psi(r)) \mid (\ell \to r) \in R_L(x) \} = \hat{R}_L \cup \{(\triangleright\psi(s_0x)\triangleleft \to \lambda)\}.$$

Then $\hat{R}_L(x)$ is a finite erasing string-rewriting system. Note that Lemma 1 can also be applied to those rules $\ell \to r$ of R_L for which $|\ell| = 4$ holds, since for these rules, ℓ starts or ends with one of the sentinels $\triangleleft$ or $\triangleright$, which are mapped onto themselves by ψ.

Lemma 2. *For each $x \in \Sigma^*$, the system $\hat{R}_L(x)$ is λ-confluent if and only if the system $R_L(x)$ is λ-confluent.*

Proof. From the definition of the encoding ψ we see that, for all $w, z \in \hat{\Delta}^*$, $\triangleright w\triangleleft \Rightarrow_{R_L} \triangleright z\triangleleft$ if and only if $\triangleright\psi(w)\triangleleft \Rightarrow_{\hat{R}_L} \triangleright\psi(z)\triangleleft$. Further, $(\triangleright\psi(s_0x)\triangleleft \to \lambda)$ is the only rule of $\hat{R}_L(x)$ with right-hand side λ, while $(\triangleright s_0x\triangleleft \to \lambda)$ is the only rule of $R_L(x)$ with right-hand side λ. It follows immediately that $\hat{R}_L(x)$ is λ-confluent, if and only if $R_L(x)$ is.   □

This yields the following result.

Lemma 3. *The following problem is not even recursively enumerable:*

INSTANCE: A word $x \in \Sigma^*$.
QUESTION: Is the erasing string-rewriting system $\hat{R}_L(x)$ λ-confluent?

Thus, we have the following undecidability result.

Theorem 3. *The problem of deciding whether a finite erasing string-rewriting system is λ-confluent is not recursively enumerable.*

Each rule of the system $\hat{R}_L$ is erasing, that is, its right-hand side is obtained from its left-hand side by erasing up to six factors (see the proof of Lemma 1).

Finally we extend the above undecidability result to finite factor-erasing string-rewriting systems by simulating the system $\hat{R}_L$ by a factor-erasing string-rewriting system $\tilde{R}_L$. Essentially, the new system will simulate the application of a rule of $\hat{R}_L$ by up to seven separate steps that each erase a single factor only. Accordingly, each rule of $\hat{R}_L$ will be replaced by finitely many factor-erasing rules that each erase one of the factors deleted by the original rule. By adding certain contexts to these rules it will be ensured that they can only be applied in the correct order, and that rules corresponding to different rules of $\hat{R}_L$ will not interfere with each other.

We define the string-rewriting system $\tilde{R}_L$ as $\tilde{R}_L = \bigcup_{i=1}^s \tilde{R}_i$, where, for $i = 1, \ldots, s$, $\tilde{R}_i$ is a finite set of factor-erasing rules that is defined from the rule $(\psi(\ell_i) \to \psi(r_i))$ of $\hat{R}_L$. Finally, for each $x \in \Sigma^*$, we will consider the system

$$\tilde{R}_L(x) = \tilde{R}_L \cup \{(\triangleright\psi(s_0x)\triangleleft \to \lambda)\}.$$

Then $\tilde{R}_L(x)$ is a finite factor-erasing string-rewriting system on Ω.

First, we define a finite set LC of *left contexts* by taking

$$\text{LC} = \{\triangleright\} \cup (\{\triangleright\} \cdot \psi(\hat{\Delta})) \cup (\{\triangleright\} \cdot \psi(\hat{\Delta}^2)) \cup \psi(\hat{\Delta}^3).$$

Then, for each rule $(\psi(\ell_i) \to \psi(r_i))$ of $\hat{R}_L$, we define a finite collection of factor-erasing rules $\tilde{R}_i$.

Actually, there are five different cases based on the form of the rule considered. Here we present the most involved case, that is, $(\ell_i \to r_i)$ is a rule of the form $\ell_i = d_{i_1}d_{i_2}d_{i_3}$ and $r_i = d_{j_1}d_{j_2}$ for some $d_{i_1}, d_{i_2}, d_{i_3}, d_{j_1}, d_{j_2} \in \hat{\Delta}$, that is,

$$\psi(\ell_i) = c^{\hat{s}}1^{\widehat{m}-i_1}0^{i_1}(c^{\hat{s}}d1^{\widehat{m}}0^{\widehat{m}})^2 c^{\hat{s}}1^{\widehat{m}-i_2}0^{i_2}(c^{\hat{s}}d1^{\widehat{m}}0^{\widehat{m}})^2 c^{\hat{s}}1^{\widehat{m}-i_3}0^{i_3}(c^{\hat{s}}d1^{\widehat{m}}0^{\widehat{m}})^2,$$

and

$$\psi(r_i) = c^{\hat{s}}1^{\widehat{m}-j_1}0^{j_1}(c^{\hat{s}}d1^{\widehat{m}}0^{\widehat{m}})^2 c^{\hat{s}}1^{\widehat{m}-j_2}0^{j_2}(c^{\hat{s}}d1^{\widehat{m}}0^{\widehat{m}})^2.$$

Then $\tilde{R}_i$ contains the following rules for all $u \in$ LC:

(1) $uc^i\underline{c^{\hat{s}-i}1^{\widehat{m}-i_1}0^{i_1}-1}0(c^{\hat{s}}d1^{\widehat{m}}0^{\widehat{m}})^2 c^{\hat{s}}1^{\widehat{m}-i_2}0^{i_2}(c^{\hat{s}}d1^{\widehat{m}}0^{\widehat{m}})^2 c^{\hat{s}}1^{\widehat{m}-i_3}0^{i_3}(c^{\hat{s}}d1^{\widehat{m}}0^{\widehat{m}})^2$
$\to u \cdot c^i0(c^{\hat{s}}d1^{\widehat{m}}0^{\widehat{m}})^2 c^{\hat{s}}1^{\widehat{m}-i_2}0^{i_2}(c^{\hat{s}}d1^{\widehat{m}}0^{\widehat{m}})^2 c^{\hat{s}}1^{\widehat{m}-i_3}0^{i_3}(c^{\hat{s}}d1^{\widehat{m}}0^{\widehat{m}})^2,$

(2) $c^i0(c^{\hat{s}}\underline{d1^{j_1}1^{\widehat{m}-j_1}0^{\widehat{m}}})(c^{\hat{s}}d1^{\widehat{m}}0^{\widehat{m}})c^{\hat{s}}1^{\widehat{m}-i_2}0^{i_2}(c^{\hat{s}}d1^{\widehat{m}}0^{\widehat{m}})^2 c^{\hat{s}}1^{\widehat{m}-i_3}0^{i_3}(c^{\hat{s}}d1^{\widehat{m}}0^{\widehat{m}})^2$
$\to c^i0(\underline{c^{\hat{s}}1^{\widehat{m}-j_1}0^{\widehat{m}}})(c^{\hat{s}}d1^{\widehat{m}}0^{\widehat{m}})c^{\hat{s}}1^{\widehat{m}-i_2}0^{i_2}(c^{\hat{s}}d1^{\widehat{m}}0^{\widehat{m}})^2 c^{\hat{s}}1^{\widehat{m}-i_3}0^{i_3}(c^{\hat{s}}d1^{\widehat{m}}0^{\widehat{m}})^2,$

(3) $c^i0c^{\hat{s}}1^{\widehat{m}-j_1}0^{j_1}\underline{0^{\widehat{m}-j_1}}(c^{\hat{s}}d1^{\widehat{m}}0^{\widehat{m}})c^{\hat{s}}1^{\widehat{m}-i_2}0^{i_2}(c^{\hat{s}}d1^{\widehat{m}}0^{\widehat{m}})^2 c^{\hat{s}}1^{\widehat{m}-i_3}0^{i_3}(c^{\hat{s}}d1^{\widehat{m}}0^{\widehat{m}})^2$
$\to c^i0c^{\hat{s}}1^{\widehat{m}-j_1}0^{j_1}(c^{\hat{s}}d1^{\widehat{m}}0^{\widehat{m}})c^{\hat{s}}1^{\widehat{m}-i_2}0^{i_2}(c^{\hat{s}}d1^{\widehat{m}}0^{\widehat{m}})^2 c^{\hat{s}}1^{\widehat{m}-i_3}0^{i_3}(c^{\hat{s}}d1^{\widehat{m}}0^{\widehat{m}})^2,$

(4) $c^i0c^{\hat{s}}1^{\widehat{m}-j_1}0^{j_1}(c^{\hat{s}}d1^{\widehat{m}}0^{\widehat{m}})\underline{c^{\hat{s}}1^{\widehat{m}-i_2}0^{i_2}}(c^{\hat{s}}d1^{\widehat{m}}0^{\widehat{m}})^2 c^{\hat{s}}1^{\widehat{m}-i_3}0^{i_3}(c^{\hat{s}}d1^{\widehat{m}}0^{\widehat{m}})^2$
$\to c^i0c^{\hat{s}}1^{\widehat{m}-j_1}0^{j_1}(c^{\hat{s}}d1^{\widehat{m}}0^{\widehat{m}})^3 c^{\hat{s}}1^{\widehat{m}-i_3}0^{i_3}(c^{\hat{s}}d1^{\widehat{m}}0^{\widehat{m}})^2,$

(5) $c^i0c^{\hat{s}}1^{\widehat{m}-j_1}0^{j_1}(c^{\hat{s}}d1^{\widehat{m}}0^{\widehat{m}})^2 c^{\hat{s}} \cdot \underline{d1^{j_2}1^{\widehat{m}-j_2}0^{\widehat{m}}}c^{\hat{s}}1^{\widehat{m}-i_3}0^{i_3}(c^{\hat{s}}d1^{\widehat{m}}0^{\widehat{m}})^2$
$\to c^i0c^{\hat{s}}1^{\widehat{m}-j_1}0^{j_1}(c^{\hat{s}}d1^{\widehat{m}}0^{\widehat{m}})^2 c^{\hat{s}}1^{\widehat{m}-j_2}0^{\widehat{m}}c^{\hat{s}}1^{\widehat{m}-i_3}0^{i_3}(c^{\hat{s}}d1^{\widehat{m}}0^{\widehat{m}})^2,$

(6) $c^i0c^{\hat{s}}1^{\widehat{m}-j_1}0^{j_1}(c^{\hat{s}}d1^{\widehat{m}}0^{\widehat{m}})^2 c^{\hat{s}}1^{\widehat{m}-j_2}0^{j_2}\underline{0^{\widehat{m}-j_2}}c^{\hat{s}}1^{\widehat{m}-i_3}0^{i_3}(c^{\hat{s}}d1^{\widehat{m}}0^{\widehat{m}})^2$
$\to c^i0c^{\hat{s}}1^{\widehat{m}-j_1}0^{j_1}(c^{\hat{s}}d1^{\widehat{m}}0^{\widehat{m}})^2 c^{\hat{s}}1^{\widehat{m}-j_2}0^{j_2}(c^{\hat{s}}d1^{\widehat{m}}0^{\widehat{m}})^2,$

(7) $\underline{c^i0}c^{\hat{s}}1^{\widehat{m}-j_1}0^{j_1}(c^{\hat{s}}d1^{\widehat{m}}0^{\widehat{m}})^2 c^{\hat{s}}1^{\widehat{m}-j_2}0^{j_2}(c^{\hat{s}}d1^{\widehat{m}}0^{\widehat{m}})^2$
$\to c^{\hat{s}}1^{\widehat{m}-j_1}0^{j_1}(c^{\hat{s}}d1^{\widehat{m}}0^{\widehat{m}})^2 c^{\hat{s}}1^{\widehat{m}-j_2}0^{j_2}(c^{\hat{s}}d1^{\widehat{m}}0^{\widehat{m}})^2.$

For all other rules of $(\ell_k \to r_k)$ of R_L, the rules of $\tilde{R}_k$ are defined analogously. In particular, for a rule of the form $(\ell_k \to r_k) = (\triangleright d_{i_1}d_{i_2}d_{i_3} \to \triangleright d_{j_1})$ of R_L, the rules in the corresponding group $\tilde{R}_k$ are defined with the restriction that only the element $\triangleright \in$ LC is used in rule (1). Then $\tilde{R}_L = \bigcup_{i=1}^{s} \tilde{R}_i$ is a finite factor-erasing string-rewriting system, and for each $x \in \Sigma^*$, so is the system $\tilde{R}_L(x) = \tilde{R}_L \cup \{(\triangleright\psi(s_0 x)\triangleleft \to \lambda)\}$.

Obviously, for all $w, z \in \hat{\Delta}^*$, if $\triangleright\psi(w)\triangleleft \Rightarrow_{\hat{R}_L}^* \triangleright\psi(z)\triangleleft$, then $\triangleright\psi(w)\triangleleft \Rightarrow_{\tilde{R}_L}^*$ $\triangleright\psi(z)\triangleleft$. It follows that $\tilde{R}_L(x)$ is not λ-confluent, if $\hat{R}(x)$ is not λ-confluent. Conversely, if $\triangleright\psi(w)\triangleleft \Rightarrow_{\tilde{R}_L}^* \triangleright\psi(z)\triangleleft$, then also $\triangleright\psi(w)\triangleleft \Rightarrow_{\hat{R}_L}^* \triangleright\psi(z)\triangleleft$. Observe that the rules of a subsystem $\tilde{R}_i$ are necessarily applied in the order given. If the i-th rule of $\hat{R}_L$ is to be simulated, then the first rule of the subsystem $\tilde{R}_i$ is used to mark the first letter of the occurrence of the left-hand side $\psi(\ell_i)$ to be rewritten by the factor $c^i 0$. The left context $u \in$ LC of this rule ensures that no still uncompleted simulation of an application of a rule of $\hat{R}_L$ is currently being executed on the factor of length at most three to the left of this position. Thus, it is possible to interleave simulations of several rules of $\hat{R}_L$, but these rules must be applied to factors of the current word that are sufficiently far apart from each other. It follows that $\tilde{R}_L(x)$ is λ-confluent, if $\hat{R}_L(x)$ is. Hence, we obtain the following result.

Lemma 4. *The following problem is not recursively enumerable:*

INSTANCE: A word $x \in \Sigma^*$.
QUESTION: Is the factor-erasing string-rewriting system $\tilde{R}_L(x)$ λ-confluent?

Thus, we have the following result.

Theorem 4. *The problem of deciding whether a finite factor-erasing string-rewriting system is λ-confluent is not recursively enumerable.*

5   Conclusion and Open Problems

It is known that lc-R-automata of type $\mathcal{R}_2$ accept the context-free languages [1]. Further, if M is an lc-R-automaton of type $\mathcal{R}_2$ such that the corresponding string-rewriting system $R(M)$ is confluent, then $L(M)$ and its reversal, $(L(M))^R$, are both deterministic context-free languages [17]. What can be said about the class of languages that are accepted by λ-confluent lc-R-automata of type $\mathcal{R}_2$?

For finite special string-rewriting systems, λ-confluence is decidable in polynomial time. Here we have seen that λ-confluence becomes undecidable (in fact, non-r.e.), if we add left and right contexts to special string-rewriting systems. However, the contexts used in the proof of Theorem 4 are rather large. Does λ-confluence remain undecidable even for finite factor-erasing string-rewriting systems with left and right contexts of length at most one? If not, what is the smallest positive integer constant k such that λ-confluence is decidable for finite factor-erasing string-rewriting systems with left and right contexts of length at most $k - 1$, but it becomes undecidable for finite factor-erasing string-rewriting systems with contexts of length up to k?

References

1. Basovník, S.: Learning restricted restarting automata using genetic algorithm. Master's thesis, Charles University, Faculty of Mathematics and Physics, Prague (2010)
2. Basovník, S., Mráz, F.: Learning limited context restarting automata by genetic algorithms. In: Dassow, J., Truthe, B. (eds.) Proc. Theorietag 2011, pp. 1–4. Otto-von-Guericke-Universität, Magdeburg (2011)
3. Beeri, C.: An improvement on Valiant decision procedure for equivalence of deterministic finite turn pushdown machines. Theoret. Comput. Sci. 3, 305–320 (1976)
4. Book, R., Otto, F.: String-Rewriting Systems. Springer, New York (1993)
5. Černo, P.: Clearing restarting automata and grammatical inference. In: Heinz, J., de la Higuera, C., Oates, T. (eds.) JMLR Workshop and Conf. Proc. ICGI 2012, vol. 21, pp. 54–68. University of Maryland, College Park (2012)
6. Černo, P., Mráz, F.: Clearing restarting automata. Fund. Inf. 104, 17–54 (2010)
7. Jančar, P., Mráz, F., Plátek, M., Vogel, J.: On monotonic automata with a restart operation. J. Autom. Lang. Comb. 4, 287–311 (1999)
8. Jančar, P., Mráz, F., Plátek, M., Vogel, J.: Restarting automata. In: Reichel, H. (ed.) FCT 1995. LNCS, vol. 965, pp. 283–292. Springer, Heidelberg (1995)
9. Jurdziński, T., Otto, F., Mráz, F., Plátek, M.: On the complexity of 2-monotone restarting automata. Theory Comput. Systems 42, 488–518 (2008)
10. Kapur, D., Krishnamoorthy, M., McNaughton, R., Narendran, P.: An $O(|T|^3)$ algorithm for testing the Church-Rosser property of Thue systems. Theor. Comput. Sci. 35, 109–114 (1985)
11. Mráz, F., Otto, F., Plátek, M.: Learning analysis by reduction from positive data. In: Sakakibara, Y., Kobayashi, S., Sato, K., Nishino, T., Tomita, E. (eds.) ICGI 2006. LNCS (LNAI), vol. 4201, pp. 125–136. Springer, Heidelberg (2006)
12. Ó'Dúnlaing, C.: Finite and Infinite Regular Thue Systems. Ph.D. thesis, University of California, Santa Barbara (1981)
13. Ó'Dúnlaing, C.: Undecidable questions related to Church-Rosser Thue systems. Theor. Comput. Sci. 23, 339–345 (1983)
14. Otto, F.: Some undecidability results for non-monadic Church-Rosser Thue systems. Theoret. Comput. Sci. 33, 261–278 (1984)
15. Otto, F.: On deciding the confluence of a finite string-rewriting system on a given congruence class. J. Comput. Syst. Sci. 35, 285–310 (1987)
16. Otto, F.: Completing a finite special string-rewriting system on the congruence class of the empty word. Appl. Algebra in Eng. Comm. and Comput. 2, 257–274 (1992)
17. Otto, F., Černo, P., Mráz, F.: Limited context restarting automata and McNaughton families of languages. In: Freund, R., Holzer, M., Truthe, B., Ultes-Nitsche, U. (eds.) Proc. Fourth Workshop on Non-Classical Models for Automata and Applications (NCMA 2012). books@ocg.at, Band 290, pp. 165–180. Oesterreichische Computer Gesellschaft, Wien (2012)
18. Valiant, L.: The equivalence problem for deterministic finite-turn pushdown automata. Information and Control 25, 123–133 (1974)

Comparing Two-Dimensional One-Marker Automata to Sgraffito Automata*

Daniel Průša[1], František Mráz[2], and Friedrich Otto[3]

[1] Czech Technical University, Faculty of Electrical Engineering
Department of Cybernetics, Center for Machine Perception
Karlovo náměstí 13, 121 35 Prague 2, Czech Republic
prusapa1@cmp.felk.cvut.cz

[2] Charles University, Faculty of Mathematics and Physics
Department of Computer Science
Malostranské nám. 25, 118 25 Prague 1, Czech Republic
frantisek.mraz@mff.cuni.cz

[3] Fachbereich Elektrotechnik/Informatik, Universität Kassel
34109 Kassel, Germany
otto@theory.informatik.uni-kassel.de

Abstract. We compare two types of automata for accepting picture languages to each other: the *two-dimensional one-marker automaton* and the *sgraffito automaton*. On the one hand, it is shown that deterministic sgraffito automata are strictly more powerful than deterministic two-dimensional one-marker automata. On the other hand, nondeterministic two-dimensional one-marker automata accept some picture languages that cannot be accepted by sgraffito automata. However, if nondeterministic two-dimensional one-marker automata were to accept all picture languages that are accepted by (deterministic) sgraffito automata, then the complexity classes NL (nondeterministic logarithmic space) and P (deterministic polynomial time) would coincide. Accordingly, it is likely that the classes of picture languages accepted by these two types of nondeterministic automata are incomparable under inclusion.

Keywords: picture languages, two-dimensional one-marker automaton, sgraffito automaton, recognizable picture languages.

1 Introduction

The *two-dimensional one-marker automaton* ($2M_1A$) is a device for accepting picture languages. It was introduced by Blum and Hewitt [1] as a four-way finite-state automaton equipped with an additional marker (*pebble*). These devices are able to recognize some important topological properties, e.g., whether a black-and-white picture contains only a single component of black pixels.

---

* The first author was supported by the Grant Agency of the Czech Republic under the project P103/10/0783 and the second author under the projects P103/10/0783 and P202/10/1333.

S. Konstantinidis (Ed.): CIAA 2013, LNCS 7982, pp. 268–279, 2013.

Ito et. al. gave a quantitative analysis of this model, studying the space required by a three-way Turing machine to simulate a $2M_1A$ [10]. They also studied two-dimensional one-marker Turing machines working in logarithmic space [8]. Also the one-dimensional variant of the $2M_1A$ that is equipped with k distinguishable markers received much attention, as this model characterizes the complexity classes L and NL of deterministic and nondeterministic logarithmic space, similar to some other related devices like the two-way finite automaton with multiple heads or linearly bounded counters [6].

On the other hand, the *sgraffito automaton* (2SA) was introduced only recently [13]. It is a bounded two-dimensional Turing machine that in each step replaces the currently scanned symbol by a symbol of smaller weight. Hence, it can be seen as a two-dimensional variant of the Hennie machine [5], visiting each of its tape positions just a bounded number of times independently of the size of the input. It is more powerful than the on-line tessellation automaton [9], which characterizes the family of recognizable picture languages REC [4]. Also it has recently been shown in [14] that the deterministic sgraffito automaton is more expressive than the *four-way alternating automaton* [12], and it accepts all *sudoku-deterministically recognizable* picture languages [2]. On the other hand, the classes of picture languages accepted by the sgraffito automaton, the one-marker automaton and the on-line tessellation automaton have the same main closure properties with respect to language operations [11,13], and all these classes collapse to the regular languages when restricted to one-dimensional inputs, that is, strings.

Here we compare the expressive power of deterministic and nondeterministic two-dimensional one-marker automata to that of deterministic sgraffito automata, and we relate it to the family REC. We first show that a deterministic 2SA can simulate a deterministic $2M_1A$, and then we use a result by Hsia and Yeh [7] on one-dimensional deterministic k-marker automata to derive a separation between the two models. Next we turn to the nondeterministic variants of these models. We provide examples of picture languages that are accepted by $2M_1A$, but that are beyond the power of the 2SA or outside of REC. Finally, we prove that the complexity classes NL (nondeterministic logarithmic space) and P (deterministic polynomial time) would coincide if $2M_1A$ could accept all picture languages that are accepted by deterministic sgraffito automata. As it is expected that NL is a proper subclass of P, this means that most likely the classes of picture languages that are accepted by $2M_1A$ and by 2SA (and in fact, by 2DSA) are incomparable under inclusion.

The paper is structured as follows. In the next section we introduce the basic notation and notions on picture languages and the sgraffito automaton in short. Then we separate the deterministic $2M_1A$ from the deterministic 2SA in Section 3, and in Section 4 we study the relation of the latter automata to the nondeterministic $2M_1A$. The paper closes with a short summary and some open problems in Section 5.

2   Preliminaries

Here we use the common notation and terms on pictures and picture languages
(see, e.g., [4]). If Σ is a finite alphabet, then $\Sigma^{*,*}$ is used to denote the set of
rectangular pictures over Σ, that is, if $P \in \Sigma^{*,*}$, then P is a two-dimensional
array of symbols from Σ. If P is of size $m \times n$, this is denoted by $P \in \Sigma^{m,n}$. For
$1 \le i \le m$ and $1 \le j \le n$, $P(i,j)$ (or shortly $P_{i,j}$) identifies the symbol located
in the i-th row and the j-th column of P.

Two (partial) binary operations are used to concatenate pictures. Let $P \in
\Sigma^{k,\ell}$ and $Q \in \Sigma^{m,n}$. The *column concatenation* $P \oplus Q$ is defined iff $k = m$, and
the *row concatenation* $P \ominus Q$ is defined iff $\ell = n$. These products are specified
by the following schemes:

$$P \oplus Q = \begin{pmatrix} P_{1,1} \cdots P_{1,\ell} & Q_{1,1} \cdots Q_{1,n} \\ \vdots \ddots \vdots & \vdots \ddots \vdots \\ P_{k,1} \cdots P_{k,\ell} & Q_{m,1} \cdots Q_{m,n} \end{pmatrix} \text{ and } P \ominus Q = \begin{pmatrix} P_{1,1} & \cdots & P_{1,\ell} \\ \vdots & \ddots & \vdots \\ P_{k,1} & \cdots & P_{k,\ell} \\ Q_{1,1} & \cdots & Q_{1,n} \\ \vdots & \ddots & \vdots \\ Q_{m,1} & \cdots & Q_{m,n} \end{pmatrix}.$$

We use a set of five special markers (*sentinels*) $\mathcal{S} = \{\vdash, \dashv, \top, \bot, \#\}$, where we
assume $\Sigma \cap \mathcal{S} = \emptyset$ for any alphabet Σ considered. In order to enable an automaton
to detect the border of P easily, we define the *boundary picture* $\widehat{P}$ over $\Sigma \cup \mathcal{S}$ of
size $(m + 2) \times (n + 2)$, which is illustrated by the following scheme:

#	$\top$	$\top$	$\cdots$	$\top$	$\top$	#
$\vdash$						$\dashv$
$\vdots$			P			$\vdots$
$\vdash$						$\dashv$
#	$\bot$	$\bot$	$\cdots$	$\bot$	$\bot$	#

Here the symbols $\vdash, \dashv, \top$ and $\bot$ uniquely identify the corresponding borders (left,
right, top, bottom) of $\widehat{P}$, while the symbol $\#$ marks the corners of $\widehat{P}$. Actually,
our automata will not be able to visit these corner elements unless in the special
case of the empty picture $P \in \Sigma^{0,0}$ (see below), but we include them anyway in
order to have a rectangular form for $\widehat{P}$.

Let $\mathcal{H} = \{R, L, D, U, Z\}$ be the set of *head movements*, where the first four
elements denote directions (right, left, down, up) and Z represents no movement.

Definition 1. A two-dimensional sgraffito automaton (2SA) *is given by a 7-
tuple* $\mathcal{A} = (Q, \Sigma, \Gamma, \delta, q_0, Q_F, \mu)$, *where* Q *is a finite set of states,* Σ *is an input
alphabet,* Γ *is a working alphabet containing* Σ, $q_0 \in Q$ *is the initial state,*
$Q_F \subseteq Q$ *is a set of final states,* $\delta : (Q \smallsetminus Q_F) \times (\Gamma \cup \mathcal{S}) \to 2^{Q \times (\Gamma \cup \mathcal{S}) \times \mathcal{H}}$ *is a
transition relation, and* $\mu : \Gamma \to \mathbb{N}$ *is a weight function, such that the following
two properties are satisfied:*

1. $\mathcal{A}$ is bounded, *that is, whenever it scans a symbol from $\mathcal{S}$, then it immediately moves to the nearest field of P without changing this symbol,*
2. $\mathcal{A}$ is weight-reducing, *that is, for all $q, q' \in Q$, $d \in \mathcal{H}$, and $a, a' \in \Gamma$, if $(q', a', d) \in \delta(q, a)$, then $\mu(a') < \mu(a)$.*

Finally, $\mathcal{A}$ is deterministic (a 2DSA), *if $|\delta(q, a)| \leq 1$ for all $q \in Q$ and $a \in \Gamma \cup \mathcal{S}$.*

The notions of configuration and computation are defined as usual. In the initial configuration on input P, the tape contains $\widehat{P}$, $\mathcal{A}$ is in state q_0, and its head scans the top-left corner of P. If P is the empty picture, then the head initially scans the bottom-right corner of $\widehat{P}$. The automaton $\mathcal{A}$ accepts P iff there is a computation of $\mathcal{A}$ on input P that finishes in a state from Q_F.

When designing a sgraffito automaton for a picture language, it suffices to describe a bounded two-dimensional Turing machine that visits each tape cell only a constant number of times (i.e., a two-dimensional Hennie machine [5]). In [13] it was shown that any such machine can be transformed into an equivalent sgraffito automaton (preserving determinism). This fact will be utilized in our constructive proofs below.

When Blum and Hewitt introduced the marker automaton, they considered two types of markers – *physical* and *abstract* markers, but they proved that k physical markers can be simulated by k abstract markers and vice versa. Accordingly, we just concentrate on physical markers. To move a physical marker, it is required that the head moves over the marker and transfers the marker by a head movement to a neighbouring position. The deterministic 2M$_1$A is denoted here as 2DM$_1$A. Note that 2M$_1$As are bounded in the same way as 2SAs.

3   Comparing the Deterministic Models

Deterministic sgraffito automata are quite powerful. During a computation, they can visit any tape field constantly many times. This makes it possible for a 2DSA-automaton to perform a *depth-first search* (DFS) on a graph represented within a two-dimensional picture P. Actually this property was used in [13] for searching the graph of a computation of a nondeterministic four-way finite-state automaton (4FA). In general, each tape field $P(i, j)$ can contain a subset of constant size of the vertices of a graph satisfying the property that all vertices adjacent to any vertex from $P(i, j)$ are located either in $P(i, j)$ or in neighbouring fields $P(i', j')$ such that $|i - i'| + |j - j'| = 1$. Hence, each tape field represents a set of vertices and edges of size limited by a constant.

A 2DSA-automaton can now perform a DFS on graphs of this form [3]. During DFS traversal the visited nodes and edges are marked. Initially, all vertices are marked as *fresh*. When a vertex is visited for the first time, it is marked as *open*. When the search backtracks from a vertex, it is marked as *closed*. Similarly, all edges of the graph are initially marked as *unexplored*. When an edge is used by the algorithm for the first time, its status is changed to *discovery*, if it leads to to a fresh vertex, otherwise it is marked as *back* edge.

During the DFS traversal through the graph, each vertex and each edge is visited only a constant number of times, and so this DFS can be implemented on a 2DSA using a finite number of symbols in its working alphabet.

In the proof below we will use a 2DSA to perform a DFS on a graph consisting of nodes of the form (i, j, q), where (i, j) is a coordinate in a picture and q is a state of a 2DM$_1$A when visiting this coordinate.

Theorem 1. $\mathcal{L}(2DM_1A)$ *is contained in* $\mathcal{L}(2DSA)$.

Proof. Let $\mathcal{M}$ be a 2DM$_1$A, let Q be its set of states, let $q_0 \in Q$ be its initial state, and let P be an input picture of size $m \times n$. W.l.o.g. we can assume that $\mathcal{M}$ is allowed to halt (and accept or reject) only when it scans the marker and when it does not move the head. Otherwise it always performs an instruction.

The key idea of the presented construction is to design a 2DSA $\mathcal{A}$ that detects all situations in which $\mathcal{M}$ leaves the marker, moves across some tape fields (without the marker) and returns back to the marker. We write $(i, j, q) \circlearrowleft_{\mathcal{M}} q'$ when there is such a traversal of $\mathcal{A}$ starting at position (i, j) in state q and ending at the same position in state q'. We define a directed graph $G_M = (V_M, E_M)$, where $V_M = \{1, \ldots, m\} \times \{1, \ldots, n\} \times Q$. We represent $(i, j, q) \circlearrowleft_{\mathcal{M}} q'$ in G_M by the edge $((i, j, q), (i, j, q'))$. Moreover, for every configuration in which $\mathcal{M}$ scans the marker and performs a transition that either moves the marker or does not move the head at all, we add the edge representing this transition.

Suppose G_M has been created. Since $\mathcal{M}$ is deterministic, the outgoing degree of each vertex is at most one. $\mathcal{A}$ can easily detect whether $\mathcal{M}$ accepts or not. It starts at $(1, 1, q_0)$ and follows the path given by outgoing edges. The path either ends in some (i, j, q) or it closes a loop onto itself (this can be detected by marking the vertices traversed). $\mathcal{A}$ accepts in the former case if $q \in Q_F$.

To construct G_M, we define an auxiliary graph $G = (V, E)$ as follows. The vertices are tuples of the form (i, j, q, r), where (i, j) is a position, $q \in Q$, and $r \in \{0, 1\}$. The bit r indicates the presence of the marker (0 – not present, 1 – present) at the time a transition is performed over the field (i, j) from state q. We define V as the union of two subsets:

1. V_1 contains each vertex (i, j, q, r) where $r = 0$;
2. V_2 contains $(i, j, q, 1)$ iff $\mathcal{M}$ in state q scanning the tape field (i, j) with the marker makes a transition moving to a neighbouring field without moving the marker.

Due to our assumption on computations of $\mathcal{M}$ above, each configuration given by $v = (i, j, q, r) \in V$ induces one transition. Let the transition move the head to a position (i', j') and let it change the state to q'. Then, this transition is represented by adding the edge $((i, j, q, r), (i', j', q', 0))$ to E. Each vertex in V_2 is thus a source (with no incoming edges). Since each vertex has one outgoing edge, each component of G has the form depicted in Figure 1(b): there is one directed cycle and possibly several trees rooted in vertices of this cycle.

If there is a directed path in G from $v_1 = (i, j, q, 1) \in V$ to $v_2 = (i, j, q', 0) \in V$, and if this path does not contain any other vertex of the form $(i, j, q'', 0)$, then

$(i, j, q) \circlearrowright_{\mathcal{M}} q'$. Note that the indicator bit in v_2 is 0 even though following the path the head of $\mathcal{M}$ will return to the marker. This is intensional for technical reasons – v_2 is the last element of the path considered, but it can be an inner element of other paths which start in different configurations. We want to represent both these situations just by a single vertex of G. Conversely, if $(i, j, q) \circlearrowright_{\mathcal{M}} q'$, then G necessary contains a directed path with the aforementioned properties.

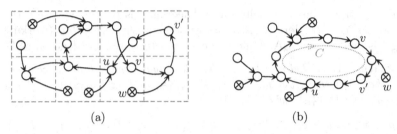

(a)   (b)

Fig. 1. An example of a component of the graph G, displayed over a portion of the tape (a), and also as an isomorphic graph without the tape (b). The vertices $\bigcirc$ and $\otimes$ are from V_1 and V_2, respectively. If the DFS on G^R is started at v and it reaches w, the stack in the corresponding field contains u and v, where u is the topmost element, and thus, the directed path from w to u in G is detected. If the DFS is started at v' and reaches in two steps w, the same path is found again, since u is contained in the walk formed of the path from w to v' and the cycle C.

Let C be the cycle of a component and let v be one of its vertices. Consider the DFS on G^R, the reversion of G, starting at v, and let $w = (i, j, q, 1) \in V_2$. Assume that the DFS detects a directed path from v to w in G^R, i.e., a directed path from w to v in G (all the vertices of the path have currently the status *open*), see Figure 1(a). Take the walk in G created by prolonging the path by adding the cycle C after v. Now, if $(i, j, q) \circlearrowright_{\mathcal{M}} q'$, then $(i, j, q', 0)$ is a vertex contained in this walk. This observation helps us to design an algorithm for $\mathcal{A}$ for finding all these situations. We use a stack $T(i, j)$ in each tape field (i, j) to record the order of vertices (from the tape field (i, j)) that appear in the walks described. This information is important to guarantee that no directed path in G that passes through the starting tape field multiple times will be used to produce an edge in G_M. This algorithm proceeds as follows:

1. Choose a vertex in G, follow its outgoing edges (and mark them) until a directed cycle C is detected. Assume this happens at a vertex v. Mark v.
2. Traverse back through C until v is encountered again. During this process, push every vertex visited at a tape field (i, j) onto $T(i, j)$.
3. Starting at v, perform a DFS on G^R. Whenever a vertex $u = (i, j, q, r)$ is visited and its status changes to *open*, push it onto $T(i, j)$. When the status is changed to *closed*, pop the vertex from $T(i, j)$. Whenever a vertex $w = (i, j, q, 1) \in V_2$ is reached, check the topmost element in $T(i, j)$ (before pushing w onto it). This is the only vertex in which the directed path in G

starting at w ends, without passing more vertices at the same tape field. If the topmost element is $u = (i, j, q', 0)$, then $(i, j, q) \circlearrowleft_{\mathcal{M}} q'$ (hence, add the induced edge to G_M).

4. Return to step 1 to find the next component of G.

The number of operations performed over each stack is bounded by a constant, since the corresponding field is visited constantly many times. Hence, all the states of a stack $T(i, j)$ can be represented in the tape field (i, j).   □

In order to prove that the inclusion in Theorem 1 is proper, we use a theorem of Hsia and Yeh [7]. Here a *two-way k-marker finite automaton* (working over strings) is denoted as $\mathsf{M_kA}$, while its deterministic variant is denoted as $\mathsf{DM_kA}$.

Proposition 1. [7] *Let* $n \geq 1$, *let* L *be a language over* Σ *that is accepted by an always halting* $\mathsf{DM_nA}$, *but not by any* $\mathsf{DM_{n-1}A}$, *and let* $\rho(L, \$)$ *be the following language over* $\Sigma \cup \{\$\}$, *where* $\$ \notin \Sigma$:

$$\rho(L, \$) = \{\, w_1 \$ w_2 \$ \dots \$ w_m \mid m \geq 2, \, |\{\, i \mid w_i \in L \,\}| = |\{\, j \mid w_j \notin L \,\}| \,\}.$$

Then $\rho(L, \$)$ *is accepted by a* $\mathsf{DM_{n+1}A}$, *but it is not accepted by any* $\mathsf{DM_nA}$.

For a string $w \in \Sigma^*$ and a symbol $a \in \Sigma$, let $S(w, a)$ denote the square picture over Σ of size $|w| \times |w|$, where the first row equals w and all remaining fields contain the symbol a. Moreover, for a (string) language $L \subseteq \Sigma^*$, let $S(L, a)$ be the picture language defined as $S(L, a) = \{\, S(w, a) \mid w \in L \,\}$.

Lemma 1. *Let* L *be a string language over* Σ, *and let* $a \in \Sigma$. *If the picture language* $S(L, a)$ *is accepted by a* $\mathsf{2DM_1A}$, *then* L *is accepted by a* $\mathsf{DM_4A}$.

Proof. Let $\mathcal{M}_1$ be a $\mathsf{2DM_1A}$ accepting $S(L, a)$. We design a $\mathsf{DM_4A}$ $\mathcal{M}_2$ that simulates $\mathcal{M}_1$ as follows. Let the four markers of $\mathcal{M}_2$ be denoted as K_1, K_2, K_3, and K_4. A configuration of $\mathcal{M}_1$ is identified by the marker position (i_M, j_M), the head position (i_H, j_H), and the state of the finite control. $\mathcal{M}_2$ uses K_1 and K_2 to represent the marker position by placing K_1 on the i_M-th field of the tape and K_2 on the j_M-th field. Analogously, the head position is represented by K_3 and K_4, while the state of $\mathcal{M}_1$ is stored within the state of $\mathcal{M}_2$. Initially, all the markers K_i are placed at the leftmost tape field. $\mathcal{M}_2$ now simulates $\mathcal{M}_1$ step by step. The symbol scanned by $\mathcal{M}_1$ is retrieved from the tape iff its head is located in the first row (i.e., K_3 is at the first tape field), otherwise $\mathcal{M}_1$ is known to scan the symbol a. Finally, $\mathcal{M}_2$ accepts, if and when $\mathcal{M}_1$ enters a final state.   □

Based on Proposition 1 we now construct a string language that is not accepted by any $\mathsf{DM_4A}$. Let $L_1 = \{\, w \in \{0, 1\}^* \mid |w|_0 = |w|_1 \,\}$, which is not accepted by any $\mathsf{DM_1A}$, as one-marker automata only accept regular (string) languages. Next we define L_2 to L_4 by taking $L_2 = \rho(L_1, \$_1)$, $L_3 = \rho(L_2, \$_2)$, and $L_4 = \rho(L_3, \$_3)$, where $\$_1$, $\$_2$, and $\$_3$ are auxiliary separators. For each $1 \leq k \leq 4$, the language L_k is not accepted by any $\mathsf{DM_kA}$, and thus, L_4 is the language we are looking for. To illustrate these languages we give some examples:

$$011\$_110\$_11001\$_100 \in L_2, \qquad\qquad u_2 = 01\$_11\$_2100\$_101 \notin L_3,$$
$$u_1 = 01\$_11\$_2101 \in L_3, \qquad u_1\$_3u_2 = 01\$_11\$_2101\$_301\$_11\$_2100\$_101 \in L_4.$$

Lemma 2. *The language $S(L_4, 0)$ is accepted by a* 2DSA.

Proof. We describe a 2DSA $\mathcal{A}$ for $S(L_4, 0)$. Let P be an input picture over $\Sigma = \{0, 1, \$_1, \$_2, \$_3\}$. $\mathcal{A}$ can easily check whether P is a square picture in which all rows except the first one only contain the symbol 0.

Let R be the first row of P, and let x be a working symbol. The idea of the computation consists in iteratively evaluating membership of substrings in R in the languages L_i. For this the following algorithm is used, starting with $i = 1$:

1. Check whether the consecutive substrings u from $\{0, 1, x\}^*$ in R (delimited by $\$_1$, $\$_2$, $\$_3$ or by the left or right sentinel, respectively) contain the same number of 0's and 1's. For each substring u, write the result into its leftmost field (1 – yes, 0 – no), and replace the content of the other fields of u by x.
2. If $i < 4$, rewrite every occurrence of the symbol $\$_i$ in R into x, increase i by 1, and return to step 1.

Figure 2(a) illustrates the stages of this iteration on input $u_1 \$_3 u_2$.

```
0 1 $₁ 1 $₂ 1 0 1 $₃ 0 1 $₁ 1 $₂ 1 0 0 $₁ 0 1        1 0 0 1 0 0 0 1 1 0
1 x x 0 $₂ 0 x x $₃ 1 x x 0 $₂ 0 x x x 1 x          x x 0 1 0 0 0 1 1 0
1 x x x x 0 x x $₃ 1 x x x x 1 x x x x x            x x x x 0 0 0 1 1 0
1 x x x x x x x x 0 x x x x x x x x x x             x x x x x 0 0 x 1 0
1 x x x x x x x x x x x x x x x x x x x             x x x x x x 0 x x 0
                    (a)                                     (b)
```

Fig. 2. (a) The first row is the input, the other rows show the result after each of the four iterations. (b) A comparison of the number of 0's and 1's in the string 1001000110. Each partial product is written into a new row. The last row indicates that there are more 0's than 1's.

It remains to show how $\mathcal{A}$ makes the evaluation in step 1. Consider a general situation when R stores a string over $\Sigma \cup \{x\}$. $\mathcal{A}$ detects the consecutive blocks of symbols in $\{0, 1, x\}$ by moving its head from left to right across R. Assume that $R' \subseteq R$ stores such a complete block u. Then $\mathcal{A}$ uses the space below this block to compare the numbers of 0's and 1's. First, it copies u to the row below. Then it scans this copy, replacing one occurrence of the symbol 0 and one occurrence of the symbol 1 by x. This subroutine is repeated until there is no occurrence of 0 or of 1 left. At this point $\mathcal{A}$ detects whether the numbers match or not. An evaluation of the string 1001000110 is demonstrated in Figure 2(b).

Since there are 4 iterations in the main algorithm and each of them visits each tape field constantly many times, the above algorithm can be realized by the 2DSA $\mathcal{A}$.   □

From the choice of L_4 and from Lemma 1 we see that $S(L_4, 0)$ is not accepted by any 2DM$_1$A. This yields the following improvement of Theorem 1.

Corollary 1. $\mathcal{L}(2DM_1A)$ *is properly contained in* $\mathcal{L}(2DSA)$.

4   Comparing the Nondeterministic Models

We define two picture languages that will allow us to separate $\mathcal{L}(2M_1A)$ from $\mathcal{L}(2SA)$ and REC. Let L_{copy} be the picture language over $\Sigma = \{\square, \boxtimes, \blacksquare\}$ that consists of all pictures $U \oplus C \oplus U$, where U is a square picture over $\{\square, \blacksquare\}$, and C is a column of symbols $\boxtimes$. Further, let L_{cols} be the subset of L_{copy} containing those pictures $U \oplus C \oplus U$ in which each column of U contains exactly one black pixel $\blacksquare$ (See Figures 3(a) and 3(b)). L_{cols} is a variant of a language used in [12] to separate the languages accepted by four-way alternating automata from REC, and L_{copy} is a widely used example of a rather complicated picture language [4]. Already in [8] Ito et. al. noticed that this language is accepted by a 2M$_1$A. Here we give our own proof, as we use the same strategy to accept L_{cols} by a 2DM$_1$A.

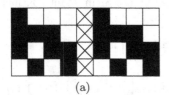   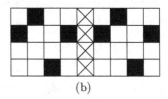

(a)   (b)

Fig. 3. (a) A picture of size 4×9 from L_{copy}. (b) A picture of size 4×9 from L_{cols}.

Lemma 3. L_{copy} *is accepted by a* 2M$_1$A.

Proof. A 2M$_1$A $\mathcal{M}$ can accept L_{copy} as follows. Let P be an input picture. First, $\mathcal{M}$ checks whether P is of size $n \times (2n + 1)$ (starting at the top-left corner, repeatedly move the head two fields to the right and one field down), marks the topmost field of the central column (which is detected by $\mathcal{M}$ when moving diagonally from the bottom-right corner), and checks whether the central column only contains the symbol $\boxtimes$, and whether the other columns do not contain this symbol (done by scanning P column by column).

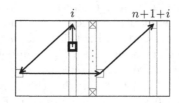

Fig. 4. Finding the corresponding column in the second half of a picture of size $n \times (2n + 1)$. The walk is a composition of vertical, horizontal and diagonal moves.

If all these tests succeed, then P is of the form $U \oplus C \oplus V$, where U and V are square pictures over $\{\square, \blacksquare\}$ and C is a column of $\boxtimes$'s. Next $\mathcal{M}$ moves its head to the top-left corner. The marker is still at this position as it has not yet been moved. Now the following algorithm is used to check whether $U = V$:

1. Let the marker be placed at a position (i, j). Memorize the symbol $U(i, j)$.
2. Leaving the marker at this position, use the movement depicted in Figure 4 to reach position $(1, n + 1 + i)$ (i.e., the topmost field of column i of V).
3. Move downwards until the border is detected. During this movement, whenever $V(r, j) = U(i, j)$ for a row r, then there is the nondeterministic choice of moving leftwards along row r. If the marker is reached, then $r = i$ and $U(i, j) = V(i, j)$, otherwise this branch of the computation fails.
4. If the marker is reached in step 3, then the marker is moved to the neighbouring field in the same row. If the symbol ⊠ is scanned, then the marker is moved to the leftmost field of the next row. If there is no next field of U, accept, otherwise continue with step 1.   □

Lemma 4. L_{cols} *is accepted by a* 2DM$_1$A.

Proof. The construction of a 2DM$_1$A $\mathcal{M}$ for L_{cols} is identical to the one given in the proof of the previous lemma. However, no nondeterministic choices are needed in step 3, as each column only contains a single black pixel.   □

For establishing some non-inclusion results below, we need a technical result on *crossing sequences* for 2SA, which were already presented in [13].

Lemma 5. *Let* L *be a picture language over* Σ *accepted by a* 2SA, *and let* $f : \Sigma^{*,*} \times \mathbb{N} \to \mathbb{N}$ *be the function that is defined by taking* $f(L, n)$ *to be the number of pictures in* L *of size* $n \times (2n + 1)$. *If* $L \subseteq L_{\text{copy}}$, *then* $f(L, n) \in 2^{\mathcal{O}(n \log n)}$.

An analogous counting argument applies to every language L in REC. The number of different placements of tiles along the border between two columns is $2^{\mathcal{O}(n)}$, which limits $f(L, n)$ to be $2^{\mathcal{O}(n)}$ [4].

Theorem 2. $\mathcal{L}(2\text{M}_1\text{A})$ *is not contained in* $\mathcal{L}(2\text{SA})$, *and* $\mathcal{L}(2\text{DM}_1\text{A})$ *is not contained in* REC.

Proof. L_{copy} contains 2^{n^2} pictures of size $n \times (2n + 1)$, and so $L_{\text{copy}} \notin \mathcal{L}(2\text{SA})$, and $L_{\text{cols}} \notin$ REC, as it contains $n^n \in 2^{\mathcal{O}(n \log n)}$ pictures of size $n \times (2n + 1)$.   □

Thus, while $\mathcal{L}(2\text{DM}_1\text{A})$ is a proper subclass of $\mathcal{L}(2\text{DSA})$ (Corollary 1), Theorem 2 shows that $\mathcal{L}(2\text{M}_1\text{A})$ is not contained in $\mathcal{L}(2\text{SA})$. Does the converse inclusion hold? Below we relate this question to the problem of whether the inclusion NL $\subseteq$ P is strict. For doing so, we first extend the simulation presented in Lemma 1 to nondeterministic marker automata and to square pictures for which the number of rows and columns is polynomial in the length of the input string. For $w \in \Sigma^*$, $a \in \Sigma$ and $k \in \mathbb{N}$, we denote by $S(w, a, k)$ the picture over Σ of the size $|w|^k \times |w|^k$ that contains w in the first row, starting at column one. The remaining fields all contain the symbol a. For a string language $L \subseteq \Sigma^*$, let $S(L, a, k) = \{ S(w, a, k) \mid w \in L \}$.

Lemma 6. *Let* L *be a string language over* Σ, *let* $a \in \Sigma$, *and* $k \in \mathbb{N}$. *If the picture language* $S(L, a, k)$ *is accepted by a* 2M$_1$A, *then* L *is accepted by a* M$_{4k}$A.

Proof. In comparison to the construction in the proof of Lemma 1, it is now necessary to represent the vertical and horizontal positions of the head and the marker, respectively, in the range from 1 to $|w|^k$. This can be done using k markers for each value, in this way representing each such value in the numeral system with base $|w|$. Hence, in total $4k$ markers are sufficient.   □

The additional space provided by the picture language $S(L, a, k)$ now allows us to simulate a polynomial time-bounded one-dimensional Turing machine for L by a 2DSA.

Lemma 7. *Let L be a string language over Σ, and let $a \in \Sigma$. If $L \in \mathsf{P}$, then there is an integer $k \in \mathbb{N}$ such that the picture language $S(L, a, k)$ is accepted by some 2DSA.*

Proof. Let $\mathcal{T}$ be a deterministic Turing machine that accepts the language $L \subseteq \Sigma^*$ in time $t(n) \in \mathcal{O}(n^k)$. We can assume that the tape of $\mathcal{T}$ is infinite only to the right. Moreover, we can assume that $t(n) \leq n^k$ for all $n \geq n_0$, where n_0 is a suitable integer.

We design a 2DSA $\mathcal{A}$ for accepting the picture language $S(L, a, k)$ as follows. Let P be an input picture. First, $\mathcal{A}$ verifies whether row one of P contains a word of the form wa^r for some $w \in \Sigma^*$ and $r \geq 0$, and whether all other rows only contain the symbol a. Secondly, it checks whether P is of the correct size $n^k \times n^k$, where $n = |w|$. Polynomials are recognizable functions [4], and so this check can be performed. If $|w| < n_0$, then $\mathcal{A}$ decides whether $\mathcal{T}$ accepts w simply by table-lookup. Otherwise, $\mathcal{A}$ simulates $\mathcal{T}$ on input w. For every $1 \leq i \leq n^k$, the i-th configuration of the computation of $\mathcal{T}$ on input w is written to the i-th row of P. There is sufficient space to represent all these configurations. Acceptance is then decided based on the last of these configurations.   □

From this technical result we can now draw the following conclusion.

Theorem 3. *If $\mathcal{L}(\text{2DSA}) \subset \mathcal{L}(\text{2M}_1\text{A})$, then the complexity classes NL and P coincide.*

Proof. Let L_1 be a problem from the complexity class P. For some input symbol a and a suitable integer k, the picture language $L_2 = S(L_1, a, k)$ is accepted by a 2DSA (Lemma 7). Hence, if $\mathcal{L}(\text{2DSA}) \subset \mathcal{L}(\text{2M}_1\text{A})$, then L_2 is accepted by some 2M$_1$A. Thus, Lemma 6 implies that L_1 is accepted by an M$_{4k}$A, which in turn yields that L_1 belongs to the complexity class NL. Thus, $\mathsf{NL} = \mathsf{P}$ follows.   □

As it is widely believed that NL is a proper subclass of P, we obtain the following conjecture:

Conjecture: The classes of picture languages $\mathcal{L}(\text{2DSA})$ and $\mathcal{L}(\text{2M}_1\text{A})$ are incomparable under inclusion.

5   Conclusion

While deterministic sgraffito automata are strictly more powerful than deterministic two-dimensional one-marker automata, it appears that in the non-deterministic case, the corresponding types of automata yield incomparable classes of picture languages. However, the question arises of whether the implication in Theorem 3 is actually a characterization, that is, does NL = P hold if and only if $\mathcal{L}$(2DSA) is contained in $\mathcal{L}$(2M$_1$A)? Actually, Lemma 7 extends to the nondeterministic case. Hence, it follows in analogy to Theorem 3 that $\mathcal{L}$(2SA) $\subset$ $\mathcal{L}$(2M$_1$A) would imply that NL even coincides with the complexity class NP (nondeterministic polynomial time). Also it remains to compare 2DSA and 2DM$_1$A for unary picture languages, that is, picture languages over a one-letter alphabet. Are 2DSA still more expressive in this setting?

References

1. Blum, M., Hewitt, C.: Automata on a 2-dimensional tape. In: Proceedings of the 8th Annual Symposium on Switching and Automata Theory (SWAT, FOCS 1967), pp. 155–160. IEEE Computer Society, Washington, DC (1967)
2. Borchert, B., Reinhardt, K.: Deterministically and sudoku-deterministically recognizable picture languages. In: Loos, R., Fazekas, S., Martín-Vide, C. (eds.) Proc. Report 35/07, LATA 2007, pp. 175–186. Research Group on Mathematical Linguistics, Universitat Rovira i Virgili, Tarragona (2007)
3. Even, S., Even, G.: Graph Algorithms. Cambridge University Press (2011)
4. Giammarresi, D., Restivo, A.: Two-dimensional languages. In: Rozenberg, G., Salomaa, A. (eds.) Handbook of Formal Languages, vol. 3, pp. 215–267. Springer, New York (1997)
5. Hennie, F.: One-tape, off-line Turing machine computations. Information and Control 8, 553–578 (1965)
6. Holzer, M., Kutrib, M., Malcher, A.: Complexity of multi-head finite automata: Origins and directions. Theoretical Computer Science 412(1-2), 83–96 (2011)
7. Hsia, P., Yeh, R.: Finite automata with markers. In: Proc. ICALP, pp. 443–451 (1972)
8. Inoue, A., Inoue, K., Ito, A., Wang, Y., Okazaki, T.: A note on one-pebble two-dimensional Turing machines. Information Sciences 162, 295–314 (2004)
9. Inoue, K., Nakamura, A.: Some properties of two-dimensional on-line tessellation acceptors. Information Sciences 13, 95–121 (1977)
10. Ito, A., Inoue, K., Takanami, I.: The simulation of two-dimensional one-marker automata by three-way Turing machines. In: Dassow, J., Kelemen, J. (eds.) IMYCS 1988. LNCS, vol. 381, pp. 92–101. Springer, Heidelberg (1989)
11. Ito, A., Inoue, K., Wang, Y.: Nonclosure properties of two-dimensional one-marker automata. Int. J. Pattern Rec. Artif. Int. 11(07), 1025–1050 (1997)
12. Kari, J., Moore, C.: New results on alternating and non-deterministic two-dimensional finite-state automata. In: Ferreira, A., Reichel, H. (eds.) STACS 2001. LNCS, vol. 2010, pp. 396–406. Springer, Heidelberg (2001)
13. Průša, D., Mráz, F.: Two-dimensional sgraffito automata. In: Yen, H.-C., Ibarra, O.H. (eds.) DLT 2012. LNCS, vol. 7410, pp. 251–262. Springer, Heidelberg (2012)
14. Průša, D., Mráz, F., Otto, F.: New results on deterministic sgraffito automata. In: Béal, M.-P., Carton, O. (eds.) DLT 2013. LNCS, vol. 7907, pp. 409–419. Springer, Heidelberg (2013)

Deterministic Counter Machines
and Parallel Matching Computations*

Stefano Crespi Reghizzi and Pierluigi San Pietro

Dipartimento di Elettronica, Informazione e Bioingegneria (DEIB)
Politecnico di Milano, Piazza Leonardo da Vinci 32, Milano I-20133
{stefano.crespireghizzi,pierluigi.sanpietro}@polimi.it

Abstract. For the classical family of languages recognized by quasi-realtime deterministic multi-counter machines of the partially blind type, we propose a new implementation by means of matching finite-state computations, within the model of consensually regular languages (recently introduced by the authors), whose properties are summarized. A counter machine computation is mapped on multiple DFA computations that must match in a precise sense. Such implementation maps the original counters onto a multiset over the states of the DFA. By carefully synchronizing and mutually excluding counter operations, we prove that the union of such counter languages is also consensual. This approach leads to a new way of specifying counter languages by means of regular expressions that define matching computations.

Keywords: formal languages, quasi-realtime partially blind counter machine, consensual language, closure under union, multiset machine.

1 Introduction

Multi Counter Machines (MCM) have been used since half a century to model computation and formal languages, yet their theory is less established than, say, the theory of pushdown machines and context-free languages. Several MCM types exist depending on the operations permitted on counters, on determinism, and on other constraints on reversals and spontaneous moves (see [5, 6, 11] among others). Other language families offer a double characterization by grammars and by machines, but only the most restricted MCM subfamilies (such as commutative semilinear and one-counter languages) enjoy some sort of generative model. We present a new approach to study MCM languages that cuts across traditional classifications and offers some promise to clarify the complex computations that have so far hindered their understanding. As a bonus we obtain the possibility to specify counter languages with regular expressions that describe the interacting computational threads in a rather perspicuous way.

This paper focuses on a rather rich family: the deterministic, quasi-realtime MCM to be named *det-QR-PBLIND*: counters are nonnegative integers that can be tested for zero value only in the final configuration, but the machine crashes whenever it tries to decrement a zero counter (so-called *partially blind* condition). Although this model is

---

* Work partially supported by MIUR project PRIN 2010LYA9RH-006.

S. Konstantinidis (Ed.): CIAA 2013, LNCS 7982, pp. 280–291, 2013.
© Springer-Verlag Berlin Heidelberg 2013

deterministic, we know that also some nondeterministic MCM's can be implemented by the approach to be presented.

We spend a few words to informally introduce the model of *consensual language* [2–4], whose initial inspiration came from the idea to model language processes that interact and re-enforce each other. Consider the alphabet (called *internal*) obtained by uniting the terminal alphabet with its marked copy. Two or more words on the internal alphabet *strongly match* (metaphorically, they provide consensus to each other) if they coincide when the marks are ignored, and in every position exactly one of the words has an unmarked character. Thus a regular set – the *base* – over the internal alphabet specifies another language over the terminal alphabet, called a *Consensually Regular Language* (CREG): a word is accepted if a corresponding set of matching words is in the base. Clearly, the family *REG* of regular languages is included in CREG.

Family CREG is known to be in NLOGSPACE, to include non-semilinear languages, and to be incomparable with the family of context-free (*CF*) languages. Moreover, some interesting results concern regular languages. First, the descriptional complexity of a language (i.e., the size of its minimal NFA) can be exponentially larger than the size of the minimal DFA for the base language. Second, family REG coincides with the consensual languages generated by a strictly locally testable [10] base.

Our main result is that every det-QR-PBLIND language is also a CREG: the proof is by simulation of the counter machine on a set of parallel threads that match. Since each thread is a word of the base language, which is recognized by a DFA, the latter machine is simultaneously active in multiple states. Therefore, the computational model for recognizing CREG is a *multiset* machine. Then the simulation creates as many copies of the original states of the det-QR-PBLIND machine, as the number of counters. The second result is that the union closure of det-QR-PBLIND languages is a CREG. The proof relies on original constructions for obtaining the closure under union, not for the whole CREG class, but only for the specific DFA's used as base languages in the simulation.

Although the path is rather technical, in our opinion the final result is valuable, because it offers a new way of specifying such MCM languages and their union by means of regular expressions (under the consensual interpretation). We hope that this novel specification style will be convenient in the areas where counter machines are used.

Paper organization: Sect. 2 lists the basic definitions and introduces the CREG model with its known properties and some examples; Sect. 3 presents the main result and its proof. The conclusion mentions open problems.

2   First Definitions and Properties

Let Σ denote a finite terminal alphabet and ϵ the empty word. For a word x, $|x|$ denotes the length of x; the i-th letter of x is $x(i)$, $1 \leq i \leq |x|$, i.e., $x = x(1)x(2)\ldots x(|x|)$.

A *finite automaton* (FA) A is a tuple $(\Sigma, Q, \delta, q_0, F)$ where: Q is a finite set of states; $\delta : Q \times \Sigma \to 2^Q$ is the state-transition function, q_0 is the initial state, and $F \subseteq Q$ is the set of final states. If, for every pair q, a, $|\delta(q, a)| \leq 1$ then A is deterministic (DFA), otherwise is nondeterministic (NFA). For a DFA we write $\delta(q, a) = q'$ instead of $\{q'\}$.

We list some standard definitions for counter machines [6, 9]. Let $\mathbf{x} \in \mathbb{N}^m$ be a vector of m nonnegative integer variables denoted by x_i and called *counters*. A *counter valuation* is a mapping giving, for every element x_i, a value in $\mathbb{N}$. Notice that x_i and **x** denote both the variables and their valuation, at no risk of confusion. For an integer m, let I^m be the set of all words y in $\{-1, 0, 1\}^m$. A word of I^m may be viewed as an *increment vector* **y** in $\mathbb{N}^m$; intuitively, **y**, defines a m-tuple of moves in the set $\{+1, -1, 0\}$ applied to m counters, such that if $y_i = 1, 0, -1$, counter i is, respectively, incremented, unchanged or decremented.

Definition 1. *A (nondeterministic) partially blind, multi-counter machine (PBLIND) is specified by a tuple* $\mathcal{M} = \langle \Sigma, S, \gamma, m, s_0, S_{fin} \rangle$ *where*

- *S is a finite set of states, $s_0 \in S$ is initial, and $S_{fin} \subseteq S$ are the final states;*
- *$m \geq 0$ is the number of counters;*
- *$\gamma \subseteq S \times (\Sigma \cup \{\epsilon\}) \to I^m \times S$ is the transition relation;*
- *A move of $\mathcal{M}$ is an element of the relation $\to_\mathcal{M} \subseteq ((\Sigma \cup \{\epsilon\}) \times S \times \mathbb{N}^m) \times (S \times \mathbb{N}^m)$ defined, $\forall a \in \Sigma \cup \{\epsilon\}, s, s' \in S, \mathbf{x}, \mathbf{x}' \in \mathbb{N}^m$, as follows:*

$$(s, \mathbf{x}) \xrightarrow{a}_\mathcal{M} (s', \mathbf{x}') \text{ if } \exists \mathbf{y} \in \mathbf{I}^m \mid (s, a, \mathbf{y}, s') \in \gamma \text{ and } \mathbf{x}' = \mathbf{x} + \mathbf{y} \quad (1)$$

$\mathcal{M}$ is called deterministic, *denoted det-PBLIND if, for all $s, s', s'' \in S, \mathbf{y}', \mathbf{y}'' \in I^m$ all the following conditions hold:*

1. *for all $a \in \Sigma$, if $(s, a, \mathbf{y}', s') \in \gamma$ then $(s, \epsilon, \mathbf{y}'', s'') \notin \gamma$;*
2. *for all $b \in \Sigma \cup \{\epsilon\}$, if $(s, b, \mathbf{y}', s') \in \gamma, (s, b, \mathbf{y}'', s'') \in \gamma$ then $(\mathbf{y}', s') = (\mathbf{y}'', s'')$.*

Notice that the domain of γ does not refer to counters, which is reason to call this machine partially blind.

A configuration is a pair in $S \times \mathbb{N}^m$. The initial configuration is $(s_0, 0^m)$, and a final configuration is an element of $S_{fin} \times 0^m$. Relation $\to_\mathcal{M}$ is extended as usual to $\Sigma^* \times S \times \mathbb{N}^m \times S \times \mathbb{N}^m$: if $w \in \Sigma^n$ for $n > 0$ then $(s, \mathbf{x}) \xrightarrow{w}_\mathcal{M} (s', \mathbf{x}')$ is:

$$(s, \mathbf{x}) \xrightarrow{w(1)}_\mathcal{M} (\xrightarrow{\epsilon}_\mathcal{M})^{k_1} \xrightarrow{w(2)}_\mathcal{M} \cdots \xrightarrow{w(n)}_\mathcal{M} (\xrightarrow{\epsilon}_\mathcal{M})^{k_n} (s', \mathbf{x}') \text{ where each } k_i \geq 0. \quad (2)$$

When $s = s_0, \mathbf{x} = 0^m$, sequence (2) is called a *run* of $\mathcal{M}$ with *label* w and *length* $|w| + k_1 + \cdots + k_n$. If also $s' \in S_{fin}, \mathbf{x}' = 0^m$ then the run is *accepting*.

The *language accepted* by a PBLIND $\mathcal{M}$ is the set $L(\mathcal{M})$ of words $w \in \Sigma^*$ such that $(s_0, 0^m) \xrightarrow{w}_\mathcal{M} (s, 0^m)$ for some $s \in S_{fin}$, i.e., the set of labels of accepting runs.

A PBLIND machine $\mathcal{M}$ works in *quasi-realtime* if there exists a constant c such that in Eq. (2) the length of each subsequence $(\xrightarrow{\epsilon}_\mathcal{M})^*$ (of so-called spontaneous moves) is bounded by c; it works in *real-time* if $c = 0$, i.e, $\gamma \subseteq S \times \Sigma \to I^m \times S$. Quasi-realtime (resp. real-time) PBLIND machines are denoted by QR-PBLIND (resp. RT-PBLIND); similarly the deterministic variants are called det-QR-PBLIND and det-RT-PBLIND.

2.1 Consensual Languages

We present the basic elements of consensual language theory following [2, 3]. Let $\underline{\Sigma}$ be the alphabet obtained by *marking* each letter $a \in \Sigma$ as $\underline{a}$. The union $\Sigma \cup \underline{\Sigma}$ is named the *internal* alphabet (because it is only used within the technical device of match functions) and denoted by $\widetilde{\Sigma}$. To express a sort of agreement or consensus between words over the internal alphabet, we first introduce a binary relation, called *match*, over $\widetilde{\Sigma}$ then extended to words. In our metaphor, such matching words provide mutual consensus on the validity of the corresponding word over Σ, thereby justifying the name "consensual" of the language family.

Definition 2. *The partial, symmetrical, and associative binary operator,* match, @ : $\widetilde{\Sigma} \times \widetilde{\Sigma} \to \widetilde{\Sigma}$, *is defined, first for all* $a \in \Sigma$, *then for all words* $w, w' \in \widetilde{\Sigma}^n, n \geq 0$ *as:*

$$\begin{cases} a@\underline{a} = \underline{a}@a = a \\ \underline{a}@\underline{a} = \underline{a} \\ undefined, \ otherwise \end{cases} \qquad \begin{cases} \epsilon @ \epsilon = \epsilon \\ w@w' = (w(1)@w'(1)) \cdot \ldots \cdot (w(n)@w'(n)) \end{cases}$$

In words, the match is undefined if $|w| \neq |w'|$, or whenever in some position i the match $w(i)@w'(i)$ is undefined, which happens when both letters are in Σ, when both are in $\underline{\Sigma}$ and differ, and when either one is marked but is not the marked copy of the other. For instance, $\underline{a}abb @ \underline{a}abb = \underline{a}abb$ while $\underline{a}a\underline{b}\underline{b} @ \underline{a}abb$ is undefined.

The *match* of a finite nonempty set of internal words $w_1, \ldots, w_m$ is denoted by $w = w_1@w_2@ \ldots @w_m$ or by $@\{w_1, w_2, \ldots, w_m\}$, and is a partially defined function. The number m is called the *degree* of the match. The match result is further qualified as *strong* if $w \in \Sigma^*$, or as *weak* otherwise. By Def. 2, if w is a strong match, in each position $1 \leq i \leq |w|$, exactly one word, say w_h, is unmarked, i.e., $w_h(i) \in \Sigma$, and $w_j(i) \in \underline{\Sigma}$ for all $j \neq h$; we say that word w_h *places* the letter at position i and the other words *consent* to it. The match operator is extended to two (or more) languages $L', L'' \subseteq \widetilde{\Sigma}^*$ by means of

$$L' @ L'' = \{w' @ w'' \mid w' \in L', w'' \in L''\}$$

and its repeated application to a language is defined by

$$\begin{cases} L^{1@} = L \\ L^{i@} = L @ L^{(i-1)@}, i \geq 2. \end{cases}$$

Definition 3 (Consensual language.). *The* closure under match, *or* @-closure, *of a language* $L \subseteq \widetilde{\Sigma}^*$ *is* $L^@ = \bigcup_{i \geq 1} L^{i@}$. *Let* $B \subseteq \widetilde{\Sigma}^*$. *The* consensual language with base B *is defined as* $\mathcal{C}(B) = B^@ \cap \Sigma^*$. *The family of* consensually regular *languages, denoted by CREG is the collection of all languages* $\mathcal{C}(B)$ *such that* B *is regular.*

We note that [3, 4] consider also context-free and context-sensitive bases, but here we only need regular bases, for which we resume the relevant known properties.

Proposition 1. *Summary of known CREG properties, proved in [3, 4].*

1. *Family comparisons. CREG includes REG, is incomparable with the context-free and deterministic context-free families, is included within the context-sensitive family, and it contains non-semilinear (in the sense of Parikh) languages.*
2. *CREG is closed under reversal, union and intersection with regular languages, and under reverse alphabetic homomorphism [8].*

3. *REG coincide with the family of consensual languages having a strictly locally testable base.*

4. *CREG is in NLOGSPACE (hence in polynomial time).*

To illustrate, Tab. 1 specifies by means of "consensual regular expressions" some typical languages recognized by counter machines of different types. The examples suggest that counting operations controlled by a finite-state control unit can be simulated by matching computations. The next section will formalize the translation from det-QR-PBLIND counter machines to CREG.

Table 1. Specification of typical counter languages by consensual regular expressions

---

1. The det-RT-PBLIND 2-counter language $\{a^n b^n c^n \mid n > 0\}$ is a CREG with base

$$\underline{a}^* a \, \underline{a}^* \underline{b}^* b \, \underline{b}^* \underline{c}^* c \, \underline{c}^*$$

   For instance, $aabbcc$ is the (strong) match of $\underline{a}\,a\,\underline{b}\,\underline{b}\,\underline{c}\,c$ and $a\,\underline{a}\,\underline{b}\,b\,\underline{b}\,c\,\underline{c}$. See also the multiset transition relation, Fig. 1.

2. The (non-deterministic) RT-PBLIND one-counter language $\{a^n b^n \mid n \geq 1\} \cup \{a^m b^{2m} \mid m \geq 1\}$ (it is the union of two det-RT-PBLIND 1-counter languages) is a CREG with base $B_1 \cup B_2 \cup B_3 \cup B_4$, where
$$\begin{cases} B_1 = \underline{a}^+ (\underline{b}\,b)^+ \\ B_2 = \underline{a}^* a\,\underline{a}^* \, (\underline{b}\,\underline{b})^* \, b\,\underline{b}\,(\underline{b}\,\underline{b})^* \\ B_3 = \underline{a}^* a\,\underline{a}^* (\underline{b}\,\underline{b})^* b\,(\underline{b}\,\underline{b})^* \cup \underline{a}^* a\,\underline{a}^* \, (\underline{b}\,\underline{b})^* \, b\,b\,\underline{b}\,(\underline{b}\,\underline{b})^* \\ B_4 = \underline{a}^* a\,\underline{a}^* \, (\underline{b}\,\underline{b})^* \, b\,b\,(\underline{b}\,\underline{b})^* \end{cases}$$
   In fact, the words consensually generated by the above sets have the form $a^r b^s$, $r, s > 0$. The case s odd is dealt with by subexpression B_3: $\{a^{2n+1} b^{2n+1} \mid n \geq 0\} = \mathcal{C}(B_3)$. The case s even is twofold. The sublanguage corresponding to $s = 2r$ is consensually obtained by $B_1 \cup B_2$: the former expression places all even-positioned (within b^+) b's at once, and the latter places one a and one odd-positioned b at a time. The sublanguage corresponding to $r = s$ is obtained by $B_2 \cup B_4$: the two expressions place respectively one odd- and one even-positioned b, as well as one a. Notice that the match of B_1 and B_4 is undefined.

3. The language $\{a^n b^n \mid n > 0\}^+$ is accepted by a 2-counter machine but not by any partially blind multi-counter machine [7]. It is a CREG with base $(\underline{a} \cup \underline{b})^* \, a\,\underline{a}^* b\,\underline{b}^* \, (\underline{a} \cup \underline{b})^*$.

4. The language $\{ba^1 ba^2 ba^3 \ldots ba^k \mid k \geq 1\}$ has a non-semilinear Parikh image. It is specified by the base $(\underline{a} \cup \underline{b})^* \, b\,\underline{a}^* a\,\underline{a}^* \, (\underline{b}\,\underline{a}^* a\,\underline{a}^*)^+$.

Recognition of Consensually Regular Languages. Consider now the DFA recognizing a base language R and a word in $\mathcal{C}(R)$, which is the strong match of some words in R. The matching words correspond to as many DFA computations, to be next formalized by means of multisets of states: the multiplicity of a state in the multiset equals the number of computations that have reached that state. In [3] a nondeterministic machine is defined, called a *multiset machine*, having as auxiliary memory a multiset of states of the DFA that recognizes the base language. To avoid confusion, we do not call such device a "counter machine" although in reality it is; also, this machine should not be confused with the "multiset" automata studied in [1]. Since the multiset cardinality is bounded by the length of the input word, using a binary encoding of multiplicities, the machine operates in (nondeterministic) logarithmic space (as stated in Proposition 1).

To define a transition relation for the multiset machine, we need from [3] some notation for multisets and multiple computations on a DFA.

A finite *multiset* over a given set Q is a total mapping $Z : Q \to \mathbb{N}$. The cardinality of multiset Z is $|Z| = \sum_{q \in Q} Z(q)$. If $Z(q) > 0$ then we say that $q \in Z$ with *multiplicity* $Z(q)$. For all multisets Z, Z' over Q, let the *underlying set* be $[\![Z]\!] = \{q \in Q \mid Z(q) > 0\}$, let the *inclusion* $Z \subseteq Z'$ hold if, for every $q \in Q$, $Z(q) \le Z'(q)$, and let the *sum* $Z \uplus Z'$ and the *difference* $Z - Z'$ be the multisets specified by the following characteristic functions, for all $q \in Q$:

$$(Z \uplus Z')\,(q) = Z(q) + Z'(q), \quad (Z - Z')\,(q) = max\,(0, Z\,(q) - Z'\,(q)).$$

Let $A = (\widetilde{\Sigma}, Q, \delta, q_0, F)$ be a DFA. In the following, assume, without loss of generality, that the transition relation of A is total, i.e, for every state q and every input symbol a, $|\delta(q, a)| = 1$. In order to define a transition relation on multisets of states, we extend the (total) state transition function δ to multisets over Q, positing for every multiset Z that $\delta(Z, a)$ is the multiset whose characteristic function is defined for every $q \in Q$ by:

$$\sum_{\substack{p \in Q \mid \\ \delta(p,a)=q}} Z(p).$$

Definition 4. *The* consensual transition relation *of A, denoted by $\leadsto_A \subseteq \mathbb{N}^Q \times \Sigma \times \mathbb{N}^Q$, is defined, for $a \in \Sigma$ and for multisets Z, Z' over Q, as:*

$$Z \overset{a}{\leadsto}_A Z' \text{ if } \exists q \in Z : \ Z' = \{\delta(q, a)\} \uplus \delta(Z - \{q\}, \underline{a}).$$

It is evident that if $Z \overset{a}{\leadsto}_A Z'$ then $|Z| = |Z'|$, i.e., the cardinality does not change. Relation $\overset{a}{\leadsto}_A$ can be extended as usual from a letter a to a word $w \in \Sigma^*$.

A special role is played by the *initial* multisets $\{(q_0)^k\}$, defined $\forall k > 0$, and by the *final* multisets Z, defined by $[\![Z]\!] \subseteq F$. The following crisp definition of CREG is obtained.

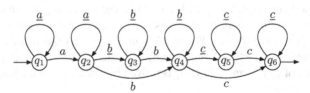

$$\{q_1, q_1\} \overset{a}{\leadsto}_A \{q_1, q_2\} \overset{a}{\leadsto}_A \{q_2, q_2\} \overset{b}{\leadsto}_A \{q_3, q_4\} \overset{b}{\leadsto}_A \{q_4, q_4\} \overset{c}{\leadsto}_A \{q_5, q_6\} \overset{c}{\leadsto}_A \{q_6, q_6\}$$

Fig. 1. DFA of the base language $\underline{a}^* a\, \underline{a}^* \underline{b}^* b\, \underline{b}^* \underline{c}^* c\, \underline{c}^*$, which defines language $\{a^n b^n c^n \mid n > 0\}$ of Table 1, Ex. 1; a run of the multiset machine for word $aabbcc$ is shown at bottom

Proposition 2. *[3] Let $R \subseteq \widetilde{\Sigma}^*$ and $A = (\widetilde{\Sigma}, Q, \delta, q_0, F)$ be a DFA recognizing R. Then $\mathcal{C}(R) = \{w \mid \exists k > 0, \exists\, a\ final\ multiset\ Z\ such\ that\ \{(q_0)^k\} \overset{w}{\leadsto}_A Z\}$.*

The transition relation and Prop. 2 are exemplified in Fig. 1.

3   From Counter Machines to Matching Computations

As seen, CREG languages are recognized by multiset machines with the following characteristics: they are nondeterministic real-time multi-counter devices that recognize a word when certain specified multiplicity counters differ from zero and the others are null; a move transfers the current multiplicity values to other counters and may increment or decrement one of the values. Such characteristics are quite different from any class (known to us) of counter machines, including in particular the partially blind, so that it is not obvious whether the latter can be simulated by such multiset machines. In the rest of the paper we prove the next theorem.

Theorem 1. *The closure under union of det-QR-PBLIND is strictly included in CREG.*

Since the proof is based on several preparatory steps and lemmas, it is preferable to outline first the overall logical path. Let M be a det-QR-PBLIND machine with m counters. The first part of the proof (Par. 3.1), to reduce the conceptual distance from multiset machines, normalizes M in several ways: real-time, at most one counter operation per step, counter operations rarefied every so many steps, and provision of an initialized extra counter. Such transformations are standard for counter machines and we do not have to prove their correctness.

The second part (Par. 3.2) considers the regular bases of CREG languages and introduces some constraints with two goals:

1. to allow a systematic simulation of any normalized det-QR-PBLIND machine, and
2. to ensure that, for any two such bases R' and R'' consensually defining det-QR-PBLIND languages L' and L'', their union $R' \cup R''$ consensually defines $L' \cup L''$.

Since CREG is not known to be closed under union, the constraints carefully synchronize the steps that *place* a letter $a \in \Sigma$ in such a way that words from R' and R'' do not make any false positive match.

After such preparation, the last part of the proof (Par. 3.3) presents the translation from the *normalized* det-QR-PBLIND m-counter machine M to the *constrained* DFA A of the base language. The key idea is simple (details apart): A's state-transition graph consists of $m + 1$ interconnected copies of M's state-transition graph. Let the j-th copy of state q be denoted by q^j. When M is in state q and the j-th counter evaluates to k, in the multiset machine controlled by A, state q^j has multiplicity k. To finish, the equivalence of $L(M)$ and $C(L(A))$ is proved by induction on the length of the words.

3.1   Normal Forms of Counter Machines

First, we restrict the set of increment vectors. For an integer m, let J^m be the set $\{1, 0, -1\}^m \cap (0^+ \cup 0^*(+1)0^* \cup 0^*(-1)0^*)$; such increment vectors specify that at most one counter is incremented or decremented and all other counters stay.

Lemma 1. *If a language is recognized by a nondeterministic (resp., deterministic) QR-PBLIND machine $\mathcal{M} = \langle \Sigma, S, \gamma, m, s_0, S_{fin} \rangle$, then it is recognized by a nondeterministic (resp., deterministic) QR-PBLIND machine such that one or more of the following restrictions apply:*

Real-time: $\gamma \subseteq S \times \Sigma \to I^m \times S$, *i.e., no ϵ move is allowed.*

Simple-operation form: $\gamma \subseteq S \times \Sigma \to J^m \times S$, *i.e., the machine increments or decrements at most one counter per move.*

(h, b)-**Rarefied operation form:** *Let $h > 0$, with $0 \le b < h$, and for all $n > 0$, let $s_0, \ldots, s_n \in S$, $\mathbf{x}_0, \ldots, \mathbf{x}_n \in \mathbb{N}^m$, $a_1, \ldots, a_n \in \Sigma$ be such that $(s_0, \mathbf{x}_0) \xrightarrow{a_1}_{\mathcal{M}}$ $(s_1, \mathbf{x}_1) \cdots \xrightarrow{a_n}_{\mathcal{M}} (s_n, \mathbf{x}_n)$ is a run; then for every $1 \le i \le n$ if $i \ne b \mod h$, $\mathbf{x}_i = \mathbf{x}_{i-1}$, i.e., the machine may alter its counters only at moves occurring at positions $b \mod h$.*

The above normal forms (for machines allowing zero-testing of counters), are stated in Th. 1.2 of [5]. The idea is that, given any $c > 0$, one can "compress" the machine: if an original counter $\mathbf{x}_i$ stores a value $k > 0$ then the compressed machine stores $\lfloor k/c \rfloor$ in $\mathbf{x}_i$ and the residue of $k - \lfloor k/c \rfloor$ in the states. In particular, compression can be applied to make a QR-PBLIND $\mathcal{M}$ into a RT-BLIND machine $\mathcal{M}'$ (by taking c as the length of the longest sequence of ϵ-moves of $\mathcal{M}$), and to make $\mathcal{M}'$ also simple-operation (by delaying the actual increment or decrement of a counter). Moreover, the (h, b)-rarefied operation form can be obtained by considering $c = h$ and using the finite memory also to count modulo h, the steps of the run, to allow increments/decrements to occur only at steps numbered b modulo h.

To simplify the proof of Lemma 3, for a m-counter machine of the types listed in Lemma 1, we introduce the following technical arrangement. Every move that increments or decrements a counter x_i, $1 \le i \le m$, must also decrement or increment another counter x_j at the same time. However, since the sum of all counters cannot change, if all counters started at zero, this kind of machine would not be able to make an increment. Hence, we assume that counter x_m, called the *initialized counter*, starts with a nonzero value: a configuration of $\mathcal{M}$ is an element of $S \times \mathbb{N}^m$, and a configuration is *initial* if it is of the form $(s_0, 0^{m-1}k)$ for some $k \ge 0$. All remaining definitions (e.g., a move) are the same of QR-PBLIND machines, with the exception of accepting runs and hence of recognized language: counter x_m must start and end with the same value $k \ge 0$. Initialized (QR-)PBLIND machines with $m + 1$ counters are equivalent to QR-PBLIND machines with m-counters.

Definition 5 (Initialized PBLIND Machine). *Given $m > 0$, define J_m^m as the set of all increment vectors y of length m in $0^+ \cup 0^*(+1)0^*(-1)0^* \cup 0^*(-1)0^*(+1)0^*$. A PBLIND machine $\mathcal{M} = \langle \Sigma, S, \gamma, m, s_0, S_{fin} \rangle$ is initialized if $\gamma \subseteq S \times (\Sigma \cup \{\epsilon\}) \to J_m^m \times S$. For every integer $k \ge 0$, for every state $s \in S_{fin}$, for every word $w \in \Sigma^+$, a run of $\mathcal{M}$ of the form $(s_0, 0^{m-1}k) \xrightarrow{w}_{\mathcal{M}} (s, 0^{m-1}k)$ is accepting for w. The initialized language of $\mathcal{M}$ is the set $L_{in}(\mathcal{M})$ of words w such that $\mathcal{M}$ has an accepting run for w.*

Given a simple-operation QR-PBLIND machine $\mathcal{M}$ with m counters, it is straightforward to define an equivalent initialized QR-PBLIND machine, *initialized($\mathcal{M}$)*, with $m + 1$ counters, by adding one counter to be incremented (resp. decremented) for every decrement (resp. increment) move of $\mathcal{M}$. Notice that if M is in (h, b)-rarefied form and/or operating in real time, then also *initialized($\mathcal{M}$)* enjoys the same properties. The fact (to be next stated) that *initialized($\mathcal{M}$)* is equivalent to $\mathcal{M}$, is trivial, since both machines have essentially the same accepting runs, provided that *initialized($\mathcal{M}$)* is initialized with a value k in counter $m + 1$, at least as large as the maximum sum of all counters during the run, so that it cannot try to go below zero: all counters $1, \ldots m$ go back to zero at the end of the run if, and only if, counter $m + 1$ goes back to k.

Proposition 3. *Let $\mathcal{M}$ be QR-PBLIND (resp. det-QR-PBLIND) machine with m counters. For all $h > 1$, $0 \leq b < h$, there exists an initialized $(m + 1)$-counter RT-PBLIND (resp. det-RT-PBLIND) $\mathcal{M}'$ in (h, b)-rarefied form such that $L(\mathcal{M}) = L_{in}(\mathcal{M}')$.*

3.2 Constrained Form of Base Languages

This section casts the base languages needed for implementing det-RT-PBLIND machines in a disciplined form that ensures closure under union. A few definitions are needed. For all $h > 1$, let R be a nonempty proper subset of $\{0, \ldots, h - 1\}$. For every $X \subset \widetilde{\Sigma}$, define:

$$R_{h,X} = \left\{ x \in \widetilde{\Sigma}^h \widetilde{\Sigma}^* \mid \forall i, 1 \leq i \leq |x| : (i - 1 \mod h) \notin R \implies x(i) \in X \right\}$$

Hence, $R_{h,X}$ is the set of words in $\widetilde{\Sigma}^+$, of length at least h, such that every position i with $i - 1 \mod h$ is in R may be any symbol in $\widetilde{\Sigma}$, while every remaining position is a symbol in X. The two cases that are considered in the following are when $X = \Sigma$, i.e., $R_{h,\Sigma}$, and when $X = \underline{\Sigma}$, i.e., $R_{h,\underline{\Sigma}}$.

Example 1. Let $h = 5$, $R = \{0, 1, 3\}$, $\Sigma = \{a\}$. We show some words in $\widetilde{\Sigma}^5$, where each position is reported, for simplicity, as an index ranging from 1 to 5: $a_1 a_2 a_3 a_4 a_5$ and $a_1 \underline{a}_2 a_3 a_4 a_5$ are in $R_{h,\Sigma}$, since 2 and 4 are in R, hence positions 3 and 5 are in Σ; $\underline{a}_1 a_2 \underline{a}_3 a_4 \underline{a}_5$ and $\underline{a}_1 a_2 \underline{a}_3 a_4 \underline{a}_5$ are in $R_{h,\underline{\Sigma}}$ since, again, 2 and 4 are in R, hence in this case positions 3 and 5 are in $\underline{\Sigma}$.

By definition, $\Sigma^h \Sigma^* \subseteq R_{h,\Sigma}$ and $\underline{\Sigma}^h \underline{\Sigma}^* \subseteq R_{h,\underline{\Sigma}}$, i.e., all words of length at least h are: in $R_{h,\Sigma}$ when they are over Σ; in $R_{h,\underline{\Sigma}}$ when they are over $\underline{\Sigma}$; moreover, $R_{h,\underline{\Sigma}}$ does not include any word over Σ, since R is a proper subset of $\{0, \ldots, h - 1\}$. A few simple properties of $R_{h,X}$ follow, with their short proofs:

I) $R_{h,\Sigma} @ R_{h,\Sigma} = \emptyset$.
   Let $x, y \in R_{h,\Sigma}$, with $|x| = |y| \geq h$. Since $R \subset \{0, \ldots, h - 1\}$, the complement of R is not empty, i.e., there exists at least a position i, $1 \leq i \leq |x|$, such that $i - 1$ mod $h \notin R$; then both $x(i), y(i) \in \Sigma$, hence $x @ y$ is undefined.

II) $R_{h,\underline{\Sigma}} @ R_{h,\Sigma} = R_{h,\Sigma}$.
   Let $x \in R_{h,\underline{\Sigma}} @ R_{h,\Sigma}$, i.e., $x = y @ z$ for some $y, z \in R_{h,\Sigma}$. For all i, $1 \leq i \leq |x|$, if $i - 1$ mod $h \notin R$ then both $y(i), z(i) \in \underline{\Sigma}$, hence also $x(i) = y(i) @ z(i) \in \underline{\Sigma}$, i.e., x is still in $R_{h,\underline{\Sigma}}$. The converse case is obvious, since $\underline{\Sigma}^h \underline{\Sigma}^* \subseteq R_{h,\underline{\Sigma}}$.

III) if $R' \subseteq R$, $R' \neq \emptyset$ and $X \subset \widetilde{\Sigma}$, then $R'_{h,X} \subseteq R_{h,X}$.
   Let $x \in R'_{h,X}$. For every i, $1 \leq i \leq |x|$, if $(i - 1 \mod h) \notin R$ then also $i - 1$ mod $h \notin R'$, hence $x(i) \in X$. But this is the definition of $R_{h,X}$.

Definition 6. *For all $h \geq 2$ and for all nonempty sets $R \subset \{0, \ldots, h - 1\}$, a language $B \subseteq \widetilde{\Sigma}^*$ is called (h, R)-counting if there exist a finite set $B_\circ \subseteq \Sigma^+$ and two sets included in $\widetilde{\Sigma}^h \widetilde{\Sigma}^+$, denoted by $B_{h,\underline{\Sigma}}$ and $B_{h,\Sigma}$, such that:*

$$B = B_\circ \cup B_{h,\underline{\Sigma}} \cup B_{h,\Sigma}, \ B_{h,\underline{\Sigma}} \subseteq R_{h,\underline{\Sigma}}, \ \text{and} \ B_{h,\Sigma} \subseteq R_{h,\Sigma}.$$

By definition, it follows that every $w \in \mathcal{C}(B)$, of length at least h, is the match of exactly one string z in $B_{h,\Sigma}$ and of zero, one or more strings in $B_{h,\underline{\Sigma}}$:

Proposition 4. *If $B \subseteq \widetilde{\Sigma}^+$ is (h, R)-counting, for some $h \geq 2, R \subset \{0, \ldots, h-1\}$, then for all $x \in \Sigma^+$, $x \in \mathcal{C}(B)$ if, and only if, $x \in B_o \cup B_{h,\Sigma}$ or $x \in B_{h,\Sigma}@(B_{h,\underline{\Sigma}})^@$.*

Lemma 2. *Let $B', B'' \subseteq \widetilde{\Sigma}^+$ be, respectively, (h, R')-counting and (h, R'')-counting, for some $h \geq 2$ and some $R', R'' \subset \{0, \ldots, h-1\}$. If $R' \cap R'' = \emptyset$ and $R' \cup R'' \subset \{0, \ldots, h-1\}$ then $B = B' \cup B''$ and $R = R' \cup R''$ are such that:*

1. *$(B'_{h,\Sigma} \cup B''_{h,\Sigma})^@ \subseteq R_{h,\Sigma}$ and $B'_{h,\Sigma}@B''_{h,\Sigma} = \emptyset$;*
2. *B is (h, R)-counting, with $B_o = B'_o \cup B''_o$, $B_{h,\Sigma} = B'_{h,\Sigma} \cup B''_{h,\Sigma}$ and $B_{h,\underline{\Sigma}} = B'_{h,\underline{\Sigma}} \cup B''_{h,\underline{\Sigma}}$;*
3. *$\mathcal{C}(B) = \mathcal{C}(B') \cup \mathcal{C}(B'')$.*

Proof. Part 1 follows immediately from above Properties (I), (II) and (III): $(B'_{h,\underline{\Sigma}} \cup B''_{h,\underline{\Sigma}})^@ \subseteq (R'_{h,\underline{\Sigma}} \cup R''_{h,\underline{\Sigma}})^@ \subseteq (R_{h,\underline{\Sigma}})^@ \subseteq R_{h,\underline{\Sigma}}$, and $B'_{h,\Sigma}@B''_{h,\Sigma} \subseteq R'_{h,\Sigma}@R''_{h,\Sigma} \subseteq R_{h,\Sigma}@R_{h,\Sigma} = \emptyset$.

Part 2 also follows immediately, since $B'_{h,\Sigma} \cup B''_{h,\Sigma} \subseteq R_{h,\Sigma}$, and $B'_{h,\underline{\Sigma}} \cup B''_{h,\underline{\Sigma}} \subseteq R'_{h,\underline{\Sigma}} \cup R''_{h,\underline{\Sigma}} \subseteq R_{h,\underline{\Sigma}}$.

Part 3: The case $\mathcal{C}(B') \cup \mathcal{C}(B'') \subseteq \mathcal{C}(B' \cup B'') = \mathcal{C}(B)$ is obvious by Def. 3. For the converse case, let $x \in \mathcal{C}(B)$. By Part 2, B is (h, R)-counting. Then, by Prop. 4, either $x \in B_o \cup B_{h,\Sigma} = B'_o \cup B''_o \cup B'_{h,\Sigma} \cup B''_{h,\Sigma}$, and hence $x \in \mathcal{C}(B') \cup \mathcal{C}(B'')$, or $x \in B_{h,\Sigma}@(B_{h,\underline{\Sigma}})^@$. Hence, there exist $w \in B_{h,\Sigma}$ and $n \leq |x|$ words with $w_1 \in B_{h,\Sigma}$ and $w_2, \ldots, w_n \in B_{h,\underline{\Sigma}}$ such that $w_1@w_2@\ldots w_n = x$. We may assume that each $w_i \notin \underline{\Sigma}^+$, $1 \leq i \leq n$, since its removal does not affect the match result. We claim that either every $w_i \in B'$ or every $w_i \in B''$, from which the thesis follows. Without loss of generality assume $w_1 \in B'_{h,\Sigma}$ (the case $w_1 \in B''_{h,\Sigma}$ being symmetrical). By contradiction, assume that there exists j, $2 \leq j \leq n$, such that $w_j \in B''$. Since $w_j \notin \underline{\Sigma}^+$, there exists a position p, $1 \leq p \leq |x|$, such that $w_j(p) \in \Sigma$. But $w_j \in B''_{h,\underline{\Sigma}} \subseteq R''_{h,\underline{\Sigma}}$: by definition of $R''_{h,\underline{\Sigma}}$, $(p-1 \mod h) \in R''$. Moreover, in order for match x to be defined, $w_1(p) \in \underline{\Sigma}$; therefore, since $w_1 \in B'_{h,\Sigma} \subseteq R'_{h,\Sigma}$ and by definition of $R'_{h,\Sigma}$, $(p-1 \mod h) \in R'$. But $R' \cap R'' = \emptyset$, which is a contradiction with $(p-1 \mod h)$ being in both R'' and R'. $\square$

3.3 From Normalized Counter Machines to Constrained Base Languages

The next lemma says that det-QR-PBLIND machines recognize only consensually regular languages and paves the way to the proof of Th. 1.

Lemma 3. *For every det-QR-PBLIND M, for all $h \geq 2$, for all $0 \leq b \leq h-1$, there exists an $(h, \{b\})$-counting language $B \subseteq \widetilde{\Sigma}^+$ such that $\mathcal{C}(B) = L(M)$.*

Proof (Sketch.). Let $\mathcal{M} = \langle \Sigma, S, \gamma, m, s_0, S_{fin} \rangle$ be a det-QR-PBLIND machine, for some $m \geq 2$, which, by Prop. 3, can be assumed to be an initialized det-RT-PBLIND machine, in (h, b)-rarefied normal form, $0 \leq b < h$, $h \geq 2$. In particular, $\mathcal{M}$ may increment or decrement a counter only at steps i such that $i = b \mod h$; counter m is the initialized counter. Let B_o be the set of words in $L(M)$ shorter than h.

For all $0 \leq i \leq m$, let S^i be a marked copy of S. If $s \in S$ then its marked copy in S^i is denoted by s^i. Define a DFA $A = (\tilde{\Sigma}, Q, \delta, init, F)$, where:

- the initial state $init$ is a new symbol and $Q = \cup_{0 \leq i \leq m} S^i \cup \{init\}$;
- The transition function δ is defined as follows, for all $a \in \Sigma$, for all $s \in S$:
  1. if $(s_0, a, 0^m, s) \in \gamma$ then $\delta(init, a) = s^0$ and $\delta(init, \underline{a}) = s^m$, else both $\delta(init, a)$ and $\delta(init, \underline{a})$ are undefined;
  2. for all $r \in S$, for all $1 \leq i, j \leq m$, if there exists $\mathbf{y} \in J_m^m$ such that $y_j = 1, y_i = -1$ and $(s, a, \mathbf{y}, r) \in \gamma$ then $\delta(s^i, a) = r^j$, else $\delta(s^i, a)$ is undefined;
  3. for all $r \in S$, if $(s, a, 0^m, r) \in \gamma$ then $\delta(s^0, a) = r^0$, else $\delta(s^0, a)$ is undefined;
  4. for all $r \in S, 0 \leq i \leq m$, if there exists $\mathbf{y} \in J_m^m$ such that $(s, a, \mathbf{y}, r) \in \gamma$ then $\delta(s^i, \underline{a}) = r^i$, else $\delta(s^i, \underline{a})$ is undefined;

The idea is that A is composed of $m + 1$ copies, numbered $0, 1, \ldots, m$ of the transition graph γ of $\mathcal{M}$. The i-th copy of γ, $1 \leq i \leq m$, is intended to simulate counter i of $\mathcal{M}$ during a computation on a multiset machine with base $L(A)$. Let a move of γ be $(s, a, r, \mathbf{y})$, i.e., from state s to state r while reading a letter a and with increment vector $\mathbf{y}$. Then, for every copy i there is a transition from s^i to r^i while reading $\underline{a}$, thus in the multiset machine the cardinality of s^i is transferred to the cardinality of r^j. If $\mathbf{y}$ increments counter i and decrements counter j (necessarily $i \neq j$), then there is a transition from s^i to r^j while reading a (i.e., the cardinality of s_i is decremented of one and the cardinality of r^j is incremented of one). If $\mathbf{y}$ does not increment or decrement, however, we include also a transition from s^0 to r^0 while reading a (since the multiset machine needs to make a strong match at every step). In this way the multiset machine associated with A is able to simulate the original det-QR-PBLIND machine: if $\mathcal{M}$ is such that $(s_0, 0^m k) \xrightarrow{w}_\mathcal{M} (s, \mathbf{x})$, for some configuration $(s, \mathbf{x})$, then $\{(init)^{k+1}\} \rightsquigarrow_A Z$, where Z is a multiset such that: $Z(s^0) = 1, Z(s^i) = x_i$, for every $i, 1 \leq i \leq m$, and for all $j, 0 \leq j \leq m, Z(r^j) = 0$ for every $r \in S, r \neq s$. The proof that $B = L(A) \cup B_o$ is $(h, \{b\})$-counting and that $\mathcal{C}(L(A)) \cup B_0 = L(\mathcal{M})$ is omitted. $\square$

We finish with the proof of the main theorem.

Proof of Th. 1. Let L be the union of $k > 0$ deterministic QR-PBLIND languages. The proof that L is consensual is by induction on k. The inductive hypothesis is that if L is the union of k det-QR-PBLIND languages then for all $h \geq k + 1$, for all $R \subset \{0, \ldots, h - 1\}$, with $|R| = k$, there exists a (h, R)-counting language $B \subseteq \tilde{\Sigma}^+$ such that $\mathcal{C}(B) = L$. The base case $k = 1$ is Lemma 3.

Let $L = L' \cup L''$, where L' is the union of k det-QR-PBLIND languages and L'' is det-QR-PBLIND. By Lemma 3 applied to L'' and by induction hypothesis applied to L', for all $h > 1$, for all $R', R'' \subseteq \{0, \ldots, h - 1\}$, with $|R'| = m$ and $|R''| = 1$, there exist a (h, R')-counting language B' and a (h, R'')-counting language B'' such that $\mathcal{C}(B') = L'$ and $\mathcal{C}(B'') = L''$. For all $h \geq k + 2$, for all $R \subset \{0, \ldots, h - 1\}$, with $|R| = k + 1$, we define a (h, r)-counting language B such that $\mathcal{C}(B) = L$. Given R, it is always possible to choose, among all R', R'' as above, two sets partitioning $R \subset \{0, \ldots, h - 1\}$ (i.e., such that $R' \cup R'' = R$ and $R' \cap R'' = \emptyset$). Let $B = B' \cup B''$;

by Lemma 2, part 3, $\mathcal{C}(B) = \mathcal{C}(B') \cup \mathcal{C}(B'')$, hence the inclusion in CREG follows. The inclusion is strict, by considering Ex. 3 of Tab. 1. $\qquad\qquad\qquad\qquad$ $\square$

4   Conclusion

The interest for multi-counter machines and their languages is almost as old as for the Chomsky's families of languages, yet they do not enjoy grammars or other forms of declarative specification, and suffer by the annoying details of low-level operations on finite-state counters and control states. Our present effort goes in the direction of specifying MCM languages by means of consensual regular expressions, a notation we (subjectively) find quite readable and amenable to language transformation and composition. The way to fully attain this goal is however still long and uncertain, after the result presented, the systematic construction of a consensual regular specification for deterministic QRT-partially-blind languages and their union. An immediate unanswered problem is whether the closure of det-QR-PBLIND languages under catenation and Kleene star is in CREG. The last two examples in Table 1 suggest that other types of counter machines, in particular the nondeterministic QRT-PBLIND and even some non-PBLIND machines, can in some cases be simulated by multiset machines since they are CREG. But a precise characterization of what counter languages belong to CREG remains to be done. In fact, the CREG family cuts across traditional classifications of languages recognized by counter machines.

References

1. Calude, C.S., Păun, G., Rozenberg, G., Salomaa, A. (eds.): Multiset Processing. LNCS, vol. 2235. Springer, Heidelberg (2001)
2. Crespi Reghizzi, S., San Pietro, P.: Consensual definition of languages by regular sets. In: Martín-Vide, C., Otto, F., Fernau, H. (eds.) LATA 2008. LNCS, vol. 5196, pp. 196–208. Springer, Heidelberg (2008)
3. Crespi Reghizzi, S., San Pietro, P.: Consensual languages and matching finite-state computations. RAIRO - Theor. Inf. and Applic. 45(1), 77–97 (2011)
4. Crespi Reghizzi, S., San Pietro, P.: Strict local testability with consensus equals regularity. In: Moreira, N., Reis, R. (eds.) CIAA 2012. LNCS, vol. 7381, pp. 113–124. Springer, Heidelberg (2012)
5. Fischer, P.C., Meyer, A.R., Rosenberg, A.L.: Counter machines and counter languages. Mathematical Systems Theory 2(3), 265–283 (1968)
6. Greibach, S.A.: Remarks on the complexity of nondeterministic counter languages. Theoretical Computer Science 1(4), 269–288 (1976)
7. Greibach, S.A.: Remarks on blind and partially blind one-way multicounter machines. Theor. Comput. Sci. 7, 311–324 (1978)
8. Hopcroft, J., Ullman, J.: Formal languages and their relation to automata. Addison-Wesley, Wokingham (1969)
9. Hromkovic, J.: Hierarchy of reversal and zerotesting bounded multicounter machines. In: Chytil, M.P., Koubek, V. (eds.) MFCS 1984. LNCS, vol. 176, Springer, Heidelberg (1984)
10. McNaughton, R., Papert, S.: Counter-free Automata. MIT Press, Cambridge (1971)
11. Minsky, M.: Recursive unsolvability of Post's problem of 'tag' and other topics in the theory of Turing machines. Annals of Mathematics 74(3), 437–455 (1961)

Early Nested Word Automata
for XPath Query Answering on XML Streams

Denis Debarbieux[1,3], Olivier Gauwin[4,5], Joachim Niehren[1,3],
Tom Sebastian[2,3], and Mohamed Zergaoui[2]

[1] Inria Lille
[2] Innovimax
[3] LIFL
[4] LaBRI
[5] University of Bordeaux

Abstract. Algorithms for answering XPATH queries on XML streams
have been studied intensively in the last decade. Nevertheless, there still
exists no solution with high efficiency and large coverage. In this paper,
we introduce early nested word automata in order to approximate earliest
query answering algorithms for nested word automata in a highly efficient
manner. We show that this approximation can be made tight in practice
for automata obtained from XPATH expressions. We have implemented
an XPATH streaming algorithm based on early nested word automata in
the FXP tool. FXP outperforms most previous tools in efficiency, while
covering more queries of the XPATHMARK benchmark.

1 Introduction

XML is a major format for information exchange besides JSON, also for RDF
linked open data and relational data. Therefore, complex event processing for
XML streams has been studied for more than a decade [12,7,19,20,5,17,13,9,18].
Query answering for XPATH is the most basic algorithmic task on XML streams,
since XPATH is a language hosted by the W3C standards XSLT and XQUERY.

Memory efficiency is essential for processing XML documents of several giga
bytes that do not fit in main memory, while high time efficiency is even more
critical in practice. Nevertheless, so far there exists no solution for XPATH query
answering on XML streams with high coverage and high efficiency. The best
coverage on the usual XPATHMARK benchmark [8] is reached by Olteanu's SPEX
[19] with 24% of the use cases. The time efficiency of SPEX, however, is only
average, for instance compared to GCX [20], which often runs in parsing time
without any overhead (since the CPU can work in parallel with file accesses to
the stream). We hope that this unsatisfactory situation can be resolved in the
near future by pushing existing automata techniques forwards [12,17,9,18].

In contrast to sliding window techniques for monitoring continuous streams
[3,15], the usual idea of answering queries on XML streams is to buffer only *alive*
candidates for query answers. These are stream elements which may be selected
in some continuation of the stream and rejected in others. All non-alive elements

S. Konstantinidis (Ed.): CIAA 2013, LNCS 7982, pp. 292–305, 2013.

should be either output or discarded from the buffer. Unfortunately, this kind of *earliest query answering* is not feasible for XPATH queries [6], as first shown by adapting counter examples from online verification [14]. A second argument is that deciding aliveness is more difficult than deciding XPATH satisfiability [9], which is CONP-hard even for small fragments of XPATH [4]. The situation is different for queries defined by deterministic *nested word automata* (NWAs) [1,2], for which earliest query answering is feasible with polynomial resources [17,10]. Many practical XPATH queries (without aggregation, joins, and negation) can be compiled into small NWAs [9], while relying on non-determinism for modeling descendant and following axis. This, however, does not lead to an efficient streaming algorithm. The problem is that a cubic time precomputation in the size of the *deterministic* NWA is needed for earliest query answering, and that the determinization of NWAs raises huge blow-ups in average (in contrast to finite automata).

Most existing algorithms for streaming XPATH evaluation approximate earliest query answering. Most prominently, SPEX's algorithm on basis of transducer networks [19], SAXON's streaming XSLT engine [13], and GCX [20] which implements a fragment of XQUERY. The recent XSEQ tool [18], in contrast, restricts XPATH queries by ruling out complex filters all over. In this way, node selection can always be decided with 0-delay [11] once having read the attributes of the node (which follow its opening event). Such queries are called begin-tag determined [5] if not relying on attributes. In this paper, we propose a new algorithm approximating earliest query answering for XPATH queries that is based on NWAs. One objective is to improve on previous approximations, in order to support earliest rejection for XPATH queries with negation, such as for instance:

//book[not(pub/text()='Springer')][contains(text(),'Lille')]

When applied to an XML document for an electronic library, as below, all books published from **Springer** can be rejected once its publisher was read:

```
<lib>...<book>...<pub> Springer </pub>
        ...<content>...Lille...</content>...</book>...</lib>
```

SPEX, however, will check for all books from Springer whether they contain the string **Lille** and detect rejection only when the closing tag **</book>** is met. This requires unnecessary processing time and buffering space.

As a first contribution, we provide an approximation of the earliest query answering algorithm for NWAs [10,17]. The main idea to gain efficiency, is that selection and rejection should depend only on the current state of an NWA but not on its current stack. Therefore, we propose *early nested word automata* (ENWAs) that are NWAs with two kinds of distinguished states: rejection states and selection states. The query answering algorithm then runs ENWAs for all possible candidates while determinizing on-the-fly, and using a new algorithm for sharing the runs of multiple alive candidates. Our stack-and-state sharing algorithm for multi-running ENWAs is original and nontrivial. As a second contribution, we show how to compile XPATH queries to ENWAs by adapting the previous translation to NWAs from [9], mainly by distinguishing selection and rejection states.

The third contribution is an implementation of our algorithms in the FXP 1.1 system, that is freely available. It covers 37% of the use cases in XPATHMARK, while outperforming most previous tools in efficiency. The only exception is GCX, which does slightly better on some queries, probably due to using C++ instead of Java. Our approximation of earliest query answering turns out to be tight for XPATH in practice: it works in an earliest manner in the above example and for all supported queries from XPATHMARK with only two exceptions. These are queries with nontrivial valid subfilter, similarly to the examples showing the hardness of aliveness [9].

Outline. Section 2 starts with preliminaries on nested word automata and earliest query answering. Section 3 introduces ENWAs. Section 4 recalls the tree logic FXP which abstracts from Forward XPATH. Section 5 sketches how to compile FXP to ENWA queries. Section 6 presents our new query answering algorithm for ENWAs with stack-and-state sharing. Section 7 sketches our implementation and exprimental results. We refer to the Appendix of the long version[1] for missing proofs and further details on constructions and experiments.

2   Preliminaries

Nested Words and XML Streams. Let Σ and Δ be two finite sets of tags and internal letters respectively. A *data tree* over Σ and Δ is a finite ordered unranked tree, whose nodes are labeled by a tag in Σ or else they are leaves containing a string in Δ^*, i.e., any data tree t satisfies the abstract grammar $t ::= a(t_1, \ldots, t_n) \mid "w"$ where $a \in \Sigma$, $w \in \Delta^*$, $n \geq 0$, and $t_1, \ldots, t_n$ are data trees. A *nested word* over Δ and Σ is a sequence of internal letters in Δ, opening tags `<a>`, and closing tags `</a>`, where $a \in \Sigma$, that is well nested so that every opening tag is closed properly. Every data tree can be linearized in left-first depth-first manner into a unique nested word. For instance, $l(b(p("ACM"), c(\ldots)), \ldots)$ becomes `<l><b><p>ACM</p><c>...</c></b>...</l>`. We will restrict ourselves to nested words that are linearizations of data trees. The positions of nested words are called *events*, of which there are three kinds: opening, closing, and internal, depending on the letter at the event. Note also, that every *node* in a data tree corresponds to a pair of an opening event and a corresponding closing event. The correspondence is established by a parser processing a nested word stream.

The XML data model provides data trees with five different types of nodes: element, text, comment, processing-instruction, and attributes.[2] The latter four are always leaf nodes. Any sequence of children of element nodes starts with a sequence of attribute nodes, followed by a sequence of nodes of the other 4 types. For an XML data tree t the "child" relation ch^t relates all element nodes to their non-attribute children. Attribute nodes are accessed by the attribute relation $@^t$, which relates all element nodes to their attribute nodes. The "next-sibling"

---

[1] The long version can be found at `http://hal.inria.fr/hal-00676178`.

[2] Attributes are nodes of data trees but *not* nodes in terms of the XML data model.

relation ns^t relates non-attributes nodes in t to their non-attribute next-sibling node. In that sense attributes in the XML data model are unordered. An XML stream contains a nested word obtained by linearization of XML data trees.

Nested Word Automata. A nested word automaton (NWA) is a pushdown automaton that runs on nested words [2]. The usage of the pushdown of an NWA is restricted: a single symbol is pushed at opening tags, a single symbol is popped at closing tags, and the pushdown remains unchanged when processing internal letters. Furthermore, the stack must be empty at the beginning of the stream, and thus it will also be empty at its end. More formally, a nested word automaton is a tuple $A = (\Sigma, \Delta, Q, I, F, \Gamma, rul)$ where Σ and Δ are the finite alphabets of nested words, Q a finite set of states with subsets $I, F \subseteq Q$ of initial and final states, Γ a finite set of stack symbols, and rul is a set of transition rules of the following three types, where $q, q' \in Q$, $a \in \Sigma$, $d \in \Delta$ and $\gamma \in \Gamma$:

(**open**) $q \xrightarrow{\langle a \rangle : \gamma} q'$ can be applied in state q, when reading the opening tag `<a>`. In this case, γ is pushed onto the stack and the state is changed to q'.

(**close**) $q \xrightarrow{\langle /a \rangle : \gamma} q'$ can be applied in state q when reading the tag `</a>` with γ on top of the stack. In this case, γ is popped and the state is changed to q'.

(**internal**) $q \xrightarrow{d} q'$ can be applied in state q when reading the internal letter d. One then moves to state q'.

A configuration of an NWA is a state-stack pair in $Q \times \Gamma^*$. A run of an NWA on a nested word over Σ and Δ must start in a configuration with some initial state and the empty stack, and then rewrites this configuration on all events of the nested word according to some rule. A run is successful if it can be continued until the end of the nested word, while reaching some final state. Note that the stack must be empty then. The *language* $\mathcal{L}(A)$ of an NWA A is the set of all data trees with some successful run on their linearization. See Figs. 1 and 2 for an example of an NWA and a successful run on the nested word of a data tree.

An NWA is called *deterministic* or a dNWA if it is deterministic as a pushdown automaton. In contrast to more general pushdown automata, NWAs can always be determinized [2], essentially, since they have the same expressiveness as bottom-up tree automata. In the worst case, the resulting automata may have $2^{|Q|^2}$ states. In experiments, we also observed huge size explosions in the average case. Therefore, we will mostly rely on on-the-fly determinization.

Automata Queries. We restrict ourselves to monadic (node selection) queries for data trees with fixed alphabets Σ and Δ. A monadic query over these alphabets is a function P that maps all data trees t over these alphabets to some subset $P(t)$ of nodes of t. We will use NWAs to define monadic queries (as usual for showing that tree automata capture MSO queries). The idea is that an NWA should only test whether a candidate node is selected by the query on a given tree, but not generate the candidate by itself. Therefore, a unique candidate node is assumed to be annotated on the input tree by some external process. We fix a single variable x for annotation and set the tag alphabet of such NWAs to $\{a, a^x \mid a \in \Sigma\}$. Letters a^x are called annotated (or "starred" in the

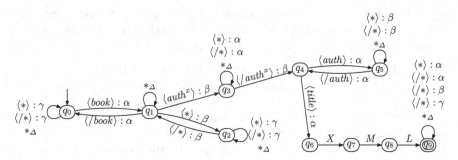

Fig. 1. An NWA for XPATH query //book[starts-with(title,'XML')]/auth, which selects all authors of book nodes having a title that starts with "XML". It runs on well-formed libraries as in the introduction, where the children of book nodes contain the sequence of authors followed by the title. We add the special symbol $*$ to Σ that captures all infinitely many other tags non-specified at the same state, and similarly a special symbol $*_\Delta$ to Δ that captures all other internal letters not mentionned there.

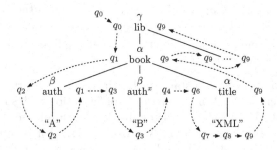

Fig. 2. An example run of the NWA in Fig. 1

terminology of [17]) while letters a are not. A monadic query P can then be defined by all NWAs that recognize the set of all trees t annotated at some single node belonging to $P(t)$. An example for a deterministic NWA is given in Fig. 1, while a successful run of this NWA is depicted in Fig. 2, on a library in which the second author of the first book is annotated by x.

Earliest Query Answering. Let P be a query, t a data tree with node π, and e an event of the nested word of t such that the opening event of π is before or equal to e. We call π *safe for selection* at e if π is selected for all data trees t' $(\pi \in P(t'))$ whose nested word is a possible continuation of the stream of t at event e, i.e., of the prefix of the nested word of t until e. We call π *safe for rejection* at e if π is rejected for all data trees t' $(\pi \notin P(t'))$ such that the nested word of t' is a possible continuation of the stream of t beyond e. We call π *alive* at e if it is neither safe for selection nor rejection at e. An earliest query answering (EQA) algorithm outputs selected nodes at the earliest event when they become safe for selection, and discards rejected nodes at the earliest event when they become safe for rejection. Indeed, an EQA algorithm buffers only alive nodes. The

problem to decide the aliveness of a node is EXPTIME-hard for queries defined by
NWAs. For dNWAs it can be reduced to the reachability problem of pushdown
machines which is in cubic time [9]. This, however, is too much in practice with
NWAs of more than 50 states, 50 stack symbols, and $4 * 50^2 = 10.000$ transition
rules, so that the time costs are in the order of magnitude of $10.000^3 = 10^{12}$.

3   Early Nested Word Automata

We will introduce early NWAs for approximating earliest query answering for
NWAs with high time efficiency. The idea is to avoid reachability problems of
pushdown machines, by enriching NWAs with selection and rejection states[3], so
that aliveness can be approximated by inspecting states, independently of the
stack. As we will see in Section 5, we can indeed distinguish appropriate selection
and rejection states when compiling XPATH queries to NWAs.

A subset Q' of states of an NWA A is called an *attractor* if any run of A that
reaches a state of Q' can always be continued and must always stay in a state
of Q'. It is easy to formalize this condition in terms of necessary and impossible
transition rules of A.

Definition 1. *An early nested word automaton (*ENWA*) is a triple $E = (A, S, R)$
where A is an* NWA, *S is an attractor of A of final states called selection states,
and R an attractor of non-final states called rejection states. The* query defined
by E is the query defined by A.

In the example NWA in Fig. 1, we can define $S = \{q_9\}$ and $R = \emptyset$. We could add
a sink state to the automaton and to the set of rejection states. Also all selection
and respectively rejection states can be merged into a single state.

An ENWA defines the same language or query as the underlying NWA. Let
us consider an ENWA E defining a monadic query and a data tree with some
annotated node π. Clearly, whenever *some* run of E on this annotated tree
reaches a selection state then π is safe for selection. By definition of attractors,
this run can always be continued until the end of the stream while staying in
selection states and thus in final states. In analogy, whenever *all* runs of E reach
a rejection state, then π is safe for rejection, since none of the many possible
runs can ever escape from the rejecting states by definition of attractors, so none
of them can be successful. For finding the first event, where all runs of E either
reach a rejection state or block, it is advantageous to assume that the underlying
NWA is deterministic. In this case, if some run reaches a rejection state or blocks,
we can conclude that all of them do.

We call an ENWA deterministic if the underlying NWA is. We next lift the
determinization procedure for NWAs to ENWAs. Let $E = (A, S, R)$ be an ENWA
and A' the determinization of A. The deterministic ENWA $E' = (A', S', R')$ is

---

[3] The semantics of selection states is identical with the semantics of final states in
the acceptance condition for NWAs in [2]. The idea of analogous rejection states,
however, is original to the present paper to the best of our knowledge.

Formulas	F ::= F ∧ F	F ∨ F	¬F	*true*	A(F)	L(F)	K(F)	O(T,s)
Axes	A ::= @	*ch*	*ch*$^+$	*ch**	*ns*	*fs*	*fo*
Labels	L ::= *x*	*a*	*nsp*$_a$
Types	K ::= *element*	*text*	*comment*	*processing-instruction*
Comparisons	O ::= *equals*	*contains*	*starts-with*	*ends-with*
Texts	T ::= *text*$_x$(F)

Fig. 3. Abstract syntax of FXP where $x \in \mathcal{V}$ is a variable, $a \in \Sigma$ is a label, attribute, or namespace, and $s \in \Delta^*$ a string data value

defined such that S' contains all those sets of pairs such that *some* of them has the second component in S, while R' contains all those sets of pairs such that *all* of them have their second component in R. From the construction of A' it is not difficult to see that S' and R' are attractors of A'. Notice that the selection delay is preserved by ENWA determinization, so that we can decide whether all runs of E reach a rejection state at event e, by running the determinized version until event e.

Lemma 1. *For any event e of the stream of a tree t, there exists a run of E going into S at event e if and only if there is a run of E' going into S' at e. Likewise all runs of E go into R at event e iff all runs of E' go into R' at e.*

4   FXP Logic

Rather than dealing with XPATH expressions directly, we first compile a fragment of XPATH into the hybrid temporal logic FXP [9]. Even though the translation from XPATH to FXP is mainly straightforward, it leads to a great simplification, mainly due to the usage of variables for node selection. We are going to compile a larger fragment of XPATH than previously [9], since supporting node types, attributes, strings data values and patterns, and all forward axes of XPATH, we also need to extend FXP accordingly. The XPATH query `//book[starts-with(title,'XML')]/auth`, for example, will be compiled to the following FXP formula with one free variable x:

$$ch^*(book(starts\text{-}with(text_y(ch(title(y(true)))), XML) \land ch(auth(x(true)))))$$

FXP formulas will talk about data trees t satisfying the XML data model based on its typed relations: attribute $@^t$, child ch^t, descendant $(ch^+)^t = (ch^t)^+$, descendant-or-self $(ch^*)^t = (ch^t)^*$, next-sibling ns^t, following-sibling $fs^t = (ns^t)^*$, and following $fo^t = ((ch^t)^*)^{-1} \circ (ns^t)^+ \circ (ch^t)^*$. The abstract syntax of FXP formulas with alphabets Σ, Δ and a set of variables $\mathcal{V}$ is given in Fig. 3. There is a single atomic formula *true*. A non-atomic formula can be constructed with the usual boolean operators, or be a test for a variable $x(F)$, a node label $a(F)$, a namespace $nsp_a(F)$, or an XML node type $K(F)$. There are also formulas $A(F)$ for navigating with any typed relation A^t supported by the XML data model. Finally there are various comparisons $O(T, s)$ between string data values $text_y(F)$

$$[\![F_1 \wedge F_2]\!]_{t,\pi,\mu} \Leftrightarrow [\![F_1]\!]_{t,\pi,\mu} \wedge [\![F_2]\!]_{t,\pi,\mu}$$
$$[\![F_1 \vee F_2]\!]_{t,\pi,\mu} \Leftrightarrow [\![F_1]\!]_{t,\pi,\mu} \vee [\![F_2]\!]_{t,\pi,\mu}$$
$$[\![\neg F]\!]_{t,\pi,\mu} \Leftrightarrow \neg [\![F]\!]_{t,\pi,\mu}$$
$$[\![true]\!]_{t,\pi,\mu} \Leftrightarrow true$$
$$[\![A(F)]\!]_{t,\pi,\mu} \Leftrightarrow \exists \pi' \in [\![A]\!]_{t,\pi,\mu} \ s.t. \ [\![F]\!]_{t,\pi',\mu}$$
$$[\![L(F)]\!]_{t,\pi,\mu} \Leftrightarrow [\![F]\!]_{t,\pi,\mu} \wedge [\![L]\!]_{t,\pi,\mu}$$
$$[\![K(F)]\!]_{t,\pi,\mu} \Leftrightarrow [\![F]\!]_{t,\pi,\mu} \wedge [\![K]\!]_{t,\pi,\mu}$$

$$[\![O(T,s)]\!]_{t,\pi,\mu} \Leftrightarrow O([\![T]\!]_{t,\pi,\mu}, s)$$
$$[\![A]\!]_{t,\pi,\mu} = A^t(\pi)$$
$$[\![x]\!]_{t,\pi,\mu} \Leftrightarrow \pi = \mu(x)$$
$$[\![a]\!]_{t,\pi,\mu} \Leftrightarrow \text{label of } \pi \text{ in } t \text{ is } a$$
$$[\![nsp_a]\!]_{t,\pi,\mu} \Leftrightarrow \text{namespace of } \pi \text{ in } t \text{ is } a$$
$$[\![K]\!]_{t,\pi,\mu} \Leftrightarrow \pi \text{ has type } K \text{ in } t$$
$$[\![text_x(F)]\!]_{t,\pi,\mu} = \text{data value of } \mu(x) \colon [\![F]\!]_{t,\pi,\mu}$$

Fig. 4. Semantics of FXP formulas F for an XML data tree t with node π and variable assignment μ to nodes of t

accessed from the y-node in the data tree and string constants $s \in \Delta^*$, but no more general comparisons as needed for join operations. The formal semantics of FXP is defined in Fig. 4. Given an XML data tree t, a node π of t, and a variable assignment μ to nodes of t, a formula F evaluates to a truth value $[\![F]\!]_{t,\pi,\mu}$. Formulas F with one free variable define monadic queries. For compiling XPATH, we restrict ourselves to formulas where all subformulas contain at most one free variable. Also there may be some bound variables y introduced by $text_y(F)$.

5 Compiler from FXP to Early Nested Word Automata

We sketch a compiler from FXP formulas to ENWAs, that follows the usual approach of compiling tree logics such as MSO into tree automata. Compared to previous compilers into NWAs in [10,17], the most novel part is the distinction of appropriate selection and rejection states. It should also be noticed, that our compiler will heavily rely on non-determinism, in order to compile formulas $A(F)$ where A is a recursive axis such as the descendant or following axis. However, we will try to preserve determinism as much as possible, so that we can compile many formulas $\neg F$ without having to determinize the ENWA for F.

The construction is by recursion on the structure of F. Given an FXP formula F with n free variables, the compiler produces an ENWA with node labels in $\Sigma \times 2^{\mathcal{V}}$ that defines the same n-ary query. With $n = 1$ as for FXP formulas obtained from XPATH, this yields ENWAs defining monadic queries by identifying $\Sigma \times 2^{\{x\}}$ with $\{a, a^x \mid a \in \Sigma\}$. Let F and F' be two formulas that were compiled to $E = (A, S, R)$ and $E' = (A', S', R')$ with state sets Q and Q' respectively. The NWA for a **conjunction** $F \wedge F'$ is the product of A and A'. We choose selection states $S \times S'$, since a node is safe for selection for $F \wedge F'$ iff it is safe for selection for both F and F'. As rejection states we choose $(R \times Q') \cup (Q \times R')$, which may lead to a proper approximation of earliest query answering. Also a large number of conjunctions may lead to an exponential blow-up of the states. The NWA of a **disjunction** $F \vee F'$ is the union of A and A'. As selection states we use $S \cup S'$ which is exact, and as rejection states $R \cup R'$. Note that we compile conjunctions and disjunctions differently, since unions may introduce non-determinism while products do not. For **negations** $\neg F$, where E

is deterministic, we simply swap the final states of E, and exchange selection and rejection states. This is correct since we maintain pseudo-completeness as an invariant (see [9]), and remains exact, since a node is safe for selection for $\neg F$ iff it is safe for rejection for F, and conversely. Otherwise, we determinize E in a first step, which is exact by Lemma 1, and second apply the previous construction. The ENWAS for **navigation** formulas $A(F)$ for the various axes A guess an A-successor of the root and then run E starting from there. The selection and rejection states remain unchanged except for the treatment of the root. A better construction preserving determinism is available for formulas $ch(F)$ under the condition that F contains only the axis ch and ch^* (see [9] again). There is also an optimized construction for **attribute** access $@(F)$ which uses internal transitions only. Further optimizations are possible based on **node typing** of the XML data model (such as that attribute children precede all children of other node types). Node **label** and **type** testers $L(F)$ and $K(F)$ work as usual. They may add new rejection states to R while preserving the selection states S. ENWAS for **string comparisons** at the root $O(text_y(y), s)$ can be obtained from a DFA with accepting and rejecting states that recognizes all strings s' such that $O(s', s)$. General string comparisons $O(text_y(F), s)$ can be reduced to the previous case, since they are equivalent to $F[y/O(text_z(z), s)]$.

6   Early Query Answering

We show how to use ENWAS for evaluating monadic queries on XML streams. Our basic algorithm generates all possible answer candidates, and runs the ENWA on them based on on-the-fly determinization. We then improve this algorithm so that configurations and runs of multiple answer candidates may be shared.

On-the-fly Determinization. Let E be an ENWA that defines a monadic query, i.e., with tag alphabet $\{a, a^x \mid a \in \Sigma\}$ where x is a fixed variable. Rather than running E, we want to run its determinization E'. This can be done while generating E' on the fly. At any time point, we store the subset of the states and transitions of E' that was used before. If a missing transition rule is needed then one we compute it from E and adds it to E'. It should be noticed that each transition can be computed in polynomial time (but not in linear time). Recall also that the states of E' are sets of pairs of states of E. For efficiency reasons, we will substitute such sets by integers, so that the known transitions of E' can be executed as efficiently as if E was deterministic at beforehand. Therefore, we will assume in the sequel that E is deterministic. We will also assume that it is pseudo-complete, so that runs can never get blocked.

Buffering Possibly Alive Candidates. Suppose we are given a stream containing a nested word of some data tree, and that we want to compute the answer of the query defined by E on this data tree in an early manner. That is, we have to find all nodes of the data tree that can be annotated by x, so that E can run successfully on the annotated data tree. At any event e of the stream, our algorithm maintains a finite set of candidates in a so called buffer. A candidate

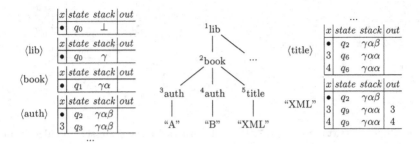

Fig. 5. Evolution of the buffer for the ENWA from Fig. 1 when answering the XPATH query `//book[starts-with(title,'XML')]/auth` on a sample document

is a triple that contains a value for x, a state of E that is neither selecting nor rejecting, and a stack of E. The value of x can either be a node opened before the current event e, or "unknown" which we denote by $\bullet$. At the start event, there exists only a single candidate in the buffer, which is $(\bullet, q_0, \bot)$ where q_0 is the unique initial state of E and $\bot$ the empty stack. At any later event, there will be at most one candidate containing the $\bullet$.

Lazy Candidate Generation. New candidates are generated lazily under supervision of the automaton. This can happen at all opening events for which there exists a bullet candidate (which then is unique). Consider the `<a>` event of some node π and let $(\bullet, q, S)$ be the bullet candidate in the buffer at this event. The algorithm then computes the unique pair (γ, q') such that $q \xrightarrow{\langle a^x \rangle : \gamma} q'$ is a transition rule of E. If q' is a selection state, then π is an answer of the query, so we can output π directly. If q' is a rejection state, then π is safe for rejection (since E is deterministic), so we can ignore it. Otherwise, π may still be alive, so we add the candidate $(\pi, q', S\gamma)$ to the buffer.

Candidate Updates. At every event, all candidates in the buffer must be updated except for those that were newly created. First, the algorithm updates the configuration of the candidate by applying the rule of E with the letter of the current event to the configuration of the candidate. If a selecting state is reached, the node of the candidate is output and discarded from the buffer. If a rejection state is reached, the candidate is also discarded from the buffer. Otherwise, the node may still be alive, so the candidate is kept in the buffer.

Example. We illustrate the basic algorithm in Fig. 5 on the ENWA from Fig. 1 and the document from Fig. 2. Initially the buffer contains a single candidate with an unknown node $\bullet$, that starts in the initial state of the ENWA with an empty stack. According to opening tags $\langle lib \rangle$ and $\langle book \rangle$ we launch open transitions and apply state and stack changes. At the opening event of node 3, i.e. when reading the open tag $\langle auth \rangle$ in state q_1, a new candidate is created. This is possible, since there exists the transition rule $q_1 \xrightarrow{\langle auth^x \rangle : \beta} q_3$ in the ENWA and since q_3 is neither a rejection nor a selection state. Similarly a new candidate

302   D. Debarbieux et al.

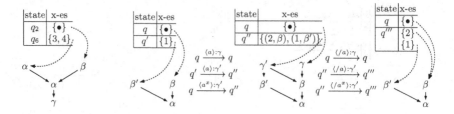

Fig. 6. Buffer of `<title>`   **Fig. 7.** Data structures for the state sharing algorithm

will be created for node 4 at its opening event. Only after having consumed the text value of the title node 5, a selection state is reached for the candidates with node 3 and 4, such that they can be output and removed from the buffer.

Stack and State Sharing. For most queries of the XPATHMARK benchmark, the buffer will contain only 2 candidates at every event, of which one is the ● candidate. It may happen though that the number of candidates grows linearly with the size of the document. An example is the XPATH query /a[following::b] on a document whose root has a large list of only a-children. There the processing time will grow quadratically in the size of the document. All candidates (of which there are $O(n)$ for documents of size n) must be touched for all following events on the stream (also $O(n)$). A quadratic processing time is unfeasible even for small documents of some megabytes, so this is a serious limitation.

We next propose a data structure for state and stack sharing, that allows to solve this issue. The idea is to share the work for all candidates in the same state, by letting their stacks evolve in common. Thereby the processing time per event for running the ENWA on all candidates will become linear in the number of states and stack symbols of the ENWA, instead of linear in the number of candidates in the buffer. In addition to this time per event, the algorithm must touch each candidate at most three times, once for creation, output, and deletion. We will use a directed acyclic graph (DAG) with nodes labeled in Γ for sharing multiple stacks in the obvious manner. In addition, we use a table $B : Q \times \Gamma \to Aggreg$ relating a state and a root of the DAG through an aggregation of nodes or ●. The shared representation of the buffer at the `<title>`-event in Fig. 5 is illustrated in Fig. 6 for instance. Here we have $B(q_6, \alpha) = \{3,4\}$, $B(q_2, \beta) = \{●\}$. In this case, the aggregations are set of candidate nodes or the ●, but this will not be enough in general (see example below). Whenever a selection state is reached in the B-table, the nodes in the aggregate of this state will be output and the aggregate will be deleted from the data structure. For rejection states, we only have to discard the aggregate. Note that rejected or selected nodes get deleted entirely from the data structure this way, since no node may appear twice in different aggregates, again due to determinism.

The precise functioning of our DAG-based buffering is illustrated by example in Fig. 7. There one has to store enough information when sharing at opening events, so that one can undo the sharing properly at closing events. From the first configuration, we reach the second with the `<a>` event for node 2, for which a new candidate will be buffered. This candidate 2 will be created from the •-candidate whose configuration has β on top of the stack, goes into state q'', and pushes γ'. However, there is also the candidate for node 1 which will go into the same state q'' while pushing the same stack symbol γ', but from a configuration with β' on top of the stack. The pairs $(2, \beta)$ and $(1, \beta')$ must be stored in the aggregation, so we define $B(q'', \gamma') = \{(2, \beta), (1, \beta')\}$. The next event has the letter </a>, where we have to undo the sharing. Now we decompose the aggregate, to update the data structure to $B(q''', \beta) = \{2\}$, $B(q''', \beta') = \{1\}$ and $B(q, \beta) = \{\bullet\}$.

Theorem 1. *For any deterministic* ENWA *E with state set Q defining a monadic query P and data tree t, the time complexity of our streaming algorithm to compute $P(t)$ is in $\mathcal{O}(|E| + |Q|\,|t|)$ and its space complexity in $O(|E| + depth(t)\,|Q| + C)$, where C is the maximal number of alive candidates of P on t at any event.*

7   Implementation and Experimental Results

The FXP tool is released under the version 1.1 and is available under the GPL licence at http://fxp.lille.inria.fr. A compiler from XPATH to FXP is freely available in the QuiXPath tool at http://code.google.com/p/quixpath. It covers a slightly larger fragment of XPATH than discussed here. In particular it supports top-level aggregations, which are reduced to earliest query answering for n-ary queries (and not only monadic queries). QuiXPath also supports backwards axes such as SPEX. We eliminated them at the cost of forward axis and regular closure. As noticed in [16] conditional regular axes are not enough.

We also implemented the static determinization algorithm for NWAs, which explodes for most practical queries, even if restricted to accessible states, but do not need it for evaluating the queries of the XPATHMARK benchmark. In contrast, on-the-fly determinization explores only small fragments of the determinized NWAs. One should also mention that we obtain high efficiency results also due to projection, where parts of the input documents are projected according to the content of the query. This precise projection algorithm is new and of interest but out of scope of this present paper.

We tested our system against the revised[4] version of XPATHMARK query set [8]. It turns out that all queries are answered in an earliest manner with two exceptions, that use valid and unsatisfiable subfilters. The query from the introduction is also treated in an earliest manner, so FXP improves on SPEX in this respect. We have also compared our FXP tools to various systems on XPATHMARK, such as SPEX, SAXON, and GCX. Input documents were produced by the XMark generator. We give in Table 1 a collection of XPATH queries, where we report for each system the *throughput* obtained on a 1.1GB XMark file. There "–" states that the query was not supported. Notice however that the GCX system competes very well. Nevertheless we believe that we obtain good results with respect to that the GCX system was done in C++, in contrast to FXP, developed in Java.

---

[4] http://users.dimi.uniud.it/~massimo.franceschet/xpathmark/index.html

Table 1. Throughput on XPATHMARK queries in millions of events per second

	A1	A2	A3	A4	A5	A6	A7	A8	B1	B2	B3	B4	B5	B6	B7	B11	B12	B13	B14	B15
FXP	2.7	2.5	2.4	2.3	3.5	3.4	3.4	2.4	2.8	2.2	3.3	3.7	2.1	2.4	1.9	1.9	2.2	2.2	2.0	1.6
SPEX	0.7	1.5	1.1	0.9	0.9	0.9	0.8	0.9	0.9	1.1	0.4	0.8	–	–	–	–	–	0.6	–	–
SAXON	1.7	1.8	1.8	–	–	1.6	–	–	–	–	–	–	–	–	–	–	–	–	–	–
GCX	2.5	3.0	2.9	–	–	–	–	–	3.3	–	–	–	–	2.3	3.3	–	–	–	–	–

Conclusion and Future Work. We have shown how to approximate earliest query answering for XPATH on XML streams by using ENWAS. An implementation of our algorithms is freely available in the FXP system. Our practical solution outperforms existing algorithms in performance and coverage. In follow-up work, we extended the coverage of our XPATH fragment by aggregate queries, arithmetic operations, and float comparisons. For this we propose *networks of automata registrations*, such that each of them can evaluate one subquery in a query decomposition. For future work we hope that we can extend this approach to cover database joins as well, and thereby reach over 90% coverage of XPATHMARK.

References

1. Alur, R., Madhusudan, P.: Visibly pushdown languages. In: 36th ACM Symposium on Theory of Computing, pp. 202–211. ACM Press (2004)
2. Alur, R., Madhusudan, P.: Adding nesting structure to words. Journal of the ACM 56(3), 1–43 (2009)
3. Barbieri, D., Braga, D., Ceri, S., Della Valle, E., Grossniklaus, M.: C-SPARQL: a continuous query language for RDF data streams. Int. J. Semantic Computing 4(1), 3–25 (2010)
4. Benedikt, M., Fan, W., Geerts, F.: XPath satisfiability in the presence of DTDs. Journal of the ACM 55(2), 1–79 (2008)
5. Benedikt, M., Jeffrey, A.: Efficient and expressive tree filters. In: Arvind, V., Prasad, S. (eds.) FSTTCS 2007. LNCS, vol. 4855, pp. 461–472. Springer, Heidelberg (2007)
6. Benedikt, M., Jeffrey, A., Ley-Wild, R.: Stream Firewalling of XML Constraints. In: ACM SIGMOD, pp. 487–498 (2008)
7. Fernandez, M., Michiels, P., Siméon, J., Stark, M.: XQuery streaming à la carte. In: ICDE, pp. 256–265 (2007)
8. Franceschet, M.: XPathMark: An XPath benchmark for the XMark generated data. In: 3rd International XML Database Symposium (2005)
9. Gauwin, O., Niehren, J.: Streamable fragments of forward XPath. In: Bouchou-Markhoff, B., Caron, P., Champarnaud, J.-M., Maurel, D. (eds.) CIAA 2011. LNCS, vol. 6807, pp. 3–15. Springer, Heidelberg (2011)
10. Gauwin, O., Niehren, J., Tison, S.: Earliest query answering for deterministic nested word automata. In: Kutyłowski, M., Charatonik, W., Gębala, M. (eds.) FCT 2009. LNCS, vol. 5699, pp. 121–132. Springer, Heidelberg (2009)
11. Gauwin, O., Niehren, J., Tison, S.: Queries on XML streams with bounded delay and concurrency. Information and Computation 209, 409–442 (2011)

12. Gupta, A.K., Suciu, D.: Stream processing of XPath queries with predicates. In: SIGMOD Conference, pp. 419–430 (2003)
13. Kay, M.: A streaming XSLT processor. In: Balisage: The Markup Conf., vol. 5 (2010)
14. Kupferman, O., Vardi, M.Y.: Model checking of safety properties. Formal Methods in System Design 19(3), 291–314 (2001)
15. Le-Phuoc, D., Dao-Tran, M., Xavier Parreira, J., Hauswirth, M.: A native and adaptive approach for unified processing of linked streams and linked data. In: Aroyo, L., Welty, C., Alani, H., Taylor, J., Bernstein, A., Kagal, L., Noy, N., Blomqvist, E. (eds.) ISWC 2011, Part I. LNCS, vol. 7031, pp. 370–388. Springer, Heidelberg (2011)
16. Ley, C., Benedikt, M.: How Big Must Complete XML Query Languages Be? In: 12th International Conference on Database Theory, pp. 183–200 (2009)
17. Madhusudan, P., Viswanathan, M.: Query automata for nested words. In: Královič, R., Niwiński, D. (eds.) MFCS 2009. LNCS, vol. 5734, pp. 561–573. Springer, Heidelberg (2009)
18. Mozafari, B., Zeng, K., Zaniolo, C.: High-performance complex event processing over XML streams. In: ACM SIGMOD, pp. 253–264 (2012)
19. Olteanu, D.: SPEX: Streamed and progressive evaluation of XPath. IEEE Trans. on Know. Data Eng. 19(7), 934–949 (2007)
20. Schmidt, M., Scherzinger, S., Koch, C.: Combined static and dynamic analysis for effective buffer minimization in streaming XQuery evaluation. In: ICDE (2007)

Invertible Transducers, Iteration and Coordinates

Klaus Sutner

Carnegie Mellon University
Pittsburgh, PA 15213, USA

Abstract. We study natural computational problems associated with iterated transductions defined by a class of invertible transducers over the binary alphabet. The transduction semigroups of these automata are known to be free Abelian groups and the orbits of words can be described as affine subspaces in a suitable geometry defined by the generators of these groups. We show how to compute the associated coordinates of words in quadratic time and how to apply coordinates to the problem of deciding whether two words generate the same orbit under a given transduction. In some special cases our algorithms can be implemented on finite state machines.

1 Invertible Transducers

We consider Mealy automata $\mathcal{A}$ where all transitions are of the form $p \xrightarrow{a/\pi(a)} q$; here $\pi = \pi_p$ is a permutation of the alphabet $\mathbf{2} = \{0,1\}$ that depends on the source p of the transition. When π is the transposition we refer to p as a *toggle state*, and as a *copy state*, otherwise. By selecting any state p in $\mathcal{A}$ as the initial state we obtain a transduction $\mathcal{A}(p) : \mathbf{2}^\star \to \mathbf{2}^\star$. A moment's thought reveals that $\mathcal{A}(p)$ is a length-preserving permutation of $\mathbf{2}^\star$. Moreover, the corresponding inverse permutation can be obtained by interchanging 0 and 1 labels in all transitions. Correspondingly, these automata are called *binary invertible transducers*. The groups generated by the collection of all transitions $\mathcal{A}(p)$ as p ranges over the state set of $\mathcal{A}$ have attracted considerable attention in the last two decades, see [1,6,7,13]. One reason invertible transducers are relevant in group theory and symbolic dynamics is that they afford very compact descriptions of surprisingly complicated groups. For example, Grigorchuk has constructed a group of intermediate growth that can be interpreted as the transduction group of a binary invertible transducer on only 5 states and with a single toggle state.

Our objective here is the study of iteration and in particular the computational complexity of problems associated with iterated transductions in an invertible transducer. Write $\mathcal{S}(\mathcal{A})$ for the semigroup generated by the basic transductions $\mathcal{A}(p)$. Each transduction f in $\mathcal{S}(\mathcal{A})$ defines its iterate $f^\star \subseteq \mathbf{2}^\star \times \mathbf{2}^\star$, a length-preserving equivalence relation on $\mathbf{2}^\star$ that we will refer to as the *orbit relation* of f: two words are related by $f^\star$ if they have the same orbits (as sets) under f.

Any equivalence relation on $\mathbf{2}^\star$ is associated with a number of natural decision problems. First, there is the recognition problem, the problem of deciding $x f^\star y$

S. Konstantinidis (Ed.): CIAA 2013, LNCS 7982, pp. 306–318, 2013.

given two words x and y. In our context we will refer to this as the *Orbit Problem*. Second, and closely related, is the first canonical form problem, the question of how hard it is to compute the *root function* $x \mapsto \min(z \in \mathbf{2}^\star \mid z \, f^\star \, x)$; here the minimum is understood to be with respect to length-lexicographical order. See [4] for a recent discussion of general complexity results relating to these questions, and [8] for results relating to rational relations. Since our equivalence relations are generated by iteration there are several other natural problems to consider. The *Iteration Problem* asks for the complexity of computing $x \, f^t$ for some transduction f, a word x and $t \geq 0$. The recognition problem has a slightly stronger variant that we refer to as the *Timestamp Problem*: given words x and y, find the least number $t \geq 0$ such that $x \, f^t = y$, or determine that no such t exists. Since f is length-preserving we only need to consider words of length k, in which case the brute-force method takes $O(k \, 2^k)$ steps for either problem; of course, we are interested in polynomial time solutions.

As Grigorchuk's example shows, even small invertible transducers with a single toggle state can produce very complicated transduction groups. In this paper we will therefore focus on a class of simple binary invertible transducers first introduced in [19] that we refer to as *cycle-cum-chord transducers*, or CCC transducers for short. These transducers have state set $\{0, 1, \ldots, n-1\}$ and transitions

$$p \xrightarrow{a/a} p - 1, \quad p > 0 \qquad \text{and} \qquad 0 \xrightarrow{0/1} n - 1, \quad 0 \xrightarrow{1/0} m - 1$$

where $1 \leq m \leq n$. We write $\mathfrak{A}_m^n$ for this transducer; the example $\mathfrak{A}_3^5$ is shown in figure 1. It is shown in [19] that the semigroups generated by CCC transducers are in fact free Abelian groups. The argument is based on a normal form for transductions proposed by Knuth. The normal form also allows one to define a natural geometry on $\mathbf{2}^\star$ that describes the orbits of words under a transduction f as affine subspaces, see section 2 below. As a consequence, it is polynomial-time decidable whether two transductions give rise to the same equivalence relation and we can construct the minimal transition system recognizing $f^\star$ in the sense of Eilenberg [3]. The reference identifies some CCC transducers where this minimal transition system is finite, so that the orbit relation is rational and thus automatic in the sense of [9,11].

In this paper we will show that for some CCC transducers timestamps can be computed by a finite state machine; more precisely, there is a transducer that computes the witness t in reverse binary. For arbitrary CCC transducers our methods produce a quadratic time algorithm. It is known that there is a natural coordinate system for the level sets $\mathbf{2}^k$ in the full binary tree $\mathbf{2}^\star$, based on the elementary maps defined by the transducer, see [19]. We will show how to compute coordinates in quadratic time algorithm in general. As before, some CCC transducers admit faster algorithms and coordinates can be computed by a suitable transducer.

This paper is organized as follows. In section 2 we provide background information on invertible transducers, introduce cycle-cum-chord transducers and describe their basic properties, using Knuth normal form as a central tool. In the next section, we discuss the complexity of the timestamp and coordinate problem for CCC transducers. Lastly, section 4 contains comments on related decision problems and mentions open problems.

2   Transduction Groups and Iteration

We consider Mealy machines of the form $\mathcal{A} = \langle Q, \mathbf{2}, \delta, \lambda \rangle$ where Q is a finite set, $\mathbf{2} = \{0, 1\}$ is the input and output alphabet, $\delta : Q \times \mathbf{2} \to Q$ the transition function and $\lambda : Q \times \mathbf{2} \to \mathbf{2}$ the output function. We can think of $\mathbf{2}^\star$ as acting on Q via δ, see [2,16,10] for background. We are here only interested in *invertible transducers* where $\lambda(p, .) : \mathbf{2} \to \mathbf{2}$ is a permutation for each state p. When this permutation is the transposition in the symmetric group $\mathfrak{S}_2$ on two letters, we refer to p as a *toggle state*, and as a *copy state*, otherwise. Fixing a state p as initial state, we obtain a transduction $\mathcal{A}(p) : \mathbf{2}^\star \to \mathbf{2}^\star$ that is easily seen to be a length-preserving permutation. If the automaton is clear from context we write $\underline{p}$ for this function; $\mathcal{S}(\mathcal{A})$ denotes the semigroup generated by all the functions $\mathcal{A}(p)$ as p ranges over Q.

If we think of $\mathbf{2}^\star$ as an infinite, complete binary tree in the spirit of [17], we can interpret our transductions as automorphisms of this tree, see [13,18]. Any automorphism f of $\mathbf{2}^\star$ can be written in the form $f = (f_0, f_1)s$ where $s \in \mathfrak{S}_2$: s describes the action of f on $\mathbf{2}$, and f_0 and f_1 are the automorphisms induced by f on the two subtrees of the root, which subtrees are naturally isomorphic to the whole tree. Write σ for the transposition in $\mathfrak{S}_2$. The automorphisms f such that $f = (f_0, f_1)\sigma$ are *odd*, the others *even*. In terms of wreath products the whole automorphism group can be written as

$$\mathsf{Aut}(\mathbf{2}^\star) \simeq \mathsf{Aut}(\mathbf{2}^\star) \wr \mathfrak{S}_2 = (\mathsf{Aut}(\mathbf{2}^\star) \times \mathsf{Aut}(\mathbf{2}^\star)) \rtimes \mathfrak{S}_2$$

with group operation $(f_0, f_1)s \, (g_0, g_1)t = (f_0 g_{s(0)}, f_1 g_{s(1)}) \, st$, see [13,18]. The collection $\mathfrak{J}$ of all maps defined by invertible transducers is easily seen to be closed under inverse and composition, and thus forms a subgroup $\mathfrak{J} \subseteq \mathsf{Aut}(\mathbf{2}^\star)$. For automata groups $G \subseteq \mathfrak{J}$ the wreath form naturally induces three maps ∂_0, ∂_1 and par such that $f = (\partial_0 f, \partial_1 f) \, \mathsf{par} f$. The parity is simply determined by the corresponding state being toggle or copy. The operations ∂_s are the *left residuals*, see [15,5,13]: for any word x, define the function $\partial_x f$ by $(x\,f)\,(z\,\partial_x f) = (xz)\,f$ for all words z (for transductions, we write function application on the right and use diagrammatic composition for consistency with relational composition). It follows that

$$\partial_{xy} f = \partial_y \partial_x f \qquad\qquad \partial_x(fg) = \partial_x f \, \partial_{xf} g$$

The transduction semigroup $\mathcal{S}(\mathcal{A})$ is naturally closed under residuals. In fact, we can describe the behavior of all the transductions by a transition system $\mathcal{C}$, much the way $\mathcal{A}$ describes the basic transductions: the states are all transductions in $\mathcal{S}(\mathcal{A})$ and the transitions are $f \xrightarrow{s/sf} \partial_s f$. Thus $\mathcal{C}$ contains $\mathcal{A}$ as a subautomaton. Of course, this system is infinite in general; it is referred to as the *complete automaton* in [13].

We will focus on a class of invertible transducers called *cycle-cum-chord trans-ducers (CCC)*: their diagrams consist of a cycle plus one chord, the source of the chord is the only toggle state, all others are copy states. More precisely, a CCC transducer has state set $\{0, 1, \ldots, n-1\}$ and wreath representation $\underline{0} = (\underline{n^-}, \underline{m^-})\sigma$ and $\underline{k} = (\underline{k^-}, \underline{k^-})$ for $0 < k < n$. Here $1 \le m \le n$ and we occasionally write p^- rather than $p-1$ to improve legibility. We will write $\mathfrak{A}_m^n$ for this transducer. The diagram of $\mathfrak{A}_3^5$ is shown in figure 1.

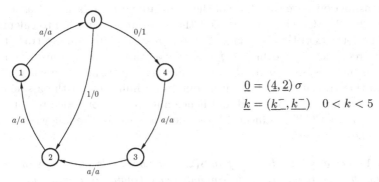

$$\underline{0} = (\underline{4}, \underline{2})\,\sigma$$
$$\underline{k} = (\underline{k^-}, \underline{k^-}) \quad 0 < k < 5$$

Fig. 1. The cycle-cum-chord transducer $\mathfrak{A}_3^5$ and its representation in the wreath form from section 2

It was shown in [19] that the semigroups associated with these transducers are in fact free Abelian groups. More precisely, in the degenerate case $n = m$, $\mathfrak{A}_m^n$ generates the Boolean group $\mathbf{2}^n$. For $n > m$ let $s = \gcd(n, m)$, then $\mathfrak{A}_m^n$ generates the free Abelian group $\mathbb{Z}^{n-s}$. As explained in [19], one only needs to be concerned with the case when n and m are coprime. Hence, from now on, we assume $s = 1$. Commutativity of $\mathcal{S}(\mathfrak{A}_m^n)$ is easily seen. The reason we obtain a group is that the following *cancellation identity* holds in the semigroup:

$$\underline{0}^2\,\underline{1}^2\ldots(\underline{m^-})^2\,\underline{m}\,\underline{m+1}\ldots\underline{n^-} = I.$$

It follows from the cancellation identity that $\mathcal{S}(\mathfrak{A}_m^n)$ is a quotient of $\mathbb{Z}^{n-1}$. To establish isomorphism one can use the *Knuth Normal Form* of a transduction suggested in [12]. To this end, we extend $\mathfrak{A}_m^n$ to an infinite transducer with additional copy states k where $k \ge n$ and transitions $\underline{k} = (\underline{k^-}, \underline{k^-})$. This extension does not change the (semi)group generated by the machine because of the *shift identities* $\underline{k}^2 = \underline{k+m}\,\underline{k+n}$. One can then show that for every transduction f there is a unique flat representation

$$f = \underline{k_1}\,\underline{k_2}\,\ldots\,\underline{k_r},$$

where $k_1 < k_2 < \ldots k_r$, the *Knuth normal form (KNF)* of f.

Thus we have two natural representations for transductions: the semigroup representation $f = \underline{0}^{e_0} \underline{1}^{e_1} \ldots \underline{n-1}^{e_{n-1}}$ where $e_i \geq 0$, and the unique group representation $f = \underline{0}^{e'_0} \underline{1}^{e'_1} \ldots \underline{n-2}^{e'_{n-s-1}}$ where $e'_i \in \mathbb{Z}$. Correspondingly, the *group representation* of f is the integer-valued vector $(e'_0, \ldots, e'_{n-s-1})$. We will refer to $\sum |e_i|$ as the *weight* of f.

2.1 Rational Orbits

Given a transduction f we can think of the associated orbit relation f^* as a language over $(\mathbf{2} \times \mathbf{2})^*$. One can then exploit the group representation to calculate Brzozowski quotients of this language. We obtain a generally infinite transition system that recognizes the orbits of f and whose states naturally are given by pairs of transductions, see [19] for details. Somewhat surprisingly, for some CCC transducers this transition system turns out to be finite for all the associated transductions. Thus, f^* is rational and hence automatic. For space reasons we focus here on the CCC Transducer $\mathfrak{A}_2^3$, see the reference for the following result and some generalizations.

Theorem 1. *For any transduction f in $\mathcal{S}(\mathfrak{A}_2^3)$, the orbit relation of f is rational. Accordingly, the root function can be computed by a length-preserving finite state transducer.*

This property is not shared by all CCC transducers; for example, the orbit relation of $\underline{0}$ in $\mathfrak{A}_3^4$ fails to be rational. It seems difficult to characterize CCC transducers with rational orbits.

For $\mathfrak{A}_2^3$, Knuth normal form has a number of interesting properties that will be important in section 3. Let $\mathsf{KNF}(f)$ denote the Knuth normal form of f. For any transduction f, write $\mathsf{sh}^s(f)$ for the transduction obtained by replacing any term $\underline{k}$ in the KNF of f by $\underline{k+s}$. In group representation, we have $\mathsf{sh}^1(a,b) = (-2b, a - 2b)$. Lastly, let $\gamma_0 = \underline{0}$, $\gamma_1 = \underline{0}\,\underline{1}$, $\gamma_2 = \underline{0}^{-1}$ and $\gamma_3 = \underline{0}^{-1}\underline{1}^{-1}$ and set $\gamma'_i = \mathsf{sh}^1(\gamma_i)$.

Lemma 1. *Let $0 \leq k$ and $0 \leq i < 4$. Then $\mathsf{KNF}(\underline{0}^{2^{4k+i}}) = \mathsf{sh}^{8k+2i}(\gamma_i)$. More generally, for $f = \underline{0}^a \underline{1}^b$, we have $\mathsf{KNF}(f^{2^{4k+i}}) = \mathsf{sh}^{8k+2i}(\mathsf{KNF}(\gamma_i^a \gamma_i'^b))$.*

Proof. A straightforward computation shows that $\mathsf{KNF}(\underline{0}^{16}) = \underline{8}$ and it follows by induction that $\mathsf{KNF}(\underline{0}^{2^{4k}}) = \underline{8k}$ for all $k \geq 0$. But then $\mathsf{KNF}(\underline{0}^{2^{4k+1}}) = \mathsf{KNF}(\underline{8k}^2) = \underline{8k+2}\,\underline{8k+3} = \mathsf{sh}^{8k+2}(\gamma_1)$. The cases $i = 2, 3$ are entirely similar. The second claim follows immediately from the first. $\square$

The existence proof for KNF is based on a weakly confluent rewrite system; other than a bound on the number of rewrite steps that is logarithmic in the weight of the given transduction there is no further information on the complexity computing the normal form. As it turns out, for $\mathfrak{A}_2^3$, rewriting is not required at all, a finite state transducer suffices to determine KNF in the following sense.

For space reasons, let us focus on determining the KNF for $\underline{0}^t$ rather than the general group elements $\underline{0}^t\underline{1}^s$. Note that we can think of the KNF of f as an ω-sequence $\kappa \in \mathbf{2}^\omega$ where $\kappa_i = 1 \iff i$ appears in the normal form of f. Since all but finitely many of the terms in κ are 0 we can also think of KNF as a finite bit-vector u such that $\kappa = u\,0^\omega$. We can pre-compute these finite bit-vectors of $\underline{0}^a$ for $0 \le a < 16$ and pad to length 8 whenever necessary:

> 00000000 10000000 00110000 1011000   000010111 100010111
> 001110111 101110111 000000111 100000111 001100111 101100111
> 000010001 100010001 001110001 101110001

All but the first 4 entries have length 9 and require a "carry" to the next block. According to lemma 1 we can now determine KNF of $\underline{0}^t$ as follows. Let T be a 0-indexed table whose entries are the 16 KNFs, right-padded or truncated to form blocks of length 8. If there is no carry, on input hex-digit d the correct output is T_d, but with a carry it is $T_{d+1 \bmod 16}$. Figure 2 shows a sketch of the appropriate transducer; input is hexadecimal, output is binary in blocks of 8 bits.

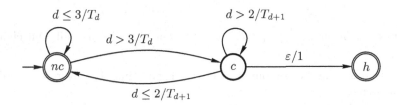

Fig. 2. A transducer that determines the Knuth normal form of a transduction $\underline{0}^a$ for CCC transducer $\mathfrak{A}_2^3$

The state nc is the no-carry state, c is carry, and h takes care of pending carries after the last input digit. For example, for $a = 3921 = (15F)_{16r}$ we get three blocks plus one 1 because of the carry:

$$T_1 T_5 T_0 T_1 = 10000000\ 10001011\ 00000000\ 1,$$

corresponding to KNF $\underline{0}\ \underline{8}\ \underline{12}\ \underline{14}\ \underline{15}\ \underline{24}$. Note that the KNF transducer can be converted into a recurrence equation for the length of $\mathsf{KNF}(f)$, but it seems difficult to obtain a closed form solution. Also, a similar construction works for general group elements, but the machinery becomes considerably more complicated since we now have to deal with both generators $\underline{0}$ and $\underline{1}$ of the transduction group.

2.2   Computing Iterates

Knuth normal form also suggests that $x f^t$ can be computed easily: we have $az\,f = a\,f(z\,\partial_a f)$ and residuation for a transduction written in KNF comes down to a left shift, except possibly for a first term $\underline{0}$. Hence, after processing an initial segment of x, the residuals of f^t will have low weight and, from then on, every single bit of x can be processed in constant time. In terms of the complete automaton $\mathcal{C}$ from section 2 this means that there are only a few non-trivial strongly connected components and every sufficiently long path winds up in one of them. For example, in the case of $\mathfrak{A}_2^3$ the complete automaton has 8 non-trivial strongly connected components the largest of which as 6 states. Technically we have the following results.

Proposition 1. *Given a transduction $f \in \mathcal{S}(\mathfrak{A}_m^n)$ of weight w we can compute its residuals in time $O(n \log w)$.*

Proof. Let $f = (e_0, \ldots, e_{n^-})$ be the semigroup representation where $\sum e_i = w$. Then the semigroup representations of the residuals are

$$\partial_0 f = (e_1, e_2, \ldots, e_m + \lfloor e_0/2 \rfloor, \ldots, e_{n-1}, \lceil e_0/2 \rceil)$$
$$\partial_1 f = (e_1, e_2, \ldots, e_m + \lceil e_0/2 \rceil, \ldots, e_{n-1}, \lfloor e_0/2 \rfloor)$$

An entirely similar argument applies to group representation as well. Our claim follows.   □

It follows that $x f$ can be computed in $O(|x|\, n \log w)$ time. However, we can do better than that.

Proposition 2. *Given a transduction $f \in \mathcal{S}(\mathfrak{A}_m^n)$ we can compute $x f$ in time linear in $|x|$, with coefficients depending on f.*

Proof. Suppose f has group representation $(e_0, e_1, \ldots, e_{n-2})$ so that $w = \sum |e_i|$. The first bit of $az\,f$ depends only on the parity of e_0. As we have seen in proposition 1, we can compute the residuals of f in $O(n \log w)$ steps, and these residuals have weight at most w. Moreover, we can express residuation as an affine operation of the form

$$\partial_s \boldsymbol{u} = \begin{cases} A \cdot \boldsymbol{u} & \text{if } \boldsymbol{u} \text{ is even,} \\ A \cdot \boldsymbol{u} - (-1)^s \boldsymbol{a} & \text{otherwise.} \end{cases}$$

where $\boldsymbol{u} \in \mathbb{Z}^{n-1}$ is the group representation of the transduction, see [14]. The spectral radius of A is less than 1, hence residuation is a contraction and after a transient part all weights are bounded by a constant depending only on n and m. Since the length of the transient is independent of x our claim follows.   □

We do not know how to obtain more precise bounds on the cost of computing $x f$. In particular there appears to be no easy way to determine the number and size of the non-trivial strongly connected components of the complete automaton, short of actual computation.

3   Timestamps and Coordinates

One can show that for any CCC transducer $\mathfrak{A}^n_m$ the group H of transductions generated by $\underline{p}$, $0 \le p < m$, acts transitively on 2^ℓ (which set of words is often referred to as a level set in connection with the infinite binary tree). For $\ell = km$, the quotient group H' obtained by factoring with respect to $\underline{i}^{2^k}$ acts simply transitively on the level set 2^ℓ. As a consequence, there is a natural coordinate system for 2^{km}: for every $\ell = km$ there is a bijection

$$2^\ell \to \mathbb{Z}/(2^k) \times \ldots \times \mathbb{Z}/(2^k)$$

where the product on the right has m terms. We will write $\langle w \rangle_\ell \in (\mathbb{Z}/(2^k))^m$ for the *coordinates* of a word w: $\langle w \rangle_\ell = (a_0, \ldots, a_{m-})$ if, and only if, $w = 0^\ell \underline{0}^{a_0} \underline{1}^{a_1} \ldots \underline{m}^{-a_{m-}}$. We use $x \equiv y$ to express the fact that two integer vectors of length m are componentwise congruent modulo 2^k. Also, for a transduction f, define the ℓ-coordinates of f by $\langle f \rangle_\ell = \langle 0^\ell f \rangle_\ell$. For example, in $\mathfrak{A}^3_2$, letting $f = \underline{0}^{-1}\underline{1}^3$ we get $\langle f \rangle_{2k} = (2^k - 1, 3)$ for $k \ge 2$. By commutativity it follows that $\langle 0^\ell f^i \rangle_\ell \equiv i \cdot \langle f \rangle_\ell$ and $\langle 0^\ell f^\star \rangle_\ell \equiv \mathbb{N} \cdot \langle f \rangle_\ell$, so that the orbit of 0^ℓ is a linear subspace of $(\mathbb{Z}/(2^k))^m$. Again by commutativity general orbits can be described as affine subspaces of $(\mathbb{Z}/(2^k))^m$:

$$\langle w f^\star \rangle_\ell \equiv \langle w \rangle_\ell + \mathbb{N} \cdot \langle f \rangle_\ell$$

Thus, it is of interest to be able to calculate coordinates. More formally, we wish to address the following problem, assuming a CCC transducer $\mathfrak{A}^n_m$ is fixed.

Problem: **Coordinate Problem**
Instance: A word $x \in 2^\ell$ where $\ell = km$.
Output: The coordinates $\langle x \rangle_\ell \in (2^k)^m$ of x.

Closely related is the question how many times a given transduction f must be applied to obtain a particular point in the orbit of a given word x. We refer to this as the Timestamp Problem:

Problem: **Timestamp Problem**
Instance: A transduction f, two words $x, y \in 2^k$.
Output: The least $t \ge 0$ such that $y = x f^t$, if it exists; NO otherwise.

Clearly the Orbit Problem reduces to the Timestamp Problem, which, as we will see shortly, in turn reduces to the Coordinate Problem. We will show that all of them can be solved in quadratic time. Let us first deal with the Timestamp Problem.

Theorem 2. *The Timestamp Problem can be solved in quadratic time: given two words x and y of length $\ell = km$ and a transduction $f \in \mathcal{S}(\mathfrak{A}^n_m)$ we can find a timestamp $t \ge 0$ such that $x f^t = y$, or determine that no such t exists, in $O(\ell^2)$ steps.*

Proof. We may safely assume that $m < n$ and $\gcd(n, m) = 1$. Also, we only need to consider the case where f is odd, since otherwise f simply copies the first k_1 bits in the input word, where $\underline{k_1}$ is the first term in the KNF of f. Write $x = x_0 \ldots x_{\ell-1}$ and $y = y_0 \ldots y_{\ell-1}$ and let $f = (e_0, \ldots, e_d) \in \mathbb{Z}^d$ in group representation where $d = n - 2$. We will determine the bits in the binary expansion of the timestamp $t = \sum t_i 2^i$, starting with the least significant digit. Initialize a symbolic vector

$$V = \sum_{i < k} t_i 2^i (e_0, \ldots, e_d).$$

with entries in the polynomial ring $\mathbb{Z}[\boldsymbol{\tau}]$, $\boldsymbol{\tau} = t_0, \ldots, t_{k-1}$. We proceed in k rounds $r = 0, \ldots, k - 1$, each consisting of m stages $s = 0, \ldots, m - 1$. In round r, stage s, perform the following actions. If $s = 0$, bind t_r to 0 or 1 so as to satisfy $V_1 + x_{mr} = y_{mr} \pmod 2$ where V_1 is the first component of V. If $s > 0$, check that $V_1 + x_{mr+s} = y_{mr+s} \pmod 2$ and exit returning NO if the test fails. In either case, finish the stage by replacing V by $\partial_{x_{mr+s}} V$. If all rounds complete successfully, return the required timestamp $t = \sum t_i 2^i$.

To see that the algorithm is correct, write $x[i]$ for the prefix of x of length i. Induction shows that if the algorithm has not returned NO by the end of round r, state s, then, letting $p = \sum_{i \le r} t_i 2^i \in \mathbb{N}$, we have

$$x[mr + s + 1] f^p = y[mr + s + 1] \quad \text{and} \quad V = \partial_{x[mr+s+1]} f^{p+\pi}$$

where π is a polynomial in variables $t_{r+1}, \ldots, t_{k-1}$. The binding of t_r at stage $s = 0$ of round r always exists since the transduction represented by V is guaranteed to be odd by lemma 14 in [19].

First assume that indeed $x f^t = y$ for some $0 \le t < 2^k$. But then the algorithm correctly determines the timestamp t: In each round another binary digit of t is determined. Given the bindings for t, V initially represents the transduction f^t, and represents the appropriate quotients during execution, so that all the tests succeed. If, on the other hand, y is not in the orbit of x, consider the longest prefix x' of x such that $x' f^p = y'$ for the corresponding prefix y' of y and some $p \ge 0$. Writing the length of x' as $mr + s$, the algorithm will run up to round r, stage s. However, at the next stage a mismatch will be found and the algorithm returns NO, as required.

As to the time complexity of the algorithm, note that all numerical coefficients have at most k bits. Since the polynomials have at most k terms, a brute-force implementation will require time cubic in ℓ. However, all entries in V during round r are of the form $c + d \sum_{i \ge r} t_i 2^{i-r}$ where $c, d \in \mathbb{N}$. Thus, there is not need to maintain the full polynomials during the computation. As a consequence, quadratic time suffices.   □

Here is an example that shows how the computation unfolds for $\mathfrak{A}_2^3$, $f = \underline{0}$ and $x = 101110010010$, $y = 001000011000$. In this case $y = x f^{43}$. There are 6 rounds with 2 stages each.

r t_i	V_1	V_2
	$t_0 + 2t_1 + 4t_2 + 8t_3 + 16t_4 + 32t_5$	0
0 $t_0 = 1$	$-2t_1 - 4t_2 - 8t_3 - 16t_4 - 32t_5$	$1 - t_1 - 2t_2 - 4t_3 - 8t_4 - 16t_5$
	$1 + t_1 + 2t_2 + 4t_3 + 8t_4 + 16t_5$	$t_1 + 2t_2 + 4t_3 + 8t_4 + 16t_5$
1 $t_1 = 1$	-1	$-1 - t_2 - 2t_3 - 4t_4 - 8t_5$
	$1 - t_2 - 2t_3 - 4t_4 - 8t_5$	2
2 $t_2 = 0$	$2 + 2t_3 + 4t_4 + 8t_5$	$1 + t_3 + 2t_4 + 4t_5$
	$-1 - t_3 - 2t_4 - 4t_5$	$-1 - t_3 - 2t_4 - 4t_5$
3 $t_3 = 1$	0	$1 + t_4 + 2t_5$
	$1 + t_4 + 2t_5$	0
4 $t_4 = 0$	$-2 - 2t_5$	$-2 - t_5$
	t_5	$1 + t_5$
5 $t_5 = 1$	2	1
	-1	-1

The technique of the last theorem can be pushed slightly to provide a fast algorithm to compute coordinates. Suppose $x \in \mathbf{2}^\ell$ where $\ell = km$. We need to compute integers $e_0, \ldots, e_{m^-}$ such that

$$x = 0^\ell \, \underline{0}^{e_0} \ldots \underline{m}^{-e_{m^-}}.$$

Let us call the transduction on the right f. Then for any $r < \ell$

$$x = 0^r f \cdot 0^{\ell-r} \partial_{0^r} f.$$

Since the first bit of $0^{\ell-r} \partial_{0^r} f$ depends only on the parity of $\partial_{0^r} f$ we can determine the coefficients of the binary expansions of the exponents e_i.

Theorem 3. *The Coordinate Problem can be solved in quadratic time: given a word x of length $\ell = km$ we can determine its coordinates in $O(\ell^2)$ steps.*

Proof. Let $x = x_0 \ldots x_{\ell 1-1}$ and write the coordinates of x as $e_i = \sum_{j<k} t_{i,j} 2^j$. Initialize a vector

$$V = \left(\sum_{j<k} t_{0,j} 2^j, \ldots, \sum_{j<k} t_{m^-,j} 2^j \right)$$

of linear polynomials, this time over the polynomial ring $\mathbb{Z}[\boldsymbol{\tau}]$ where $\boldsymbol{\tau} = (t_{i,j} \mid i, j)$. Again we proceed in k rounds $r = 0, \ldots, k - 1$, each consisting of m stages $s = 0, \ldots, m - 1$. In round r, stage s, perform the following actions. Bind $t_{s,r}$ to 0 or 1 so as to make sure that $V_1 = x_{mr+s} \pmod 2$ where V_1 is the first component of V. Then replace V by $\partial_0 V$ where we interpret the arithmetic operations involved in ∂_0 in the obvious way on V.

Correctness and running time analysis are entirely analogous to the argument in the preceding theorem.   □

Given the algorithm for the Coordinate Problem one can also tackle the Timestamp Problem via a reduction.

Proposition 3. *The Timestamp Problem reduces to the Coordinate Problem in time $O(\ell \log w + \log^2 k)$ where w is the weight of the transduction and km is the length of the words.*

Proof. We are given a transduction f and words x, y. We may safely assume that the given words have length $\ell = km$, otherwise we can simply pad and ignore some of the linear equations below. We need to find the least t such that $x f^t = y$. As we have seen,

$$\langle x f^t \rangle_\ell \equiv \langle x \rangle_\ell + t \cdot \langle f \rangle_\ell.$$

Thus, given the ℓ-coordinates of x, y and f we can simply solve a system of modular equations. The computation of $0^\ell f$ takes $O(\ell \log w)$ steps where w is the weight of f. The linear system has m equations and all coefficients have at most k bits. $\square$

For some CCC transducers the quadratic bounds from the last few results can be improved upon: finite state machines sometimes suffice to calculate coordinates and timestamps. As an example, consider again $\mathfrak{A}_2^3$. The following algorithm solves the Coordinate Problem in this case. Given a word x (here assumed to be 0-indexed) we calculate its coordinates in reverse binary as follows. The γ_i are as in section 2.1.

```
// coordinate algorithm for CCC transducer 𝔄₂³
    h = (0,0);
    for r = 0, ..., n - 1 do
        s_r = h₁ + x_{2r} mod 2;          // phase 1: bind s_r
        h = ∂₀(h + s_r · γ_r);
        t_r = h₁ + x_{2r+1} mod 2;        // phase 2: bind t_r
        h = ∂₀(h + t_r · γ_r);
    return (s,t);
```

As stated, the algorithm appears to require quadratic time. However, it can be implemented on a finite state machine because of the contraction property of residuals spelled out in section 2.

Theorem 4. *The Coordinate Problem for $\mathfrak{A}_2^3$ can be solved by a transducer that computes the coordinates in reverse binary.*

Proof. Given a word x of length $2n$ the algorithm determines a transduction $f = \underline{0}^s \underline{1}^t$ where $0 \le s, t < 2^n$. We will show by induction on n that $f(0^{|x|}) = x$ and $\partial_{0^{|x|}} f = h$. We only present the step from length $8n$ to $8n + 2$ during one round of the algorithm, the other cases are entirely similar and will be omitted. During a particular round we denote s', f' and h' the new values of s, f and h after the first phase in the execution of the algorithm, and t'', f'' and h'' for the second phase. Write $\mathbf{0}$ for 0^{8n} and consider an extension $u = x\,ab$ of x. The following argument relies on lemma 1.

In phase 1, if $00\,f = xa$ then $f' = f$ and we have $00\,f' = 00\,f = xa$. Also, $\partial_{00}f' = \partial_0\partial_0 f = \partial_0 h = h'$. Otherwise $s' = s + 2^{4n}$ and $f' = f\,\underline{0}^{2^{4n}}$. Then $00\,f' = 00\,f\underline{0}^{2^{4n}} = x\overline{a}\,\underline{0}^{2^{4n}} = x\,(\overline{a})\,\underline{0} = xa$. Furthermore, $\partial_{00}f' = \partial_0(\partial_0 f\underline{0}^{2^{4n}}) = \partial_0(h\partial_x\underline{0}^{2^{4n}}) = h'$ by lemma 1.

For phase 2 first consider the case $f'(000) = xab$. Then $t'' = t$ and we have $000\,f'' = 000\,f' = xab$. Also, $\partial_{000}f'' = \partial_0(\partial_{00}f') = \partial_0 h' = h''$. In the remaining case $t'' = t + 2^{4n}$ and $f'' = f'\underline{1}^{2^{4n}}$. Then $000\,f'' = 000\,f'\underline{1}^{2^{4n}} = xa\overline{b}\,\underline{1}^{2^{4n}} = x\,(a\overline{b})\,\underline{1} = xa(\overline{b})\underline{0} = xab$. Furthermore, $\partial_{000}f'' = \partial_0(\partial_{00}f'\underline{1}^{2^{4n}}) = \partial_0(h'\partial_{xa}\underline{1}^{2^{4n}}) = h''$, again by the lemma. □

It is straightforward to modify this algorithm to deal with timestamps.

4 Open Problems

We have characterized the complexity of various computational problems associated with a class of invertible binary transducers that relate to iteration of transductions. Specifically, we have shown that for a cycle-cum-chord transducers iterates, time-stamps and coordinates can be computed quickly. The Knuth normal form of a transduction is a critical technical device in all these arguments. We do not know in general when Knuth normal form can be computed by a finite state transducer as in section 2.1. It appears that this property is quite rare but we are currently unable to characterize the corresponding CCC transducers. The situation is similar with respect to the rationality of the orbit relation; some cases are discussed in [19] but no general characterization exists. In fact, we do not even know whether orbit rationality is decidable for CCC transducers, much less for arbitrary invertible transducers, even when the number of toggle states is restricted to just one.

As already pointed out, our CCC transducers generate free Abelian groups. Of course, there are other automata that also generate these groups. One well-known example are the so-called "sausage automata" in [13], given in wreath notation by $\underline{0} = (I, \underline{n})\,\sigma$ and $\underline{k} = (\underline{k-1}, \underline{k-1})$ for $2 \le k \le n$. Here we ignore the identity I, as customary. In fact, [14] contains a detailed characterization of invertible automata associated with free Abelian groups. It is natural to ask whether and to what degree our results carry over to these automata.

It is straightforward to check whether $\mathcal{S}(\mathcal{A})$ is commutative, using standard automata-theoretic methods. We do not know whether it is decidable whether $\mathcal{S}(\mathcal{A})$ is a group, though this property is obviously semidecidable. Unsurprisingly, many other decidability questions regarding transduction semigroups or groups of invertible transducers are also open, see [7, chap. 7] for an extensive list.

References

1. Bartholdi, L., Silva, P.V.: Groups defined by automata. CoRR, abs/1012.1531 (2010)
2. Berstel, J.: Transductions and context-free languages (2009), http://www-igm. univ-mlv.fr/~berstel/LivreTransductions/LivreTransductions.html

3. Eilenberg, S.: Automata, Languages and Machines, vol. A. Academic Press (1974)
4. Fortnow, L., Grochow, J.A.: Complexity classes of equivalence problems revisited. Inf. Comput. 209(4), 748–763 (2011)
5. Gluškov, V.M.: Abstract theory of automata. Uspehi Mat. Nauk. 16(5(101)), 3–62 (1961)
6. Grigorchuk, R., Šunić, Z.: Self-Similarity and Branching in Group Theory. In: Groups St. Andrews 2005. London Math. Soc. Lec. Notes, vol. 339. Cambridge University Press (2007)
7. Grigorchuk, R.R., Nekrashevich, V.V., Sushchanski, V.I.: Automata, dynamical systems and groups. Proc. Steklov Institute of Math. 231, 128–203 (2000)
8. Howard Johnson, J.: Rational equivalence relations. Theoretical Computer Science 47, 167–176 (1986)
9. Khoussainov, B., Nerode, A.: Automatic presentations of structures. In: Leivant, D. (ed.) LCC 1994. LNCS, vol. 960, pp. 367–392. Springer, Heidelberg (1995)
10. Khoussainov, B., Nerode, A.: Automata Theory and its Applications. Birkhäuser (2001)
11. Khoussainov, B., Rubin, S.: Automatic structures: overview and future directions. J. Autom. Lang. Comb. 8(2), 287–301 (2003)
12. Knuth, D.: Private communication (2010)
13. Nekrashevych, V.: Self-Similar Groups. In: Math. Surveys and Monographs, vol. 117. AMS (2005)
14. Nekrashevych, V., Sidki, S.: Automorphisms of the binary tree: state-closed subgroups and dynamics of 1/2-endomorphisms. Cambridge University Press (2004)
15. Raney, G.N.: Sequential functions. J. Assoc. Comp. Mach. 5(2), 177–180 (1958)
16. Sakarovitch, J.: Elements of Automata Theory. Cambridge University Press (2009)
17. Serre, J.-P.: Arbres, Amalgames, SL_2. Astérisque, vol. 46. Société Mathématique de France, Paris (1977)
18. Sidki, S.: Automorphisms of one-rooted trees: Growth, circuit structure, and acyclicity. J. Math. Sciences 100(1), 1925–1943 (2000)
19. Sutner, K., Lewi, K.: Iterating invertible binary transducers. In: Kutrib, M., Moreira, N., Reis, R. (eds.) DCFS 2012. LNCS, vol. 7386, pp. 294–306. Springer, Heidelberg (2012)

Compressed Automata for Dictionary Matching

Tomohiro I[1,2], Takaaki Nishimoto[1], Shunsuke Inenaga[1],
Hideo Bannai[1], and Masayuki Takeda[1]

[1] Department of Informatics, Kyushu University, Japan
{tomohiro.i,takaaki.nishimoto,inenaga,bannai,takeda}@inf.kyushu-u.ac.jp
[2] Japan Society for the Promotion of Science (JSPS)

Abstract. A variant of the dictionary matching problem is addressed
where the dictionary is given in an SLP-compressed form. An Aho-
Corasick automata-based algorithm is presented which pre-processes the
compressed dictionary $\mathcal{D}$ in $O(n^4 \log n)$ time using $O(n^2 \log N)$ space and
recognizes all occurrences of the patterns in $\mathcal{D}$ in amortized $O(h + m)$
running time per character, where n and N are, respectively, the com-
pressed and uncompressed sizes of $\mathcal{D}$, and h is the height of $\mathcal{D}$, and m is
the number of patterns in the dictionary.

1 Introduction

The classical pattern matching problem is, given two strings called the pattern
and the text, to find all occurrences of the pattern within the text. The *(fully)
compressed pattern matching problem* [5] is the pattern matching problem where
both the pattern and the text are given in compressed form. A variant of this
problem where the text is given in compressed form while the pattern is given
in uncompressed form, has been extensively studied for various compression
formats (see, e.g. [7]).

In this paper, we introduce a new, yet another variant of the problem, where
the pattern is given in compressed form while the text is given in uncompressed
form. In particular, we are interested in a setting where a set of patterns (called
the *dictionary*) is given in compressed form in advance, and the text is given in
a streaming fashion. A typical application would be an SDI (Selective Dissemination of Information) service.

A straight-line program (SLP) is a context-free grammar in the Chomsky
normal form which generates a single string. It is well known that outputs of
various grammar-based compression algorithms (e.g., [10,8]), as well as those of
dictionary-based compression algorithms (e.g., [16,14,15,12]), can be regarded
as, or be quickly transformed to, SLPs [11]. We use an SLP to represent a
dictionary consisting of m patterns, by designating m variables in the SLP as
the start symbols.

Given a compressed dictionary $\mathcal{D}$ represented as an SLP of size n, we consider
how to efficiently construct an Aho-Corasick (AC) automaton [1] for $\mathcal{D}$. Since
the total length N of patterns in $\mathcal{D}$ can be as large as $\Theta(2^n)$, a naïve method
which decompresses $\mathcal{D}$ takes exponential time and space in the worst case. By

S. Konstantinidis (Ed.): CIAA 2013, LNCS 7982, pp. 319–330, 2013.

exploiting some combinatorial properties on SLP-compressed dictionaries, we present a *compressed representation* of AC automata which requires $O(n^2 \log N)$ space. Hence, our representation is useful when the patterns in the dictionary are compressible. This representation allows us to recognize all occurrences of the patterns in $\mathcal{D}$ in amortized $O(h + m)$ running time per character, where h is the height of the derivation tree of the SLP representing $\mathcal{D}$, and m is the number of patterns in $\mathcal{D}$. We also show how to construct our compressed AC automata in $O(n^4 \log n)$ time using $O(n^2 \log N)$ space.

A *succinct* representation of AC automaton has been proposed [2], which requires $S(\log \sigma + 3.443 + o(1)) + m(3 \log(N/m) + O(1))$ bits of space, where S is the number of states in the AC automaton and σ is the alphabet size. Using this succinct AC automaton one can conduct dictionary matching for a given text t in $O(|t| + occ)$ time, where occ is the output size. To the best of our knowledge, our data structure is the first which uses grammar-based string compression to reduce space requirement of AC automata.

2   Preliminaries

2.1   Strings

Let Σ be a finite *alphabet*. An element of Σ^* is called a *string*. The length of a string w is denoted by $|w|$. The empty string ε is a string of length 0, namely, $|\varepsilon| = 0$. Strings x, y and z are, respectively, called a *prefix, substring*, and *suffix* of the string $w = xyz$. A prefix (suffix) of a string w is said to be *proper* if it is shorter than w. The i-th character of a string w is denoted by $w[i]$, where $1 \le i \le |w|$. For a string w and two integers i, j with $1 \le i \le j \le |w|$, let $w[i..j]$ denote the substring of w that begins at position i and ends at position j, that is, $w[i..j] = w[i] \cdots w[j]$.

For any strings $p, t \in \Sigma^+$, let $Occ(p, t)$ denote the set of all positions of t at which an occurrence of P begins, that is, $Occ(p, t) = \{k \mid k \in [1..|t| - |p| + 1], p = t[k..k + |p| - 1]\}$.

2.2   Periods and Runs of Strings

A *period* of a string w is a positive integer p such that $w[p] = [i + p]$ for every $i \in [1..|w|]$ with $i + p \le |w|$. A *run* in a string w is an interval $[i..j]$ with $1 \le i \le j \le |w|$ such that:

- the smallest period p of $w[i..j]$ satisfies $2p \le j - i + 1$.
- the interval can be extended neither to the left nor the right, without violating the above condition, that is, $w[i-1] \ne w[i+p-1]$ and $w[j-p+1] \ne w[j+1]$, provided that respective symbols exist.

Lemma 1 (Periodicity Lemma (see [4])). *Let p and q be two periods of a string x. If $p + q - \gcd(p, q) \le |x|$, then $\gcd(p, q)$ is also a period of x.*

Lemma 2 ([4]). *The periods of any $x \in \Sigma^+$ are partitioned into $O(\log |x|)$-arithmetic progressions.*

2.3   Aho-Corasick Automata

The Aho-Corasick automaton (AC automaton for short) [1] is a finite state machine which simultaneously recognizes all occurrences of multiple patterns in a single pass through a text. The AC automaton for a dictionary Π consists of three functions: *goto*, *failure*, and *output*. Fig. 1 displays an example of the AC automata.

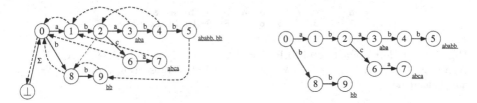

Fig. 1. On the left the Aho-Corasick automaton for $\Pi = \{\texttt{aba}, \texttt{ababb}, \texttt{abca}, \texttt{bb}\}$ is displayed, where the circles denote states, the solid and the broken arrows represent the goto and the failure functions, respectively, and the underlined strings adjacent to states mean the outputs from them. On the right the g-trie for Π is shown.

The *g-trie* for a dictionary Π is a trie representing Π. There is a natural one-to-one correspondence between the states (nodes) of the g-trie and the pattern prefixes. State q is said to *represent* string u if the path from the initial state 0 to q spells out u. For example, the initial state 0 represents the empty string ε and the state 4 represents the string $\texttt{abab}$ in Fig. 1. Let Q denote the set of states of the g-trie, and let $\perp$ be an auxiliary state not in Q. The g-trie defines the goto function g so that every edge q to r labeled c implies $g(q,c) = r$. In addition, we set $g(\perp, a) = 0$ for all $a \in \Sigma$.

The output function λ and the failure function f are defined as follows.

Definition 1. *Let q be any state. Suppose q represents string u. Then $\lambda(q)$ is the set of patterns in Π that are suffixes of u.*

Definition 2. *Let q be any state with $q \neq 0$. Suppose q represents string u. Then state $f(q)$ represents the longest proper suffix of u that is also a prefix of some pattern.*

Let $\delta : Q \times \Sigma \to Q$ be the state-transition function defined by:

$$\delta(q,a) = \begin{cases} g(q,a), & \text{if } g(q,a) \text{ is defined;} \\ \delta(f(q),a), & \text{otherwise.} \end{cases}$$

We extend δ to the domain $Q \times \Sigma^*$ in the standard way. Then we have:

Lemma 3 ([1]). *For any string $w \in \Sigma^*$, $\delta(0,w)$ is the state that represents the longest suffix of w that is also a prefix of some pattern. The number of goto and failure transitions required in computing $\delta(0,w)$ is at most $2|w|$.*

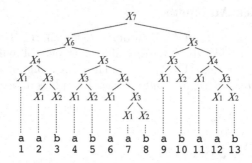

Fig. 2. The derivation tree of SLP $\mathcal{D} = \{X_1 \to$ a, $X_2 \to$ b, $X_3 \to X_1 X_2$, $X_4 \to X_1 X_3$, $X_5 \to X_3 X_4$, $X_6 \to X_4 X_5$, $X_7 \to X_6 X_5\}$, representing string $val(X_7) =$ aababaababaab

We say that a state is *branching* if it is of out-degree ≥ 2, and *terminating* if it represents some pattern. We say that a state is *explicit* if it is branching or terminating, and *implicit* otherwise.

Lemma 4. *The number of explicit states is at most* $2|\Pi|$.

2.4 Straight Line Programs

A *straight-line program (SLP)* is a set of assignments $\mathcal{D} = \{X_1 \to expr_1, X_2 \to expr_2, \dots, X_n \to expr_n\}$, where each X_i is a variable and each $expr_i$ is an expression, where $expr_i = a$ $(a \in \Sigma)$, or $expr_i = X_{\ell(i)} X_{r(i)}$ $(i > \ell(i), r(i))$. It is essentially a context-free grammar in the Chomsky normal form, that derives a single string. Let $val(X_i)$ denote the string derived from variable X_i. To ease notation, we sometimes identify $val(X_i)$ with X_i, and denote $|val(X_i)|$ as $|X_i|$, and $val(X_i)[b..e]$ as $X_i[b..e]$ for any interval $[b..e]$. An SLP $\mathcal{D}$ *represents* the string $s = val(X_n)$. The *size* of $\mathcal{D}$, denoted by $|\mathcal{D}|$, is the number n of assignments in $\mathcal{D}$. Note that $N = |s|$ can be as large as $\Theta(2^n)$.

Our model of computation is the word RAM: We shall assume that the computer word size is at least $\log_2 N$, and hence, standard operations on values representing lengths and positions of string s can be manipulated in constant time. Space complexities will be determined by the number of computer words (not bits).

We will use the following result.

Lemma 5 ([9]). *We can pre-process an SLP* $\mathcal{D} = \{X_i \to expr_i\}_{i=1}^{n}$ *in* $O(n^3)$ *time to answer the following query in* $O(n^2)$ *time: given two variables* X_i *and* X_j *$(1 \leq i, j \leq n)$, compute the length of the longest common prefix of* $val(X_i)$ *and* $val(X_j)$.

The *derivation tree* of an SLP $\mathcal{D} = \{X_i \to expr_i\}_{i=1}^{n}$ is a labeled ordered binary tree where each internal node is labeled with a non-terminal variable in $\{X_1, \dots, X_n\}$, and each leaf is labeled with a terminal character in Σ. The root node has label X_n. Fig. 2 displays an example derivation tree. Let $height(X_i)$ denote the height of derivation tree of X_i, and let $height(\mathcal{D}) = height(X_n)$.

For each variable X_i we store the length $|X_i|$ of the string derived by X_i, which can be computed in a total of $O(n)$ time using $O(n)$ space by a simple dynamic programming algorithm.

The *sorted index* of an SLP $\mathcal{D} = \{X_i \to expr_i\}_{i=1}^n$ is the permutation σ of $[1..n]$ such that the strings $val(X_{\sigma(1)}), \ldots, val(X_{\sigma(n)})$ are arranged in the lexicographical order.

Lemma 6. *The sorted index σ of an SLP of size n can be computed in $O(n^3 \log n)$ time.*

Proof. We compute the length ℓ of the longest common prefix of two variables X_i and X_j using Lemma 5. Then, comparing $val(X_i)$ and $val(X_j)$ reduces to comparing the $(\ell+1)$-th leaves of the derivation trees of X_i and X_j, which can be done in $O(n)$ time using the length of the string that each variable derives (note that the case where $\ell = min\{|X_i|, |X_j|\}$ is easier). Hence the sorted index σ can be computed in $O(n^3 + n^3 \log n) = O(n^3 \log n)$ time using any $O(n \log n)$-time comparison sort. ☐

A variable X with $X_i \to X_l X_r \in \mathcal{D}$ is said to *stab* an interval $[b..e] \subseteq [1..|X_i|]$ if $b \in [1..|X_l|]$ and $e \in [|X_l| + 1..|X_i|]$. For any $p \in \Sigma^+$, let $Occ(p, X_i)$ denote $Occ(p, val(X_i))$, and let $Occ^\xi(p, X_i)$ be the set of positions $\alpha \in Occ(p, X_i)$ such that the interval $[\alpha..\alpha + |p| - 1]$ is stabbed by X_i.

Lemma 7 ([9]). $Occ^\xi(p, X_i)$ *forms an arithmetic progression.*

We will also use the following result:

Lemma 8 ([3]). *Given an SLP $\mathcal{D} = \{X_i \to expr_i\}_{i=1}^n$ that represents a string T of length N, it is possible to pre-process $\mathcal{D}$ in $O(n)$ time using $O(n)$ space, so that any substring $T[i..i+m-1]$ of length m of T can be computed in $O(\log N + m)$ time.*

3   Problem and Compressed AC Automata

3.1   Problem Formulation

A *dictionary* is a non-empty, finite subset of Σ^+. We extend SLPs so as to represent dictionaries as follows: A *dictionary SLP* (DSLP) is an ordered pair $\langle \mathcal{D}, m \rangle$ of an SLP $\mathcal{D} = \{X_i \to expr_i\}_{i=1}^n$ and a positive integer $m \in [1..n]$. The last m variables $X_{n-m+1}, \ldots, X_n$ of $\mathcal{D}$ are designated as the *start variables*. Let $\Pi_{\langle \mathcal{D}, m \rangle}$ denote the dictionary consisting of the strings derived from the start variables. That is,

$$\Pi_{\langle \mathcal{D}, m \rangle} = \{val(X_i) \mid i \in [n - m + 1..n]\}.$$

We note that DSLP $\langle \mathcal{D}, 1 \rangle$ is equivalent to SLP $\mathcal{D}$.

Problem 1. Given a DSLP $\langle \mathcal{D}, m \rangle$ of size n, build in polynomial time and space w.r.t. n an automaton that recognizes all occurrences of patterns in $\Pi_{\langle \mathcal{D}, m \rangle}$ within an arbitrary (uncompressed) string with polynomial time delay.

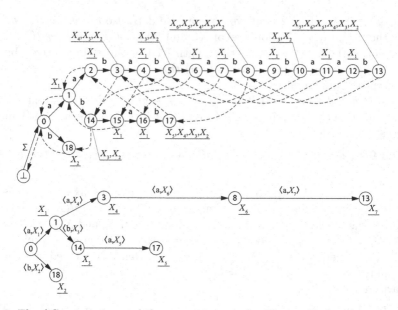

Fig. 3. The AC automaton and the compact g-trie for $\Pi_\mathcal{D}$ are displayed on the upper and on the lower, respectively, where $\mathcal{D}$ is identical to the SLP of Fig. 2

3.2 Compressed AC Automata

We consider the AC automaton for $\Pi_\mathcal{D} = \Pi_{\langle \mathcal{D},n \rangle} = \{ val(X_i) \mid i \in [1..n] \}$, not for $\Pi_{\langle \mathcal{D},m \rangle}$. Independently of $m \in [1..n]$, we use the goto and the failure functions of this automaton, and adjust the output function appropriately for $\Pi_{\langle \mathcal{D},m \rangle}$.

For a compact representation of the g-trie, we can adopt the so-called path compaction technique like the suffix trees [13]. The *compact g-trie* for $\mathcal{D} = \{X_i \to expr_i\}_{i=1}^n$ is the path-compacted trie obtained from the g-trie for $\{ val(X_i) \mid i \in [1..n]\}$ by removing the implicit states, where every edge e from q to r is labeled by $\langle a, X_i \rangle$ such that r represents string uv with $v \in \Sigma^+$, $a = v[1]$, $X_i[1..|uv|] = uv$ and X_i stabs $[1..|uv|]$. The next lemma directly follows from Lemma 4.

Lemma 9. *There are at most $2n$ states in the compact g-trie for $\mathcal{D}$ of size n.*

Fig. 3 displays the AC automaton and the compact g-trie for $\Pi_\mathcal{D}$ where $\mathcal{D}$ is identical to the example SLP of Fig. 2.

An implicit state q' on edge $e = (q, r)$ can be specified by an integer $h \geq 1$ such that q' represents the string $X_i[1..|u| + h]$ and X_i stabs $[1..|u| + h]$, where q represents string u and e is labeled by $\langle a, X_i \rangle$.

Lemma 10. *The compact g-trie can be constructed in $O(n^3 \log n)$ time using $O(n)$ space so that for any state q and any character c, $g(q,c)$ can be determined in $O(\log N)$ time.*

Proof. We can compute in $O(n^3 \log n)$ time the sorted index σ of $\mathcal{D}$ and an array storing the longest common prefix length of $val(X_{\sigma(i)})$ and $val(X_{\sigma(i+1)})$ for all $i \in [1..n-1]$. Thus the compact g-trie can be constructed in $O(n^3 \log n)$ time.

When q is an explicit state, we can find the edge $e = (q, r)$ labeled by $\langle c, X_i \rangle$ for some variable X_i in $O(\log |\Sigma|)$ time, if such e exists, and we thus determine $g(q, c)$ in $O(\log |\Sigma|)$ time. When q is an implicit state on edge e specified by integer h, we can compute the $(h + 1)$-th character in the string spelled out by e in $O(\log N)$ time by using the technique of Lemma 8, and then compare it with c to determine $g(q, c)$. $\qquad\square$

Thus, we can represent the goto function compactly. A naive implementation of the failure function, however, requires exponential space. In the following section, we describe how to represent the failure and the output functions in polynomial space with respect to n.

4   Compact Representation of AC Automaton for DSLP

Theorem 1. *Given any DSLP $\langle \mathcal{D}, m \rangle$ of size n that represents dictionary $\Pi_{\langle \mathcal{D}, m \rangle}$ of total length N, it is possible to build in $O(n^4 \log n)$ time using $O(n^2 \log N)$ space an automaton that recognizes all occurrences of patterns in $\Pi_{\langle \mathcal{D}, m \rangle}$ within an arbitrary string with $O(height(\mathcal{D}) + m)$ amortized running time per character.*

Theorem 1 follows from Lemma 9, Lemma 10, Lemma 11, Lemma 13 and Lemma 17, some of which will be proved in the following two subsections.

4.1   Compact Representation of Failure Function

As stated in the previous section we can represent any implicit state of the compact g-trie as a pair of an edge $e = (q, r)$ and an integer h. Here, we show another representation of states in the compact g-trie: A *reference-pair* of explicit/implicit state q is defined to be $\langle X_i, h \rangle$ such that q represents string $X_i[1..h]$ and X_i stabs $[1..h]$.

Lemma 11. *A mutual conversion between the two state representations can be performed in $O(\log n)$ time using some data structure of size $O(n^2)$.*

Proof. Let q be any state that represents string u. Suppose q is an explicit state. If q is terminating, let X_i be the variable corresponding to q, and otherwise, let X_i be the variable such that some out-going edge e from q is labeled by $\langle a, X_i \rangle$. Then, $\langle X_i, |X_i| \rangle$ gives a reference-pairs of q. Suppose q' is an implicit state on edge $e = (q, r)$ specified by integer h, and e is labeled by $\langle a, X_i \rangle$. Then, $\langle X_i, |u| + h \rangle$ gives a reference-pairs of q.

Conversely, suppose we are given a reference-pair $\langle X_i, h \rangle$ of some state q'. Then, it is possible to determine in $O(\log n)$ time the explicit state q that is the nearest ancestor of q', by using a simple binary search over the lengths of strings represented by the explicit states on the path from the initial state to the terminating state for X_i. $\qquad\square$

Let $Prefix(\mathcal{D})$ denote the set of prefixes of $val(X_i)$ for all variables X_i in $\mathcal{D}$. For any variable $X_i \to X_l X_r \in \mathcal{D}$, an *f-interval* of X_i is a maximal element in the set $\{[b,e] \mid 1 < b \le |X_l| < e \le |X_i|, X_i[b..e] \in \mathit{Prefix}(\mathcal{D})\}$ with respect to the set inclusion relation $\subseteq$. The *f-interval sequence* of X_i, denoted $\mathcal{F}(X_i)$, is defined to be the sequence $\{[b_k..e_k]\}_{k=1}^{s}$ of all f-intervals of X_i arranged in the increasing order of b_k. By definition $e_1, \ldots, e_s$ are also arranged in the increasing order of e_k.

The set of f-interval sequences represents the failure function f as follows:

Lemma 12. *Let q be any state. Suppose q represents string $X_i[1..h]$. If $h = 1$, then $f(q)$ is the initial state. Suppose $h \ge 2$. Choose X_i so that X_i stabs $[1..h]$. Let $\{[b_k..e_k]\}_{k=1}^{s}$ be the f-interval sequence of X_i, and let $k' \in [1..h]$ be the smallest integer such that $h \in [b_{k'}..e_{k'}]$. Then, the state $f(s)$ represents the string $X_i[b_{k'}..h]$. If no such k' exist, then $f(q)$ represents the string $X_r[1..h-|X_l|]$ where $X_i \to X_l X_r \in \mathcal{D}$.*

A naive way of encoding the f-interval sequence $\{[b_k..e_k]\}_{k=1}^{s}$ of a variable X_i is to have a linear-list of triples of $\langle b_k, e_k, X_j \rangle$ such that $X_i[b_k..e_k] = X_j[1..e_k - b_k + 1]$ and X_j stabs $[1..e_k - b_k + 1]$. The list length s can, however, be exponential with respect to n.

Example 1. Consider the SLP $\mathcal{D} = \{X_1 \to \mathtt{a}\} \cup \{X_i \to X_{i-1}X_{i-1}\}_{i=2}^{n-3} \cup \{X_{n-2} \to \mathtt{b}, X_{n-1} \to X_{n-2}X_{n-3}, X_n \to X_{n-1}X_{n-3}\}$. Then there are $2^{n-4} - 1$ f-intervals of X_n. See Fig. 4.

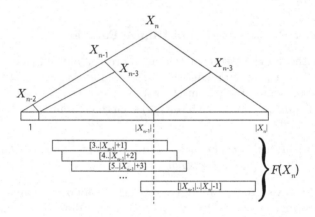

Fig. 4. The f-interval sequence $\mathcal{F}(X_n)$ of length $2^{n-4} - 1$ in Example 1 is illustrated

Fortunately, we can prove the following lemma:

Lemma 13. *The failure function f can be implemented in $O(n^4 \log n)$ time using $O(n^2 \log N)$ space so that given reference-pair of any state q, a reference-pair of the state $f(q)$ can be computed in $O(\log n)$ time.*

In order to achieve Lemma 13, we focus on cyclic structures on f-intervals.

For any variable $X_i \to X_l X_r \in \mathcal{D}$ and any f-interval $[b..e] \in \mathcal{F}(X_i)$, if there is a run $[\alpha..\beta]$ with period p such that $\alpha \le b < e \le \beta$ and $e - b + 1 \ge 2p$, we say that the run $[\alpha..\beta]$ *subsumes* the f-interval $[b..e]$. Note that if such run exists, p is the smallest period of $X_i[b..e]$ and the run is unique with respect to $[b..e]$. If a run $[\alpha..\beta]$ subsumes two distinct f-intervals $[b..e]$ and $[b'..e']$ such that $X_i[b..e] = X_i[b'..e']$ and $b < b' \leq |X_l| - p$, $[\alpha..\beta]$ is said to be *f-rich*.

Lemma 14. *For any variable $X_i \to X_l X_r$ in $\mathcal{D}$, there is at most one f-rich run.*

Proof. The existence of an f-rich run $[\alpha..\beta]$ with period p implies that $u = X_i[|X_l| - p + 1..|X_l|] = X_i[|X_l| + 1..|X_l| + p]$. Also, from the definition of f-rich run, there must exist an f-interval $[b..e]$ such that $[b..e] \supseteq [|X_l| - p + 1..|X_l| + p]$.

Assume on the contrary that there is another f-rich run $[\alpha'..\beta']$ with p' (w.l.o.g. assume $p' < p$). Since $X_i[|X_l| - p' + 1..|X_l|] = X_i[|X_l| + 1..|X_l| + p']$, $p - p'$ is a period of u. Since any interval contained in $[b..e]$ cannot be an f-interval, at least one of $\alpha' < b \leq |X_l| - p + 1$ or $|X_l| + p \leq e < \beta'$ must hold. In either case, we can see that u has a period p'. It follows from the periodicity lemma that $\gcd(p - p', p')$ is a period of u, which means that p is not the smallest period of $X_i[\alpha..\beta]$, a contradiction. □

Lemma 15. *Let $X_i \to X_l X_r$ be any variable in $\mathcal{D}$. Let $[b..e]$ and $[b'..e']$ be the first and the last f-intervals subsumed by a run $[\alpha..\beta]$ with period p, respectively. For any d with $p \leq d < d+p < b' - b$, $[b+d..e''] \in \mathcal{F}(X_i) \iff [b+d+p..e''+p] \in \mathcal{F}(X_i)$.*

Proof. We remark that $X_i[b..e']$ has period p.

Firstly, we show that $[b + d..e''] \in \mathcal{F}(X_i) \implies [b + d + p..e'' + p] \in \mathcal{F}(X_i)$. It is clear that $X_i[b + d..e''] \in Prefix(\mathcal{D})$. Note that $e'' < e'' + p < e'$ holds, since otherwise $X_i[b+d+p..e']$ is a prefix of $X_i[b+d..e'']$ and in $Prefix(\mathcal{D})$, which implies that $[b'..e']$ is not an f-interval. Assume on the contrary that $[b+d+p..e''+p] \notin \mathcal{F}(X_i)$, i.e., at least one of $X_i[b+d+p..e''+p+1] \in Prefix(\mathcal{D})$ or $X_i[b+d+p-c..e''+p] \in Prefix(\mathcal{D})$ with some $c > 0$ holds. If $X_i[b+d+p..e''+p+1] \in Prefix(\mathcal{D})$, we get $X_i[b+d+p..e''+p+1] = X_i[b+d..e''+1] \in Prefix(\mathcal{D})$, which contradicts that $[b + d..e'']$ is an f-interval. If $X_i[b + d + p - c..e'' + p] \in Prefix(\mathcal{D})$, we consider two cases: When $c > p$, we get $[b+d..e''] \subset [b+d+p-c..e''+p]$, which contradicts that $[b+d..e''] \in \mathcal{F}(X_i)$. When $c \leq p$, we get $X_i[b+d+p-c..e''+p] = X_i[b+d-c..e''] \in Prefix(\mathcal{D})$, a contradiction. Therefore $[b + d..e''] \in \mathcal{F}(X_i) \implies [b + d + p..e'' + p] \in \mathcal{F}(X_i)$ holds.

Next we show that $[b + d..e''] \in \mathcal{F}(X_i) \impliedby [b + d + p..e'' + p] \in \mathcal{F}(X_i)$. Note that $e < e''$, since otherwise $X_i[b..e''] = X_i[b + p..e'' + p]$ is in $Prefix(\mathcal{D})$, which implies that $[b + d + p..e'' + p]$ is not in $\mathcal{F}(X_i)$. Assume on the contrary that $[b + d..e''] \notin \mathcal{F}(X_i)$, i.e., at least one of $X_i[b + d..e'' + 1] \in Prefix(\mathcal{D})$ or $X_i[b+d-c..e''] \in Prefix(\mathcal{D})$ with some $c > 0$ holds. If $X_i[b+d..e''+1] \in Prefix(\mathcal{D})$, we get $X_i[b+d..e'' + 1] = X_i[b+d+p..e''+p+1] \in Prefix(\mathcal{D})$, which contradicts that $[b+d+p..e''+p]$ is an f-interval. If $X_i[b+d-c..e''] \in Prefix(\mathcal{D})$, we consider two cases: When $c > d$, we get $[b..e] \subset [b+d-c..e'']$, which contradicts that $[b..e] \in \mathcal{F}(X_i)$. When $c \leq d$, we get $X_i[b+d-c..e''] = X_i[b+d+p-c..e''+p] \in Prefix(\mathcal{D})$, a contradiction. Therefore $[b + d..e''] \in \mathcal{F}(X_i) \impliedby [b + d + p..e'' + p] \in \mathcal{F}(X_i)$ holds. □

Lemma 15 implies that f-intervals subsumed by the f-rich run are cyclic except for $O(n)$ f-intervals around the first and the last f-intervals in the run. Then we consider storing cyclic f-intervals in a different way from the naive list of $\mathcal{F}(X_i)$. Since information of f-intervals for the first period is enough to compute failure function for any state within the cyclic part, it can be stored by an $O(n)$-size list $L_c(X_i)$. Let $L(X_i)$ denote the list storing $\mathcal{F}(X_i)$ other than cyclic f-intervals.

Lemma 16. *For any $X_i \rightarrow X_l X_r \in \mathcal{D}$, the size of $L(X_i)$ is bounded by $O(n \log N)$.*

Proof. Let X_j be any variable and let $c_0, \ldots, c_s$ ($c_0 < \cdots < c_s$) be the positions of $val(X_l)$ at which a suffix of $val(X_l)$ overlaps with a prefix of $val(X_j)$. We note that each c_k is a candidate for the beginning position of an f-interval of X_i. It follows from Lemma 2 that $c_0, \ldots, c_s$ can be partitioned into at most $O(\log|X_l|)$ disjoint segments such that each segment forms an arithmetic progression.

Let $0 \leq k < k' \leq s$ be integers such that $C = c_k, \ldots, c_{k'}$ is represented by one arithmetic progression. Let d be the step of C, i.e., $c_{k'} = c_{k'-1} + d = \cdots = c_k + (k' - k)d$. We show that if more than two of C are related to the beginning positions of f-intervals of X_i, the f-rich run subsumes all those f-intervals but the last one.

Suppose that for some $k \leq h_1 < h_2 < h_3 \leq k'$, $c_{h_1}, c_{h_2}, c_{h_3} \in C$ are corresponding to f-intervals, namely, $[c_{h_1}..e], [c_{h_2}..e'], [c_{h_3}..e''] \in \mathcal{F}(X_i)$ with $e - c_{h_1} + 1 = LCP(X_i[c_{h_1}..|X_i|], X_j)$, $e' - c_{h_2} + 1 = LCP(X_i[c_{h_2}..|X_i|], X_j)$ and $e'' - c_{h_3} + 1 = LCP(X_i[c_{h_3}..|X_i|], X_j)$. It is clear that d is the smallest period of $X_i[c_{h_1}..|X_i|]$ and $|X_l| - c_{h_1} + 1 > 2d$. Let β be the largest position of $val(X_i)$ such that $X_i[c_{h_1}..\beta]$ has period d, i.e., there is a run $[\alpha, \beta]$ with $\alpha \leq c_{h_1} < |X_l| < \beta$. Let β' be the largest position of $val(X_j)$ such that $X_j[1..\beta']$ has period d.

- If $\beta < e''$. Note that this happens only when $\beta - c_{h_3} + 1 = \beta'$. Consequently, $LCP(X_i[c_{h_1}..|X_i|], X_j) = LCP(X_i[c_{h_2}..|X_i|], X_j) = \beta'$.
- If $\beta \geq e''$. It is clear that $\beta' < e'' - c_{h_2} + 1$, since otherwise $[c_{h_3}..e'']$ would be contained in $[c_{h_2}..e']$. Then, $LCP(X_i[c_{h_1}..|X_i|], X_j) = LCP(X_i[c_{h_2}..|X_i|], X_j) = \beta'$.

In either case $X_i[c_{h_1}..e] = X_i[c_{h_2}..e'] = X_j[1..\beta']$ holds, which means that except for at most one f-interval $[c..e]$ satisfying $\beta < e$ the others are all subsumed by the f-rich run $[\alpha..\beta]$.

Since in each segment there are at most two f-intervals which are not subsumed by the f-rich run, the number of such f-intervals can be bounded by $O(\log N)$. Considering every variable X_j, we can bound the size of $L(X_i)$ by $O(n \log N)$. □

We are now ready to prove Lemma 13.

Proof of Lemma 13. Karpinski et al. considered in [6] a compressed overlap table OV for an SLP of size n such that for any pair of variables X and Y, $OV(X, Y)$ contains $O(\log N)$-size representation of overlaps between suffixes of $val(X)$ and prefixes of $val(Y)$. They showed how to compute OV in $O(n^4 \log n)$

time. Actually, their algorithm can be used to compute $L(X_i)$ and $L_c(X_i)$ for all variable $X_i \in \mathcal{D}$ in $O(n^4 \log n)$ time. From Lemma 15 and Lemma 16, the total size for $L(X_i)$ and $L_c(X_i)$ for all variable $X_i \in \mathcal{D}$ is bounded by $O(n^2 \log N)$.

Using $L(X_i)$ and $L_c(X_i)$, we can compute $f(q)$ for any state $q = \langle X_i, h \rangle$ in $O(\log n)$ time. If q is not in cyclic part of f-intervals, we conduct binary search on $L(X_i)$, otherwise on $L_c(X_i)$ with proper offset. It takes $O(\log(n \log N)) = O(\log n)$ time. $\square$

4.2   Compact Representation of Output Function

Lemma 17. *The output function λ can be implemented in $O(n^3 \log n)$ time using $O(nm)$ space so that given any state $q = \langle X_i, h \rangle$ we can compute $\lambda(q)$ in $O(height(X_i) + m)$ time.*

Proof. First we construct a tree with nodes $\Pi \cup \{\varepsilon\}$ such that for any $p \in \Pi_{\langle \mathcal{D}, m \rangle}$ the parent of p is the longest element of $\Pi_{\langle \mathcal{D}, m \rangle} \cup \{\varepsilon\}$ which is also a suffix of $X_i[1..h]$. The tree can be constructed in $O(n^3 \log n)$ time in a similar way to the construction of the compact g-trie. Note that $\lambda(q)$ can be computed by detecting the longest member p of $\Pi_{\langle \mathcal{D}, m \rangle}$ which is also a suffix of $X_i[1..h]$, and outputting all patterns on the path from p to the root of the tree. In addition, we compute in $O(n^3)$ time a table of size $O(nm)$ such that for any pair of $p \in \Pi_{\langle \mathcal{D}, m \rangle}$ and variable X_j the table has $Occ^\xi(p, X_j)$ in a form of one arithmetic progression.

Now we show how to compute the longest member of $\Pi_{\langle \mathcal{D}, m \rangle}$ which is also a suffix of $X_i[1..h]$. We search for it in descending order of pattern length. We use three variables p', i' and h', which are initially set to the longest pattern in $\Pi_{\langle \mathcal{D}, m \rangle}$, i and h, respectively. We omit the case when $|p'| = 1$ or $|p'| > h$ since it is trivial. If the end position of $X_i[1..h]$ is contained in $X_{r(i')}$ and $|p'| > h' - |X_{\ell(i')}|$, using arithmetic progression of $Occ^\xi(p', X_{i'})$, we can check if p' is a suffix of $X_i[1..h]$ or not in constant time by simple arithmetic operations. If the above condition does not hold, we traverse the derivation tree of $X_{i'}$ toward the end position of $X_i[1..h]$ updating i' and h' properly until meeting the above situation, where h' is updated to be the length of the overlapped string between $X_{i'}$ and $X_i[1..h]$.

It is not difficult to see that the total time is $O(height(X_i) + m)$. $\square$

5   Discussion

Our method of Section 4 builds the goto and the failure functions of the AC automaton for the dictionary $\Pi_{\langle \mathcal{D}, n \rangle}$ independently of m. This introduces redundant states and edges into the compact g-trie, and unnecessary failure transitions.

Another possible solution would be to divide the input DSLP into m SLPs and then build KMP-type automata for them, respectively. Concerning such KMP-type automata, we can prove:

Theorem 2. *For an SLP $\mathcal{D}$ of size n representing string P of length N, it is possible to build in $O(n^4 \log n)$ time using $O(n \log N)$ space a KMP-type automaton that recognizes all occurrences of pattern P within an arbitrary string with $O(height(\mathcal{D}))$ amortized running time per character.*

Proof. Some technique similar to the proof of Lemma 13 can be used. The details are omitted.   □

Constructing m KMP-type automata for the m SLPs takes $O(n^4 \log n)$ time and $O(mn \log N)$ space, with $O(m \, height(\mathcal{D}))$ amortized running time per character. Notice that the solution proposed in Section 4 is more efficient than this one.

References

1. Aho, A.V., Corasick, M.: Efficient string matching: An aid to bibliographic search. Comm. ACM 18(6), 333–340 (1975)
2. Belazzougui, D.: Succinct dictionary matching with no slowdown. In: Amir, A., Parida, L. (eds.) CPM 2010. LNCS, vol. 6129, pp. 88–100. Springer, Heidelberg (2010)
3. Bille, P., Landau, G.M., Raman, R., Sadakane, K., Satti, S.R., Weimann, O.: Random access to grammar-compressed strings. In: Proc. SODA 2011, pp. 373–389 (2011)
4. Crochemore, M., Rytter, W.: Text Algorithms. Oxford University Press, New York (1994)
5. Gąsieniec, L., Karpinski, M., Plandowski, W., Rytter, W.: Efficient algorithms for Lempel-Ziv encoding. In: Karlsson, R., Lingas, A. (eds.) SWAT 1996. LNCS, vol. 1097, pp. 392–403. Springer, Heidelberg (1996)
6. Karpinski, M., Rytter, W., Shinohara, A.: An efficient pattern-matching algorithm for strings with short descriptions. Nordic Journal of Computing 4, 172–186 (1997)
7. Kida, T., Shibata, Y., Takeda, M., Shinohara, A., Arikawa, S.: Collage system: A unifying framework for compressed pattern matching. Theor. Comput. Sci. 298(1), 253–272 (2003)
8. Larsson, N.J., Moffat, A.: Offline dictionary-based compression. In: Proc. DCC 1999, pp. 296–305. IEEE Computer Society (1999)
9. Miyazaki, M., Shinohara, A., Takeda, M.: An improved pattern matching algorithm for strings in terms of straight-line programs. In: Hein, J., Apostolico, A. (eds.) CPM 1997. LNCS, vol. 1264, pp. 1–11. Springer, Heidelberg (1997)
10. Nevill-Manning, C.G., Witten, I.H., Maulsby, D.L.: Compression by induction of hierarchical grammars. In: Proc. DCC 1994, pp. 244–253 (1994)
11. Rytter, W.: Application of Lempel-Ziv factorization to the approximation of grammar-based compression. Theor. Comput. Sci. 302(1–3), 211–222 (2003)
12. Storer, J., Szymanski, T.: Data compression via textual substitution. J. ACM 29(4), 928–951 (1982)
13. Weiner, P.: Linear pattern-matching algorithms. In: Proc. of 14th IEEE Ann. Symp. on Switching and Automata Theory, pp. 1–11. Institute of Electrical Electronics Engineers, New York (1973)
14. Welch, T.A.: A technique for high performance data compression. IEEE Computer 17, 8–19 (1984)
15. Ziv, J., Lempel, A.: A universal algorithm for sequential data compression. IEEE Transactions on Information Theory IT-23(3), 337–349 (1977)
16. Ziv, J., Lempel, A.: Compression of individual sequences via variable-length coding. IEEE Transactions on Information Theory 24(5), 530–536 (1978)

Enhancing Approximations
for Regular Reachability Analysis

Aloïs Dreyfus, Pierre-Cyrille Héam, and Olga Kouchnarenko

FEMTO-ST CNRS 6174, University of Franche-Comté & Inria/CASSIS, France
`firstname.name@femto-st.fr`

Abstract. This paper introduces two mechanisms for computing over-approximations of sets of reachable states, with the aim of ensuring termination of state-space exploration. The first mechanism consists in over-approximating the automata representing reachable sets by merging some of their states with respect to simple syntactic criteria, or a combination of such criteria. The second approximation mechanism consists in manipulating an auxiliary automaton when applying a transducer representing the transition relation to an automaton encoding the initial states. In addition, for the second mechanism we propose a new approach to refine the approximations depending on a property of interest. The proposals are evaluated on examples of mutual exclusion protocols.

1 Introduction and Problem Statement

Reachability analysis is a challenging issue in formal software verification. Since the reachability problem is in general undecidable in most formalisms, several ad-hoc approaches have been developed, such as symbolic reachability analysis using finite representations of infinite sets of states. *Regular model checking* (RMC for short) – a symbolic approach using regular sets to represent sets of states – tackles undecidability in either of two ways: pointing out classes of regulars sets and relations for which the reachability problem is decidable (see for instance [21]), or developing semi-algorithmic and/or approximation-based approaches (see for instance [15,16]) to semi-decide the reachability problem.

In this paper we present new approximation techniques for RMC, with the aim of providing quite efficient (semi-)algorithms. The first technique consists in over-approximating the automata representing reachable sets by merging some of their states with respect to simple syntactic criteria, or a combination of such criteria (Section 2). The second approximation technique consists in using an auxiliary automaton when applying a transducer representing the transition relation to an automaton encoding the initial states (Section 3). Moreover, for the second technique we develop a new approach to refine the approximations, close to the well-known CEGAR technique (Section 4). The proposals are evaluated on examples of mutual exclusion protocols.

Omitted proofs are available online[1].

---

[1] http://disc.univ-fcomte.fr/~adreyfus/ciaa13/version_longue.pdf

S. Konstantinidis (Ed.): CIAA 2013, LNCS 7982, pp. 331–339, 2013.
© Springer-Verlag Berlin Heidelberg 2013

Related Work. Regular model-checking remains an active research domain in computer science (see [14] and [4] for a thorough overview). In [23] the authors propose to use regular sets of strings to represent states of parametrized arrays of processes, and to represent the effect of performing an action by a predicate transformer (transducer). In this work only transducers representing the effect of a single application of a transition are considered, and consequently the reachability analysis does not terminate for a lot of protocols. To bypass this problem and still reach a fixpoint, the principal methods are acceleration (providing exact computations) [22,11,15,16,3,8], widening (extrapolating) [11,25,24], and automata abstraction [10]. Recently, new results in RMC have been obtained for specific protocols (i.e., CLP [19], communicating systems [20], tree language [1,12], or relational string verification using multi-track automata [26]), using domain-specific techniques [7]. Our contributions aim at improving the generic method in [10] by giving means to build over-approximations by merging abstract states of the system (and not of the transducer, which is never modified).

Unlike [11,10], our proposals do not require the subset-construction, minimization and determinization of the obtained automaton at each RMC step.

Formal Background. We assume the reader familiar with basic notions of language theory. (Q, Σ, E, I, F) where Q is the finite set of *states*, $E \subseteq Q \times \Sigma \times Q$ is the set of *transitions*, $I \subseteq Q$ is the set of *initial states* and $F \subseteq Q$ is the set of *final states*. We define the size of $\mathcal{A}$ by $|\mathcal{A}| = |Q| + |E|$. An automaton is *deterministic* [resp. *complete*] if I is a singleton and for each $(q, a) \in Q \times \Sigma$ there is at most [resp. at least] one $p \in Q$ such that $(q, a, p) \in E$. A path in $\mathcal{A}$ is a (possibly empty) finite sequence of transitions $(p_1, a_1, q_1) \ldots (p_n, a_n, q_n)$ such that for each i, $q_i = p_{i+1}$. The integer n is the length of the path and the word $a_1 \ldots a_n$ is its label. A path is *successful* if p_1 is initial and p_n is final. A word w is *accepted* by $\mathcal{A}$ if w is the label of a successful path. The set of words accepted by $\mathcal{A}$ is denoted $L(\mathcal{A})$. If $\mathcal{A}$ is deterministic and complete, for every state q and every word w, there exists a unique state of $\mathcal{A}$, denoted $q \cdot_{\mathcal{A}} w$ reachable from q by reading a path labeled by w. If there is no ambiguity on $\mathcal{A}$, it is simply denoted $q \cdot w$. By convention, $q \cdot \varepsilon = \{q\}$. A state q is *accessible* [resp. *co-accessible*] if there exists a path from an initial state to q [resp. if there exists a path from q to a final state]. An automaton whose states are all both accessible and co-accessible is called *trim*. If $\mathcal{A}$ is not a trim automaton, removing from $\mathcal{A}$ all states that are not both accessible and co-accessible together with all related transitions provides an equivalent trim automaton. Let $\mathcal{A}_1 = (Q_1, \Sigma, E_1, I_1, F_1)$ and $\mathcal{A}_2 = (Q_2, \Sigma, E_2, I_2, F_2)$ be two automata over the same alphabet, the product of $\mathcal{A}_1$ and $\mathcal{A}_2$ is the automaton $(Q_1 \times Q_2, \Sigma, E, I_1 \times I_2, F_1 \times F_2)$, denoted $\mathcal{A}_1 \times \mathcal{A}_2$, where $E = \{((p_1, p_2), a, (q_1, q_2)) \mid (p_1, a, q_1) \in E_1 \wedge (p_2, a, q_2) \in E_2\}$. By definition, $L(\mathcal{A}_1 \times \mathcal{A}_2) = L(\mathcal{A}_1) \cap L(\mathcal{A}_2)$. Let $\hat{\mathcal{A}} = (\hat{Q}, \Sigma, \hat{E}, \hat{I}, \hat{F})$ be the trim automaton obtained from $\mathcal{A}$, given an equivalence relation $\sim \subseteq Q \times Q$, $\mathcal{A}/_\sim$ denotes the automaton $(\hat{Q}/_\sim, \Sigma, E', \hat{I}/_\sim, \hat{F}/_\sim)$ where $E' = \{(\tilde{p}, a, \tilde{q}) \mid \exists p \in \tilde{p} \text{ and } \exists q \in \tilde{q} \text{ s.t. } (p, a, q) \in \hat{E}\}$. One can easily check that $L(\hat{\mathcal{A}}) \subseteq L(\mathcal{A}/_\sim)$. Two automata $\mathcal{A}_1 = (Q_1, \Sigma, E_1, I_1, F_1)$ and $\mathcal{A}_2 = (Q_2, \Sigma, E_2, I_2, F_2)$ are isomorphic if there exists a one-to-one function $f : Q_1 \to Q_2$ satisfying $(p, a, q) \in E$ iff $((f(p), a, f(q)) \in E$, and $f(I_1) = I_2$, $f(F_1) = F_2$ when lifted to sets. Informally, two automata are isomorphic if they are equal up to state names.

Let Σ_1 and Σ_2 be two alphabets, a *transducer* on Σ_1, Σ_2 is an automaton on $\Sigma_1 \times \Sigma_2$. Each transducer $\mathcal{T}$ on Σ_1, Σ_2 induces a relation $R_{\mathcal{T}}$ on $\Sigma_1^* \times \Sigma_2^*$ defined by: for the a_i's in Σ_1 and the b_j's in Σ_2, $(a_1 \ldots a_n, b_1 \ldots b_m) \in R_{\mathcal{T}}$ iff $n = m$ and the word $(a_1, b_1) \ldots (a_n, b_n)$ is accepted by $\mathcal{T}$. The reflexive transitive closure of $R_{\mathcal{T}}$ is denoted $R_{\mathcal{T}}^*$. Let $\mathcal{A} = (Q_1, \Sigma, E_1, I_1, F_1)$ be an automaton on Σ_1, and $\mathcal{T} = (Q_2, \Sigma_1 \times \Sigma_2, E_2, I_2, F_2)$ a transducer on $\Sigma_1 \times \Sigma_2$, we denote by $\mathcal{T}(\mathcal{A})$ the automaton $(Q_1 \times Q_2, \Sigma_2, E, I_1 \times I_2, F_1 \times F_2)$ on Σ_2 where $E = \{((p_1, p_2), b, (q_1, q_2)) \mid (p_1, a, q_1) \in E_1 \wedge (p_2, (a, b), q_2) \in E_2\}$. By definition, $L(\mathcal{T}(\mathcal{A}))$ is the set of words v satisfying $(u, v) \in R_{\mathcal{T}}$ for some words $u \in L(\mathcal{A})$. If $\mathcal{T} = (Q_2, \Sigma_1 \times \Sigma_2, E_2, I_2, F_2)$ is a transducer, we denote by $\mathcal{T}^{-1}$ the transducer $(Q_2, \Sigma_2 \times \Sigma_2, E_2', I_2, F_2)$ with $E_2' = \{(p, (a, b), q) \mid (p, (b, a), q) \in E_2\}$. One can check that $(u, v) \in R_{\mathcal{T}}$ iff $(v, u) \in R_{\mathcal{T}^{-1}}$.

Regular Reachability Problem. The following regular reachability problem – central for RMC – is known to be undecidable in general.

Input: Two finite automata $\mathcal{A}$ and $\mathcal{B}$ on Σ, and a transducer $\mathcal{T}$ on $\Sigma \times \Sigma$.

Output: **1** if $R_{\mathcal{T}}^*(L(\mathcal{A})) \cap L(\mathcal{B}) = \emptyset$, and **0** otherwise.

Since the problem is concerned with the reflexive-transitive closure, we may assume without loss of generality that for every $u \in \Sigma^*$, $(u, u) \in R_{\mathcal{T}}$. In the rest of the paper, all considered relations contain the identity.

2   Quotient-Based Approximations

This section introduces the first mechanism for computing over-approximations of sets of reachable states, which consists in over-approximating the automata representing reachable sets by merging some of their states. For doing this, basic elementary policies as well as their combinations are introduced.

Given an automaton $\mathcal{A}$, we define an *approximation* as a function mapping each automaton $\mathcal{A}$ to an equivalence relation $\sim_{\mathcal{A}}$ over the states of $\mathcal{A}$. The approximation function $\mathfrak{F}$ is *isomorphism-compatible* if for every pair of automata $\mathcal{A}_1$ and $\mathcal{A}_2$, every isomorphism φ from $\mathcal{A}_1$ to $\mathcal{A}_2$, $p \sim_{\mathcal{A}_1} q$ iff $\varphi(p) \sim_{\mathcal{A}_2} \varphi(q)$. We denote $\mathfrak{F}[\mathcal{A}]$ the automaton $\hat{\mathcal{A}}/_{\mathfrak{F}(\hat{\mathcal{A}})}$, where $\hat{\mathcal{A}}$ is the trim automaton obtained from $\mathcal{A}$. We inductively define $\mathfrak{F}^n[\mathcal{A}]$ by $\mathfrak{F}^0[\mathcal{A}] = \mathcal{A}$, and $\mathfrak{F}^n[\mathcal{A}] = \mathfrak{F}[\mathfrak{F}^{n-1}[\mathcal{A}]]$.

Let us now introduce two isomorphism-compatible approximation functions. They are easily computable, and represent simple criteria naturally used by the specifier, as for example in [10] for computing equivalence relations, or in [5] for monitoring LTL properties. The function $\mathfrak{Left}$ maps each automaton (Q, Σ, E, I, F) to the reflexive-transitive closure of the relation R_{left}, defined by $p R_{\text{left}} q$ iff $L(Q, \Sigma, E, I, \{p\}) \cap L(Q, \Sigma, E, I, \{q\}) \neq \emptyset$. The function $\mathfrak{Right}$ maps each automaton (Q, Σ, E, I, F) to the reflexive-transitive closure of the relation R_{right}, defined by $p R_{\text{right}} q$ iff $L(Q, \Sigma, E, \{p\}, F) \cap L(Q, \Sigma, E, \{q\}, F) \neq \emptyset$.

Semi-Algorithm FixPoint
Input: $\mathcal{A}$, $\mathcal{T}$, $\mathcal{B}$, $\mathfrak{F}$
  If $L(C_{\mathfrak{F}}(\mathcal{T}(\mathcal{A}))) \cap L(\mathcal{B}) \neq \emptyset$ then
    return *Inconclusive*
  EndIf
  If $L(C_{\mathfrak{F}}(\mathcal{T}(\mathcal{A}))) = L(\mathcal{A})$ then
    return *Safe*
  EndIf
  Return FixPoint($C_{\mathfrak{F}}(\mathcal{T}(\mathcal{A}))$, $\mathcal{T}$,$\mathcal{B}$,$\mathfrak{F}$)

(a) FixPoint

Semi-Algorithm FixPointT
Input: $\mathcal{A}$, $\mathcal{T}$, $\mathcal{B}$, C
Variable: k
  k:=0
  While $(L(\mathcal{T}_C^{k+1}(\mathcal{A})) \neq L(\mathcal{T}_C^k(\mathcal{A})))$ do
    $k := k + 1$
  EndWhile
  If $(L(\mathcal{T}_C^k(\mathcal{A})) \cap L(\mathcal{B}) = \emptyset)$ then
    Return *Safe*
  Else
    Return *Inconclusive*
  EndIfElse

(b) FixPointT

Fig. 1. Fixpoint algorithms

Proposition 1. *For each automaton $\mathcal{A}$, if $\mathfrak{F}$ is an isomorphism-compatible approximation function, then the sequence $(\mathfrak{F}^n[\mathcal{A}])_{n \in \mathbb{N}}$ is ultimately constant, up to isomorphism. Let $C_{\mathfrak{F}}(\mathcal{A})$ denote the limit of $(\mathfrak{F}^n[\mathcal{A}])_{n \in \mathbb{N}}$. Moreover, if for each automaton $\mathcal{A}$ and each pair of states p, q of $\mathcal{A}$, one can check in polynomial time whether $p \sim_{\mathcal{A}} q$, then $C_{\mathfrak{F}}(\mathcal{A})$ can be computed in polynomial time as well.*

In the FixPoint algorithm depicted in Fig. 1(a), given a finite automaton $\mathcal{A}$, a transducer $\mathcal{T}$, a finite automaton $\mathcal{B}$, and an isomorphism-compatible function $\mathfrak{F}$, the first check (emptiness) can be performed in polynomial time. Then, unfortunately, the equality of the languages cannot be checked in polynomial time, since the involved automata are not deterministic. Nevertheless, recently developed algorithms [17,2,9] allow solving this problem very efficiently. Note also that the equality test can be replaced by another test – e.g., isomorphism or (bi)simulation – implying language equality or inclusion, as $L(\mathcal{A}) \subseteq L(C_{\mathfrak{F}}(\mathcal{T}(\mathcal{A})))$) by construction.

Proposition 2. *The FixPoint semi-algorithm is correct: if it returns Safe, then $R_{\mathcal{T}}^*(L(\mathcal{A})) \cap L(\mathcal{B}) = \emptyset$.*

Given two approximation functions $\mathfrak{F}$ and $\mathfrak{G}$, we denote $\mathfrak{F}.\mathfrak{G}$ the approximation function defined by $(\mathfrak{F}.\mathfrak{G})(\mathcal{A}) = \mathfrak{F}(\mathcal{A}) \cap \mathfrak{G}(\mathcal{A})$ for every automaton $\mathcal{A}$. In addition, the approximation function $\mathfrak{F} + \mathfrak{G}$ is defined by: for every automaton $\mathcal{A}$, $(\mathfrak{F} + \mathfrak{G})(\mathcal{A})$ is the smallest equivalence relation containing both $\mathfrak{F}(\mathcal{A})$ and $\mathfrak{G}(\mathcal{A})$. Then using several approximation functions and combining them allow us to obtain new – stronger or weaker – approximations.

The approach has been experimented on classical mutual exclusion protocols: the Bakery algorithm by Lamport, the token ring algorithm, Dijkstra's, and Burns protocols. Using combinations of the $\mathfrak{Left}$, $\mathfrak{Right}$ approximations functions and two other functions corresponding to 1-(in, out)-simulations, all these protocols have been proven safe in few computation steps (at most 6).

3   Transducer-Based Approximations

This section introduces another approximation mechanism consisting in reasoning about the application of k copies of a transducer representing the transition relation to an automaton representing the initial states. The states reached in the transducers are encoded as a finite word, and an additional automaton is used for specifying what are the combinations of transducer states that have to be merged. This technique is inspired by an automata theoretic construction in [11], with the difference concerning the equivalence relation, and the use of automata at step k (the transducer is never modified).

Let $\mathcal{A} = (Q, \Sigma, E, I, F)$ be a finite automaton, $\mathcal{T} = (Q_T, \Sigma \times \Sigma, E_T, I_T, F_T)$ a transducer, and $\mathcal{C} = (Q_C, Q_T, E_C, \{q_{\text{init}}\}, \emptyset)$ a deterministic complete finite automaton on Q_T (i.e., the transitions of $\mathcal{C}$ are labeled with states of $\mathcal{T}$). Let φ_k be a one-to-one mapping from the set $(((Q \times Q_T) \times Q_T) \ldots \times Q_T)$ of states of $\mathcal{T}^k(\mathcal{A})$ to $Q \times Q_T^k$, where Q_T^k is the set of words of length k on Q_T. We set a relation $\sim_C$ on states of $\mathcal{T}^k(\mathcal{A})$ as follows: if p and q are states of $\mathcal{T}^k(\mathcal{A})$ such that $\varphi_k(p) = (p_0, w_p)$ and $\varphi_k(q) = (q_0, w_q)$, then $p \sim_C q$ iff $p_0 = q_0$ and $q_{\text{init}} \cdot w_p = q_{\text{init}} \cdot w_q$. The automaton $\mathcal{T}^k(\mathcal{A})/\sim_C$ is denoted $\mathcal{T}_C^k(\mathcal{A})$. One can easily check that $\sim_C$ is an equivalence relation.

Proposition 3. *An automaton isomorphic to $\mathcal{T}^k(\mathcal{A})/\sim_C$ can be computed in polynomial time in k and in the sizes of $\mathcal{A}$, $\mathcal{T}$ and $\mathcal{C}$.*

Now, given a finite automaton $\mathcal{B}$, we can use the computed automata when applying the FixPointT semi-algorithm described in Fig. 1(b). It may provide an over-approximation of reachable states: if FixPointT stops on a not too coarse approximation we can deduce that $\mathcal{R}_T^*(L(\mathcal{A})) \cap L(\mathcal{B}) = \emptyset$.

Proposition 4. *The FixPointT semi-algorithm is correct: if it returns safe then $\mathcal{R}_T^*(L(\mathcal{A})) \cap L(\mathcal{B}) = \emptyset$.*

4   Refining Transducer-Based Approximations

In this section we propose to refine transducer-based approximations when the approximate iteration is inconclusive. Intuitively, this happens when the sequence of approximations is too coarse: the result intersects with the set of bad states after k steps while the backward iteration of k copies of the transducer from the bad states does not intersect with the initial states. Our algorithm can be seen as a kind of CEGAR algorithms – the paradigm introduced in [13] and intensively studied during the last decade (see for example [10,6]), with the aim of obtaining finer approximations/abstractions by exploiting counter-examples.

Proposition 5. *If $L(\mathcal{T}_C^k(\mathcal{A})) \cap L(\mathcal{B}) \neq \emptyset$, then either $L(\mathcal{A}) \cap L(\mathcal{T}^{-k}(\mathcal{B})) \neq \emptyset$, or there exists j, $0 \leq j \leq k$ such that $L(\mathcal{T}_C^j(\mathcal{A})) \cap L(\mathcal{T}^{j-k}(\mathcal{B})) \neq \emptyset$ and $L(\mathcal{T}(\mathcal{T}_C^{j-1}(\mathcal{A}))) \cap L(\mathcal{T}^{j-k}(\mathcal{B})) = \emptyset$.*

Algorithm Split
Input: $S = (Q_S, Q_T, E_S, \{q_0\}, \emptyset)$ a deterministic automaton, $p, q \in Q_S$ and $\alpha, \beta \in Q_T$
such that $p \cdot_S \alpha = q \cdot_S \beta$
$\quad Q'_S := Q_S \cup \{r\}$ where $r \notin Q_S$; $I'_S := \{q_0\}$; $E'_S := E_S \setminus \{(q, \beta, q \cdot_S \beta)\}$;
$\quad E'_S := E'_S \cup \{(q, \beta, r)\} \cup \{(r, a, s) \mid (p \cdot \alpha, a, s) \in E_S$ and $s \in Q_S \setminus \{p \cdot_S \alpha\}\}$
$\quad E'_S := E'_S \cup \{(r, a, r) \mid (p \cdot \alpha, a, p \cdot \alpha) \in E_S\}$
$\quad$**Return** $(Q'_S, Q_T, E'_S, I'_S, \emptyset)$

Fig. 2. Algorithm Split

Algorithm Refine
Input: $\mathcal{T}$ (transducer), $\mathcal{C}$ a deterministic automaton, $S = (Q_S \times Q_C, Q, E, \{q_0\}, F_S)$ a
finite automaton, a relation $\equiv$ such that $\equiv \subseteq \sim_c$ and $L(\mathcal{T}_C(\mathcal{A}))/_\equiv \cap L(\mathcal{T}^{-1}(\mathcal{B})) = \emptyset$
$\quad$**While** $(\sim_c \not\subseteq \equiv)$ **do**
$\quad\quad$**Choose** (p, q, α) **and** (p, q', α') states of $\mathcal{T}(S)$ **such that**
$\quad\quad\quad (p, q, \alpha) \sim_c (p, q', \alpha')$ but $(p, q, \alpha) \not\equiv (p, q', \alpha')$
$\quad\quad \mathcal{C} := $Split$(\mathcal{C}, q, \alpha, q', \alpha')$
$\quad$**EndWhile**
$\quad$**Return** $\mathcal{C}$

Fig. 3. Algorithm Refine

Assume that $L(\mathcal{T}_\mathcal{C}^j(\mathcal{A})) \cap L(\mathcal{T}^{j-k}(\mathcal{B})) \neq \emptyset$ and $L(\mathcal{T}(\mathcal{T}_\mathcal{C}^{j-1}(\mathcal{A}))) \cap L(\mathcal{T}^{j-k}(\mathcal{B})) = \emptyset$.
As it is classically done in the CEGAR framework, one can compute a relation $\equiv$
on $\mathcal{T}_\mathcal{C}^j(\mathcal{A})$ such that $\equiv \subseteq \sim_c$ and $L(\mathcal{T}_\mathcal{C}^j(\mathcal{A}))/_\equiv \cap L(\mathcal{T}^{k-j}(\mathcal{B})) = \emptyset$. The existence of
$\equiv$ is trivial since the results hold for the identity relation. However, when using
the CEGAR approach, our goal is to compute a relation $\equiv$ as large as possible,
with the aim of ensuring termination of state-space exploration.

To achieve this goal, several heuristics may be used. Instead of computing
the $\equiv$ relation, building the corresponding $\mathcal{T}_\mathcal{C}^j(\mathcal{A})/_\equiv$ automaton, and then performing the fixpoint computation, we propose to use a dynamic approach. More
precisely, we prefer to modify $\mathcal{C}$ according to $\equiv$ to avoid similar states merging
which may lead to a coarser over-approximation. To modify $\mathcal{C}$ according to $\equiv$,
we propose to use the algorithms in Figs. 2 and 3. The Split algorithm modifies
the given deterministic automaton to provide a weaker abstraction. Its idea is
quite natural: if two equivalent states must be distinguished, the automaton $\mathcal{C}$
is refined to take this constraint into account.

Proposition 6. *The* Refine *algorithm always terminates.*

If $L(\mathcal{T}_\mathcal{C}^k(\mathcal{A})) \cap L(\mathcal{B}) \neq \emptyset$ and $L(\mathcal{A}) \cap L(\mathcal{T}^{-k}(\mathcal{B})) = \emptyset$, then we denote by
$J(\mathcal{A}, \mathcal{B}, \mathcal{C}, \mathcal{T}, k)$ the maximal integer j such that $0 \leq j \leq k$ and $L(\mathcal{T}_\mathcal{C}^j(\mathcal{A})) \cap$
$L(\mathcal{T}^{j-k}(\mathcal{B})) \neq \emptyset$ and $L(\mathcal{T}(\mathcal{T}_\mathcal{C}^{j-1}(\mathcal{A}))) \cap L(\mathcal{T}^{j-k}(\mathcal{B})) = \emptyset$. Now, the Reach-CEGAR
semi-algorithm in Fig. 4 encodes the whole approach: each time a too strong
approximation is detected, it is refined. This semi-algorithm may terminate
by returning *Safe* if an over-approximation of accessible states that does not

Semi-Algorithm Reach-CEGAR
Input: $\mathcal{A}, \mathcal{B}$ finite automata, $\mathcal{T}$ (transducer), $\mathcal{C}$ a deterministic automaton, an integer ℓ
Variables: integers j, k, and equivalence relation $\equiv$
> $k := \ell$
> **While** $(L(\mathcal{T}_{\mathcal{C}}^k(\mathcal{A})) \cap L(\mathcal{B}) = \emptyset$ **and** $L(\mathcal{T}_{\mathcal{C}}^{k+1}(\mathcal{A})) \neq L(\mathcal{T}_{\mathcal{C}}^k(\mathcal{A}))$) **do**
> > $k := k + 1$
> **EndWhile**
> **If** $(L(\mathcal{T}_{\mathcal{C}}^{k+1}(\mathcal{A})) = L(\mathcal{T}_{\mathcal{C}}^k(\mathcal{A}))$ **and** $L(\mathcal{T}_{\mathcal{C}}^k(\mathcal{A})) \cap L(\mathcal{B}) = \emptyset$) **then**
> > **Return** *Safe*
> **EndIf**
> **If** $L(\mathcal{A}) \cap L(\mathcal{T}^{-k}(\mathcal{B})) \neq \emptyset$ **then**
> > **Return** *Unsafe*
> **EndIf**
> $j := \mathrm{J}(\mathcal{A}, \mathcal{B}, \mathcal{C}, \mathcal{T}, k)$
> **Let** $\equiv$ be such that $\equiv \subseteq \sim_{\mathcal{C}}$ and $L(\mathcal{T}_{\mathcal{C}}^j(\mathcal{A}))/_{\equiv} \cap L(\mathcal{T}^{k-j}(\mathcal{B})) = \emptyset$
> **Return** Reach-CEGAR$(\mathcal{A}, \mathcal{T}^{-k}(\mathcal{B}), \mathcal{T}, \texttt{Refine}(\mathcal{T}, \mathcal{C}, \mathcal{T}^j(\mathcal{A}), \equiv), j)$

Fig. 4. Semi-algorithm Reach-CEGAR

contain any bad states. It may also terminate by returning *Unsafe* if it detects a reachable bad state. It may also diverge if the computed approximations have to be refined again and again.

The approach has been experimented using a three-state-automaton for $\mathcal{C}$. The Bakery algorithm by Lamport, the token ring algorithm, Dijkstra's, and Burns protocols have been proven safe in few steps. The obtained automata have sizes similar to the sizes of the input automata: there is no state explosion.

5   Conclusion

Developing efficient approximation-based techniques is a critical challenging issue to tackle reachability problems when exact approaches do not work. In this paper two new approximation techniques for the regular reachability problem have been presented. Our techniques use polynomial time algorithms, provided that recent algorithms for checking automata equivalence are used; the only exception being language inclusion testing as in [17,2,9]. As a future direction, we plan to upgrade our refinement approach, both on the precision of the approximations and on computation time. Another possible direction is to generalize our approximation mechanisms and to apply them to other RMC applications, e.g., counter systems or push-down systems.

References

1. Abdulla, P.A., Jonsson, B., Mahata, P., d'Orso, J.: Regular tree model checking. In: Brinksma, E., Larsen, K.G. (eds.) CAV 2002. LNCS, vol. 2404, pp. 555–568. Springer, Heidelberg (2002)

338   A. Dreyfus, P.-C. Héam, and O. Kouchnarenko

2. Abdulla, P.A., Chen, Y.-F., Holík, L., Mayr, R., Vojnar, T.: When simulation meets antichains. In: Esparza, Majumdar (eds.) [18], pp. 158–174
3. Abdulla, P.A., Jonsson, B., Nilsson, M., d'Orso, J.: Algorithmic improvements in regular model checking. In: Hunt Jr., W.A., Somenzi, F. (eds.) CAV 2003. LNCS, vol. 2725, pp. 236–248. Springer, Heidelberg (2003)
4. Baier, C., Katoen, J.P., Ebrary, I.: Principles of model checking, vol. 950. MIT Press (2008)
5. Bauer, A., Falcone, Y.: Decentralised LTL monitoring. In: Giannakopoulou, D., Méry, D. (eds.) FM 2012. LNCS, vol. 7436, pp. 85–100. Springer, Heidelberg (2012)
6. Boichut, Y., Courbis, R., Héam, P.-C., Kouchnarenko, O.: Finer is better: Abstraction refinement for rewriting approximations. In: Voronkov, A. (ed.) RTA 2008. LNCS, vol. 5117, pp. 48–62. Springer, Heidelberg (2008)
7. Boigelot, B.: Domain-specific regular acceleration. STTT 14(2), 193–206 (2012)
8. Boigelot, B., Legay, A., Wolper, P.: Iterating transducers in the large. In: Hunt Jr., W.A., Somenzi, F. (eds.) CAV 2003. LNCS, vol. 2725, pp. 223–235. Springer, Heidelberg (2003)
9. Bonchi, F., Pous, D.: Checking NFA equivalence with bisimulations up to congruence. Technical report, 13 p. (January 2012)
10. Bouajjani, A., Habermehl, P., Vojnar, T.: Abstract regular model checking. In: Alur, R., Peled, D.A. (eds.) CAV 2004. LNCS, vol. 3114, pp. 372–386. Springer, Heidelberg (2004)
11. Bouajjani, A., Jonsson, B., Nilsson, M., Touili, T.: Regular model checking. In: Emerson, E.A., Sistla, A.P. (eds.) CAV 2000. LNCS, vol. 1855, Springer, Heidelberg (2000)
12. Bouajjani, A., Touili, T.: Widening techniques for regular tree model checking. STTT, 1–21 (2011)
13. Clarke, E.M., Grumberg, O., Jha, S., Lu, Y., Veith, H.: Counterexample-guided abstraction refinement. In: Emerson, E.A., Sistla, A.P. (eds.) CAV 2000. LNCS, vol. 1855, pp. 154–169. Springer, Heidelberg (2000)
14. Clarke, E.M., Grumberg, O., Peled, D.: Model Checking, 2000. MIT Press (2000)
15. Dams, D.R., Lakhnech, Y., Steffen, M.: Iterating transducers. In: Berry, G., Comon, H., Finkel, A. (eds.) CAV 2001. LNCS, vol. 2102, pp. 286–297. Springer, Heidelberg (2001)
16. Dams, D., Lakhnech, Y., Steffen, M.: Iterating transducers. Journal of Logic and Algebraic Programming 52, 109–127 (2002)
17. Doyen, L., Raskin, J.-F.: Antichain algorithms for finite automata. In: Esparza, Majumdar (eds.) [18], pp. 2–22
18. Esparza, J., Majumdar, R. (eds.): TACAS 2010. LNCS, vol. 6015. Springer, Heidelberg (2010)
19. Fioravanti, F., Pettorossi, A., Proietti, M., Senni, V.: Program specialization for verifying infinite state systems: An experimental evaluation. Logic-Based Program Synthesis and Transformation, 164–183 (2011)
20. Le Gall, T., Jeannet, B.: Lattice automata: A representation for languages on infinite alphabets, and some applications to verification. In: Riis Nielson, H., Filé, G. (eds.) SAS 2007. LNCS, vol. 4634, pp. 52–68. Springer, Heidelberg (2007)
21. Gómez, A.C., Guaiana, G., Pin, J.-É.: When does partial commutative closure preserve regularity? In: Aceto, L., Damgård, I., Goldberg, L.A., Halldórsson, M.M., Ingólfsdóttir, A., Walukiewicz, I. (eds.) ICALP 2008, Part II. LNCS, vol. 5126, pp. 209–220. Springer, Heidelberg (2008)

22. Jonsson, B., Nilsson, M.: Transitive closures of regular relations for verifying infinite-state systems. In: Graf, S. (ed.) TACAS/ETAPS 2000. LNCS, vol. 1785, pp. 220–235. Springer, Heidelberg (2000)
23. Kesten, Y., Maler, O., Marcus, M., Pnueli, A., Shahar, E.: Symbolic model checking with rich assertional languages. In: Grumberg, O. (ed.) CAV 1997. LNCS, vol. 1254, Springer, Heidelberg (1997)
24. Legay, A.: Extrapolating (omega-) regular model checking. STTT 14(2), 119–143 (2012)
25. Touili, T.: Regular model-checking using widening techniques. In: VEPAS. ENTCS, vol. 50, pp. 342–356 (2001)
26. Yu, F., Bultan, T., Ibarra, O.: Relational string verification using multi-track automata. IJFCS 22, 290–299 (2011)

Generating Small Automata
and the Černý Conjecture

Andrzej Kisielewicz* and Marek Szykuła**

Department of Mathematics and Computer Science, University of Wrocław

Abstract. We present a new efficient algorithm to generate all nonisomorphic automata with given numbers of states and input letters. The generation procedure may be restricted effectively to strongly connected automata. This is used to verify the Černý conjecture for all binary automata with $n \leq 11$ states, which improves the results in the literature. We compute also the distributions of the length of the shortest reset word for binary automata with $n \leq 10$ states, which completes the results reported by other authors.

Keywords: Černý conjecture, synchronizing word, nonisomorphic automata.

We consider deterministic finite automata $A = \langle Q, \Sigma, \delta \rangle$, where Q is the set of the states, Σ is the input alphabet, and $\delta : Q \times \Sigma \to Q$ is the (complete) transition function. The cardinality $n = |Q|$ is the *size* of A, and if $k = |\Sigma|$ then A is called k-*ary*.

If there exists a w such that the image of Q by w consists of a single state, then w is called a *reset* (or *synchronizing*) word for A, and A itself is called *synchronizing*. The length of a shortest reset word of A is called its *reset length*.

The Černý conjecture states that every synchronizing automaton A with n states has a reset word of length $\leq (n-1)^2$. This conjecture was formulated by Černý in 1964, and is considered the longest-standing open problem in combinatorial theory of finite automata. So far, the conjecture has been proved only for a few special classes of automata and a cubic upper bound has been established (see Volkov [19] for an excellent survey). It is known (and not difficult to prove) that to verify the conjecture it is enough to consider only *strongly connected* automata, that is, those whose underlying digraph is strongly connected.

Trahtman [17,18] reports that, using a computer program, he has verified the Černý conjecture for all strongly connected k-ary automata of size n with $k = 2$ and $n \leq 10$, $k \leq 4$ and $n \leq 7$, and $k = 3$ and $n = 8$. Unfortunately, no method of generating such automata is described and no details of computations are given. There are 10^{20} binary automata of size $n = 10$, and it is out of reach of the present computer technology to generate all of them, so some methods to generate only

---

* Supported in part by Polish MNiSZW grant N N201 543038.
** Supported in part by NCN grant number 2011/01/D/ST6/07164, 2011–2014, and by a scholarship co-financed by an ESF project *Human Capital*.

S. Konstantinidis (Ed.): CIAA 2013, LNCS 7982, pp. 340–348, 2013.

strongly connected automata (or a restricted class containing all the strongly connected automata) must be used. Such a method is described in [1], the authors restrict themselves to the class of *initially-connected* automata (with each state reachable from a single start state), using a special string representation for such automata and parallel programming. With these tools, they are able to verify the Černý conjecture only for binary automata with $n \leq 9$ states. (For 9 states, there are about 700 billions initially-connected automata with 2 input letters.)

The theoretical part of Trahtman's work [17,18] is devoted mainly to the problem of efficiently finding the shortest (or a short) reset word. A number of good algorithms are known at present for solving this problem (see [7] and references given therein). We found however that the main problem arising in verifying the Černý conjecture for small automata is to overcome somehow the huge number of automata involved rather than to compute the reset length fast. Ideally, one would like to consider only all nonisomorphic strongly connected automata for such verification, but no efficient method to generate only automata from this class is known. There are formulas enumerating the number of nonisomorphic automata (see [6] and [5,11,13]), and methods to enumerate nonisomorphic strongly connected automata ([8,14]) Unfortunately, the ways they approach the problem do not seem useful in the task of efficient generation of the objects.

In this paper we present a new algorithm to generate efficiently all nonisomorphic automata with given numbers of states and letters, and to compute the reset length for them. The method can be extended to generate only specific classes of automata without much additional cost. In particular, a version of the algorithm generates all nonisomorphic strongly connected automata. While the algorithm still produces isomorphic copies (and some not strongly connected automata for the second version), it greatly reduces the number of considered automata as well as the overall computation cost. Also we are able to speed-up computation of reset length making use of the specific properties of the generating method.

Our method allows us us to verify and extend the known computational results. In particular, we prove that the Černý conjecture is true for all binary automata with $n \leq 11$ states. We obtain complete distributions of the reset length for all automata of size $n \leq 10$. For $n = 11$ a new gap in the distribution is observed, leading to a new conjecture concerning reset lengths.

1   Generating Automata

The algorithm is recursive. Given $n > 1$, we use known lists of all nonisomorphic automata of size n and arity 1 (which are equivalent to certain digraphs). For $k \geq 2$, having two lists of all nonisomorphic automata of size n and of arity $k - 1$ and 1, respectively, our algorithm generates a list of automata of size n and arity k. To this aim, for each pair of automata, A from the first list, and B from the second list, a special procedure, called Permutation procedure, is applied. It (1) takes as an input the pair of automata A and B, from the first and the second list, respectively, (2) generates all automata isomorphic to B (by permutations of the states of B), and (3) matches each resulting automaton with A. In this way, we obtain all the automata of arity k whose restriction to the first $k-1$ letters is isomorphic to A, while the restriction to the last letter is isomorphic to B. Matching all the pairs A, B, we obtain all nonisomorphic k-ary automata of size n. Yet, many of these automata may appear in isomorphic copies.

Using more specific ideas we design a few variant of the algorithm with different task. They generate all nonisomorphic automata of a given size either without isomorphic pairs or (for lower computational cost) with the number of such pairs relatively small. We also show how the generation process can be restricted effectively to strongly connected automata. The latter is used to verify the Černý conjecture for automata of a given size. Because of the space limit, in this paper, we describe only the theoretical aspects of the procedure, called Permutation procedure, which is designed to skip efficiently permutations of B leading to isomorphic copies. Other variants and the details of the algorithm will be given in the extended version of the paper.

1.1   Permutation Procedure

We say that two automata $A = \langle Q_A, \Sigma_A, \delta_A \rangle$ and $B = \langle Q_B, \Sigma_B, \delta_B \rangle$, are *isomorphic*, if there exist two bijections $\phi : Q_A \to Q_B$ and $\psi : \Sigma_A \to \Sigma_B$ such that for all $q \in Q_A$ and $a \in \Sigma_A$

$$\phi(\delta_A(q,a)) = \delta_B(\phi(q), \psi(a)). \tag{1}$$

In other words, isomorphic automata are equal up to renaming the states and the letters. In particular, two isomorphic automata have the same reset lengths, and the classes of shortest reset words differ only up to renaming the letters (given by ψ).

We note that various authors use various terminology here. For example, Harrison [6] calls such automata *equivalent with respect to input permutations*, and reserves the term "isomorphic automata" for the situation when the bijection ψ in (1) is the identity. If $\Sigma_A = \Sigma_B$, and A and B are isomorphic with ψ being the identity, we will say that A and B over the same alphabet are *strongly isomorphic*. Then the bijection ϕ itself is called a *strong isomorphism* or simply *isomorphism* (meaning that it forms an isomorphism itself with the second bijection being the identity). In the case, when $A = B$, ϕ is called an *automorphism*.

We consider now an automorphism that fixes the states in a given set. For an automaton $A = \langle Q, \Sigma, \delta \rangle$, and a subset S of Q, we say that the states $u, v \in Q$, $u, v \notin S$ are *conjugate under S*, and write $u \overset{S}{\simeq} v$, if there exists a (strong) automorphism $\phi : Q \to Q$ such that

$$\begin{aligned} \phi(w) &= w \text{ for each } w \in S, \\ \phi(u) &= v. \end{aligned} \tag{2}$$

Figure 1 shows an example of an automaton over a one-letter alphabet with some states conjugate under $S = \emptyset$ and $S = \{5\}$. Namely, we have $1 \overset{S}{\simeq} 2$ and $3 \overset{S}{\simeq} 4$ for $S \subseteq \{5\}$. Note that there are no two different conjugate states under any other S in this automaton.

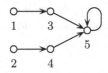

Fig. 1. An unary automaton with nontrivial automorphisms and conjugate states

The following facts are routine to prove.

Lemma 1. *For each $S \subseteq Q$, the conjugation under S is an equivalence relation. Moreover, if $R \subseteq S$, then the relation $\overset{R}{\simeq} \supseteq \overset{S}{\simeq}$.*

Checking whether two states are S-conjugate may be done by computing the corresponding group of automorphisms (fixing the states from S) or by generating and checking permutations with suitably prescribed images for S and u. During the generation of permutations, natural conditions for a permutation to be an automorphism, such as equality of indegrees, may be taken into account. Although it has an exponential cost in the worst case, our experiments show that for most of automata of small size, it works pretty fast.

In our procedure, to be able to skip superfluous permutations effectively we do some preprocessing. We assume that the set of states of the automata on both the input lists is $Q = \{1, 2, \ldots, n\}$. Before running the procedure, the following structures are created for each pair A and B of automata (with $k-1$ and 1 letter alphabet, respectively):

1. The structure **PrevB**. For each of the 2^n subsets $S \subseteq Q$ and for each $j \in Q \setminus S$, **PrevB**$[S][j]$ contains *true*, if and only if there exists some state $h \in Q \setminus S$ $(1 \leq h < j)$ for which $h \overset{S}{\simeq} j$ in B.
2. The structure **PrevA**. For each i $(1 \leq i \leq n)$ the entry **PrevA**$[i]$ contains the largest index h $(1 \leq h < i)$ such that $i \overset{S_h}{\simeq} h$ in A with $S_h = \{1, \ldots, h-1\}$. It is possible that the index does not exist.

The first structure requires computing automorphisms of B for as many as $2^n \binom{n}{2}$ conditions (in the worst case) fixing a set S and unordered pair $\{i, j\}$ with $i, j \notin S$. For each automaton B (which is of arity 1) we compute it only once and then we process all the pairs with B. The second structure requires computing automorphisms only for $\binom{n}{2}$ pairs of states (determining the set of fixed elements). For small n, this preprocessing can be done quickly and takes only a negligible amount of time compared with processing the resulting automata.

Let $A = \langle Q, \Sigma_A, \delta_A \rangle$ and $B = \langle Q, \Sigma_B, \delta_B \rangle$ be two automata with $Q = \{1, 2, \ldots, n\}$, $|\Sigma_A| = k - 1$ for some $k > 1$, and $\Sigma_B = \{b\}$ for some $b \notin \Sigma_A$. Let π be a permutation of Q. Then, by $U(A, B, \pi)$ we denote the automaton $\langle Q, \Sigma_A \cup \Sigma_B, \delta \rangle$, where δ is an extension of δ_A given by

$$\delta(q, b) = \pi^{-1}(\delta_B(\pi(q), b)). \tag{3}$$

We call it the (disjoint) *union of A and B under permutation* π. The condition $b \notin \Sigma_A$ is purely technical, so we may assume it always without further mention. Note that this construction may be viewed as identifying each state q in B with the state $\pi^{-1}(q)$ in A. An example is given in Figure 2 (loops are omitted).

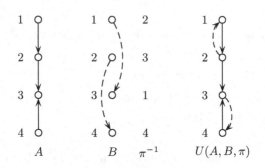

Fig. 2. The union $U(A, B, \pi)$ with $\pi = (1, 3, 2)$

The main part of our algorithm is the PERMUTB procedure presented as Algorithm 1. It takes as the input two automata A and B on $Q = \{1, 2, \ldots, n\}$, with alphabets of arity $k - 1$ and 1, respectively. It starts from the empty (partial) permutation $\pi_0 = \emptyset$ and extends it in a recursive manner. A partial permutation π_i is an injective function from $\{1, \ldots, i\}$ to $\{1, \ldots, n\}$. For each complete permutation π_n the automaton $U(A, B, \pi_n)$ is generated. The permutations are generated in the lexicographical order subject to two restrictions (reducing the number of isomorphic automata):

1. Let $h = \mathbf{PrevA}[i+1]$. If such h exists and $\pi_i(h) > j$ then matching $(i+1) \to j$ is skipped, since (as we prove below) a suitable isomorphic automaton has been generated earlier. This results in starting the corresponding "for loop" from $m = \pi_i(h) + 1$.
2. For each j, if $\mathbf{PrevB}[S_B][j]$ is true then matching $(i + 1) \to j$ is skipped. Again, we will prove that a suitable isomorphic automata have been already generated.

In the theorem below we use the notation $A|_\Gamma$ for the automaton obtained from $A = \langle Q, \Sigma_A, \delta_A \rangle$ by restricting its alphabet to a subset Γ of Σ.

Theorem 1. *Let* $A = \langle Q, \Sigma_A, \delta_A \rangle$ *and* $B = \langle Q, \Sigma_B, \delta_B \rangle$ *be two automata with disjoint alphabets* Σ_A *and* Σ_B *(where* $|\Sigma_B| = 1$*) and the same set of states* $Q = \{1, 2, \ldots, n\}$*. Then, for each automaton* C *over the alphabet* $\Sigma_A \cup \Sigma_B$ *such that* $C|_{\Sigma_A} \cong A$ *and* $C|_{\Sigma_B} \cong B$*,* PERMUTB$(0, \emptyset, \emptyset)$ *generates at least one isomorphic copy of* C*.*

The proof will be given in the extended version of the paper.

Algorithm 1. Permutation Procedure

Require: A, B – the input automata.
Require: PrevA, PrevB – preprocessed structures.
1: **procedure** PERMUTB(i,π_i,S)
2:     **if** $i = n$ **then**
3:         Report automaton $U(A, B, \pi_n)$ – the union of A and B under π_n
4:     **else**
5:         $m \leftarrow \pi_i(\textbf{PrevA}[i + 1]) + 1$, or $m \leftarrow 1$ if **PrevA**[i + 1] does not exist.
6:         **for** $j = m, \ldots, n$ **do**
7:             **if** $j \notin S$ and not **PrevB**[S][j] **then**
8:                 Extend π_i to $\pi_{i+1}(x)$ putting $\pi_{i+1}(i + 1) = j$.
9:                 PERMUTB($i + 1,\pi_{i+1},S \cup \{j\}$)
10:            **end if**
11:        **end for**
12:    **end if**
13: **end procedure**

---

2 Some Experimental Results

The problem of computing the reset length of an automaton is computationally hard (see [12], and [3] for approximating hardness). In spite of this the exponential algorithms used so far can work efficiently enough. Yet, they can vary in efficiency for different automata (see [7,9,15,17,19]).

To compute the reset length for each of the generated automata we use the standard BFS algorithm in the power automaton with storing visited subsets of states in an array (see [15,19,9]), and with preprocessing transitions (computing the images of subsets) allowing faster computations for a huge number of automata. We have found this the fastest method for considered small n values when using bit-vector encoding for sets, allowing to represent them as integers. It can also report that an automaton is not synchronizing, without separately using the standard synchronization checking algorithm on the pair automaton ([4,17]). Further technical improvements applied are described in the extended version.

We have computed the exact numbers of all nonisomorphic binary automata and those strongly connected and/or synchronizing for sizes $n \leq 10$. Also complete distributions of the reset length in this range are computed. Our results confirm all the results reported in [1] and particular facts formulated in [17]. For $n = 11$ we have computed a partial distribution proving, in particular, that all binary DFA of size 11 satisfy the Černý conjecture. We plan also to perform similar computations for $k > 2$.

2.1 The Number of Nonisomorphic Automata

The results up to 10 states are shown in Table 1. The total number of DFA is known due to the formula in [6], and we have obtained computationally exactly the same numbers. We have computed also the numbers of synchronizing DFA, strongly connected, and the number of synchronizing strongly connected DFA. The numbers of nonisomorphic strongly connected DFA on 2 labeled letters (up to strong isomorphism) have been considered in [10] (up to $n \leq 6$). They are about 2 times larger than those with unlabeled (for example, there are 658,885 such DFA for $n = 6$).

We can see that the fraction of synchronizing DFA to all DFA grows, and we may conjecture that it tends to 1 as it has been conjectured for the labeled model (P. Cameron and [16]). This growth is more rapid in strongly connected DFA; the corresponding fraction here is about 0.999 for $n = 10$.

Table 1. The exact numbers of nonisomorphic binary DFA of size n in the classes of all, synchronizing, strongly connected, and strongly connected synchronizing DFA. In the last column there is the fraction of the number of synchronizing DFA to all DFA.

n	Total	Synchronizing	Strongly connected	S. c. and synchronizing	Synch./Total
2	7	4	4	2	0.57
3	74	51	29	21	0.69
4	1,474	1,115	460	395	0.76
5	41,876	34,265	10,701	10,180	0.82
6	1,540,696	1,318,699	329,794	322,095	0.86
7	68,343,112	60,477,844	12,310,961	12,194,323	0.88
8	3,540,691,525	3,210,707,626	538,586,627	536,197,356	0.91
9	209,612,916,303	193,589,241,468	26,959,384,899	26,904,958,363	0.92
10	13,957,423,192,794	13,070,085,476,528	1,518,185,815,760	1,516,697,994,964	0.94

Let us compare our method of generating all strongly connected DFA with that of [1,2] by generating of all ICDFA (initially connected DFA). There are about 7×10^{11} and 4.4×10^{13} of ICDFA with $n = 9$ and $n = 10$ states, respectively. In our method we have generated only about 3×10^{10} and 1.7×10^{12} DFA in these cases. In fact there are about 2.7×10^{10} and 1.5×10^{12} nonisomorphic strongly connected DFA, so in our method the relative number of extra generated DFA is really low. This is confirmed by statistics we have made.

2.2   The Distribution of Reset Lengths for $n = 10$.

Since generating automata for each pair in Algorithm 1 can be computed independently we performed paralleled computations on a small computer grid. Our computations have been done on 16 computers with Intel(R) Core(TM) i7-2600 CPU 3.40GHz 4 cores and 16GB of RAM. Computing the complete distribution for all DFA with $n = 10$ states took above 800 days of total CPU time

Table 2. The exact numbers $N(\ell)$ of all and $N_{sc}(\ell)$ of strongly connected nonisomorphic binary automata of size 10 with the shortest reset word of length $\ell \geq 56$

ℓ	56	57	58	59	60	61	62	63	64	65	66	67	68	69	70	71	72	73	74	75	76	77	78	79	80	81
$N(\ell)$	607	369	168	49	18	10	8	9	106	21	3	0	0	0	0	0	2	1	1	0	0	0	0	0	0	1
$N_{sc}(\ell)$	343	160	58	38	18	10	8	9	18	10	3	0	0	0	0	0	2	1	1	0	0	0	0	0	0	1

($\sim$ 13 days of paralleled computations). Restricting to the class of strongly connected DFA reduced this time to about 80 days of CPU ($\sim$ 2 days of paralleled computations).

2.3   The Distribution of Reset Lengths for $n = 11$.

In order to verify the Černý conjecture for all binary DFA of size $n = 11$ it is sufficient to restrict the tested class of DFA to strongly connected. We have not obtained the complete distribution of reset lengths because of the huge number of DFA. In this case, we were performing the isomorphism test only for DFA with long reset length. We have also excluded the automata with a single synchronizing letter. The number of remaining strongly connected DFA we have to check was 79,246,008,127,339. The total CPU time of this experiment was above 4 years ($\sim$ 25 days of parallel computations). Note that for $n = 11$ there are about 3×10^{15} of ICDFA, so we really needed a different method than that used in [1].

Table 3. The exact numbers $N_{sc}(\ell)$ of strongly connected nonisomorphic binary automata of size 11 with the reset length $\ell \geq 76$

ℓ	76	77	78	79	80	81	82	83	84	85	86	87	88	89	90	91	92	93	94	95	96	97	98	99	100
$N_{sc}(\ell)$	3	2	0	0	9	22	12	2	1	0	0	0	0	0	3	2	1	0	0	0	0	0	0	0	1
Classes			–	–			$\mathscr{D}_n$	$\mathscr{H}_n$ $\ldots$ $\ddot{\mathscr{H}}_n$	$\mathscr{G}_n$	–	–	–	–	–	$\mathscr{E}_n$ $\mathscr{D}''_n$ $\mathscr{B}_n$	$\mathscr{W}_n$ $\mathscr{F}_n$	$\mathscr{D}'_n$	–	–	–	–	–	–	–	$\mathscr{C}_n$

Table 3 presents the obtained exact numbers of all nonisomorphic binary DFA of size $n = 11$ with large reset lengths. Also some slowly synchronizing DFA classes are presented in the table. The notation here follows [1] (this topic is discussed in more detail in the extended version of the paper). The most interesting observation is a gap between $\ell = 77$ and 80: there exist no binary automaton of size $n = 11$ with the reset length equal to 78 or 79. First, Trahtman [17] noted that the reset length $(n - 1)^2$ corresponding to the class of the Černý automata is separated from the second large reset length by a gap (in the classes of considered DFA of small size). Then the authors of [1] observed that there is a second gap in the distribution for $n = 9$. They called the DFA between the two gaps *slowly synchronizing*. There is no other gap for $n \leq 10$.

We suppose that this kind of irregularity in the upper part of the reset length distributions occur also for larger numbers of states and that more gaps for larger number of states appear. We state the following:

Gap Conjecture. *For any natural number $g \geq 1$, there exists a big enough natural number n such that there are at least g gaps in the distribution of the reset length of all binary automata of size n.*

References

1. Ananichev, D., Gusev, V., Volkov, M.: Slowly synchronizing automata and digraphs. In: Hliněný, P., Kučera, A. (eds.) MFCS 2010. LNCS, vol. 6281, pp. 55–65. Springer, Heidelberg (2010)
2. Ananichev, D., Gusev, V., Volkov, M.: Primitive digraphs with large exponents and slowly synchronizing automata. In: Zapiski Nauchnyh Seminarov POMI (Kombinatorika i Teorija Grafov. IV), vol. 402, pp. 9–39 (2012) (In Russian)
3. Berlinkov, M.: Approximating the minimum length of synchronizing words is hard. In: Ablayev, F., Mayr, E.W. (eds.) CSR 2010. LNCS, vol. 6072, pp. 37–47. Springer, Heidelberg (2010)
4. Eppstein, D.: Reset sequences for monotonic automata. SIAM Journal on Computing 19, 500–510 (1990)
5. Harary, F., Palmer, E.M.: Graphical Enumeration. Academic Press (1973)
6. Harrison, M.: A census of finite automata. Canadian Journal of Mathematics 17, 100–113 (1965)
7. Kisielewicz, A., Kowalski, J., Szykuła, M.: A Fast Algorithm Finding the Shortest Reset Words. In: Du, D.-Z., Zhang, G. (eds.) COCOON 2013. LNCS, vol. 7936, pp. 182–196. Springer, Heidelberg (2013)
8. Koršunov, A.D.: On the number of non-isomorphic strongly connected finite automata. Journal of Information Processing and Cybernetics 22(9), 459–462 (1986)
9. Kudałcik, R., Roman, A., Wagner, H.: Effective synchronizing algorithms. Expert Systems with Applications 39(14), 11746–11757 (2012)
10. Liskovets, V.A.: Enumeration of non-isomorphic strongly connected automata. Vesci Akad. Navuk BSSR Ser. Fiz.-Téhn. Navuk 3, 26–30 (1971)
11. Liskovets, V.A.: Exact enumeration of acyclic deterministic automata. Discrete Applied Mathematics 154(3), 537–551 (2006)
12. Olschewski, J., Ummels, M.: The complexity of finding reset words in finite automata. In: Hliněný, P., Kučera, A. (eds.) MFCS 2010. LNCS, vol. 6281, pp. 568–579. Springer, Heidelberg (2010)
13. Read, R.C.: A note on the number of functional digraphs. Mathematische Annalen 143, 109–110 (1961)
14. Robinson, R.W.: Counting strongly connected finite automata. In: Graph Theory with Applications to Algorithms and Computer Science, pp. 671–685 (1985)
15. Sandberg, S.: Homing and synchronizing sequence. In: Broy, M., Jonsson, B., Katoen, J.-P., Leucker, M., Pretschner, A. (eds.) Model-Based Testing of Reactive Systems. LNCS, vol. 3472, pp. 5–33. Springer, Heidelberg (2005)
16. Skvortsov, E., Tipikin, E.: Experimental study of the shortest reset word of random automata. In: Bouchou-Markhoff, B., Caron, P., Champarnaud, J.-M., Maurel, D. (eds.) CIAA 2011. LNCS, vol. 6807, pp. 290–298. Springer, Heidelberg (2011)
17. Trahtman, A.N.: An efficient algorithm finds noticeable trends and examples concerning the Černy conjecture. In: Královič, R., Urzyczyn, P. (eds.) MFCS 2006. LNCS, vol. 4162, pp. 789–800. Springer, Heidelberg (2006)
18. Trahtman, A.N.: Modifying the upper bound on the length of minimal synchronizing word. In: Owe, O., Steffen, M., Telle, J.A. (eds.) FCT 2011. LNCS, vol. 6914, pp. 173–180. Springer, Heidelberg (2011)
19. Volkov, M.V.: Synchronizing automata and the Černý conjecture. In: Martín-Vide, C., Otto, F., Fernau, H. (eds.) LATA 2008. LNCS, vol. 5196, pp. 11–27. Springer, Heidelberg (2008)

Incomplete Transition Complexity
of Basic Operations on Finite Languages[*]

Eva Maia[**], Nelma Moreira, and Rogério Reis

CMUP & DCC, Faculdade de Ciências da Universidade do Porto
Rua do Campo Alegre, 4169-007 Porto, Portugal
{emaia,nam,rvr}@dcc.fc.up.pt

Abstract. The state complexity of basic operations on finite languages
(considering complete DFAs) has been extensively studied in the literature. In this paper we study the incomplete (deterministic) state and
transition complexity on finite languages of boolean operations, concatenation, star, and reversal. For all operations we give tight upper bounds
for both descriptional measures. We correct the published state complexity of concatenation for complete DFAs and provide a tight upper bound
for the case when the *right* automaton is larger than the *left* one. For
all binary operations the tightness is proved using family languages with
a variable alphabet size. In general the operational complexities depend
not only on the complexities of the operands but also on other refined
measures.

1 Introduction

Descriptional complexity studies the measures of complexity of languages and
operations. These studies are motivated by the need to have good estimates of
the amount of resources required to manipulate the smallest representation for
a given language. In general, having succinct objects will improve our control
on software, which may become smaller and more efficient. Finite languages are
an important subset of regular languages with many applications in compilers,
computational linguistics, control and verification, etc. [9,1,8,3]. In those areas
it is also usual to consider deterministic finite automata (DFA) with partial
transition functions. As an example we can mention the manipulation of compact natural language dictionaries using Unicode alphabets. This motivates the
study of the transition complexity of DFAs (not necessarily complete), besides
the usual state complexity. The operational transition complexity of basic operations on regular languages was studied by Gao *et al.* [4] and Maia *et al.* [7].
In this paper we continue that line of research by considering the class of finite

---

[*] This work was partially funded by the European Regional Development Fund
through the programme COMPETE and by the Portuguese Government through
the FCT under projects PEst-C/MAT/UI0144/2011 and CANTE-PTDC/EIA-
CCO/101904/2008.

[**] Eva Maia is funded by FCT grant SFRH/BD/78392/2011.

S. Konstantinidis (Ed.): CIAA 2013, LNCS 7982, pp. 349–356, 2013.

languages. For finite languages, Salomaa and Yu [10] showed that the state complexity of the determinization of a nondeterministic automaton (NFA) with m states and k symbols is $\Theta(k^{\frac{m}{1+\log k}})$ (lower than 2^m as it is the case for general regular languages). Câmpeanu et al. [2] studied the operational state complexity of concatenation, Kleene star, and reversal. Finally, Han and Salomaa [5] gave tight upper bounds for the state complexity of union and intersection on finite languages. In this paper we give tight upper bounds for the state and transition complexity of all the above operations, for non necessarily complete DFAs with an alphabet size greater than 1. For the concatenation, we correct the upper bound for the state complexity of complete DFAs [2], and show that if the *right* automaton is larger than the *left* one, the upper bound is only reached using an alphabet of variable size. The transition complexity results are all new, although the proofs are based on the ones for the state complexity and use techniques developed by Maia et al. [7]. Table 1 presents a comparison of the transition complexity on regular and finite languages, where the new results are highlighted. Note that the values in the table are obtained using languages for which the upper bounds are reached. All the proofs not presented in this paper can be found in an extended version of this work[1].

Table 1. Incomplete transition complexity for regular and finite languages, where m and n are the (incomplete) state complexities of the operands, $f_1(m,n) = (m-1)(n-1)+1$ and $f_2(m,n) = (m-2)(n-2)+1$. The column $|\Sigma|$ indicates the minimal alphabet size for each the upper bound is reached.

Operation	Regular	$|\Sigma|$	Finite	$|\Sigma|$
$L_1 \cup L_2$	$2n(m+1)$	2	**3(mn-n-m) +2**	$f_1(m,n)$
$L_1 \cap L_2$	nm	1	$(\mathbf{m-2})(\mathbf{n-2})(\mathbf{2} + \sum_{i=1}^{\min(m,n)-3}(\mathbf{m-2-i})(\mathbf{n-2-i})) + \mathbf{2}$	$f_2(m,n)$
L^C	$m+2$	1	$\mathbf{m+1}$	1
$L_1 L_2$	$2^{n-1}(6m+3) - 5$, if $m, n \geq 2$	3	$\mathbf{2^n(m-n+3) - 8}$, if $m+1 \geq n$	2
			See Theorem 3 **(4)**	$n-1$
L^*	$3.2^{m-1} - 2$, if $m \geq 2$	2	$\mathbf{9 \cdot 2^{m-3} - 2^{m/2} - 2}$, if m is odd	3
			$\mathbf{9 \cdot 2^{m-3} - 2^{(m-2)/2} - 2}$, if m is even	
L^R	$2(2^m - 1)$	2	$\mathbf{2^{p+2} - 7}$, if $m = 2p$	2
			$\mathbf{3 \cdot 2^p - 8}$, if $m = 2p - 1$	

2   Preliminaries

We assume that the reader is familiar with the basic notions about finite automata and regular languages. For more details, we refer the reader to the

---
[1] http://www.dcc.fc.up.pt/Pubs/TReports/TR13/dcc-2013-02.pdf

standard literature [6,12,11]. In this paper we consider DFAs to be not necessarily complete, *i.e.* with partial transition functions. The *state complexity* of L ($sc(L)$) is equal to the number of states of the minimal complete DFA that accepts L. The *incomplete state complexity* of a regular language L ($isc(L)$) is the number of states of the minimal DFA, not necessarily complete, that accepts L. Note that $isc(L)$ is either equal to $sc(L)-1$ or to $sc(L)$. The *incomplete transition complexity*, $itc(L)$, of a regular language L is the minimal number of transitions over all DFAs that accepts L. We omit the term *incomplete* whenever the model is explicitly given. A τ-*transition* is a transition labeled by $\tau \in \Sigma$. The τ-*transition complexity* of L, $itc_\tau(L)$ is the minimal number of τ-transitions of any DFA recognizing L. It is known that $itc(L) = \sum_{\tau \in \Sigma} itc_\tau(L)$ [4,7]. For determining the transition complexity of an operation, we also consider the following measures and refined numbers of transitions. Let $A = ([0, n - 1], \Sigma, \delta, 0, F)$ be a DFA, $\tau \in \Sigma$, and $i \in [0, n-1]$. We define $f(A) = |F|$, $f(A, i) = |F \cap [0, i-1]|$, $t_\tau(A, i)$ as 1 if exist a τ-transition leaving i and 0 otherwise, and $\bar{t}_\tau(a, i)$ as its complement. Let $s_\tau(A) = t_\tau(A, 0)$, $e_\tau(A) = \sum_{i \in F} t_\tau(A, i)$, $t_\tau(A) = \sum_{i \in Q} t_\tau(A, i)$, $t_\tau(A, [k, l]) = \sum_{i \in [k, l]} t_\tau(A, i)$, and the respective complements $\bar{s}_\tau(A) = \bar{t}_\tau(A, 0)$, $\bar{e}_\tau(A) = \sum_{i \in F} \bar{t}_\tau(A, i)$, etc. We denote by $in_\tau(A, i)$ the number of transitions reaching i, $a_\tau(A) = \sum_{i \in F} in_\tau(A, i)$ and $c_\tau(A, i) = 0$ if $in_\tau(A, i) > 0$ and 1 otherwise. Whenever there is no ambiguity we omit A from the above definitions. All the above measures, can be defined for a regular language L, considering the measure values for its minimal DFA. We define $s(L) = \sum_{\tau \in \Sigma} s_\tau(L)$ and $a(L) = \sum_{\tau \in \Sigma} a_\tau(L)$. Let A be a minimal DFA accepting a finite language, where the states are assumed to be topologically ordered. Then, $s(\mathcal{L}(A)) = 0$ and there is exactly one final state, denoted π and called *pre-dead*, such that $\sum_{\tau \in \Sigma} t_\tau(\pi) = 0$. The *level* of a state i is the size of the shortest path from the initial state to i, and never exceeds $n - 1$. The level of A is the level of π.

3   Union and Intersection

Given two incomplete DFAs $A = ([0, m - 1], \Sigma, \delta_A, 0, F_A)$ and $B = ([0, n - 1], \Sigma, \delta_B, 0, F_B)$ adaptations of the classical cartesian product construction can be used to obtain DFAs accepting $\mathcal{L}(A) \cup \mathcal{L}(B)$ and $\mathcal{L}(A) \cap \mathcal{L}(B)$ [7].

Theorem 1. *For any two finite languages L_1 and L_2 with $isc(L_1) = m$ and $isc(L_2) = n$, one has:*

1. $isc(L_1 \cup L_2) \leq mn - 2$ and

$$itc(L_1 \cup L_2) \leq \sum_{\tau \in \Sigma} (s_\tau(L_1) \boxplus s_\tau(L_2) - (itc_\tau(L_1) - s_\tau(L_1))(itc_\tau(L_2) - s_\tau(L_2)))$$
$$+ n(itc(L_1) - s(L_1)) + m(itc(L_2) - s(L_2)),$$

where for x, y boolean values, $x \boxplus y = \min(x + y, 1)$.

2. $isc(L_1 \cap L_2) \leq mn - 2m - 2n + 6$ and

$$itc(L_1 \cap L_2) \leq \sum_{\tau \in \Sigma} (s_\tau(L_1)s_\tau(L_2) + (itc_\tau(L_1) - s_\tau(L_1) - a_\tau(L_1))(itc_\tau(L_2) - s_\tau(L_2) - a_\tau(L_2)) + a_\tau(L_1)a_\tau(L_2)).$$

All the above upper bounds are tight but can only be reached with an alphabet of size depending on m and n.

4   Concatenation

Câmpeanu *et al.* [2] studied the state complexity of the concatenation of a m-state complete DFA A with a n-state complete DFA B over an alphabet of size k and proposed the upper bound

$$\sum_{i=0}^{m-2} \min \left\{ k^i, \sum_{j=0}^{f(A,i)} \binom{n-2}{j} \right\} + \min \left\{ k^{m-1}, \sum_{j=0}^{f(A)} \binom{n-2}{j} \right\}, \qquad (1)$$

which was proved to be tight for $m > n - 1$. It is easy to see that the second term of (1) is $\sum_{j=0}^{f(A)} \binom{n-2}{j}$ if $m > n - 1$, and k^{m-1}, otherwise. The value k^{m-1} indicates that the DFA resulting from the concatenation has states with level at most $m - 1$. But that is not always the case, as we can see by the example[2] in Figure 2. This implies that (1) is not an upper bound if $m < n$. With these changes, we have

Theorem 2. *For any two finite languages L_1 and L_2 with $sc(L_1) = m$ and $sc(L_2) = n$ over an alphabet of size $k \geq 2$, one has*

$$sc(L_1L_2) \leq \sum_{i=0}^{m-2} \min \left\{ k^i, \sum_{j=0}^{f(L_1,i)} \binom{n-2}{j} \right\} + \sum_{j=0}^{f(L_1)} \binom{n-2}{j}. \qquad (2)$$

Given two incomplete DFAs $A = ([0, m-1], \Sigma, \delta_A, 0, F_A)$ and $B = ([0, n-1], \Sigma, \delta_B, 0, F_B)$, that represent finite languages, the algorithm by Maia *et al.* for the concatenation of regular languages can be applied to obtain a DFA $C = (R, \Sigma, \delta_C, r_0, F_C)$ accepting $\mathcal{L}(A)\mathcal{L}(B)$. The set of states of C is contained in the set $([0, m-1] \cup \{\Omega_A\}) \times 2^{[0,n-1]}$, the initial state r_0 is $(0, \emptyset)$ if $0 \notin F_A$, and is $(0, \{0\})$ otherwise; $F_C = \{(i, P) \in R \mid P \cap F_B \neq \emptyset\}$, and for $\tau \in \Sigma$, $i \in [0, m-1]$, and $P \subseteq [0, n-1]$, $\delta_C((i, P), \tau) = (i', P')$ with $i' = \delta_A(i, \tau)$, if $\delta_A(i, \tau) \downarrow$ or $i' = \Omega_A$ otherwise, and $P' = \delta_B(P, \tau) \cup \{0\}$ if $i' \in F_A$ and $P' = \delta_B(P, \tau)$ otherwise. For the incomplete state and transition complexity we have

---

[2] Note that we are omitting the dead state in the figures.

Theorem 3. *For any two finite languages L_1 and L_2 with $isc(L_1) = m$ and $isc(L_2) = n$ over an alphabet of size $k \geq 2$, and making $\Lambda_j = \binom{n-1}{j} - \binom{\bar{t}_\tau(L_2) - \bar{s}_\tau(L_2)}{j}$, $\Delta_j = \binom{n-1}{j} - \bar{s}_\tau(L_2)\binom{\bar{t}_\tau(L_2) - \bar{s}_\tau(L_2)}{j}$ one has*

$$isc(L_1 L_2) \leq \sum_{i=0}^{m-1} \min\left\{ k^i, \sum_{j=0}^{f(L_1,i)} \binom{n-1}{j} \right\} + \sum_{j=0}^{f(L_1)} \binom{n-1}{j} - 1. \quad (3)$$

and

$$itc(L_1 L_2) \leq k \sum_{i=0}^{m-2} \min\left\{ k^i, \sum_{j=0}^{f(L_1,i)} \binom{n-1}{j} \right\} +$$

$$+ \sum_{\tau \in \Sigma} \left(\min\left\{ k^{m-1} - \bar{s}_\tau(L_2), \sum_{j=0}^{f(L_1)-1} \Delta_j \right\} + \sum_{j=0}^{f(L_1)} \Lambda_j \right). \quad (4)$$

Proof. The τ-transitions of the DFA C accepting $\mathcal{L}(A)\mathcal{L}(B)$ have three forms: (i, β) where i represents the transition leaving the state $i \in [0, m-1]$; $(-1, \beta)$ where -1 represents the absence of the transition from state π_A to Ω_A; and $(-2, \beta)$ where -2 represents any transition leaving Ω_A. In all forms, β is a set of transitions of DFA B. The number of τ-transitions of the form (i, β) is at most $\sum_{i=0}^{m-2} \min\{k^i, \sum_{j=0}^{f(L_1,i)} \binom{n-1}{j}\}$ which corresponds to the number of states of the form (i, P), for $i \in [0, m-1]$ and $P \subseteq [0, n-1]$. The number of τ-transitions of the form $(-1, \beta)$ is $\min\{k^{m-1} - \bar{s}_\tau(L_2), \sum_{j=0}^{f(L_1)-1} \Delta_j\}$. We have at most k^{m-1} states in this level. However, if $s_\tau(B, 0) = 0$ we need to remove the transition $(-1, \emptyset)$ which leaves the state $(m-1, \{0\})$. On the other hand, the size of β is at most $f(L_1) - 1$ and we know that β has always the transition leaving the initial state by τ, if it exists. If this transition does not exist, *i.e.* $\bar{s}_\tau(B, 0) = 1$, we need to remove the sets with only non-defined transitions, because they originate transitions of the form $(-1, \emptyset)$. The number of τ-transitions of the form $(-2, \beta)$ is $\sum_{j=0}^{f(L_1)} \Lambda_j$ and this case is similar to the previous one.

To prove that the bounds are reachable, we consider two cases depending whether $m + 1 \geq n$ or not.

Case 1: $m+1 \geq n$ The witness languages are the ones presented by Câmpeanu *et al.* (see Figure 1).

Fig. 1. DFA A with m states and DFA B with n states

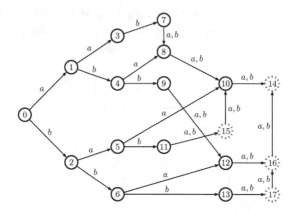

Fig. 2. DFA resulting of the concatenation of DFA A with $m = 3$ and DFA B with $n = 5$, of Fig. 1. The states with dashed lines have level > 3 and are not accounted for by formula (1).

Theorem 4. *For any two integers $m \geq 2$ and $n \geq 2$ such that $m + 1 \geq n$, there exist an m-state DFA A and an n-state DFA B, both accepting finite languages, such that any DFA accepting $\mathcal{L}(A)\mathcal{L}(B)$ needs at least $(m - n + 3)2^{n-1} - 2$ states and $2^n(m - n + 3) - 8$ transitions.*

Case 2: $m + 1 < n$ Let $\Sigma = \{b\} \cup \{a_i \mid i \in [1, n - 2]\}$. Let $A = ([0, m - 1], \Sigma, \delta_A, 0, [0, m - 1])$ where $\delta_A(i, \tau) = i + 1$, for any $\tau \in \Sigma$. Let $B = ([0, n - 1], \Sigma, \delta_B, 0, \{n - 1\})$ where $\delta_B(i, b) = i + 1$, for $i \in [0, n - 2]$, $\delta_B(i, a_j) = i + j$, for $i, j \in [1, n - 2]$, $i + j \in [2, n - 1]$, and $\delta_B(0, a_j) = j$, for $j \in [2, n - 2]$. Note that A and B are minimal DFAs.

Theorem 5. *For any two integers $m \geq 2$ and $n \geq 2$, with $m + 1 < n$, there exist an m-state DFA A and an n-state DFA B, both accepting finite languages over an alphabet of size depending on m and n, such that the number of states and transitions of any DFA accepting $\mathcal{L}(A)\mathcal{L}(B)$ reaches the upper bounds.*

Proof. We need to show that the DFA C accepting $\mathcal{L}(A)\mathcal{L}(B)$ is minimal, *i.e.*, (i) every state of C is reachable from the initial state; (ii) each state of C defines a distinct equivalence class. To prove (i), we first show that all states $(i, P) \subseteq R$ with $i \in [1, m - 1]$ are reachable. The following facts hold for the automaton C: 1) every state of the form $(i + 1, P')$ is reached by a transition from a state (i, P) (by the construction of A) and $|P'| \leq |P| + 1$, for $i \in [1, m - 2]$; 2) every state of the form (Ω_A, P') is reached by a transition from a state $(m - 1, P)$ (by the construction of A) and $|P'| \leq |P| + 1$; 3) for each state (i, P), $P \subseteq [0, n - 1]$, $|P| \leq i + 1$ and $0 \in P$, $i \in [1, m - 1]$; 4) for each state (Ω_A, P), $\emptyset \neq P \subseteq [0, n - 1]$, $|P| \leq m$ and $0 \notin P$.

Suppose that for a $1 \leq i \leq m - 2$, all states (i, P) are reachable. The number of states of the form $(1, P)$ is $m - 1$ and of the form (i, P) with $i \in [2, m - 2]$

is $\sum_{j=0}^{i} \binom{n-1}{j}$. Let us consider the states $(i + 1, P')$. If $P' = \{0\}$, then $\delta_C((i, \{0\}), a_1) = (i+1, P')$. Otherwise, let $l = \min(P' \setminus \{0\})$ and $S_l = \{s - l \mid s \in P' \setminus \{0\}\}$. Then, $\delta_C((i, S_l), a_l) = (i+1, P')$, if $2 \leq l \leq n-2$; $\delta_C((i, \{0\} \cup S_1), a_1) = (i + 1, P')$, if $l = n - 1$; and $\delta_C((i, S_1), b) = (i + 1, P')$, if $l = 1$.. Thus, all $\sum_{j=0}^{i+1} \binom{n-1}{j}$ states of the form $(i+1, P')$ are reachable. Let us consider the states (Ω_A, P'). P' is always an non empty set by construction of C. Let $l = \min(P')$ and $S_l = \{s - l \mid s \in P'\}$. Thus, $\delta_C((m - 1, S_l), a_l) = (\Omega_A, P')$, if $2 \leq l \leq n - 2$; $\delta_C((m-1, \{0\} \cup S_1), a_1) = (\Omega_A, P')$, if $l = n-1$; and $\delta_C((m-1, S_1), b) = (\Omega_A, P')$, if $l = 1$ Thus, all $\sum_{j=0}^{m} \binom{n-1}{j} - 1$ states of the form (Ω_A, P') are reachable. To prove (ii), consider two distinct states $(i, P_1), (j, P_2) \in R$. If $i \neq j$, then $\delta_C((i, P_1), b^{n+m-2-i}) \in F_C$ but $\delta_C((j, P_2), b^{n+m-2-i}) \notin F_C$. If $i = j$, suppose that $P_1 \neq P_2$ and both are final or non-final. Let $P_1' = P_1 \setminus P_2$ and $P_2' = P_2 \setminus P_1$. Without loss of generality, let P_1' be the set which has the minimal value, let us say l. Thus $\delta_C((i, P_1), a_1^{n-1-l}) \in F_C$ but $\delta_C((i, P_2), a_1^{n-1-l}) \notin F_C$. The proof corresponding to the number of transitions is similar to the proof of Theorem 3.

Theorem 6. *The upper bounds for state and transition complexity of concatenation cannot be reached for any alphabet with a fixed size for $m \geq 0$, $n > m+1$.*

Proof. Let $S = \{(\Omega_A, P) \mid 1 \in P\} \subseteq R$. A state $(\Omega_A, P) \in S$ has to satisfy the following condition:

$$\exists i \in F_A \exists P' \subseteq 2^{[0,n-1]} \exists \tau \in \Sigma : \delta_C((i, P' \cup \{0\}), \tau) = (\Omega_A, P).$$

The maximal size of S is $\sum_{j=0}^{f(A)-1} \binom{n-2}{j}$, because by construction $1 \in P$ and $0 \notin P$. Assume that Σ has a fixed size $k = |\Sigma|$. Then, the maximal number of words that reach states of S from r_0 is $\sum_{i=0}^{f(A)} k^{i+1}$ since the words that reach a state $s \in S$ are of the form $w_A \sigma$, where $w_A \in L(A)$ and $\sigma \in \Sigma$. As $n > m$, for some $l \geq 0$ we have $n = m + l$. Thus for an l sufficiently large $\sum_{i=0}^{f(A)} k^{i+1} \ll \sum_{j=0}^{f(A)-1} \binom{m+l-2}{j}$, which is an absurd. The absurd resulted from supposing that k is fixed.

5 Star and Reversal

Given an incomplete DFA $A = ([0, m - 1], \Sigma, \delta_A, 0, F_A)$ accepting a finite language, we obtain a DFA accepting $\mathcal{L}(A)^*$ using an algorithm similar to the one for regular languages [7] and a DFA that accepts $\mathcal{L}(A)^R$, reversing all transitions of A and then determinizing the resulting NFA. Note that if $f(A) = 1$ then the minimal DFA accepting $\mathcal{L}(A)^*$ has also m states. Thus, for the Kleene star operation, we will consider DFAs with at least two final states.

Theorem 7. *For any finite language L with $isc(L) = m$ one has*

1. *if $f(L) \geq 2$, $isc(L^*) \leq 2^{m-f(L)-1} + 2^{m-2} - 1$ and*

$$itc(L^*) \leq 2^{m-f(L)-1} \left(k + \sum_{\tau \in \Sigma} 2^{e_\tau(L)} \right) - \sum_{\tau \in \Sigma} 2^{n_\tau} - \sum_{\tau \in X} 2^{n_\tau},$$

where $n_\tau = \bar{t}_\tau(L) - \bar{s}_\tau(L) - \bar{e}_\tau(L)$ and $X = \{\tau \in \Sigma \mid s_\tau(L) = 0\}$.

2. *if* $m \geq 3$, $k \geq 2$, *and* l *is the smallest integer such that* $2^{m-l} \leq k^l$, $isc(L^R) \leq \sum_{i=0}^{l-1} k^i + 2^{m-l} - 1$ *moreover if* m *is odd,*

$$itc(L^R) \leq \sum_{i=0}^{l} k^i - 1 + k2^{m-l} - \sum_{\tau \in \Sigma} 2^{\sum_{i=0}^{l-1} \bar{t}_\tau(L,i)+1},$$

and, if m *is even,*

$$itc(L^R) \leq \sum_{i=0}^{l} k^i - 1 + k2^{m-l} - \sum_{\tau \in \Sigma} \left(2^{\sum_{i=0}^{l-2} \bar{t}_\tau(L,i)+1} - c_\tau(L,l) \right).$$

6   Final Remarks

In this paper we studied the incomplete state and transition complexity of basic regularity preserving operations on finite languages. Note that for the complement operation these descriptional measures coincide with the ones on regular languages. Table 1 summarizes some of those results. For unary finite languages the incomplete transition complexity is equal to the incomplete state complexity of that language, which is always equal to the state complexity of the language minus one.

References

1. Beesley, K.R., Karttunen, L.: Finite State Morphology. CSLI Publications, Stanford University (2003)
2. Câmpeanu, C., Culik, K., Salomaa, K., Yu, S.: State complexity of basic operations on finite languages. In: Boldt, O., Jürgensen, H. (eds.) WIA 1999. LNCS, vol. 2214, pp. 60–70. Springer, Heidelberg (2001)
3. Cassandras, C.G., Lafortune, S.: Introduction to discrete event systems. Springer (2006)
4. Gao, Y., Salomaa, K., Yu, S.: Transition complexity of incomplete DFAs. Fundam. Inform. 110(1-4), 143–158 (2011)
5. Han, Y.S., Salomaa, K.: State complexity of union and intersection of finite languages. Int. J. Found. Comput. Sci. 19(3), 581–595 (2008)
6. Hopcroft, J.E., Ullman, J.D.: Introduction to Automata Theory, Languages and Computation. Addison-Wesley (1979)
7. Maia, E., Moreira, N., Reis, R.: Incomplete transition complexity of some basic operations. In: van Emde Boas, P., Groen, F.C.A., Italiano, G.F., Nawrocki, J., Sack, H. (eds.) SOFSEM 2013. LNCS, vol. 7741, pp. 319–331. Springer, Heidelberg (2013)
8. Maurel, D., Guenthner, F.: Automata and Dictionaries. College Publications (2005)
9. Owens, S., Reppy, J.H., Turon, A.: Regular-expression derivatives re-examined. J. Funct. Program. 19(2), 173–190 (2009)
10. Salomaa, K., Yu, S.: NFA to DFA transformation for finite languages over arbitrary alphabets. J. of Aut., Lang. and Comb. 2(3), 177–186 (1997)
11. Shallit, J.: A Second Course in Formal Languages and Automata Theory. CUP (2008)
12. Yu, S.: Regular languages. In: Rozenberg, G., Salomaa, A. (eds.) Handbook of Formal Languages, vol. 1, pp. 41–110. Springer (1997)

Author Index